D0860490

Contemporary Electronics Circuits Deskbook

Contemporary Electronics Circuits Deskbook

Compiled by **Harry L. Helms**

McGraw-Hill Book Company

New York St. Louis San Francisco Auckland Bogotá Hamburg
Johannesburg London Madrid Mexico Montreal New Delhi
Panama Paris São Paulo Singapore Sydney Tokyo Toronto

Library of Congress Cataloging-in-Publication Data

Helms, Harry L.
Contemporary electronics circuits deskbook.

Includes index.
1. Electronic circuits—Handbooks, manuals, etc.
I. Title. II. Title: Contemporary electronics circuits deskbook.
TK7867.H44 1986 621.3815'3 85-19721
ISBN 0-07-027980-2

1234567890 HALHAL 89321098765

ISBN 0-07-027980-2

The editors for this book were Richard Krajewski and James T. Halston, the designer was Mark E. Safran, and the production supervisor was Sara L. Fliess. It was set in Univers Medium by Bi-Comp Inc.

Printed and bound by Halliday Lithograph Corporation.

For Tina . . .
with gratitude for the patience and support

Contents

Preface

Contemporary Electronics Circuits Deskbook is a compilation of circuit designs and applications which have appeared in recent electronics magazines, applications notes, and databooks. As can be seen, these circuits cover the entire range of electronics technology. Each circuit includes type numbers or values for all significant components, an identifying title, a brief description, and a citation of the original source of the circuit. Each circuit diagram has been reproduced directly from the original source, and this accounts for the differences in schematic style between the various circuits.

I hope this compilation saves you hours of searching through the literature to locate a specific circuit application. This book can also serve as a useful starting point for your own circuit designs. One of the most interesting points about this book is the number of circuit designs created by amateur radio operators; many of them meet the most exacting professional standards. Such resourcefulness and creativity is one reason why I am proud to be part of the worldwide hobby of amateur radio.

Acknowledgments

This book would not have been possible without the cooperation of the following companies and publications. At the end of each circuit description, there is a citation identifying the source of the circuit. All circuits in this book are copyrighted by the source cited and used with the written permission of the copyright holder.

Databooks and applications notes may be obtained from the company concerned or their sales representatives. Back issues of magazines may be found in university or technical libraries. In addition, some magazines may have back issues available for sale or will make photocopies of an article for a fee.

Analog Devices, Route 1 Industrial Park, P.O. Box 280, Norwood, MA, 02062

***CQ* Magazine,** 76 North Broadway, Hicksville, NY, 11801

***Ham Radio* Magazine,** Greenville, NH, 03048

Intersil, 10710 North Tantau Ave., Cupertino, CA, 95014

QST, 225 Main St., Newington, CT, 06111

Mostek, 1215 West Crosby Road, Carrollton, TX, 75006

Motorola Semiconductor Products, Box 20912, Phoenix, AZ 85036

Plessey Solid State, 3 Whatney, Irvine, CA, 92714

RCA Solid State Division, Route 202, Somerville, NJ, 08876

73 Magazine, 80 Pine St., Peterborough, NH, 03458

Signetics, P.O. Box 409, 811 E. Arques Ave., Sunnyvale, CA, 94086

Solid State Micro Technology for Music, 2076B Walsh Ave., Santa Clara, CA, 95050

Standard Microsystems, P.O. Box 1228, Delray Beach, FL, 33444

Teledyne Semiconductor, 1300 Terra Bella Ave., Mountain View, CA, 94043

HARRY HELMS

Abbreviations

A ampere
AC alternating current
AC/DC AC or DC
A/D analog-to-digital
ADC analog-to-digital converter
AF audio frequency
AFC automatic frequency control
AFSK audio frequency shift keying
AGC automatic gain control
AM amplitude modulation
ASCII American Standard Code for Information Interchange
AVC automatic volume control
BCD binary-coded decimal
BFO beat frequency oscillator
b/s bits per second
C capacitor or capacitance
cm centimeter
CMOS complimentary MOS
COR carrier-operated relay
CT center-tapped or continuous turns
CW continuous wave (Morse code transmission)
D/A digital-to-analog
DAC digital-to-analog converter
dB decibel(s)
DC direct current
DIP dual in-line package
DSB double sideband
DTMF dual-tone multi-frequency
DVM digital voltmeter
EMI electromagnetic interference
EPROM erasable PROM
F farad
FM frequency modulation
FSK frequency shift keying
GHz gigahertz
H henry
Hz hertz
IC integrated circuit
IF intermediate frequency
I/O input/output
JFET junction field-effect transistor
K kilohm
kb kilobyte
kHz kilohertz
KW kilowatt
LCD liquid crystal display
LED light emitting diode
LF low frequency
LSB lower sideband or least significant bit
M megohm
MHz megahertz
MOS metal oxide semiconductor
MOSFET metal oxide semiconductor FET
mm millimeter
ms millisecond
MSB most significant bit
mV millivolt
mW milliwatt
μA microampere
μF microfarad
μH microhenry
μs microsecond
μV microvolt
μW microwatt
PC printed circuit, personal computer, pulse-coded
PCM pulse-coded modulation
PEP peak envelope power
pF picofarad
PIN special type of diode
PIV peak inverse voltage
PLL phase-locked loop
PM phase modulation
PMOS p-channel MOS
PPM parts per million
P-P peak-to-peak
PROM programmable read-only memory
PTT push to talk
PWM pulse-width modulation
Q quality power
QRP low power transmitting
RF radio frequency
RFI radio frequency interference
RIAA Recording Industry Association of America
RMS root-mean square
RTTY radioteleprinter
SCA subsidiary communications authorization
SCR silicon controlled rectifier
S-meter signal strength meter

S/N signal to noise ratio

SPDT single-pole double-throw switch

SSB single sideband

SSTV slow-scan television

SWR standing-wave ratio

THD total harmonic distortion

TTL transistor-transistor logic

TVI television interference

UART universal asynchronous receiver-transmitter

UHF ultra high frequency

V volt

VCA voltage controlled amplifier

VCO voltage controlled oscillator

VFO variable frequency oscillator

VHF very high frequency

VLF very low frequency

VMOS vertical MOS

VOM volt-ohm meter

VOX voice operated

VTVM vacuum tube voltmeter

VXO variable crystal oscillator

W watt

WPM words per minute

Contemporary Electronics Circuits Deskbook

1

Active Filters

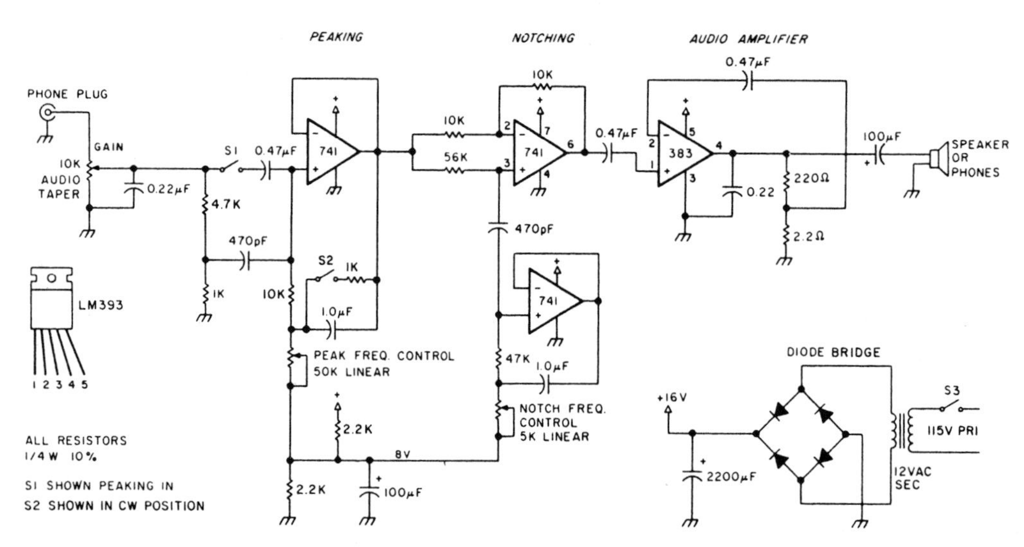

NOTCH-PEAKING AUDIO FILTER—Combines AF peaking and notching functions in single design. The notch frequency can be shifted from 500 to 3000 Hz. The bandwidth at the −3-dB point of the notch frequency is approximately 200 Hz with a rejection of greater than 20 dB.—J. Pepper, The Magical Audio Filter, *73 Magazine,* November 1983, pp. 14–16.

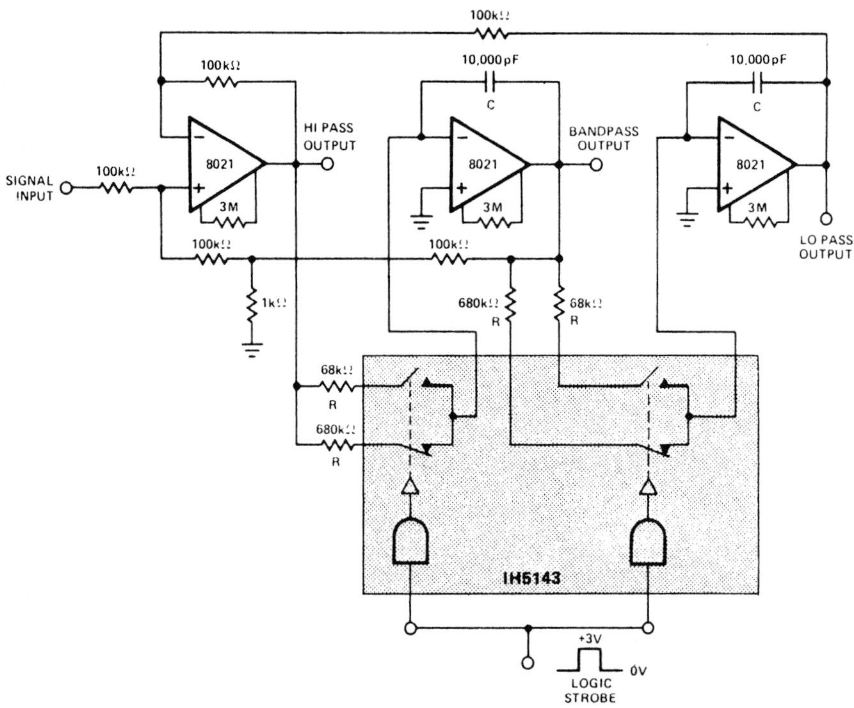

DIGITALLY TUNED LOW-PASS ACTIVE FILTER—Provides simultaneous low-pass, high-pass, and bandpass functions. With the component values shown, the center frequency is 235 Hz for a high logic input and 23.5 Hz for a low logic input. The gain and Q both equal 100.—"Intersil Data Book," Intersil, Cupertino, CA, 1981, p. 3-133.

THREE-POLE ACTIVE LOW-PASS BUTTERWORTH FILTER—Offers essentially flat response to 1 kHz, at which point response drops 60 dB/decade to no response at all at approximately 3 kHz.—"Signetics Linear LSI Data and Applications Manual," Signetics, Sunnyvale, CA, 1985, p. 6-36.

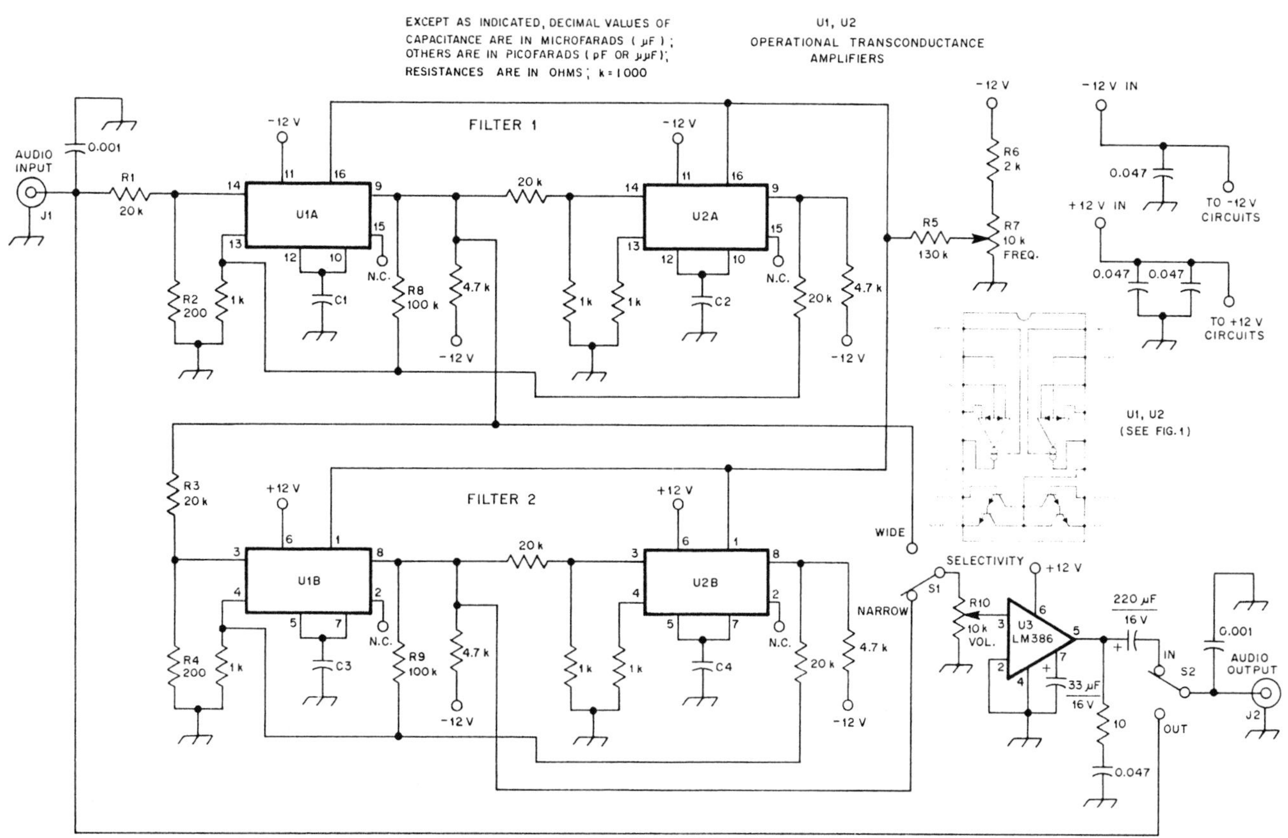

TUNABLE CW FILTER—Cascaded bandpass filters with voltage controlled center frequencies give high selectivity and low "ringing." The Q of each filter section is determined by R8 and R9, with larger values giving higher Q. U1 and U2 are each an LM13600.—R. Nelson, A Tunable CW Filter, *QST,* October 1983, pp. 14–16.

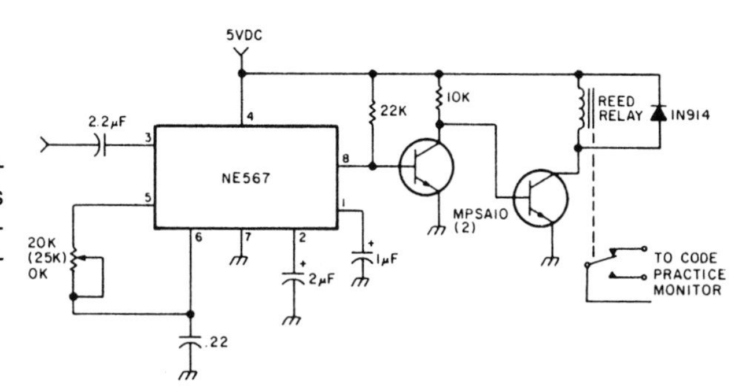

CW FILTER—Features 40-Hz bandwidth with no ringing. The filter is connected to a code practice oscillator.—R. Folkert, CW Filter, *73 Magazine,* November 1982, p. 109.

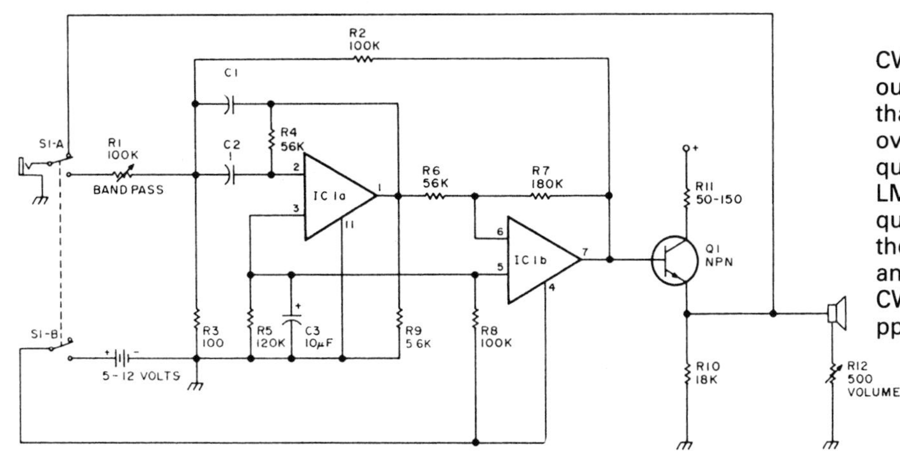

CW FILTER—Circuit has continuously adjustable bandpass from less than 30 Hz to fully open and a Q of over 25 at maximum. The center frequency is 700 Hz and IC used is an LM324 quad op amp. The center frequency can be altered by changing the values of R4 and R6.—J. Hyde and M. Minchen, The Very, Very Best CW Filter?, *73 Magazine,* July 1982, pp. 56–58.

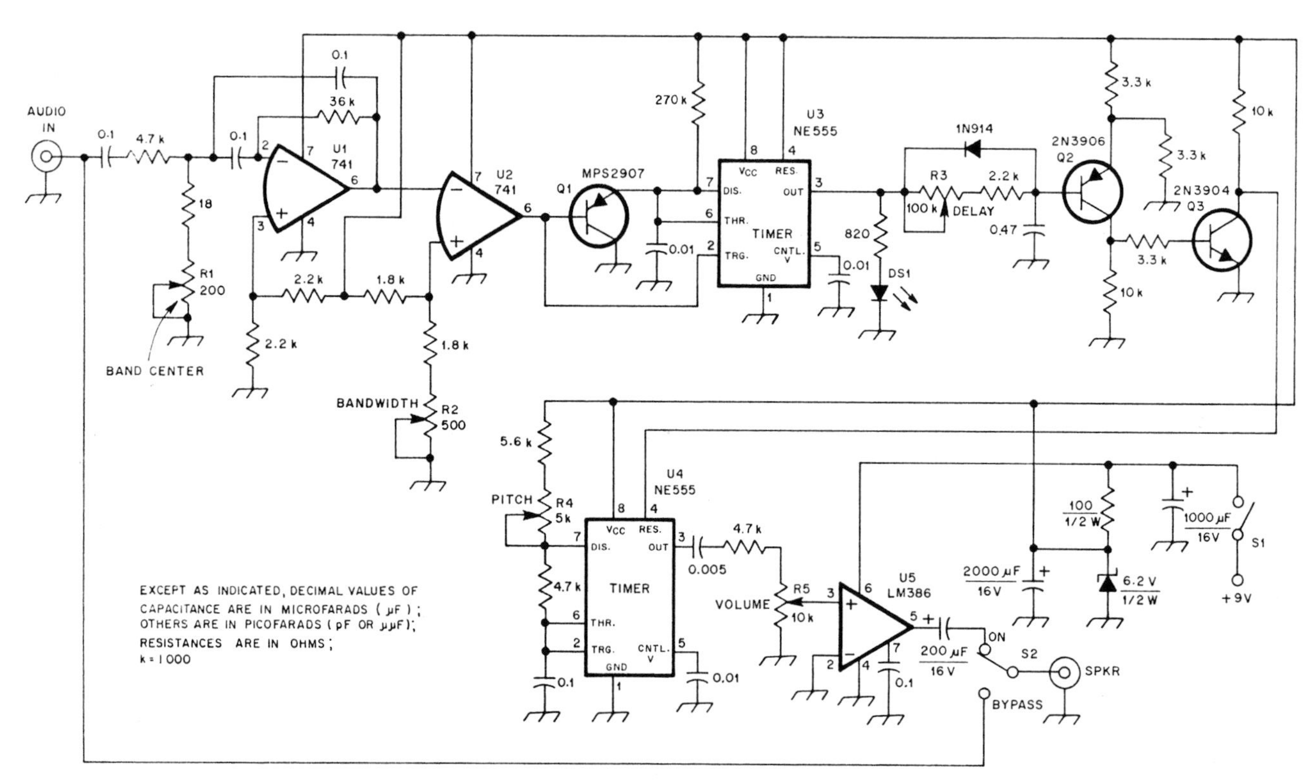

CW AUDIO FILTER—Audio input is fed to one 741, which acts as a bandpass filter. The second 741 is a comparator. The MPS2907 and 555 serve as an envelope detector and are followed by a delay unit. The second 555 is a tone generator whose output is amplified by the 386. Delay is adjustable from 0 to 16 ns.—D. Jagerman, The KC2FR QRM Fighter, *QST,* July 1982, pp. 28–30.

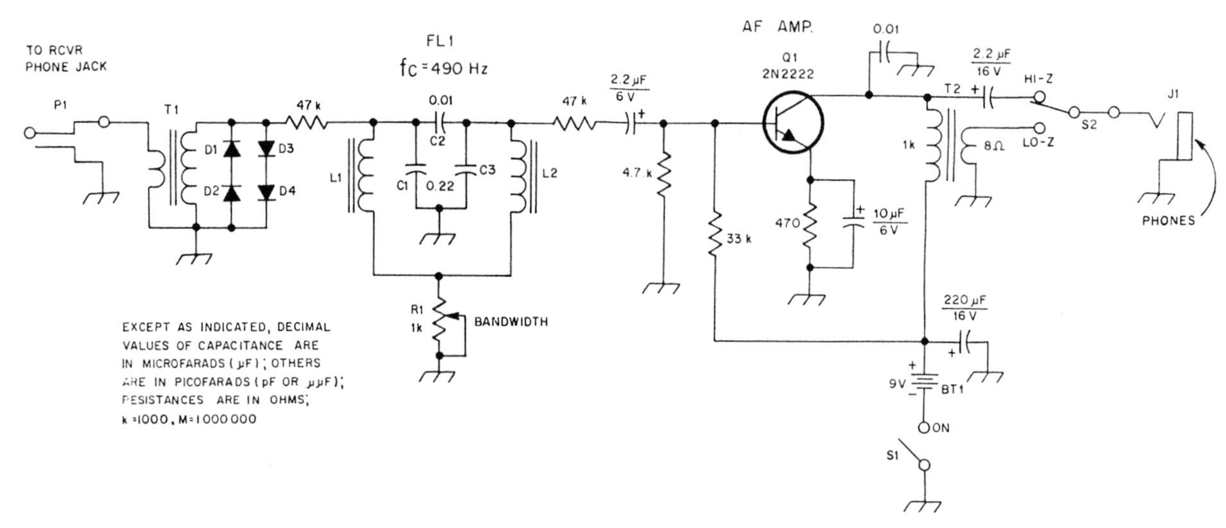

CW AUDIO FILTER—Double-tuned LC design filter features variable peak frequency and adjustable Q. The output is designed for high-impedance headphones. D1–D4 are all 1N914. T1 and T2 are miniature audio output transformers.—D. DeMaw, A Second Look at Magnetic Cores, *QST,* June 1984, pp. 15–20.

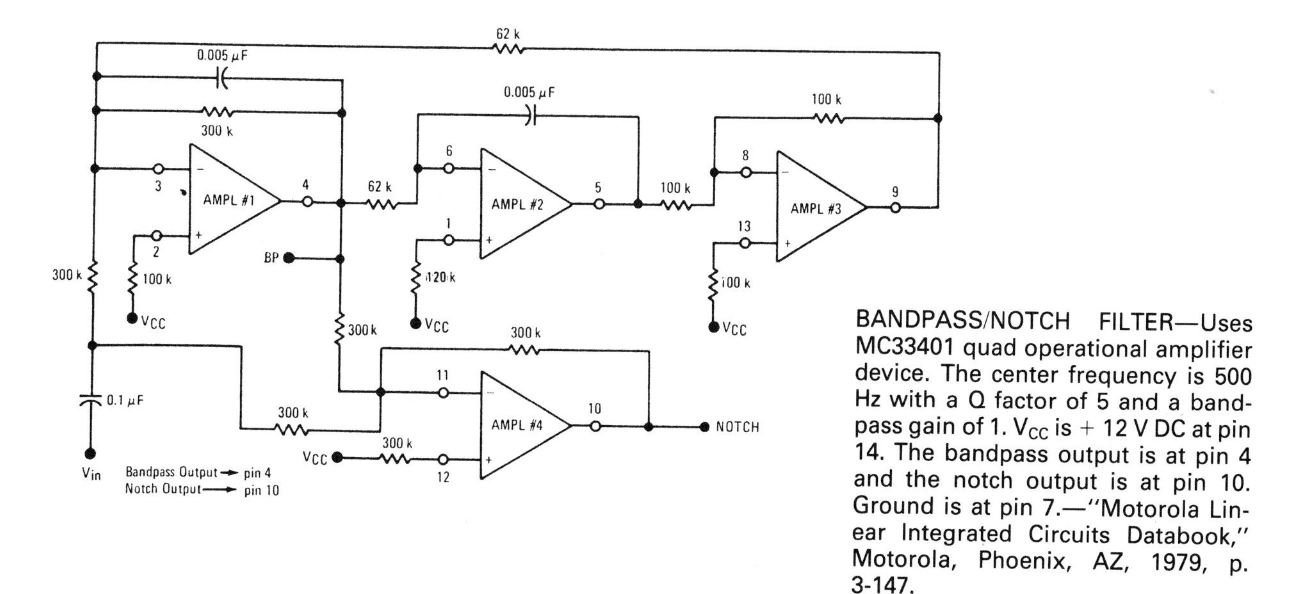

BANDPASS/NOTCH FILTER—Uses MC33401 quad operational amplifier device. The center frequency is 500 Hz with a Q factor of 5 and a bandpass gain of 1. V_{CC} is + 12 V DC at pin 14. The bandpass output is at pin 4 and the notch output is at pin 10. Ground is at pin 7.—"Motorola Linear Integrated Circuits Databook," Motorola, Phoenix, AZ, 1979, p. 3-147.

ACTIVE FILTER—Two-stage filter has center frequency of 800 Hz and a bandwidth of 100 Hz. The gain is near unity at the passband center. The filter should be added in a low-level section of the audio amplifier if a 12-V power supply is used.—P. Clower, Two-Stage Active Filter, *73 Magazine,* December 1982, p. 139.

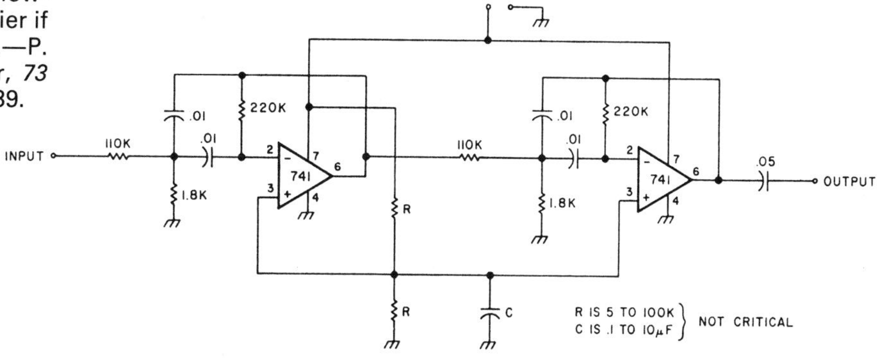

2

Amplifier Circuits

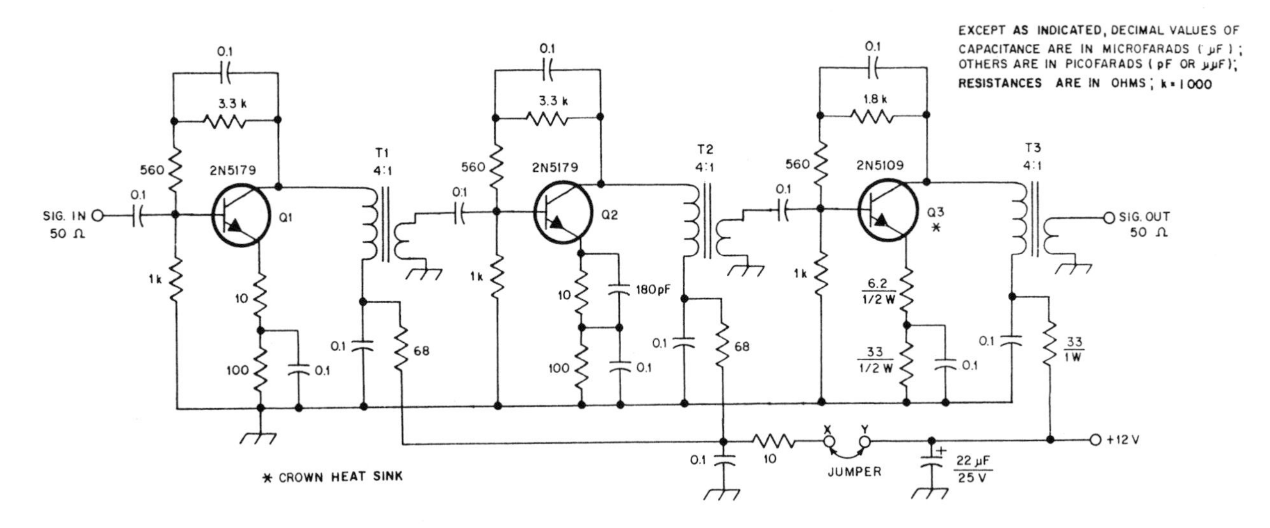

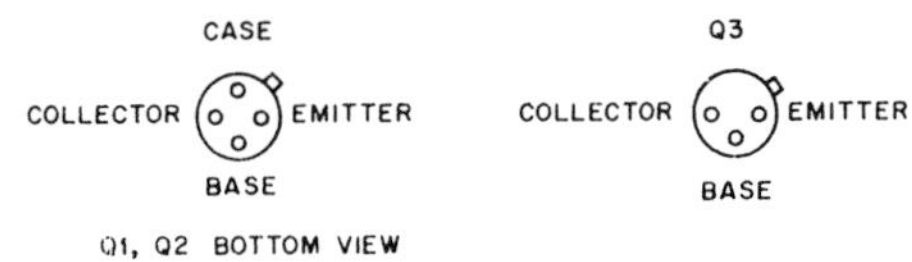

BROADBAND LINEAR AMPLIFIER—Can be used as an instrument amplifier, low-level RF amplifier section in a receiver, or as a receiving-loop preamplifier. T1 and T2 consist of primaries of 15 turns of No. 28 wire and secondaries of seven turns of No. 28 wire on FT37-43 toroid forms. T3 is wound on a FT50-43 toroid form with a primary of 12 turns of No. 26 wire and a secondary of six turns of No. 26 wire.—D. DeMaw, Broadband and Narrow-Band Amplifiers, *QST,* May 1984, pp. 26–30.

All resistor values are in ohms.

ABSOLUTE VALUE AMPLIFIER—Delivers positive output voltage for either polarity input. For positive signals, it acts as a noninverting amplifier. For negative signals, it acts as an inverting amplifier. However, accuracy is poor for input voltages under 1.0 V.—"Signetics Linear LSI Data and Applications Manual," Signetics, Sunnyvale, CA, 1985, p. 6-71.

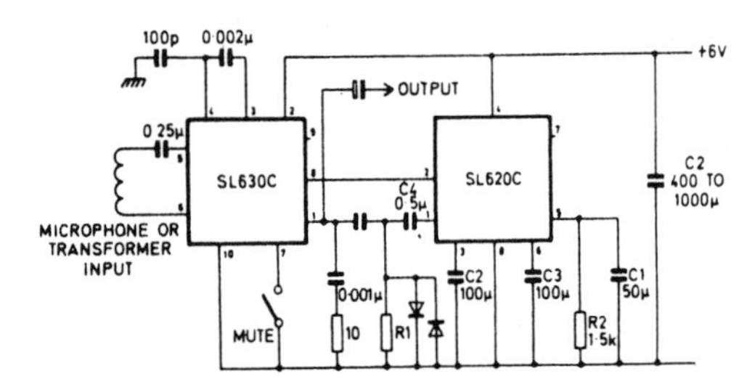

AUDIO AMPLIFIER WITH AGC—SL620C AGC generator IC controls gain of SL630C audio amplifier. The time constant of AGC depends on the values of C1, C2, and C3.—"Plessey Integrated Circuits Databook," Plessey Semiconductor, Irvine, CA, 1983, pp. 131–134.

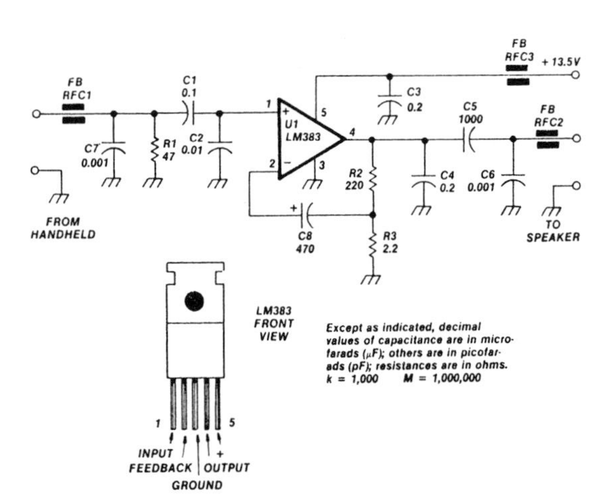

AUDIO AMPLIFIER—Capable of delivering up to 5 W and a gain of 40 dB. The circuit was designed as a remote speaker for use in high-noise areas. All capacitors are ceramic except for C2, C3, and C4, which are metalized film types. RFC1 through RFC3 are jumbo ferrite beads (VHF types).—D. Blakeslee, An Audio Amplifier for Your Handheld Transceiver, *Ham Radio,* July 1981, pp. 38–40.

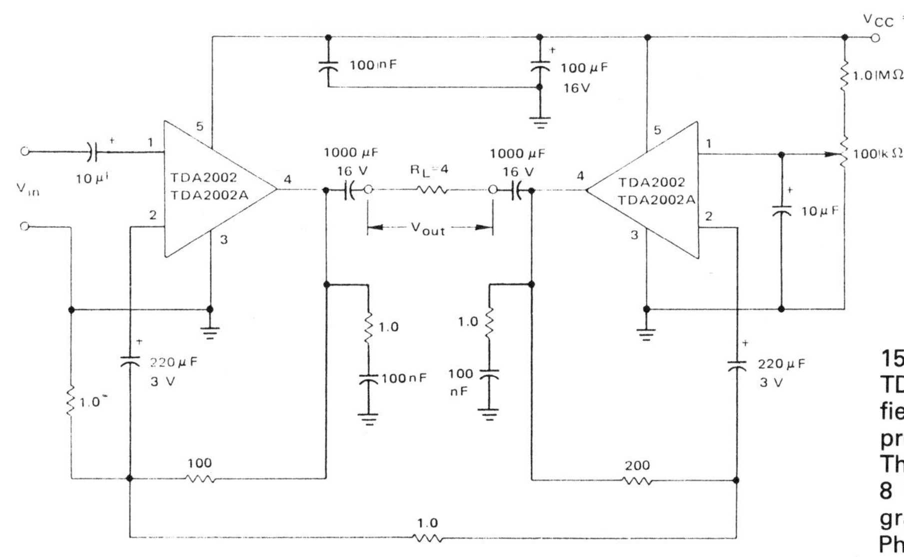

15-W AUDIO AMPLIFIER—Uses two TDA2002 class B audio power amplifier ICs in a bridge configuration to produce 15-W output into a 4-Ω load. The supply voltage may range from 8 to 18 V.—"Motorola Linear Integrated Circuits Databook," Motorola, Phoenix, AZ, 1979, p. 5-169.

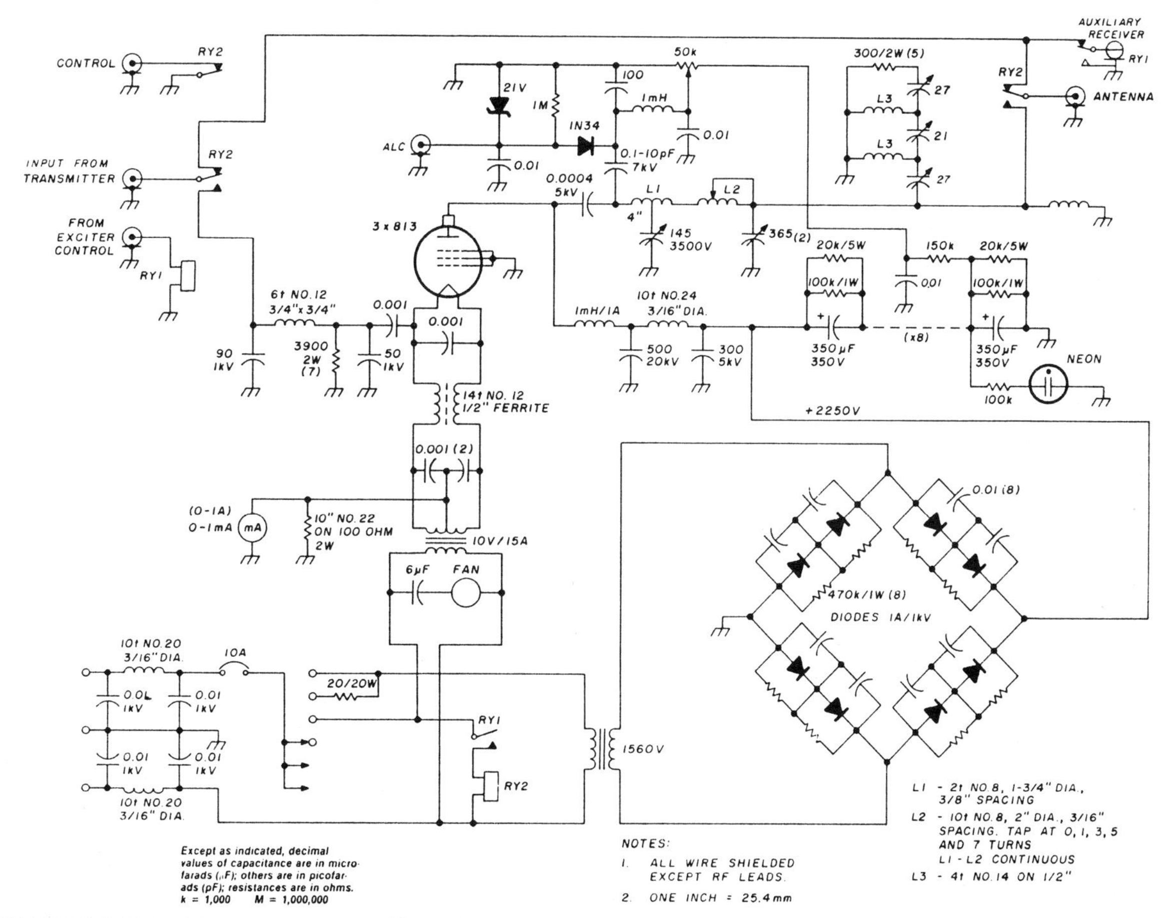

KILOWATT LINEAR RF AMPLIFIER—Designed for use in the 40-meter (7000- to 7300-kHz) through 10-meter (28,000- to 29,700-kHz) amateur radio bands with SSB and CW signals. Linearity is good throughout the range. Drive for the amplifier can be supplied by two 6146s or the equivalent. The design also features separate antenna tuning and a low-pass filter. No parasitic suppressors are used in the plate circuit.—R. P. Haviland, Low Cost Linear Design and Construction, *Ham Radio,* December 1982, pp. 12–22.

12.5-W WIDEBAND POWER AMPLIFIER—Delivers 12.5 W into a 4-Ω load with less than 1% THD to 10 kHz. The MPS-A12 transistor should be heatsinked. All the pins not shown are not connected.—"Motorola Linear Integrated Circuits Databook," Motorola, Phoenix, AZ, 1979, p. 3-94.

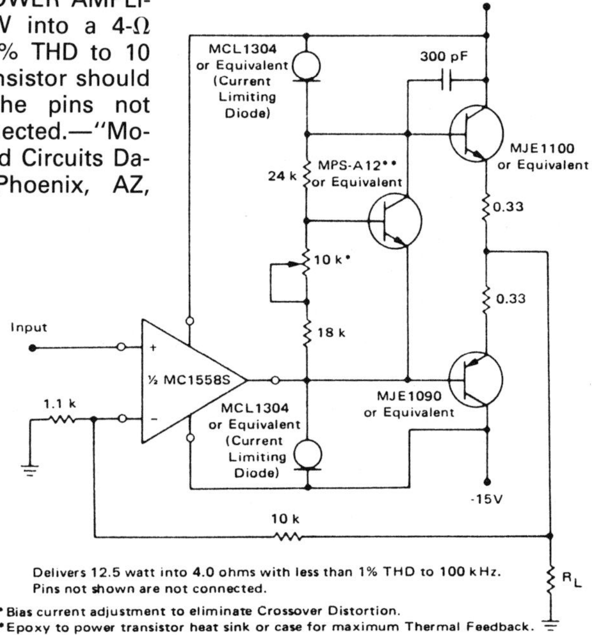

INVERTING LOGARITHMIC AMPLIFIER—With $I_{in} = I_{REF}$, the output will be zero and will increase at 1 V/decade as I_{in} reduces. The 10-V input range optimizes dynamic range in +15-V systems but can be changed proportionally by adjusting R_{in}. Adjusting R_1 will alter the scale factor, while adjustment of R_{ref} will alter the output offset at a given V_{in}. C_1, C_2, and C_3 provide phase compensation for the system.—"SSM 2100 Monolithic Log/Antilog Amplifier," Solid State Micro Technology for Music, Santa Clara, CA, 1982, SSM 2100.

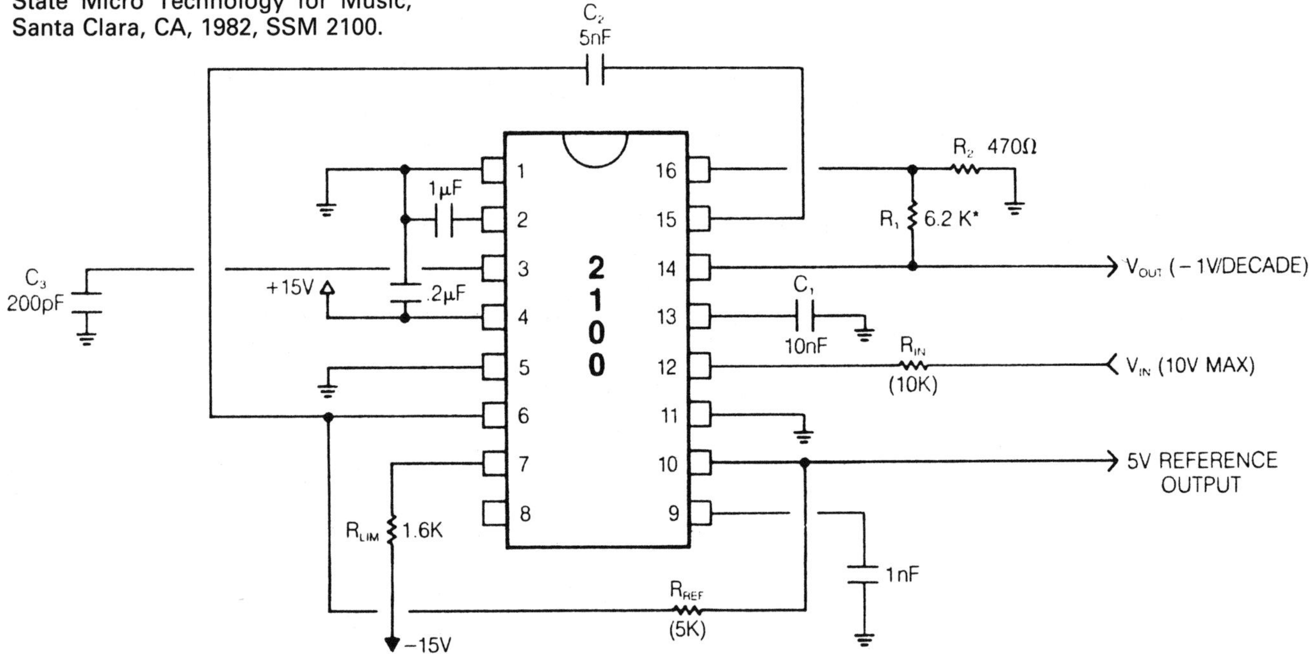

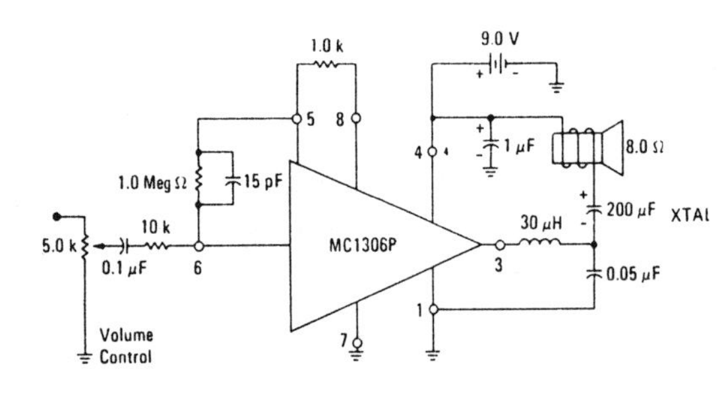

0.5-W AUDIO AMPLIFIER—Designed around MC1306P complementary power amplifier for use in portable radios, tape recorders, phonographs, and intercoms. Distortion is typically 0.5% at 250 mW. 0.5-W output is obtained at 12 V into an 8-Ω load.—"Motorola Linear Integrated Circuits Databook," Motorola, Phoenix, AZ, 1979, p. 5-17.

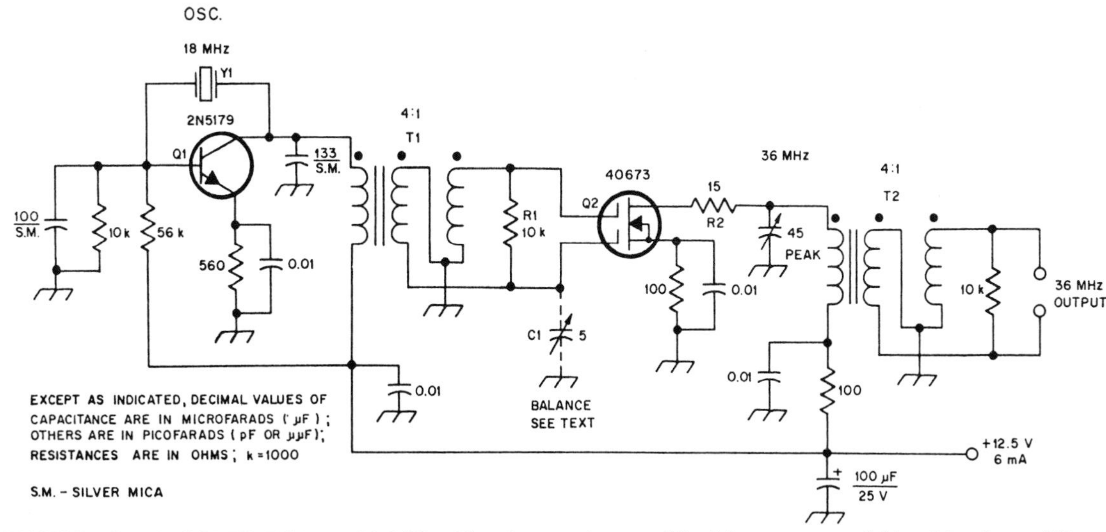

40673 PUSH-PUSH MOSFET DOUBLER—Theoretically as efficient as a straight-through class C amplifier, circuit offers output purity. This design takes the output of the 18-MHz oscillator section and gives output at 36 MHz. Circuit may be modified for higher or lower frequencies, but balance will be more difficult to obtain if the operating frequency is increased into the VHF range. T1 consists of 12 turns of No. 26 wire, trifilar wound, on a FT050-61 form. T2 has a 0.65 µH trifilar winding consisting of 13 turns of No. 26 wire on a T50-6 core.—D. DeMaw, Another Use for the 40673, *QST,* April 1984, p. 50.

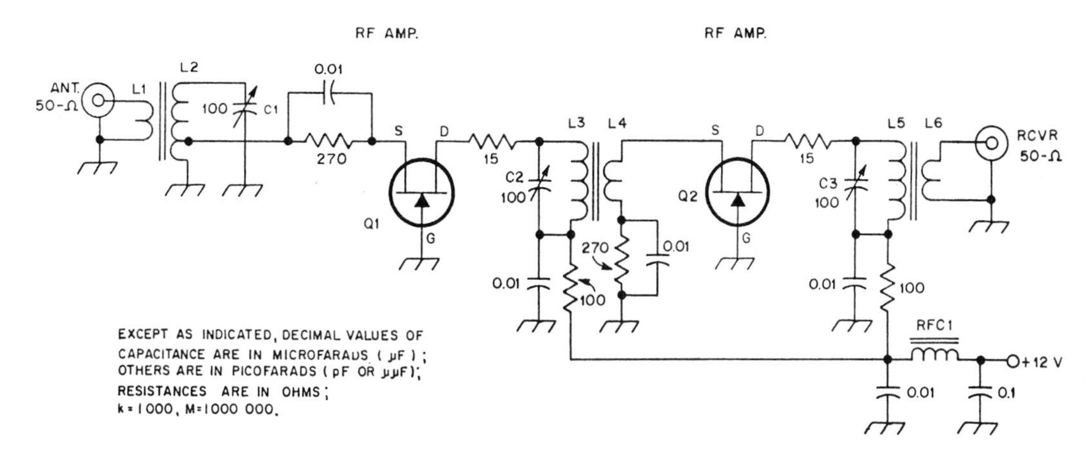

HF AMPLIFIER—Capable of providing up to a 25-dB gain from 1.5 to 30 MHz depending upon values of L1–L6 and C1–C3. Q1 and Q2 are both MPF102 or equivalent FET transistors. RFC1 is a large ferrite bead with four turns of No. 28 wire wound through it. Current configuration is as a receiving amplifier, but may be modified to serve as an instrumentation amplifier, receiver RF amplifier section, etc.—D. DeMaw, Receiver Preamps and How to Use Them, *QST,* April 1984, pp. 19–23.

Frequency (MHz)	*L1, L6*	*L2, L3, L5*	*L4*	*C1-C3 (pF max.)*
1.8-2.0	5 ts no. 28 enam. wire over main winding.	28 µH. 52 ts no. 28 enam. wire on T50-1 toroid core.† Tap L2 at 13 ts above gnd.	13 ts no. 28 enam. wire over L3 winding.	280
3.5-4.0	4 ts no. 26 enam wire over main winding.	9 µH. 42 ts no. 26 enam. wire on T50-2 toroid core. Tap L2 at 9 ts above gnd.	9 ts no. 26 enam. wire over L3 winding.	280
7.0-7.3	3 ts no. 26 enam. wire over main winding.	6 µH. 35 ts no. 26 enam. wire on T50-2 toroid core. Tap L2 at 8 ts above gnd.	8 ts no. 26 enam. wire over L3 winding.	100
10.1-10.150	3 ts no. 26 enam wire over main winding.	4 µH. 32 ts no. 26 enam. wire on T50-6 toroid core. Tap L2 at 7 ts above gnd.	7 ts no. 26 enam. wire over L3 winding.	100
14.0-14.350	2 ts no. 24 enam. wire over main winding.	2 µH. 22 ts no. 24 enam. wire on T50-6 toroid core. Tap L2 at 5 ts above gnd.	5 ts no. 24 enam. wire over L3 winding.	100
21.0-21.450	2 ts no. 24 enam. wire over main winding	1.5 µH. 19 ts no. 24 enam. wire on T50-6 toroid core. Tap L2 at 4 ts above gnd.	4 ts no. 24 enam. wire over L3 winding.	60
28.0-29.7	2 ts no. 24 enam. wire over main winding	1.0 µH. 16 ts no. 24 enam. wire on T50-6 toroid core. Tap L2 at 3 ts above gnd.	3 ts no. 24 enam. wire over L3 winding.	60

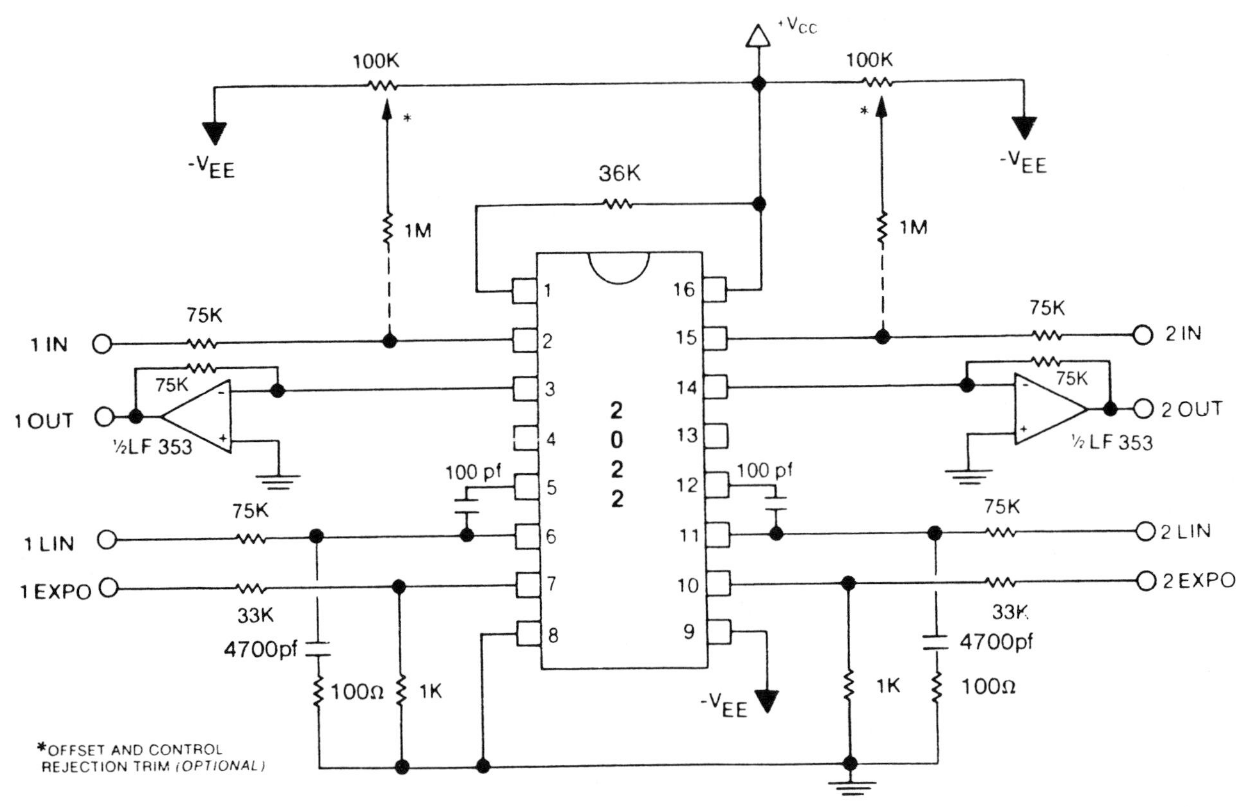

DUAL VOLTAGE CONTROLLED AMPLIFIER—Signal inputs (pins 2 and 15) are virtual ground summing nodes. The reference current established at pin 1 determines the level at which the input signal will clip. With a reference current of 0.8 mA, this will be 200 μA. The input resistors should be selected to give a signal current of ±100 μA at the input voltage peaks; the 75-K resistor shown will satisfy this condition for a +1.75-V signal, with clipping taking place at twice this level.—"Dual Linear-Antilog Voltage Controlled Amplifier," Solid State Micro Technology for Music, Santa Clara, CA, 1982, SSM 2022.

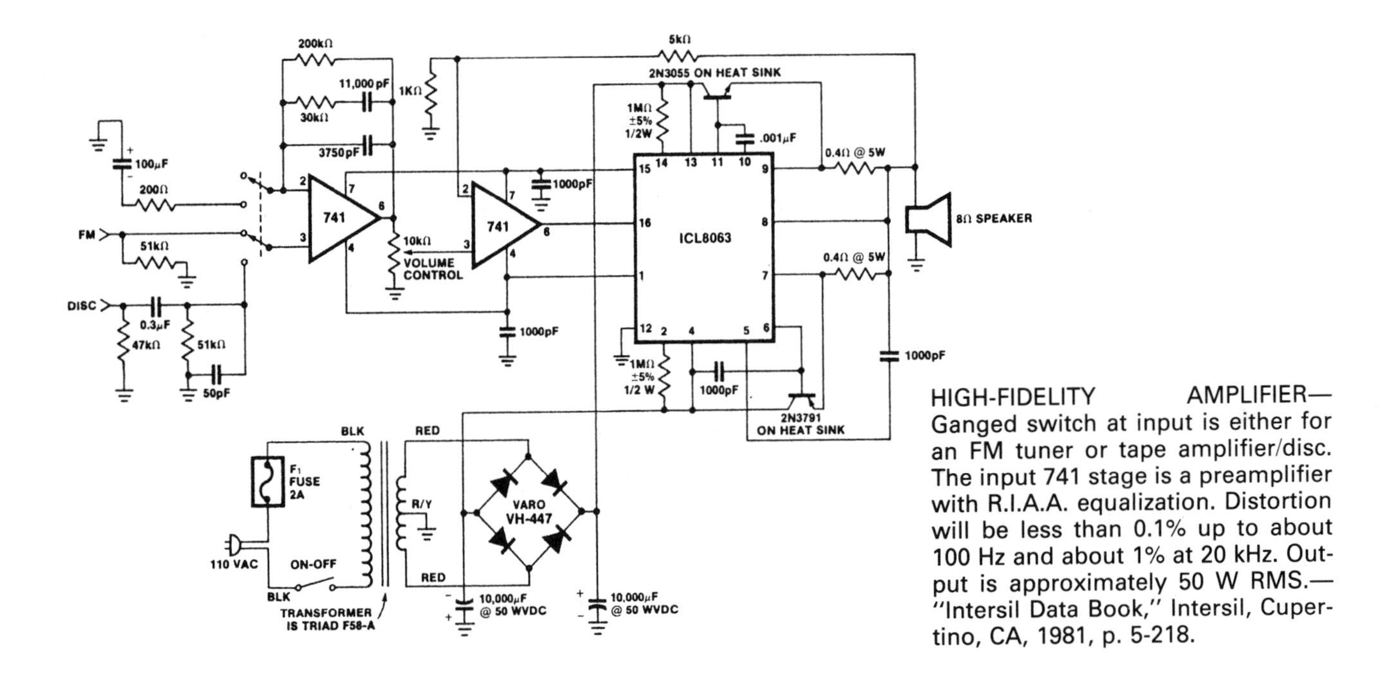

HIGH-FIDELITY AMPLIFIER—Ganged switch at input is either for an FM tuner or tape amplifier/disc. The input 741 stage is a preamplifier with R.I.A.A. equalization. Distortion will be less than 0.1% up to about 100 Hz and about 1% at 20 kHz. Output is approximately 50 W RMS.—"Intersil Data Book," Intersil, Cupertino, CA, 1981, p. 5-218.

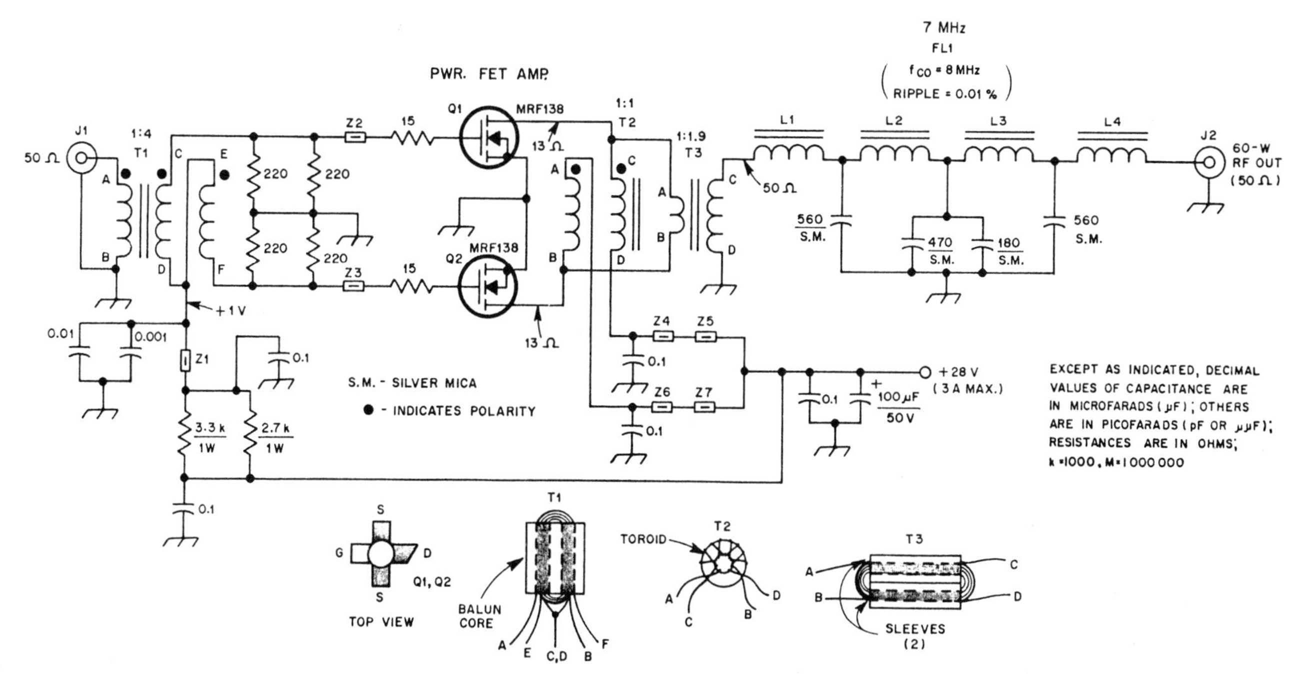

BROADBAND PUSH-PULL CLASS B AMPLIFIER—Capable of producing 60-W output with input of only 228 mW on frequencies from 1800 kHz to 29.7 MHz. The circuit requires an output filter for the desired operating frequency. The efficiency of the circuit in class B service is 72.6% with gate bias on 1 V. Parasitic suppressors are located at the gates of Q1 and Q2. L1 and L4 are formed of 13 turns of No. 22 wire on a T68-6 toroid core. L2 and L3 are formed of 19 turns of No. 22 wire on a T68-6 toroid core. T1 is a 4 : 1 balun transformer formed from 10 trifilar turns of No. 28 wire through Fair-Rite balun core No. 2843000302 or 12 trifilar turns of No. 26 wire on a FT50-43 toroid form. T2 is 12 turns of No. 22 wire on two stacked FT50-43 toroid forms. T3 is a broadband conventional transformer; it has a primary of two turns of No. 18 plastic-insulated wire and a secondary of three turns of the same wire. Both windings are wound through the holes of two S43-621-1 ferrite sleeves. Z2 and Z3 are miniature 900-mu ferrite beads (FB-43-101 or equivalent). Z1 and Z4–Z7 are large 900-mu ferrite beads (FB-43-801 or equivalent).—D. DeMaw, Go Class B or C with Power MOSFETs, *QST,* March 1983, pp. 25–29.

4-W AUDIO AMPLIFIER WITH DC VOLUME CONTROL—Applying a DC voltage of 3.5–8 V to pin 7 of TDA1013A IC controls volume of amplifier over a range of 80 dB. The output is typically 4.5 W into an 8-Ω load with a THD of 0.5%.—"Signetics Linear LSI Data and Applications Manual," Signetics, Sunnyvale, CA, 1985, p. 6-91.

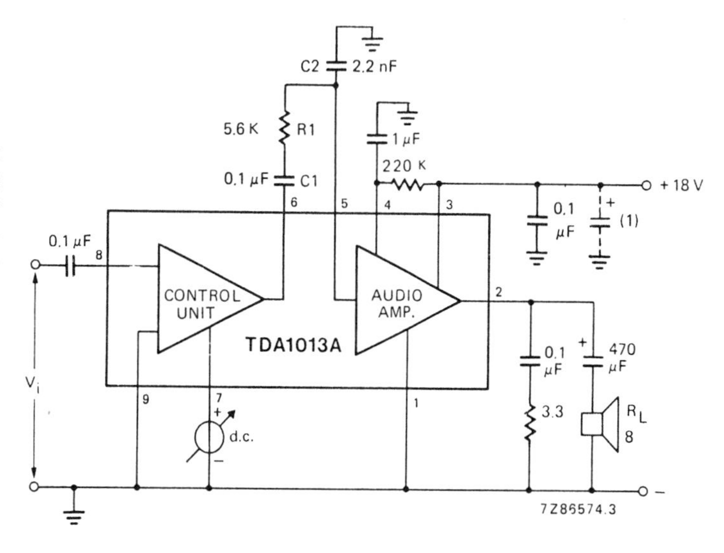

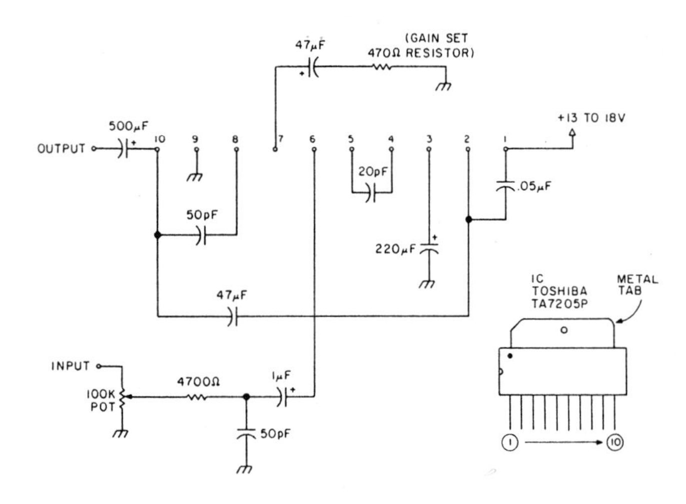

GENERAL-PURPOSE AUDIO AMPLIFIER—Can be used in numerous applications where audio amplification is needed. The 100-K variable resistor allows the circuit to handle a variety of input levels.—A. Joffe, Everyman's Audio Amplifier, *73 Magazine,* November 1982, p. 90.

3

Antenna Circuits

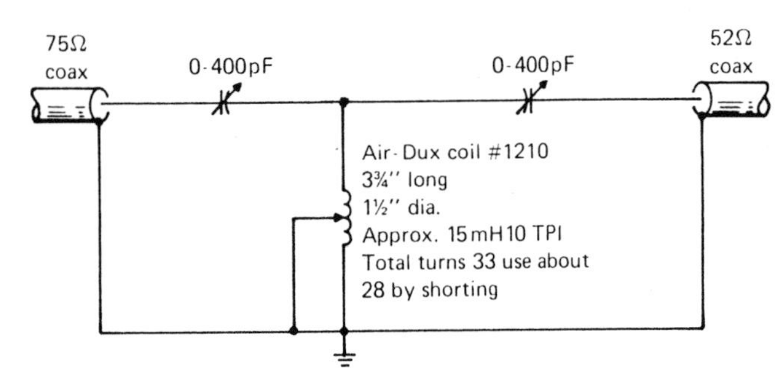

160-METER T MATCH ANTENNA TUNER—Designed for use on the 160-meter (1800- to 2000-kHz) amateur radio band. The tuner is designed to properly load the 75-Ω output of a transmitter into the 52-Ω load presented by a base-loaded vertical antenna for 160 meters.—E. Marriner, A T Match For 160 Meters, *CQ,* December 1982, pp. 60–61.

NOISE BRIDGE—Used for impedance measurement and antenna tuning. Circuit works by generating RF noise over the HF range. A receiver is used as the detector. All transistors are 2N5129 or equivalent. The transformer is wound on a T-50-2 or CF102 form. It consists of four 5-inch lengths of No. 28 wire twisted together along their entire length. The "quadrifilar" winding is obtained by taking the twisted wire bunch and winding it on the core to produce four to five turns.—*73 Magazine* Staff, QRM-Free Antenna Tuning, *73 Magazine,* August 1981, pp. 40–42.

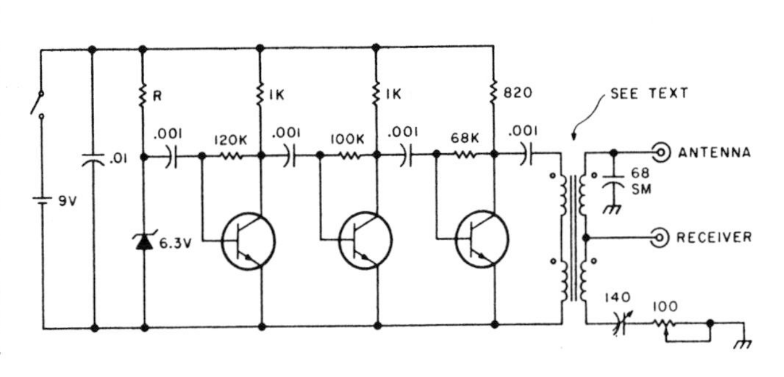

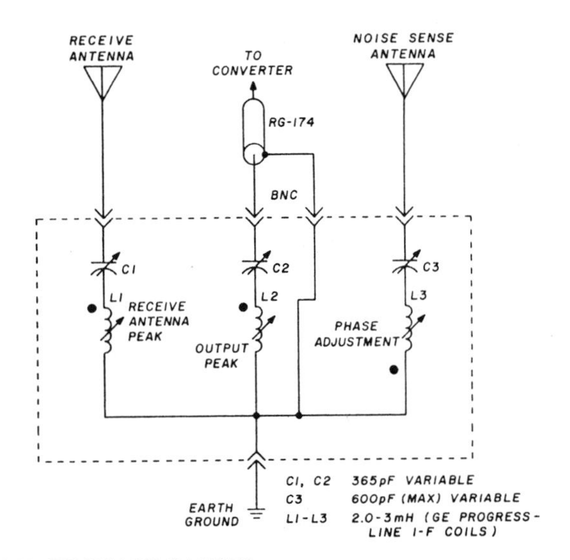

160- TO 190-kHz RF PHASE CANCELING CIRCUIT—Used to decrease electrical noise interference found in 160- to 190-kHz range by use of separate "noise sensing" antenna and altering phase of its signal input and that of normal receiving antenna. Careful adjustment can reduce electrical noise by several S units. All coils should be spaced 0.5 in apart.—S. DeFrancesco, Noise Cancellation Circuit, *Ham Radio,* March 1984, pp. 75–76.

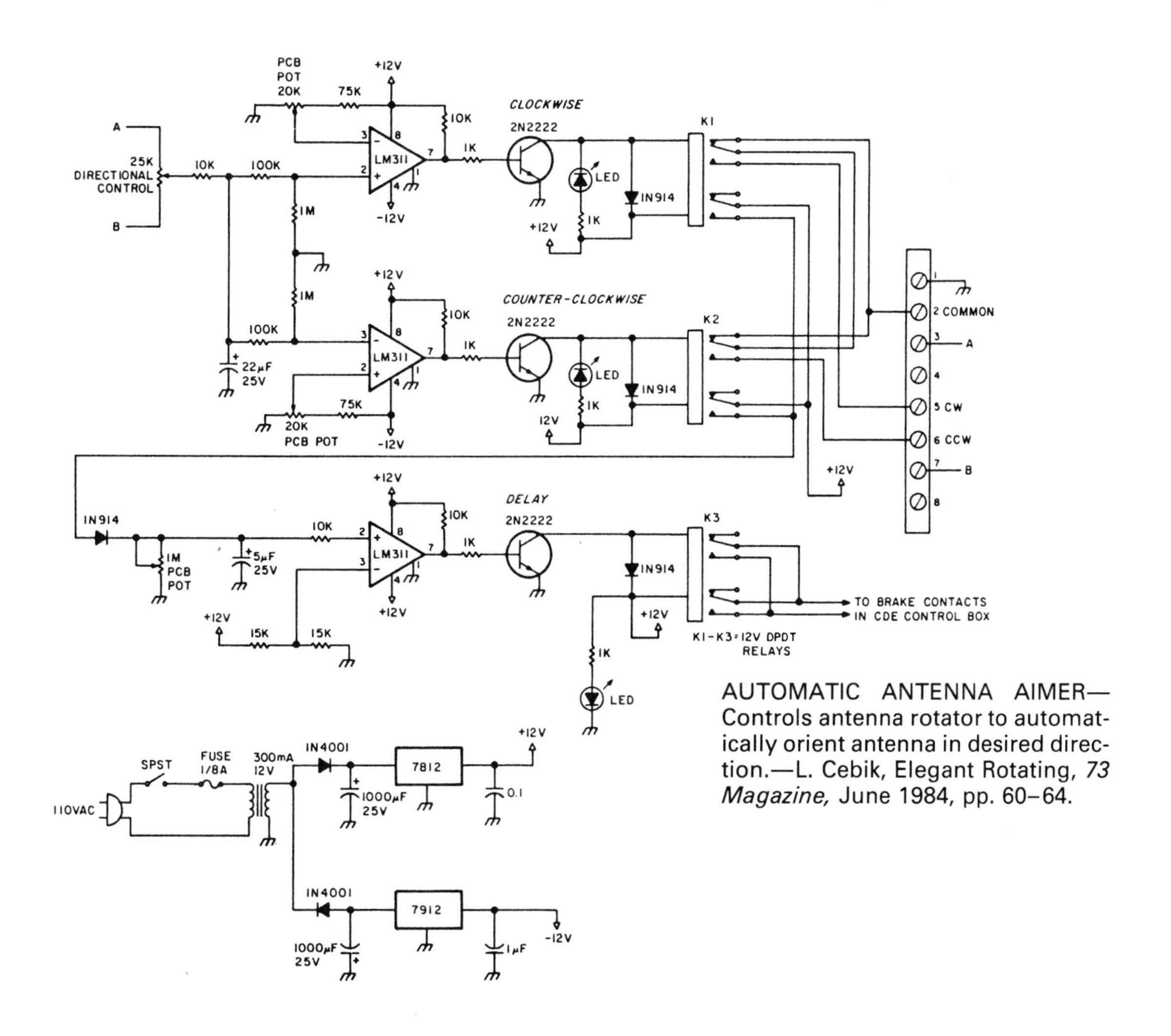

AUTOMATIC ANTENNA AIMER—Controls antenna rotator to automatically orient antenna in desired direction.—L. Cebik, Elegant Rotating, *73 Magazine,* June 1984, pp. 60–64.

UNIVERSAL ANTENNA TUNER—Allows matching of 50-Ω transmitter output into virtually any antenna load over range of 1.8–30 MHz. The circuit is capable of handling up to 1000-W DC input safely. Both C1 and C2 are 250 pF rated at 4 kV each, L is 18 to 28 μH and wound from No. 10 or No. 12 wire, and S1 and S2 are Centralab PA-2000 series or equivalent. Both C1 and C2 need to be "above" ground; plastic mounting screws and plates should be used. The SWR meter immediately following the RF input may be a commercial SWR meter or the accompanying circuit may be used.—*73 Magazine* Staff, A Tuner for Antenna Fanatics, *73 Magazine,* November 1982, pp. 42–44.

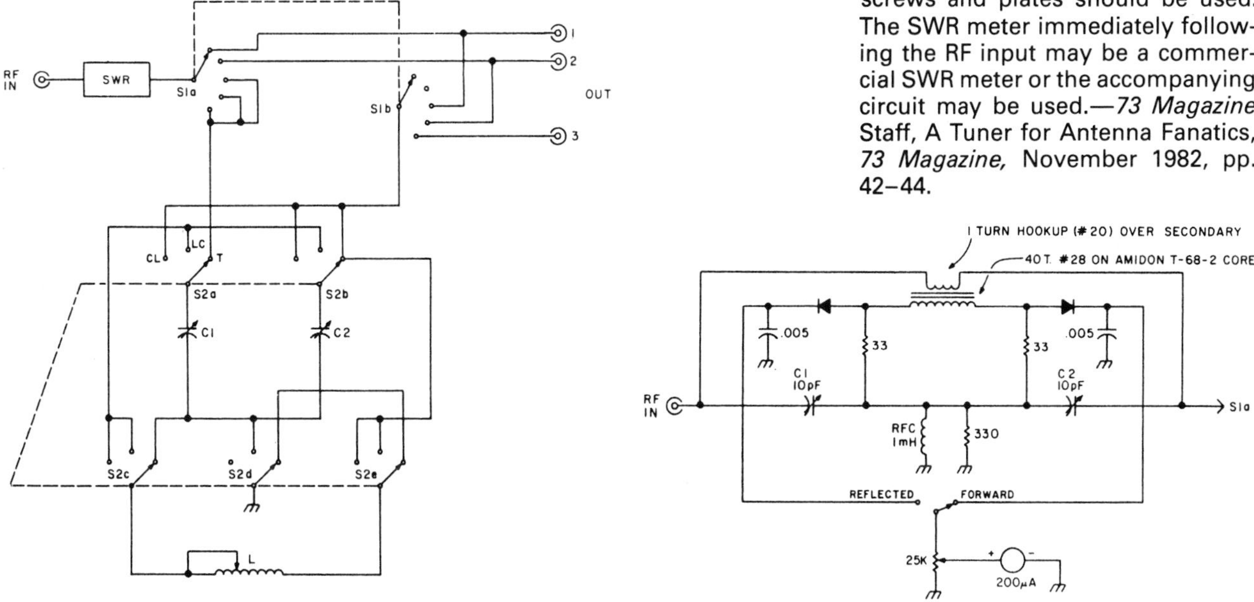

ANTENNA SWITCH/TUNER—Allows selection of up to six different antennas and matching of them to 50-Ω transmitter output from 3500 kHz to 29.7 MHz. The tuner can handle 1000 W of direct current. C1 and C2 are two-gang broadcast receiver-type capacitors, 600 pF per gang. C3 and C4 are 500-pF ceramic or mica transmitting capacitors. The tuner is not suitable for use with an open-wire line.—P. Swearingen, The Coax Matcher, *73 Magazine,* October 1982, pp. 84–88.

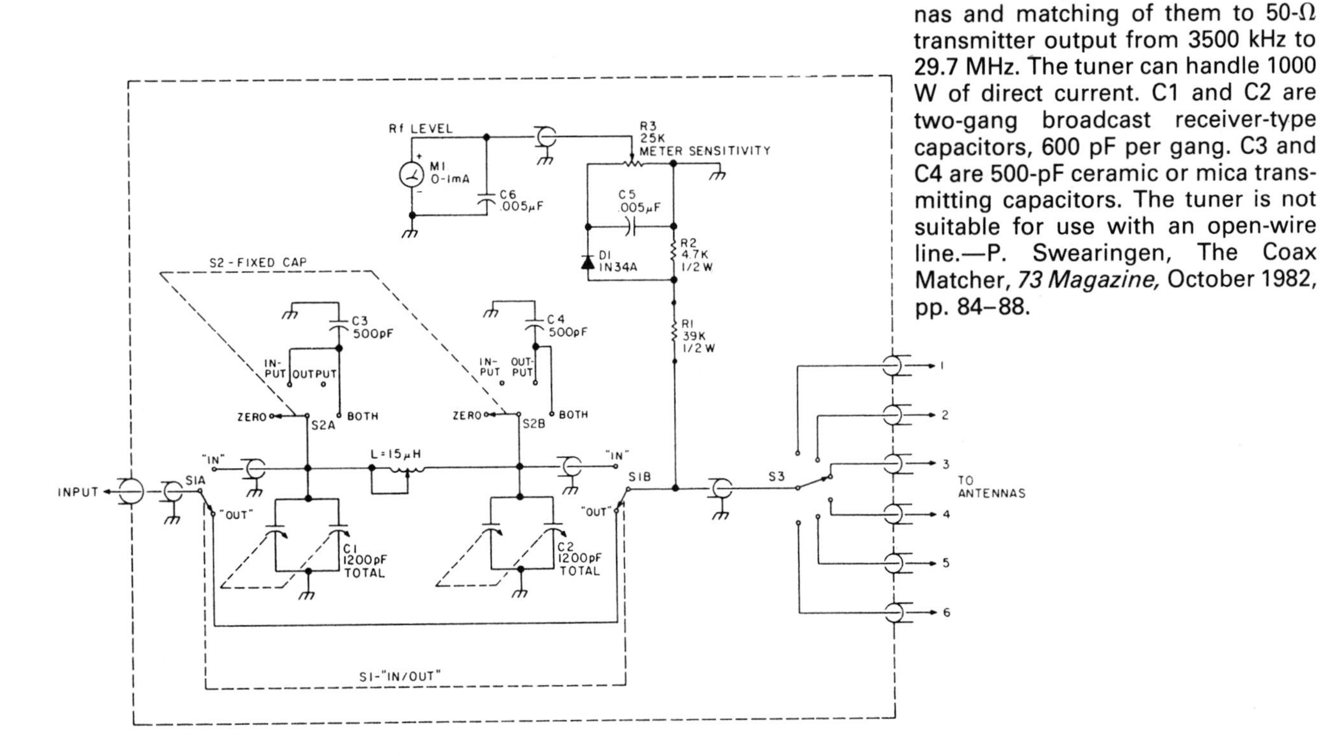

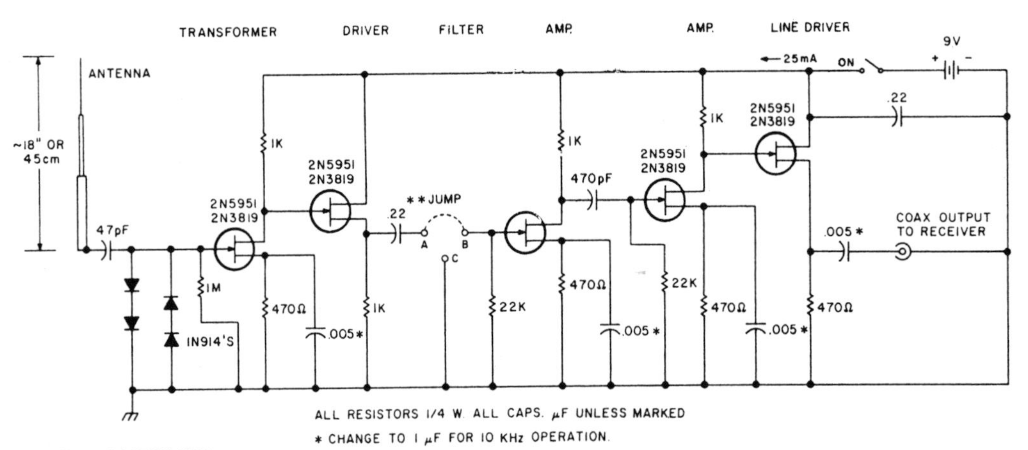

ALL RESISTORS 1/4 W. ALL CAPS. μF UNLESS MARKED

* CHANGE TO 1 μF FOR 10 KHz OPERATION.

** SEE FIG. 2 FOR QRM FILTERS.

ALL-BAND ACTIVE ANTENNA—Gives reception from 540 kHz to 30 MHz with short indoor whip antenna. Interference filters may be added at points A, B, and C in the schematic if needed. All five transistors should be 2N5951 if possible; 2N3819 may be substituted. Circuit may be modified to work down to 10 kHz if all 0.005-μF capacitors are changed to 1 μF. However, antenna will be more prone to power-line noise.—R. Wilson, The Incredible Antenna Mark 2, *73 Magazine,* October 1982, pp. 44–45.

A 10pF B 4.7K C

AC BUZZ

A 39pF B 100μH C

BC QRM MILD

DIP AT 948KHz   DIP AT 1371KHz

A 47Ω 30μH 933pF 10.4μH 1300pF 4.6μH 3.8μH 5.7μH 100Ω B C

BC QRM BAD

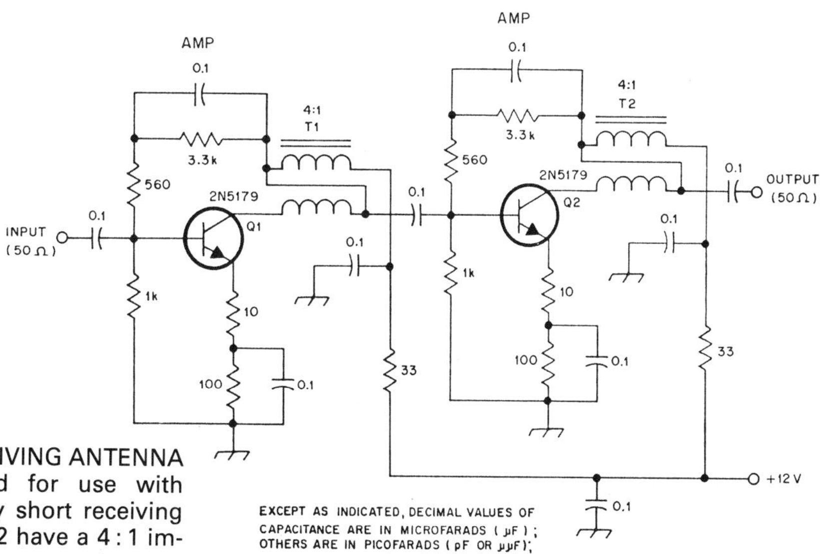

BROADBAND RECEIVING ANTENNA PREAMP—Designed for use with loops or electrically short receiving antennas. T1 and T2 have a 4 : 1 impedance ratio and are wound on FT-50-61 toroid cores which have a μH of 125. They contain 12 turns of No. 24 wire bifilar wound. All capacitors are disc ceramic.—D. DeMaw, Maverick Trackdown, *QST,* July 1980, pp. 22–25.

5'
SPLIT SHIELD
RG-58/U
5'
500pF
MINIBOX
COAX TO RECEIVER (ANY LENGTH)

1500- TO 2000-kHz SHIELDED RECEIVING LOOP—Constructed of RG-58/U coaxial cable mounted on a 6.5-foot wooden dowel, this antenna provides figure 8 reception pattern which allows for rejection of interfering stations located at right angles to plane of loop. The antenna should be mounted on a base so that it can be easily rotated. The antenna may need a preamplifier to allow the reception of weaker signals.—J. Geist, Top-Notch for Top Band, *73 Magazine,* May 1982, pp. 26–30.

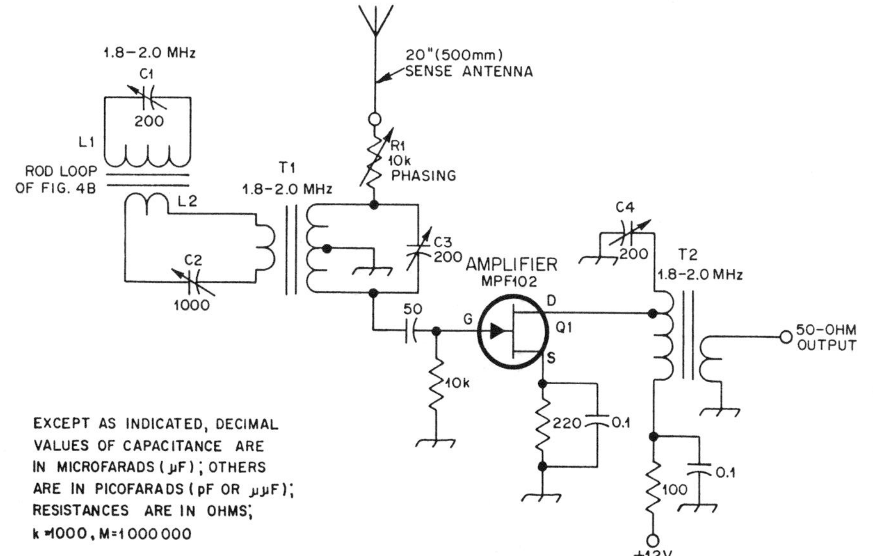

CARDIOID RESPONSE FERRITE ROD ANTENNA—Sense antenna gives essentially a unidirectional response from ferrite loop. T1 and T2 are both wound on T68-2 or T68-6 cores. The circuit can tune 500–2000 kHz, depending on values of L1 and L2; they should be values resonant with C1 and C2 at the midpoint of the desired receiving range. The rod antenna should be wound with Litz wire for best results.—D. DeMaw, Maverick Trackdown, *QST,* July 1980, pp. 22–25.

4

Audio, High Fidelity, and Music Circuits

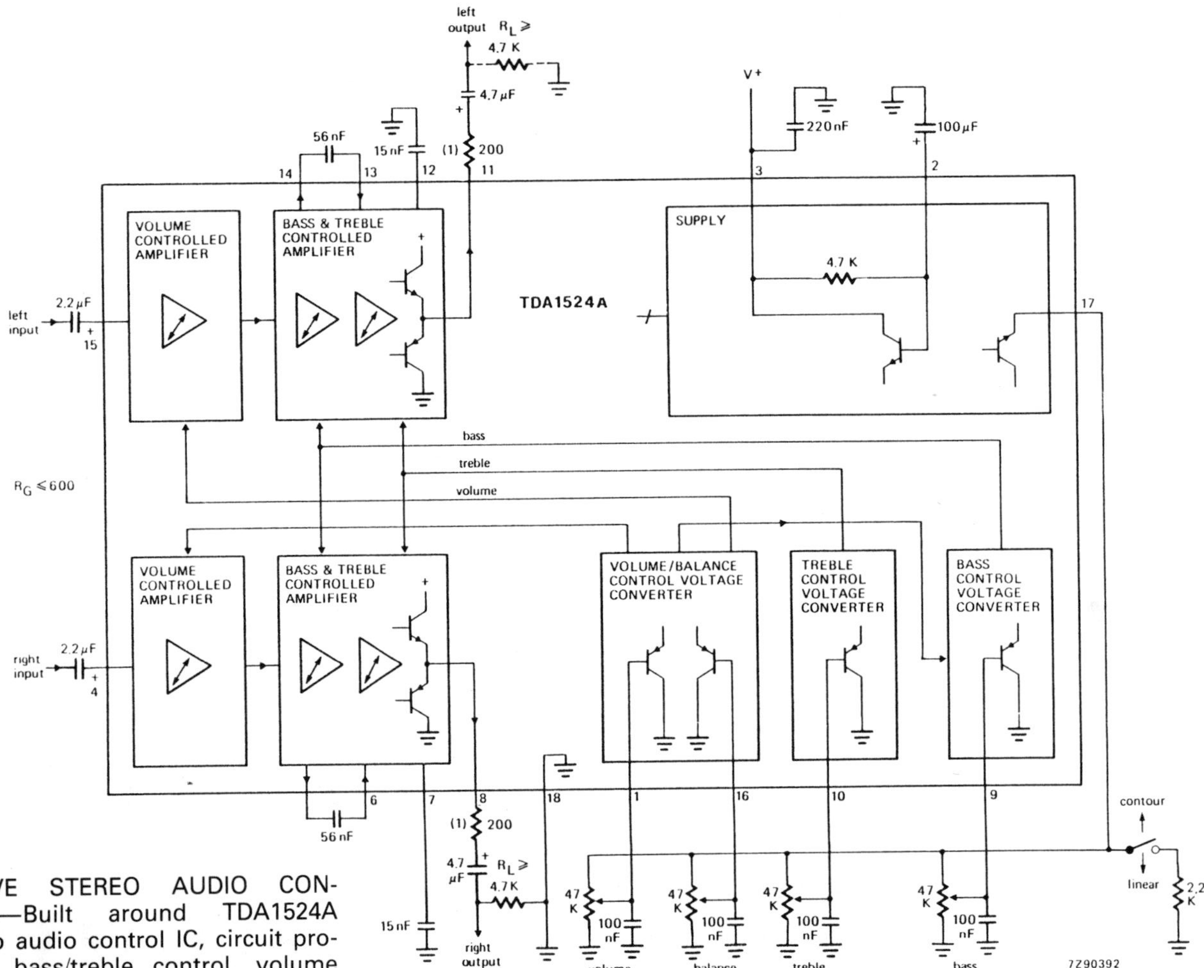

ACTIVE STEREO AUDIO CONTROL—Built around TDA1524A stereo audio control IC, circuit provides bass/treble control, volume control, and balance functions which can be controlled by potentiometers or DC voltages. Bass emphasis can be increased by a double-pole low-pass filter. Supply voltage can be 3–18 V DC (12 V nominal) and current requirements are typically 35 mA. Noise is low because of the internal gain.—"Signetics Linear LSI Data and Applications Manual," Signetics, Sunnyvale, CA, 1985, p. 5-61.

RIAA EQUALIZATION PHONOGRAPH PREAMPLIFIER—Circuit has phono cartridge output of 5 mV and gain of 40 dB at 1000 Hz. Distortion is less than 0.01% across the audio spectrum. The maximum design input to the preamplifier is 25 mV RMS.—"Signetics Analog Applications Manual," Signetics, Sunnyvale, CA, 1979, pp. 209–210.

* 1/2 OF DUAL OPERATIONAL AMPLIFIER

NOTE
All resistors are 1% metal film and are valued in ohms.

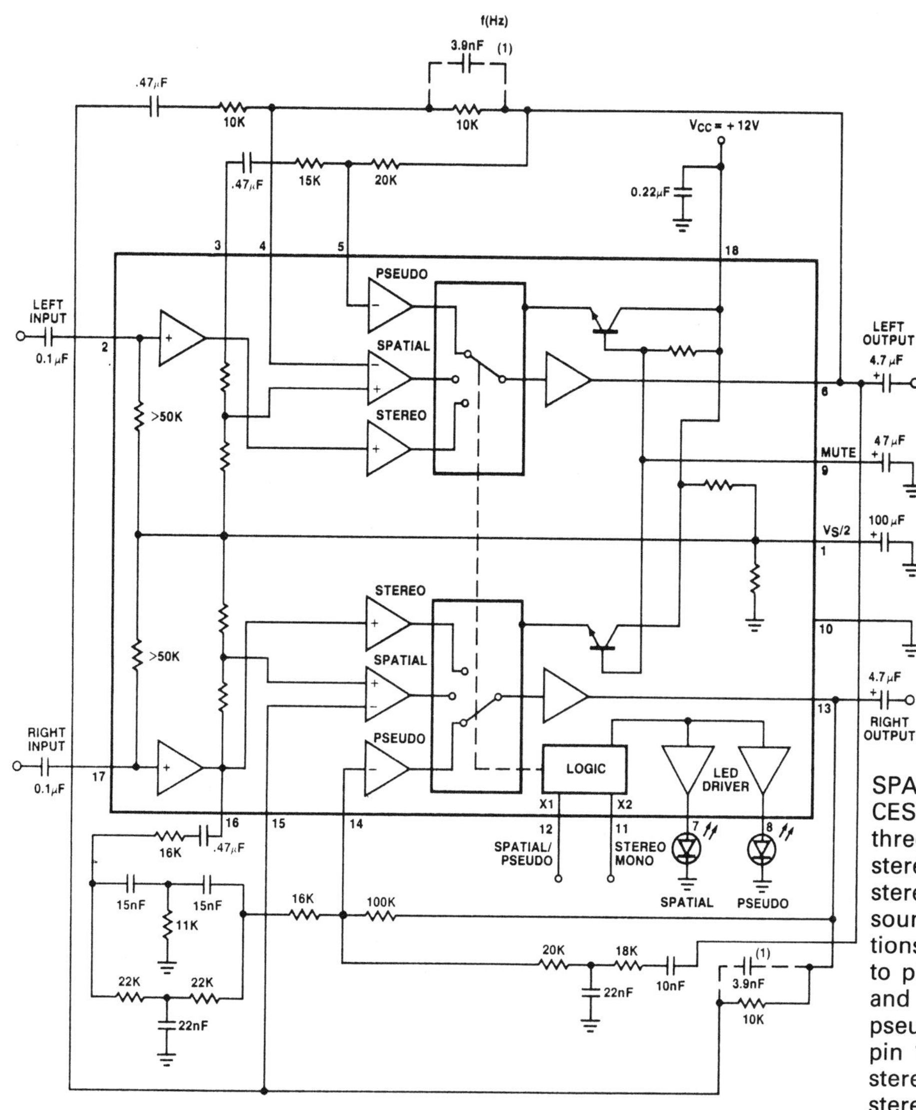

SPATIAL/PSEUDO-STEREO PROCESSOR—TDA3810 IC provides three functions: spatial sound from a stereo source, stereo sound from a stereo source, and pseudo-stereo sound from a mono source. Functions are selected by signals applied to pins 11 and 12. If pin 11 is high and pin 12 is low, the output is pseudo-stereo. If pin 11 is high and pin 12 is high, the output is spatial stereo. If pin 11 is low, the output is stereo regardless of whether pin 12 is high or low. A low signal is defined as 0 to 0.8 V and high is 2 V to V_{CC}. LEDs at pins 7 and 8 indicate operating status; if both are off, circuit is operating in stereo mode. Capacitors in dashed lines may be necessary for best high-frequency response.—"Signetics Linear LSI Data and Applications Manual," Signetics, Sunnyvale, CA, 1985, p. 5-76.

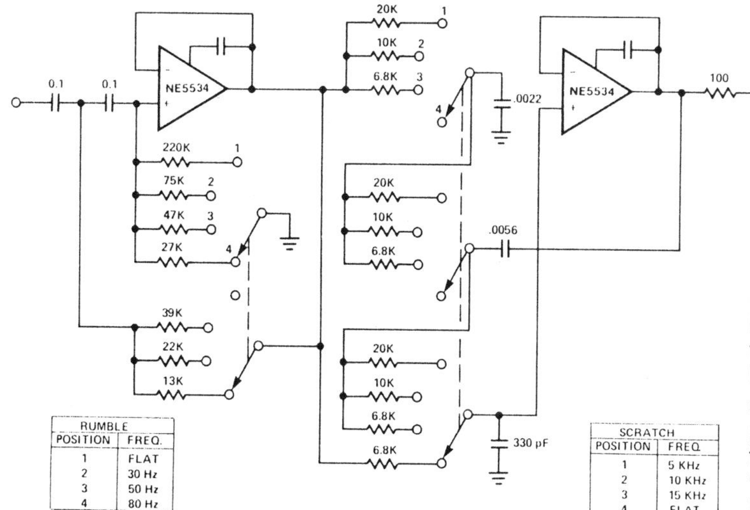

RUMBLE/SCRATCH FILTER—Follows amplifier stage of an audio system to minimize extraneous noise at low and high ends of frequency spectrum. The circuit uses the two-pole Butterworth approach and features switch-selected breakpoints.—"Signetics Analog Applications Manual," Signetics, Sunnyvale, CA, 1979, p. 35.

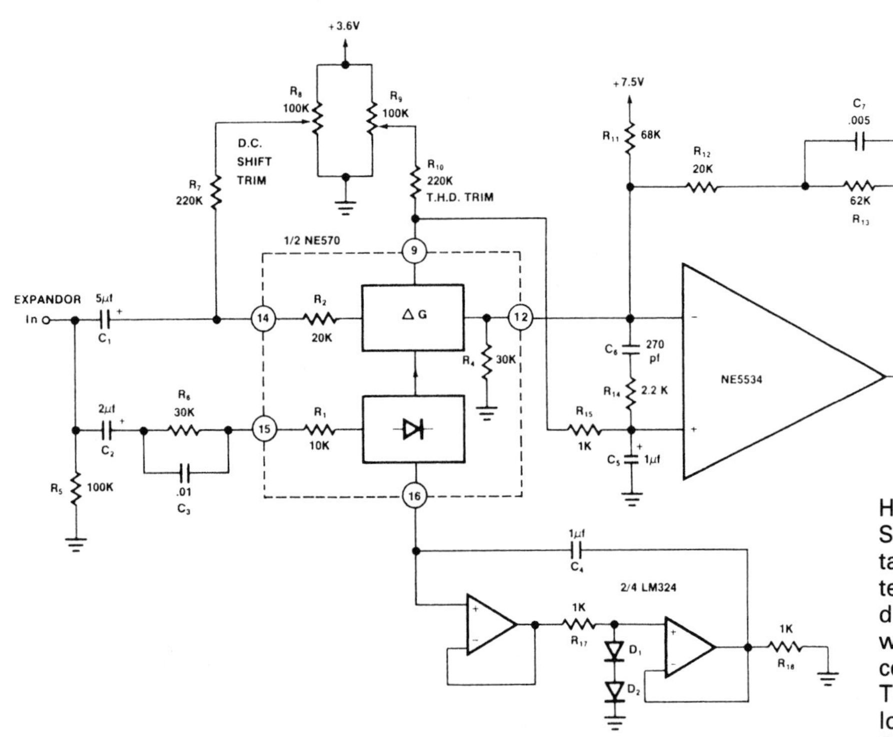

HI-FI EXPANDOR WITH DE-EMPHASIS—Used for noise reduction in tape recorders, transmission systems, bucket brigade delay lines, and digital audio systems when used with compandor. Compandor-processed audio is required at the input. The gain is unity.—"Signetics Analog Applications Manual," Signetics, Sunnyvale, CA, 1979, pp. 235–237.

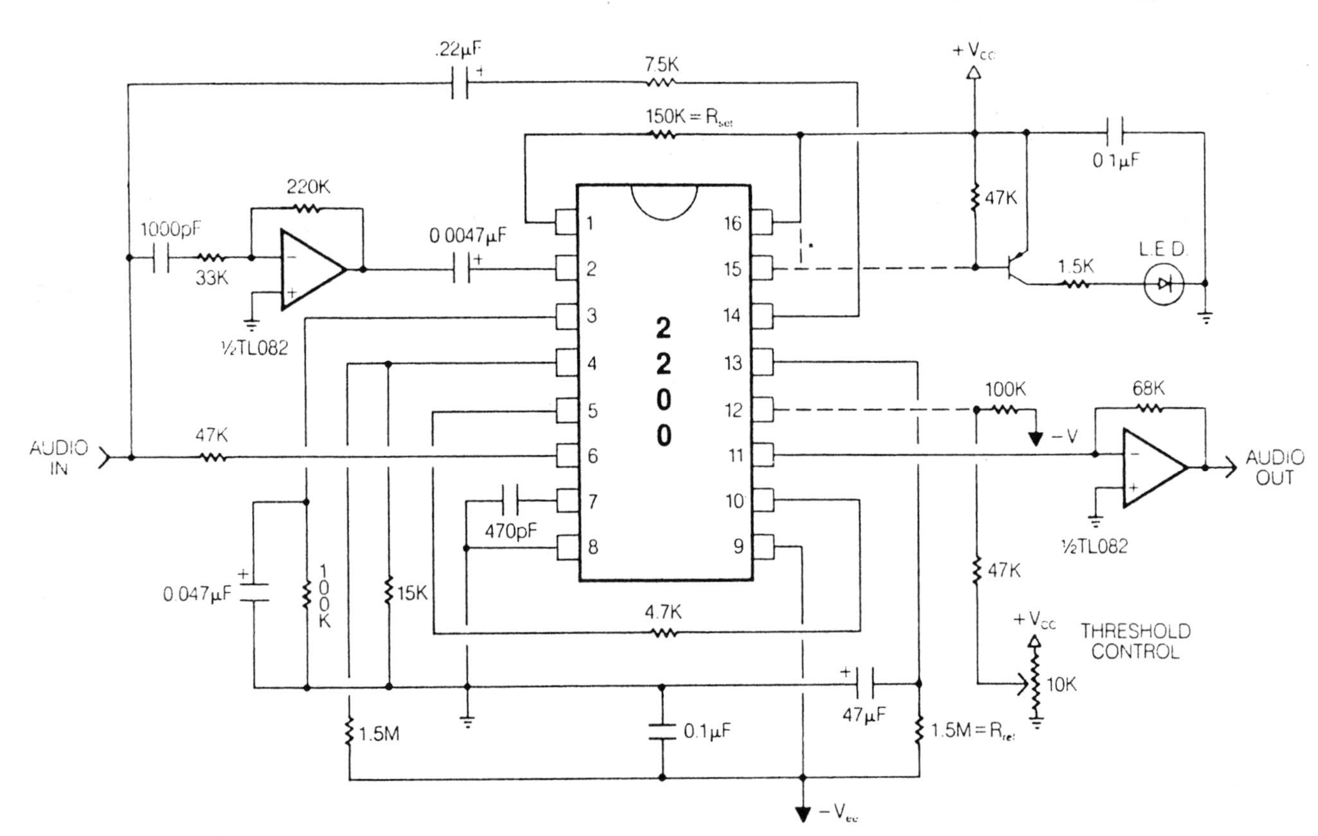

NOISE REDUCTION SYSTEM—Provides up to 30 dB of noise reduction using the Dynafex™ system developed by MICMIX Audio Products, Inc. The system combines dynamic filtering and a downward expander. In the circuit shown, the LED lights at the onset of downward expansion. The threshold control is used to tailor the system's response for best performance with varying source material and signal-to-noise ratios. The circuit shown requires 15 V DC.—"Dynafex™ Noise Reduction System," Solid State Micro Technology for Music, Santa Clara, CA, 1984, SSM 2200.

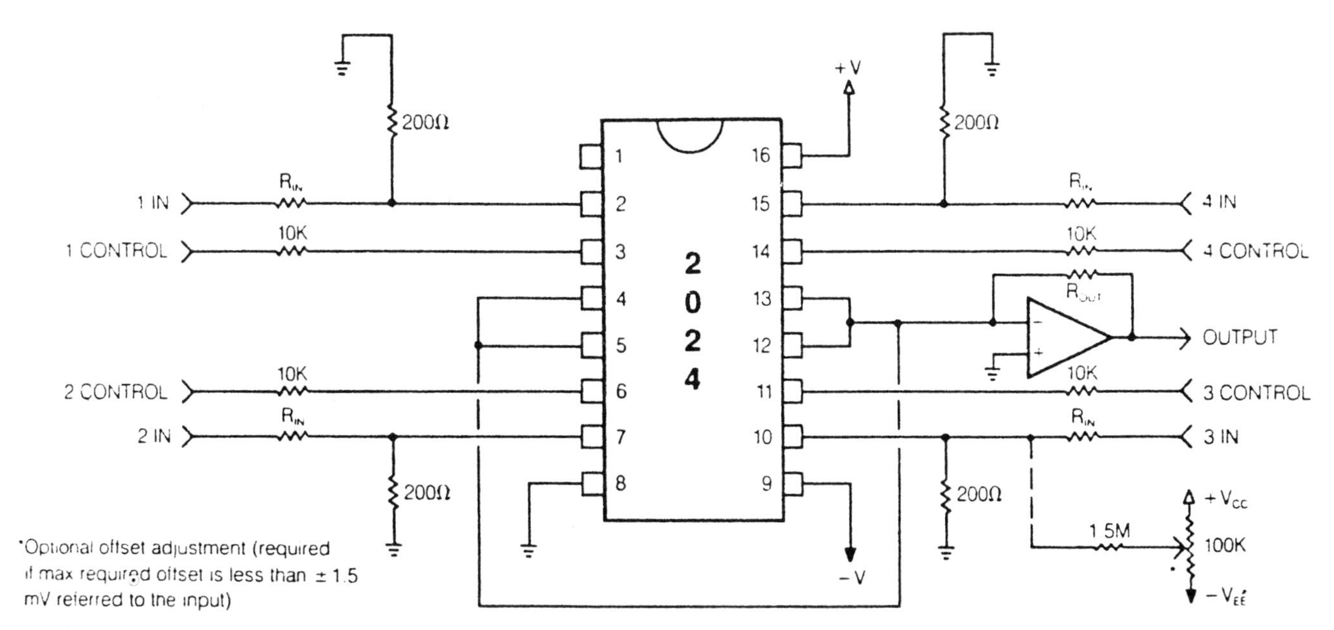

FOUR-CHANNEL MIXER—Built around SSM 2024 quad current controlled amplifier IC which has four independent VCA sections. It is a current output device. Device operates class A from ±15-V DC supplies.—"Quad Current Controlled Amplifier," Solid State Micro Technology for Music, Santa Clara, CA, 1984, SSM 2024.

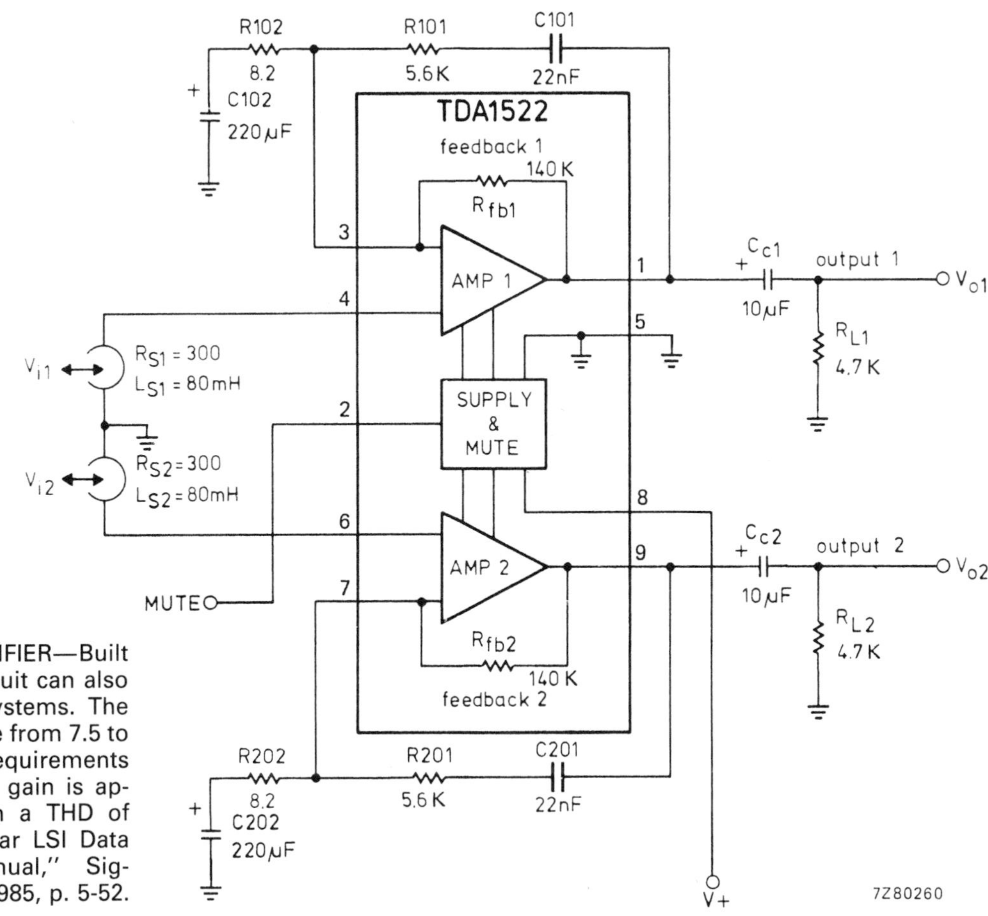

CASSETTE PREAMPLIFIER—Built around TDA1522 IC; circuit can also be used in car radio systems. The supply voltage can range from 7.5 to 23 V DC and current requirements are typically 5 mA. The gain is approximately 30 dB with a THD of 0.05%.—"Signetics Linear LSI Data and Applications Manual," Signetics, Sunnyvale, CA, 1985, p. 5-52.

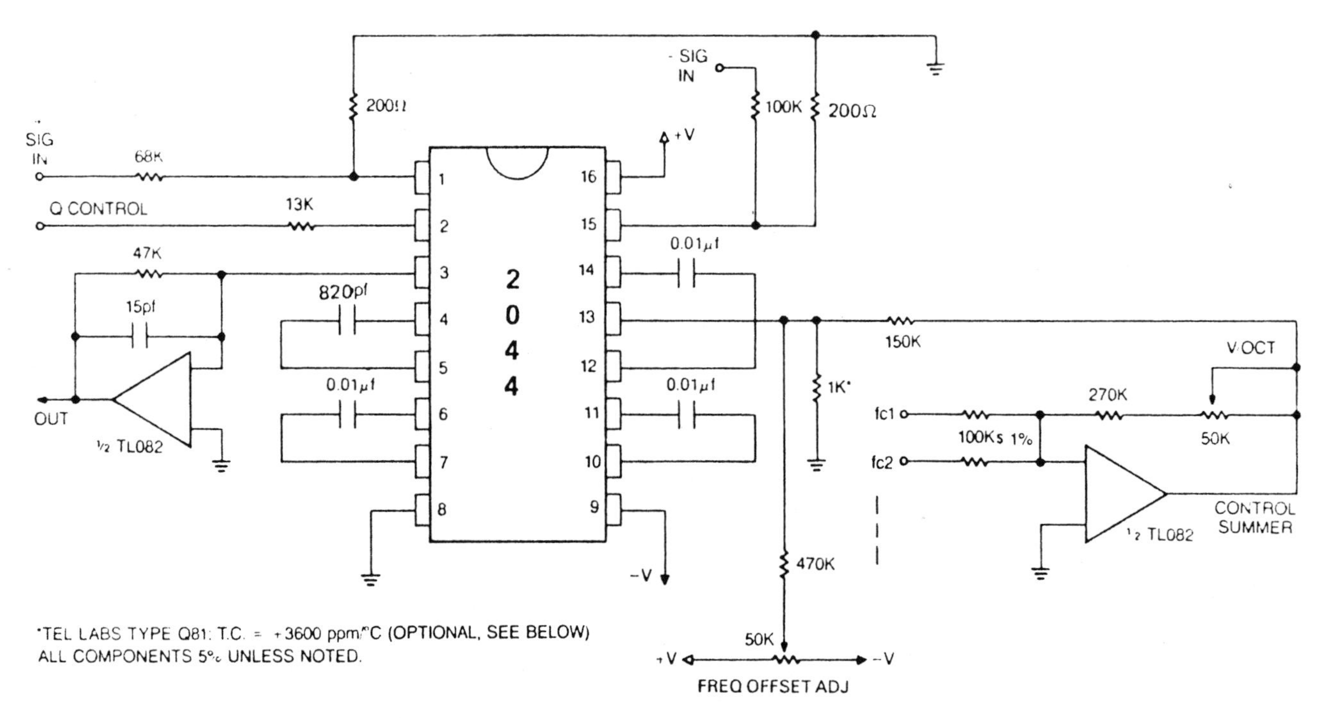

FOUR-POLE ELECTRONIC MUSIC FILTER—Built around SSM 2044 four-pole voltage controlled filter IC which features on-chip resonance control. In the circuit shown, the differential signal inputs will accept signals up to ±18 V P-P.—"4-Pole Voltage Controlled Filter," Solid State Micro Technology for Music, Santa Clara, CA, 1981, SSM 2044.

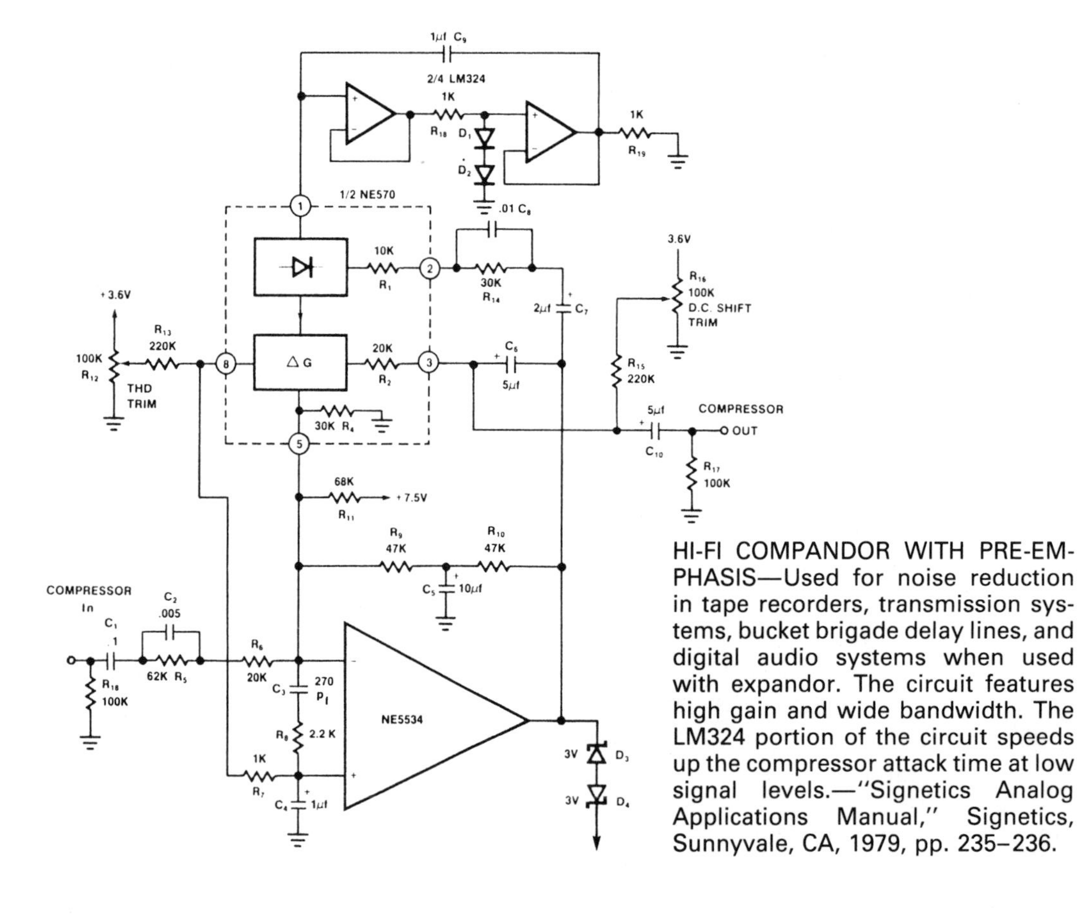

HI-FI COMPANDOR WITH PRE-EMPHASIS—Used for noise reduction in tape recorders, transmission systems, bucket brigade delay lines, and digital audio systems when used with expandor. The circuit features high gain and wide bandwidth. The LM324 portion of the circuit speeds up the compressor attack time at low signal levels.—"Signetics Analog Applications Manual," Signetics, Sunnyvale, CA, 1979, pp. 235–236.

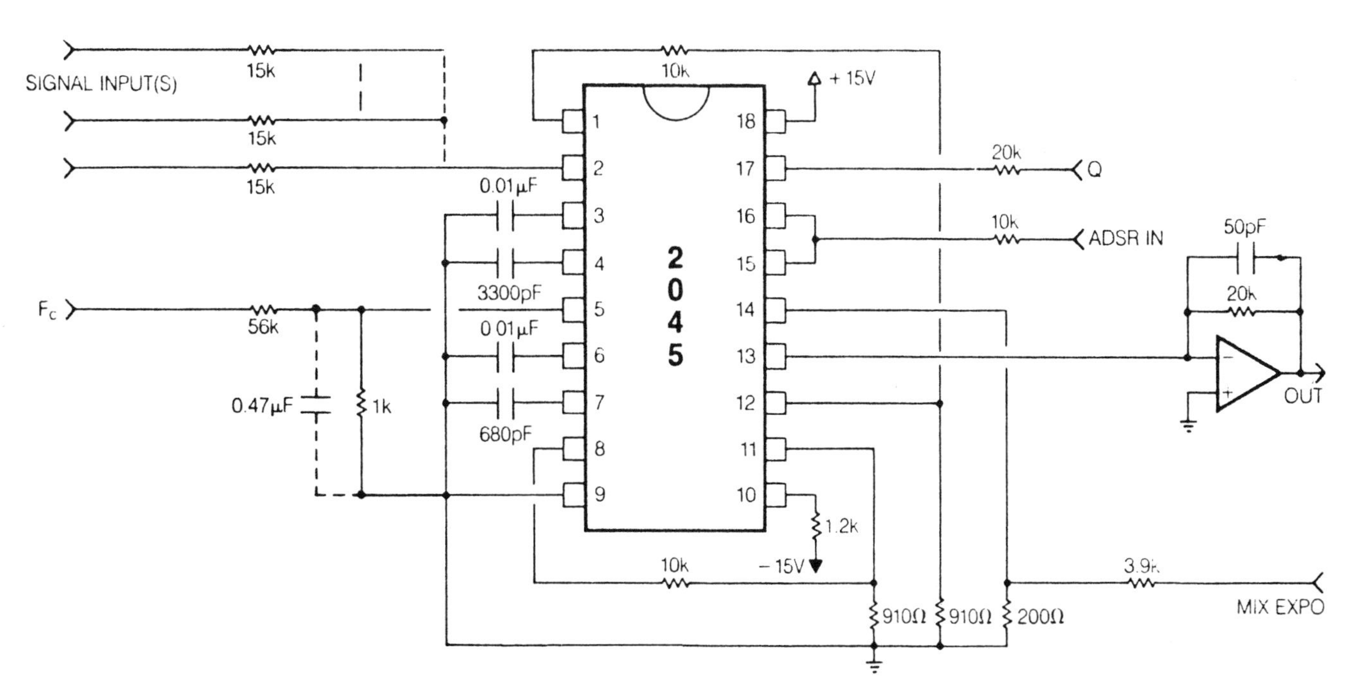

MUSIC VOICING SYSTEM—SSM 2045 provides on-chip voltage controlled filter with both two-pole and four-pole low-pass outputs and uncommitted mixer/VCA combination. The mixer/VCA section can be connected either to the filter input for waveform mixing or to the outputs for mixing between the two-pole and four-pole responses. The system as a whole delivers a "fat" open-loop sound.—"Music Voicing System," Solid State Micro Technology for Music, Santa Clara, CA, 1983, SSM 2045.

5

Automotive Circuits

AUTOMOBILE WARNING SYSTEM—Provides audible warning when headlights are left on or oil pressure light is lit. It also provides an audible turn signal indicator.—K. Barrigar, Automobile Early-Warning System, *73 Magazine,* January 1984, p. 96.

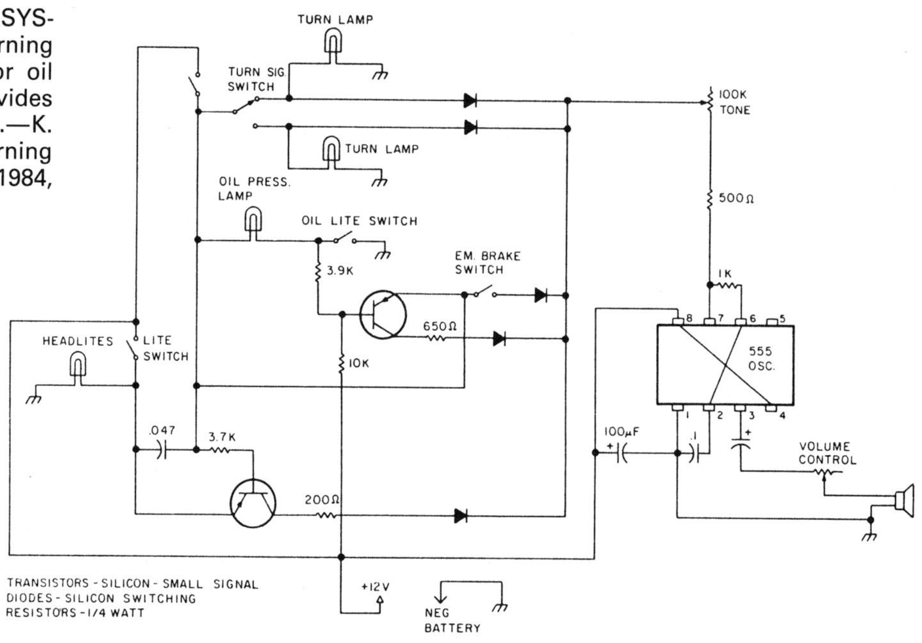

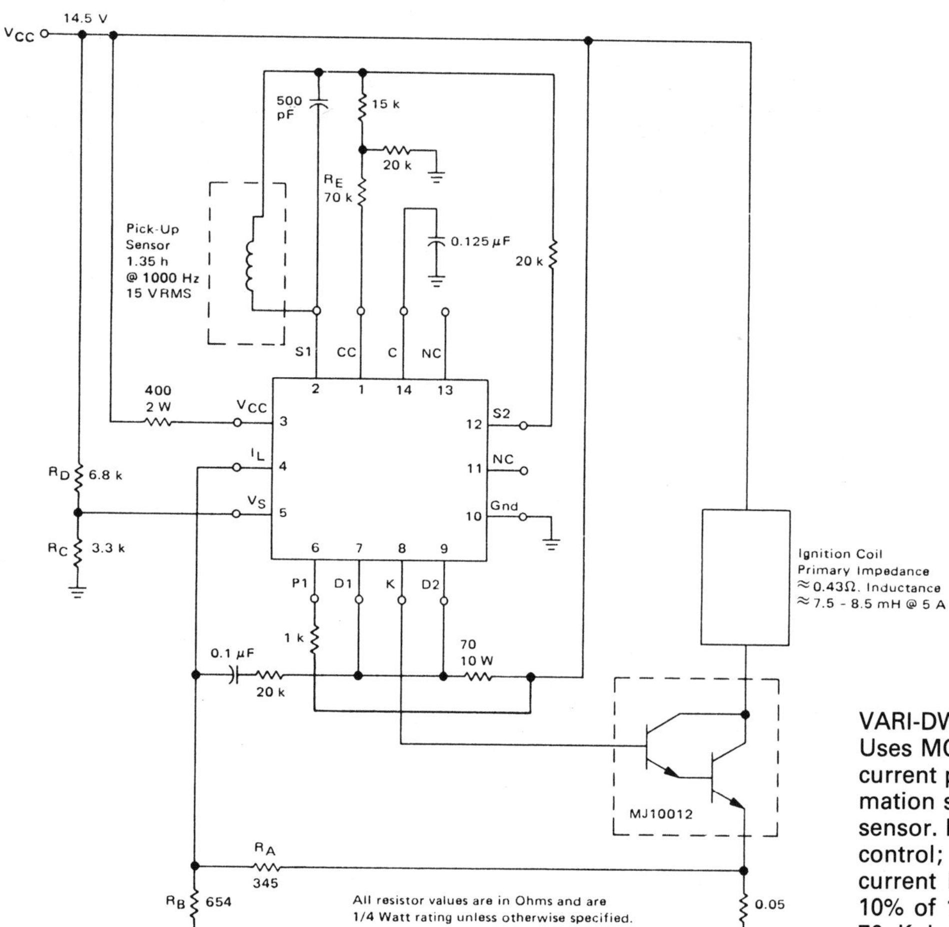

VARI-DWELL IGNITION CIRCUIT—Uses MC333 IC to provide regulated current pulses to the coil from information supplied by a flux averaging sensor. RE is the active region dwell control; RE = 70 K results in output current limit time of approximately 10% of 1000 RPM. Values less than 70 K lengthen this limit time and greater values shorten the time.—"Motorola Linear Integrated Circuits Databook," Motorola, Phoenix, AZ, 1979, p. 5-128.

BURGLAR ALARM—Alarm was designed for automotive applications and resets automatically when triggered by door switch. The alarm provides up to 90 seconds of "grace" to enter and leave the car. The time delay for U1 is adjusted by R1 and C1.—D. Sanderson, The Burglar Alarm That Resets Automatically, *QST,* July 1981, pp. 28–29.

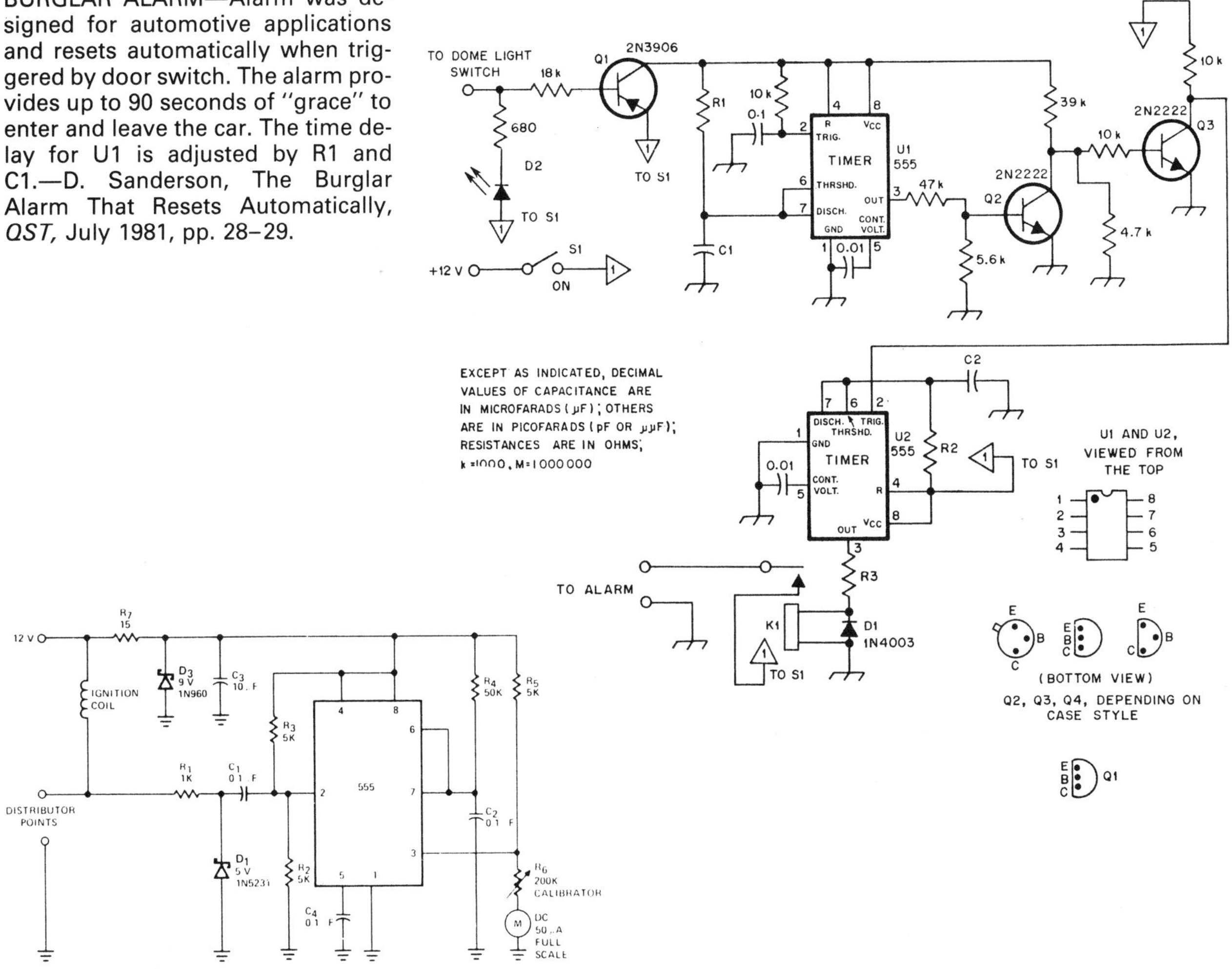

TACHOMETER—555 timer receives pulses from the distributor points. Meter M receives a calibrated current through R_6 when the timer output is high. After time-out, the meter receives no current for that part of the duty cycle. The integration of the variable duty cycle by the meter movement provides a visible indication of engine speed.—"Signetics Analog Applications Manual," Signetics, Sunnyvale, CA, 1979, pp. 156–157.

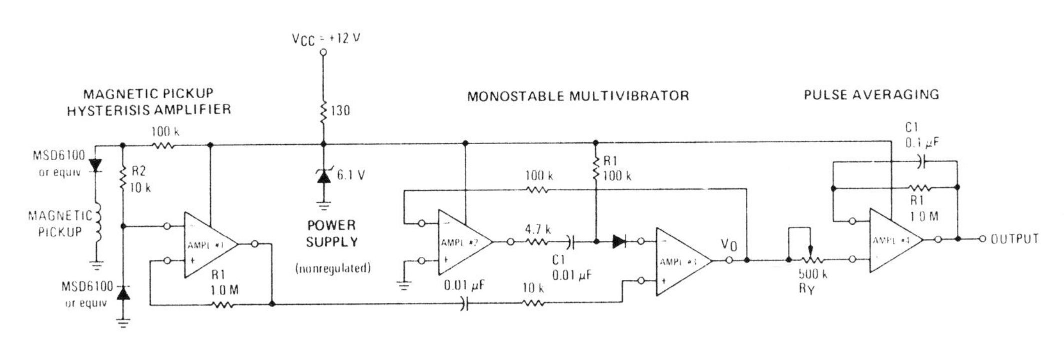

TACHOMETER CIRCUIT—Built around the MC3301 quad op amp, which can operate from a single supply source from +4 to +28 V DC. All four amplifiers operate independently of each other. Tachometer pickup is magnetic and fed into an hysteresis amplifier using one amplifier. Tachometer calibration is accomplished using the 500-K potentiometer.—"Motorola Linear Integrated Circuits Databook," Motorola, Phoenix, AZ, 1979, p. 3-138.

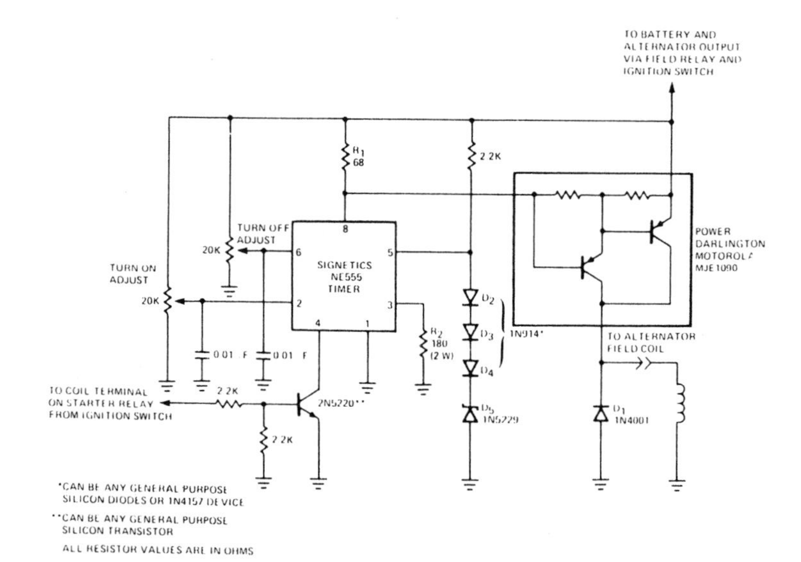

AUTOMOTIVE VOLTAGE REGULATOR—555 timer is "heart" of circuit. When the timer is off so that its output (pin 3) is low, the power Darlington transistor (MJE1090 or equivalent) is off. If the battery voltage becomes too low (less than 14.4 V), the timer turns on and the Darlington pair conducts.—"Signetics Analog Applications Manual," Signetics, Sunnyvale, CA, 1979, p. 163.

6

Conversion Circuits

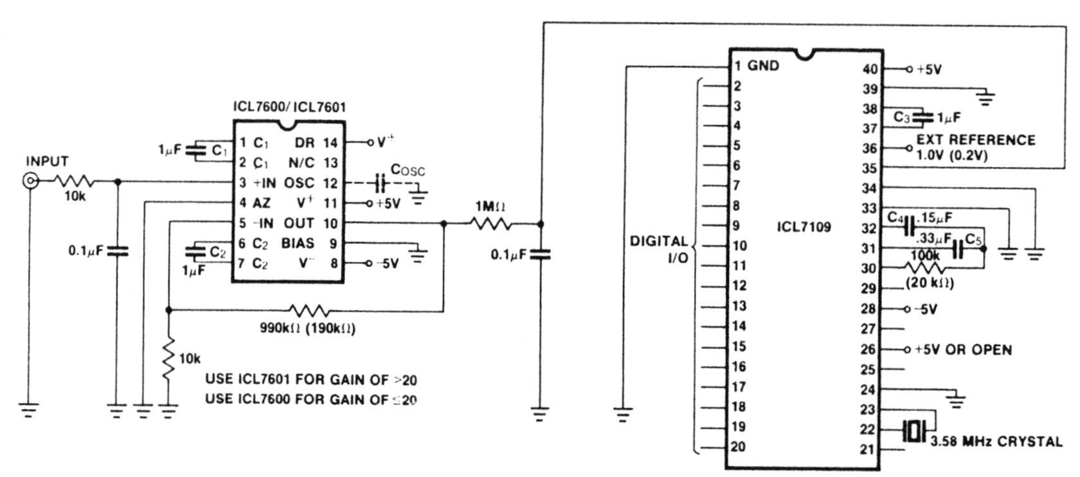

DUAL-SLOPE ANALOG-TO-DIGITAL CONVERTER—Directly interfaces to microprocessors and offers 5-μV resolution. The circuit uses a ICL7600/ICL7601 CAZ AMP preamplifier and an ICL7109 dual-slope A/D converter device; power consumption is typically 2.5 mA. The low-pass filter between the output of the CAZ op amp and the input of the ICL7109 can be used to improve the signal to noise ratio of the system by reducing bandwidth.—"Intersil Data Book," Intersil, Cupertino, CA, 1981, p. 5-127.

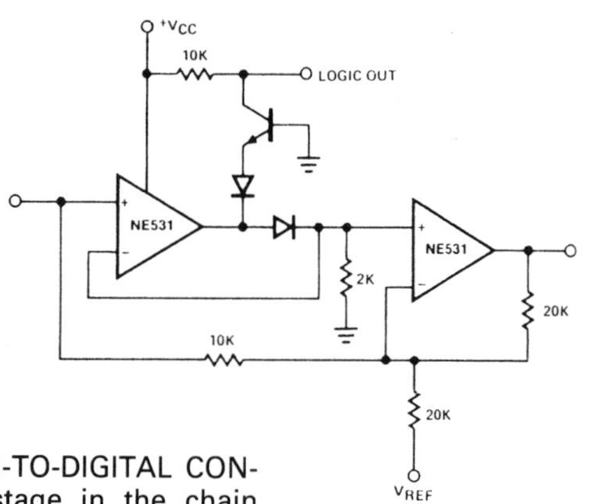

CYCLIC ANALOG-TO-DIGITAL CONVERTER—Each stage in the chain senses the polarity of the input. The stage then subtracts V_{REF} from the input and doubles the remainder if the polarity was correct. A chain of these stages gives the Gray code equivalent of the input voltage in digitized form related to the magnitude of V_{REF}.—"Signetics Analog Applications Manual," Signetics, Sunnyvale, CA, 1979, p. 37–40.

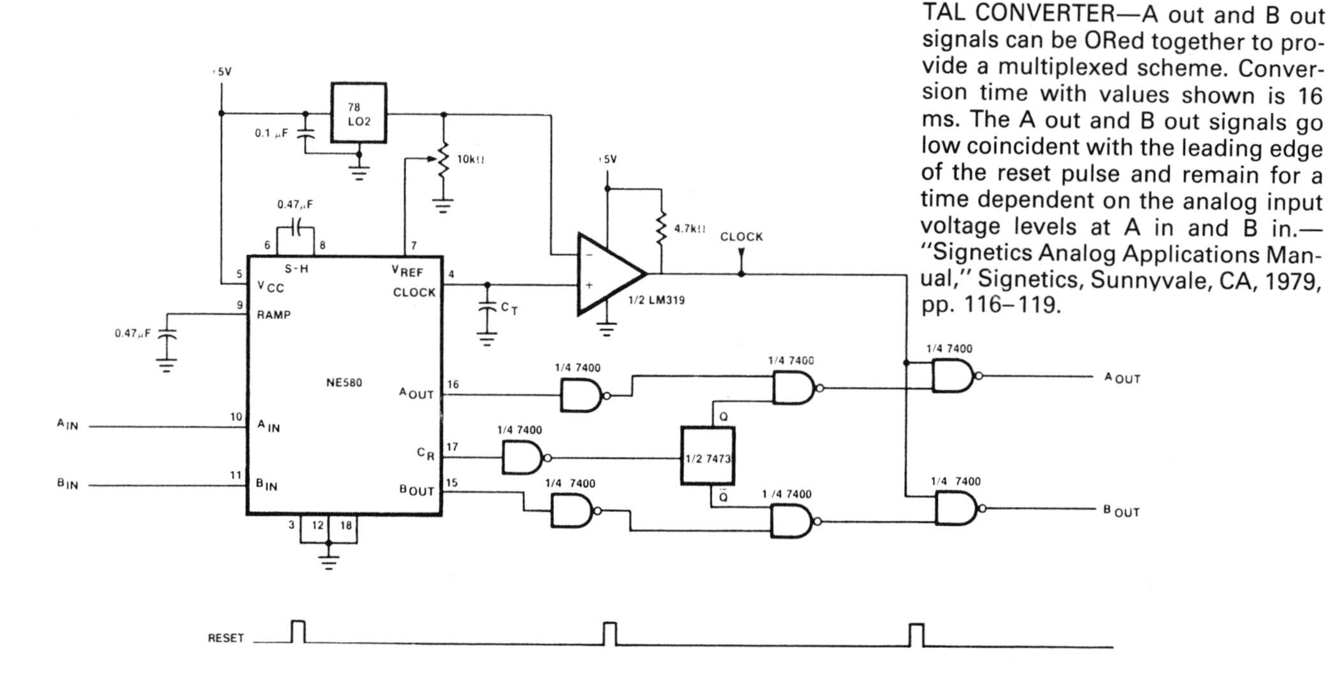

DUAL-CHANNEL ANALOG-TO-DIGITAL CONVERTER—A out and B out signals can be ORed together to provide a multiplexed scheme. Conversion time with values shown is 16 ms. The A out and B out signals go low coincident with the leading edge of the reset pulse and remain for a time dependent on the analog input voltage levels at A in and B in.—"Signetics Analog Applications Manual," Signetics, Sunnyvale, CA, 1979, pp. 116–119.

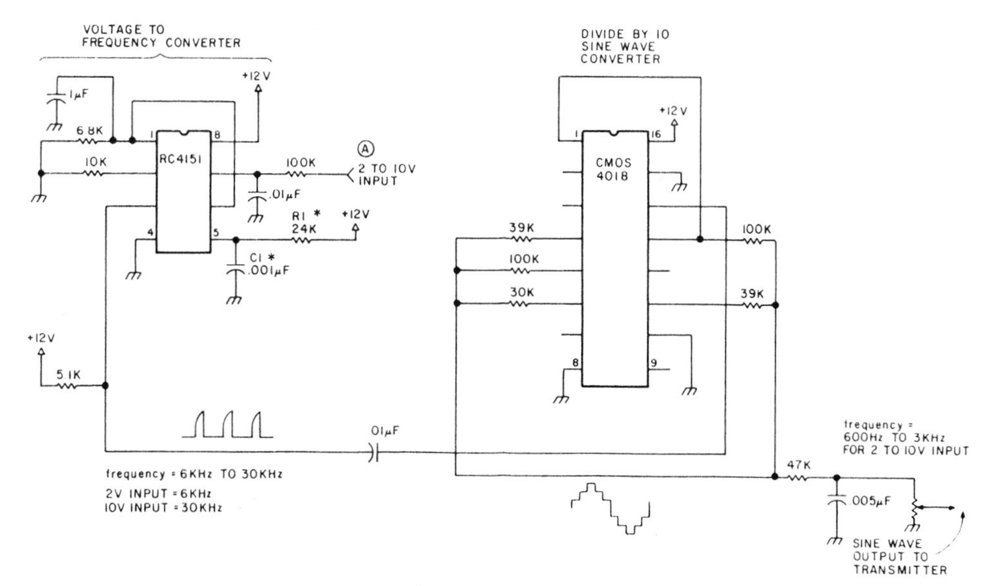

VOLTAGE-TO-FREQUENCY CONVERTER—Gives output of 600 to 3000 Hz for 2- to 10-V input. Originally designed as a telemetry device, the circuit can easily drive a transmitter. For the best results, low-temperature-coefficient, high-tolerance components should be used.—G. Allen, Amateur Telemetry, *73 Magazine,* July 1981, pp. 72–76.

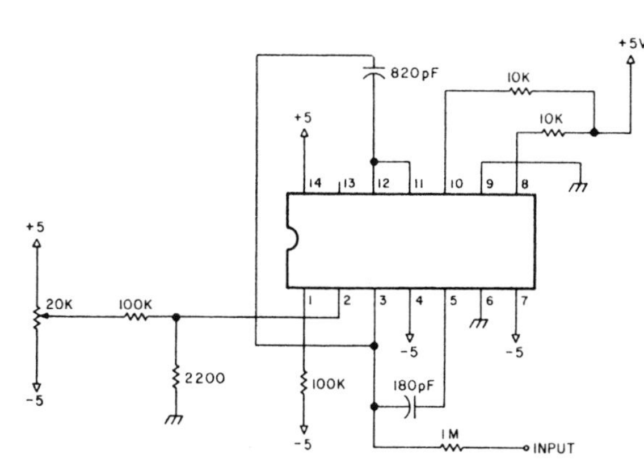

VOLTAGE-TO-FREQUENCY CONVERTER—Produces a frequency from 10 Hz to 10 kHz from a DC or AC input. An input of approximately 1.3 V will produce a 1000-Hz output.—A. Joffe, 1001 Uses for the 9400, *73 Magazine,* August 1983, pp. 88–89.

VOLTAGE-TO-FREQUENCY CONVERTER — Linear-voltage-to-frequency converter uses an LM301A op amp, and 555 timer gives 0.2% accuracy over the 0- to −10-V range. Outputs are TTL-compatible.—"Signetics Analog Applications Manual," Signetics, Sunnyvale, CA, 1979, pp. 159–161.

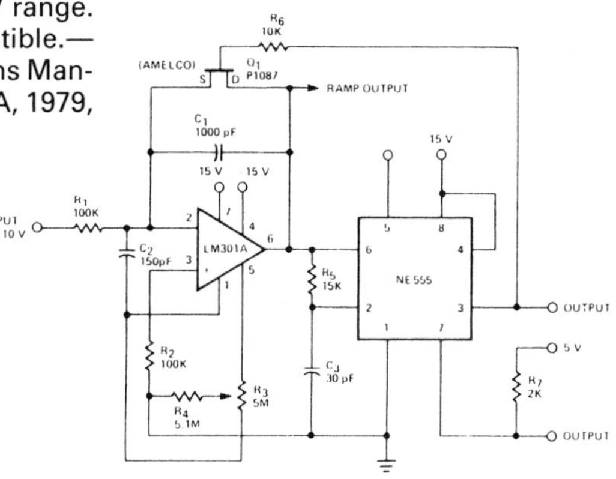

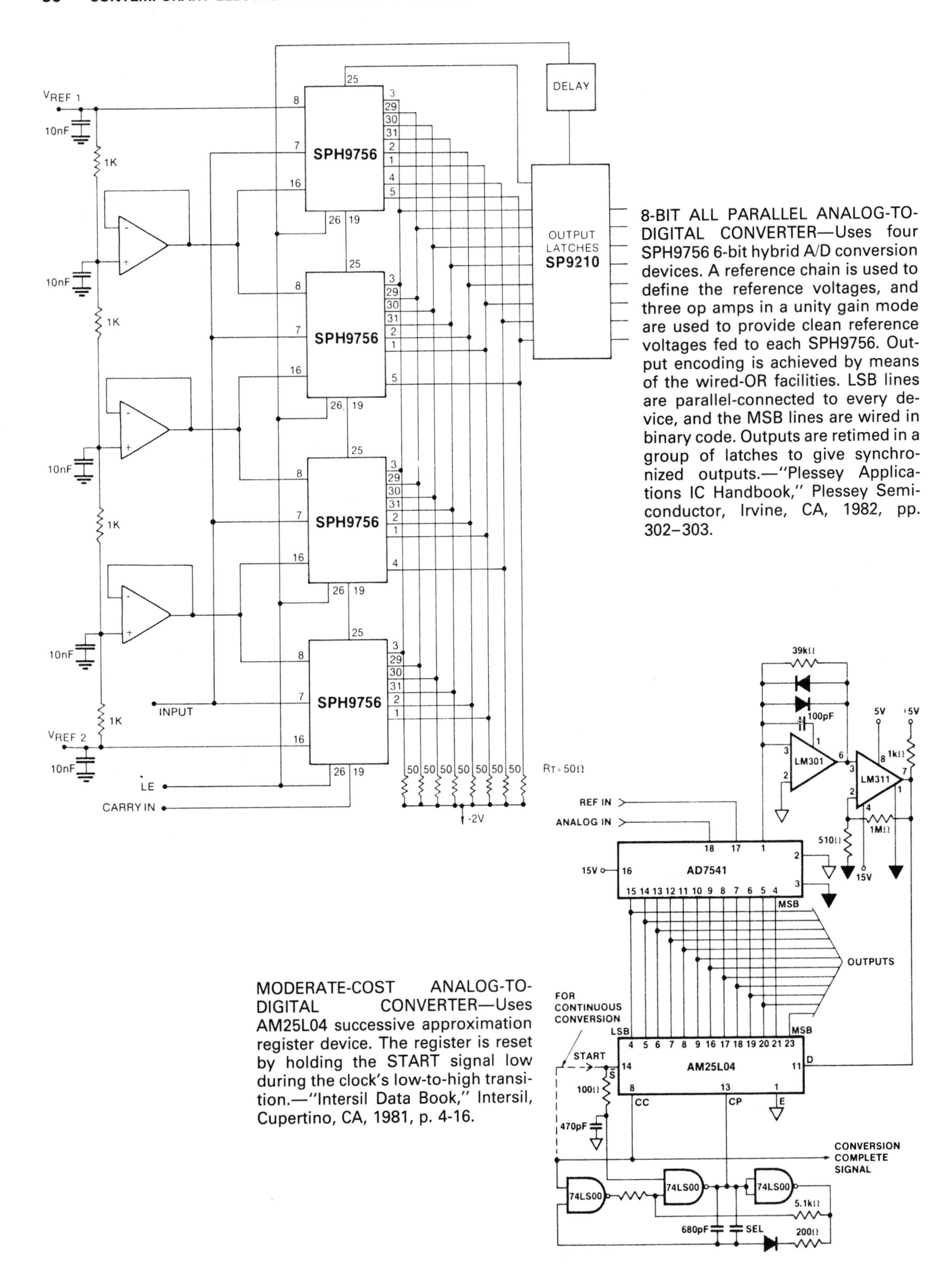

8-BIT ALL PARALLEL ANALOG-TO-DIGITAL CONVERTER—Uses four SPH9756 6-bit hybrid A/D conversion devices. A reference chain is used to define the reference voltages, and three op amps in a unity gain mode are used to provide clean reference voltages fed to each SPH9756. Output encoding is achieved by means of the wired-OR facilities. LSB lines are parallel-connected to every device, and the MSB lines are wired in binary code. Outputs are retimed in a group of latches to give synchronized outputs.—"Plessey Applications IC Handbook," Plessey Semiconductor, Irvine, CA, 1982, pp. 302–303.

MODERATE-COST ANALOG-TO-DIGITAL CONVERTER—Uses AM25L04 successive approximation register device. The register is reset by holding the START signal low during the clock's low-to-high transition.—"Intersil Data Book," Intersil, Cupertino, CA, 1981, p. 4-16.

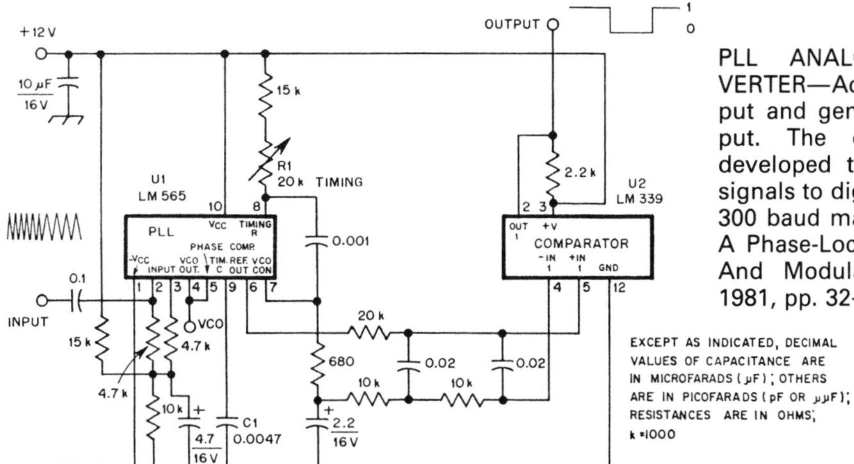

PLL ANALOG-TO-DIGITAL CONVERTER—Accepts audio tones as input and generates a TTL-level output. The circuit was originally developed to convert audio ASCII signals to digital. Data rates of up to 300 baud may be used.—R. Colton, A Phase-Locked Loop Demodulator And Modulator, *QST,* September 1981, pp. 32–33.

EXCEPT AS INDICATED, DECIMAL VALUES OF CAPACITANCE ARE IN MICROFARADS (μF); OTHERS ARE IN PICOFARADS (pF OR μμF); RESISTANCES ARE IN OHMS; k = 1000

VOLTAGE-TO-PULSE-DURATION CONVERTER—Voltage levels are converted to pulse durations by combining 741 op amp and 555 timer. Accuracies to better than 1% are obtained with this circuit, and the output signals still retain the original frequency independent of the input voltage.—"Signetics Analog Applications Manual," Signetics, Sunnyvale, CA, 1979, pp. 156–159.

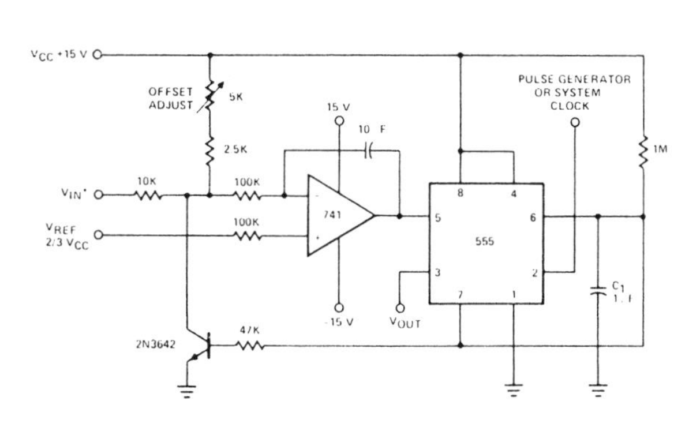

ALL RESISTOR VALUES IN OHMS

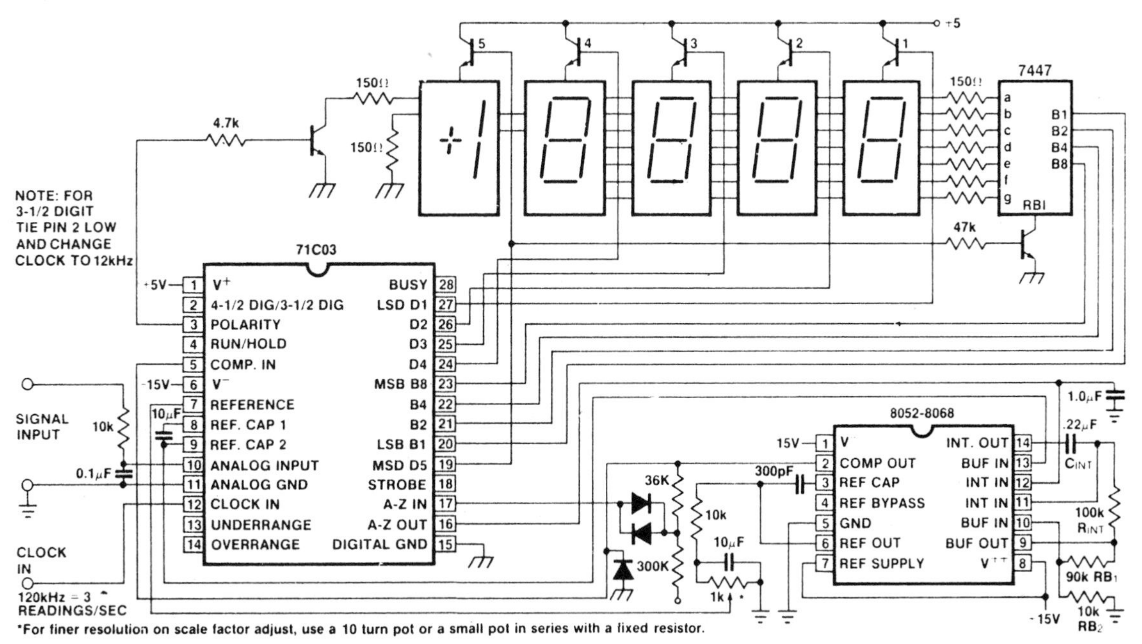

4.5-DIGIT ANALOG-TO-DIGITAL CONVERTER WITH LED DISPLAY—Uses internal reference of 8052-8068. If external reference is used, the reference supply (pin 7) should be connected to ground and the 300-pF reference capacitor deleted. The time constant of the RC input filter may be altered as needed or the filter may be deleted altogether.—"Intersil Data Book," Intersil, Cupertino, CA, 1981, p. 4-112.

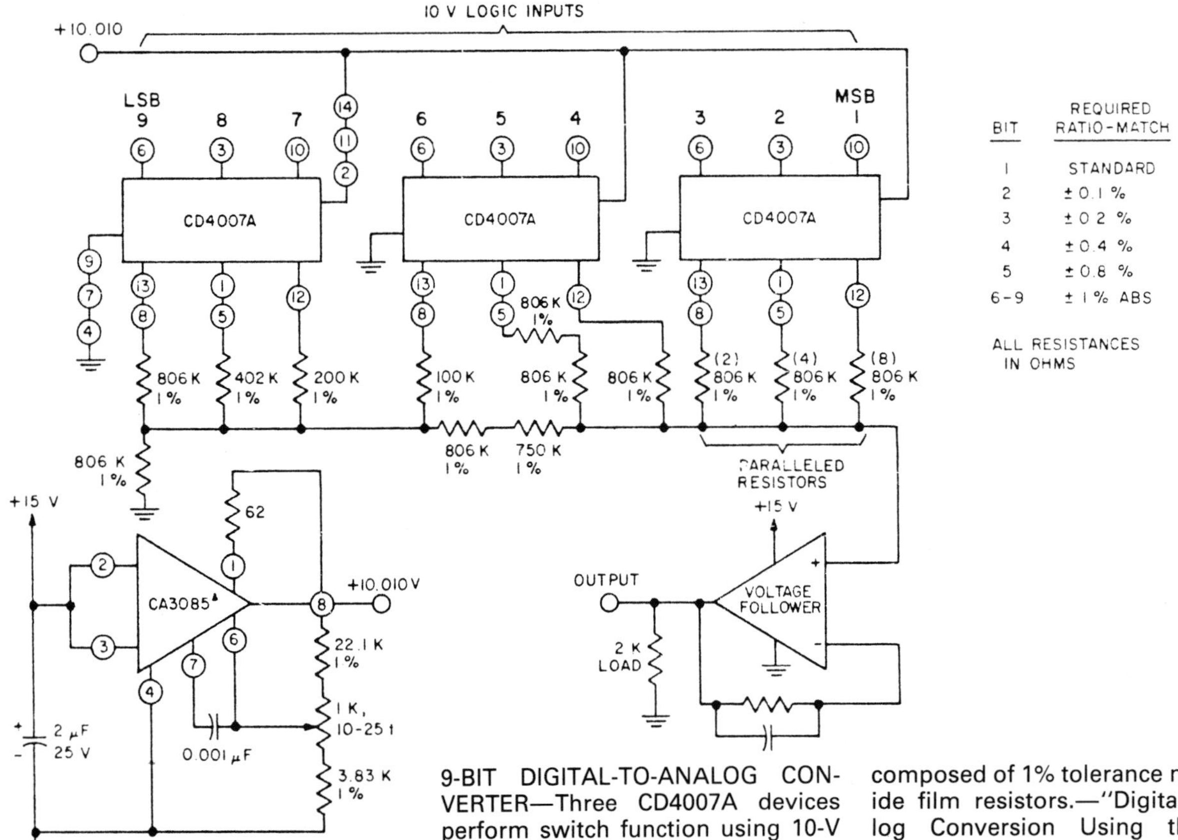

9-BIT DIGITAL-TO-ANALOG CONVERTER—Three CD4007A devices perform switch function using 10-V logic level. A single 15-V supply provides a positive bus for the follower amplifier and feeds the CA3085 voltage regulator. The resistor ladder is composed of 1% tolerance metal-oxide film resistors.—"Digital-to-Analog Conversion Using the RCA CD4007A COS/MOS IC," RCA Solid State Division, Somerville, NJ, 1973, Application Note ICAN-6080.

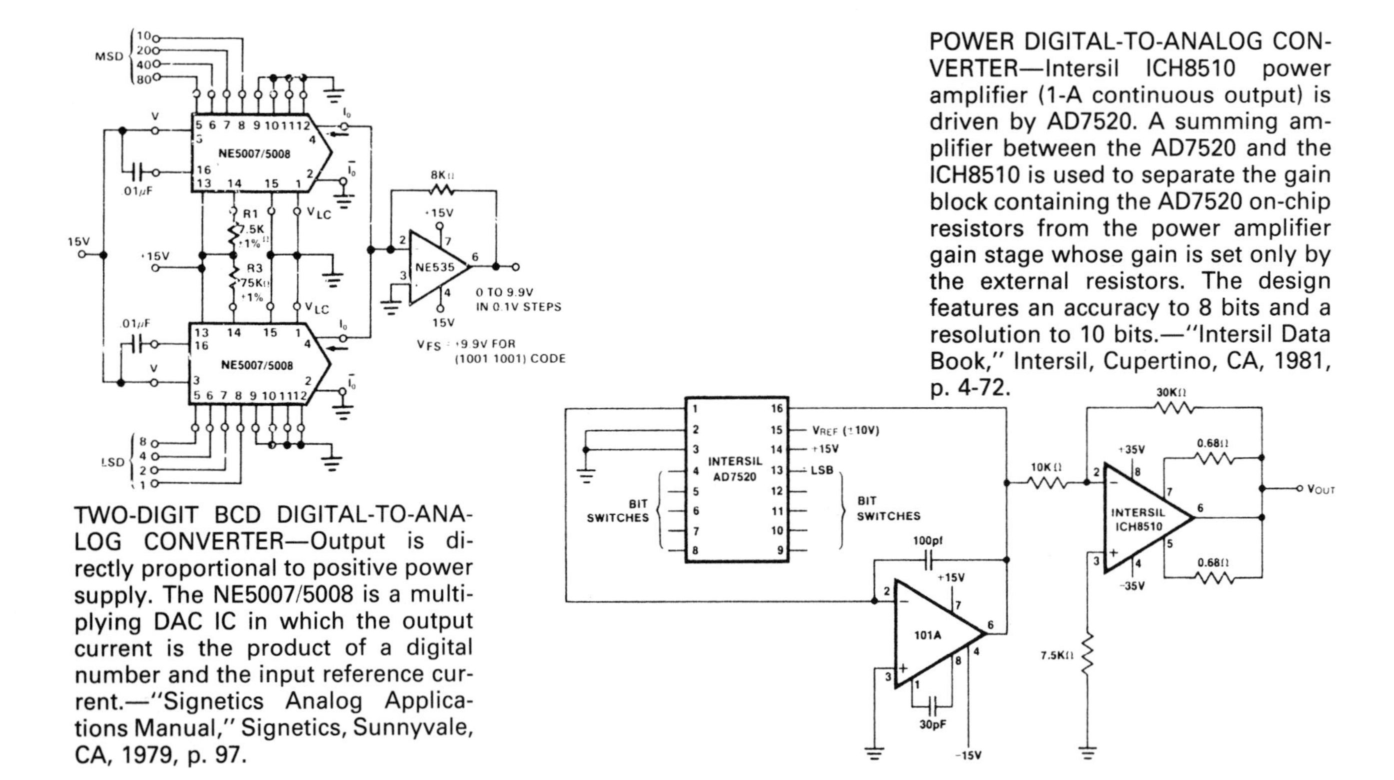

TWO-DIGIT BCD DIGITAL-TO-ANALOG CONVERTER—Output is directly proportional to positive power supply. The NE5007/5008 is a multiplying DAC IC in which the output current is the product of a digital number and the input reference current.—"Signetics Analog Applications Manual," Signetics, Sunnyvale, CA, 1979, p. 97.

POWER DIGITAL-TO-ANALOG CONVERTER—Intersil ICH8510 power amplifier (1-A continuous output) is driven by AD7520. A summing amplifier between the AD7520 and the ICH8510 is used to separate the gain block containing the AD7520 on-chip resistors from the power amplifier gain stage whose gain is set only by the external resistors. The design features an accuracy to 8 bits and a resolution to 10 bits.—"Intersil Data Book," Intersil, Cupertino, CA, 1981, p. 4-72.

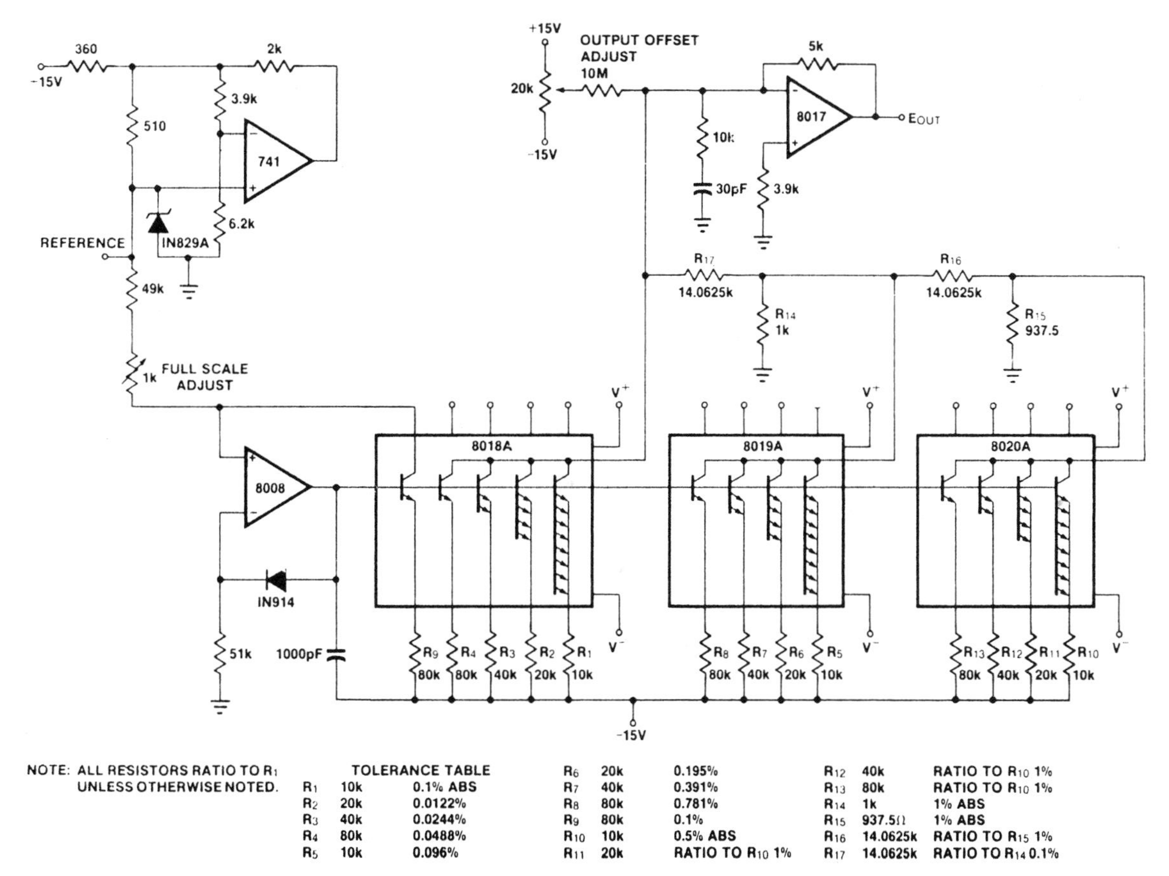

TOLERANCE TABLE

Resistor	Value	Tolerance
R_1	10k	0.1% ABS
R_2	20k	0.0122%
R_3	40k	0.0244%
R_4	80k	0.0488%
R_5	10k	0.096%
R_6	20k	0.195%
R_7	40k	0.391%
R_8	80k	0.781%
R_9	80k	0.1%
R_{10}	10k	0.5% ABS
R_{11}	20k	RATIO TO R_{10} 1%
R_{12}	40k	RATIO TO R_{10} 1%
R_{13}	80k	RATIO TO R_{10} 1%
R_{14}	1k	1% ABS
R_{15}	937.5Ω	1% ABS
R_{16}	14.0625k	RATIO TO R_{15} 1%
R_{17}	14.0625k	RATIO TO R_{14} 0.1%

12-BIT DIGITAL-TO-ANALOG CONVERTER—Based upon Intersil ICL8018A/8019A/8020A quad current switch devices. The temperature-compensated Zener diode is stabilized using an op amp as a regulated supply, and the circuit provides a very stable, precise voltage reference for the converter. The 16 : 1 and 256 : 1 resistor divider values are shown for a straight binary system; for a BCD system the dividers would be 10 : 1 and 100 : 1.—"Intersil Data Book," Intersil, Cupertino, CA, 1981, p. 4-93.

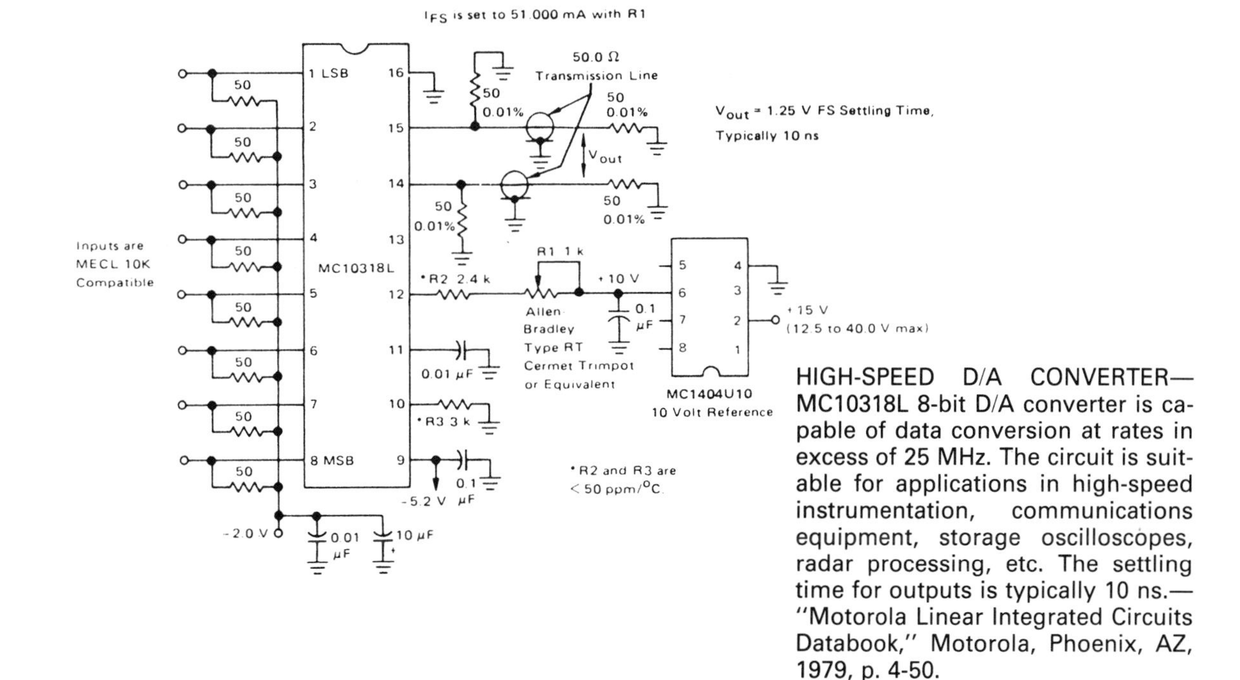

HIGH-SPEED D/A CONVERTER—MC10318L 8-bit D/A converter is capable of data conversion at rates in excess of 25 MHz. The circuit is suitable for applications in high-speed instrumentation, communications equipment, storage oscilloscopes, radar processing, etc. The settling time for outputs is typically 10 ns.—"Motorola Linear Integrated Circuits Databook," Motorola, Phoenix, AZ, 1979, p. 4-50.

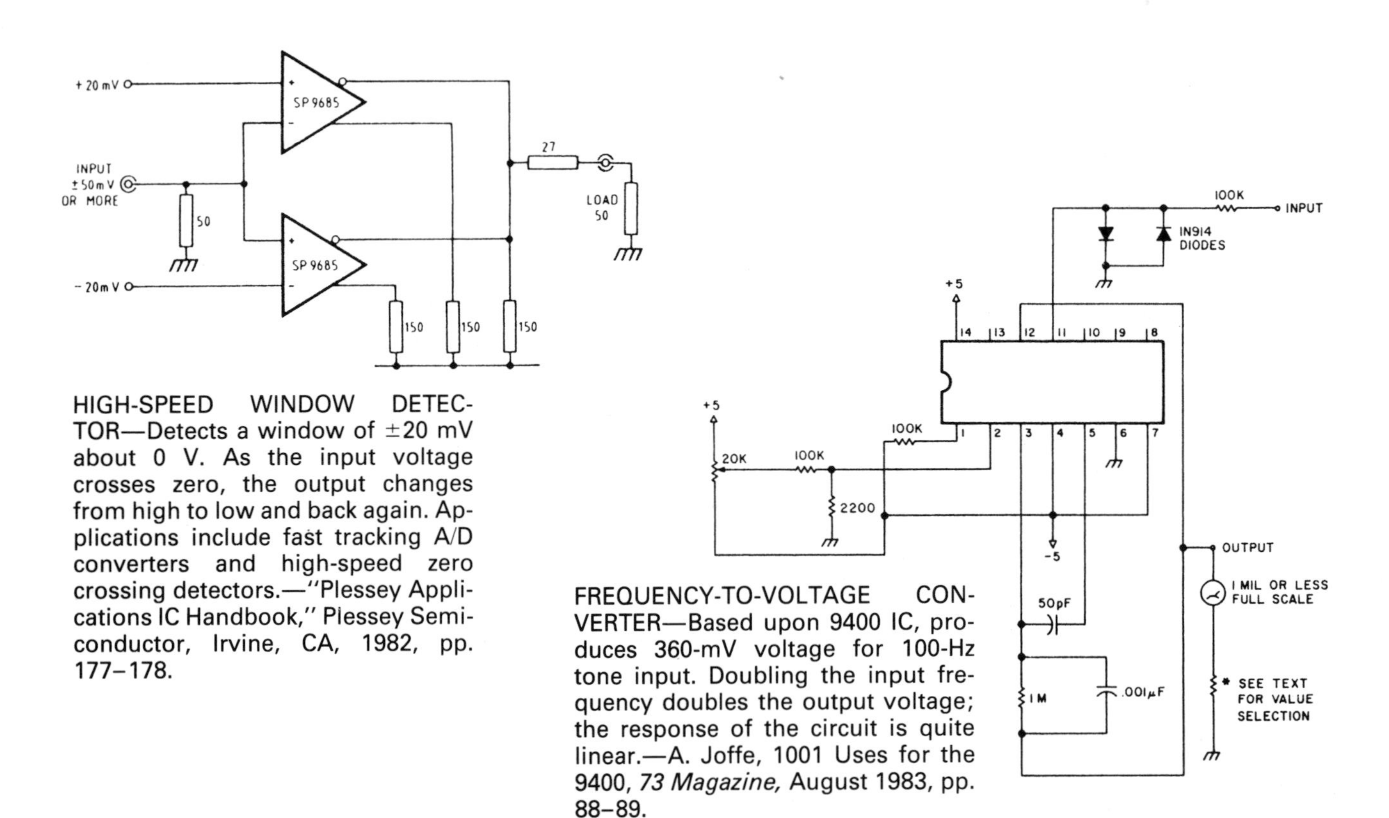

HIGH-SPEED WINDOW DETECTOR—Detects a window of ±20 mV about 0 V. As the input voltage crosses zero, the output changes from high to low and back again. Applications include fast tracking A/D converters and high-speed zero crossing detectors.—"Plessey Applications IC Handbook," Plessey Semiconductor, Irvine, CA, 1982, pp. 177–178.

FREQUENCY-TO-VOLTAGE CONVERTER—Based upon 9400 IC, produces 360-mV voltage for 100-Hz tone input. Doubling the input frequency doubles the output voltage; the response of the circuit is quite linear.—A. Joffe, 1001 Uses for the 9400, *73 Magazine,* August 1983, pp. 88–89.

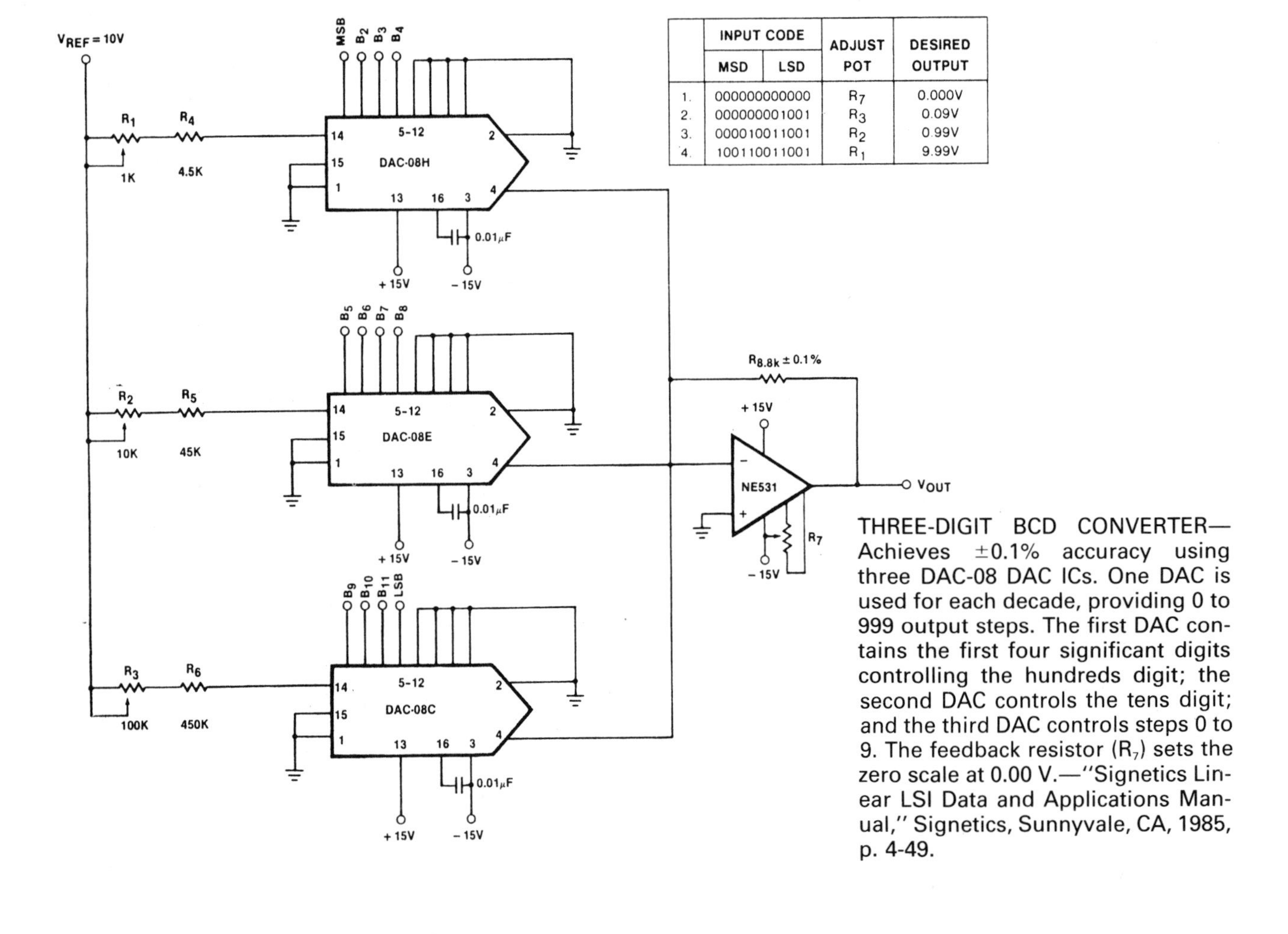

	INPUT CODE		ADJUST	DESIRED
	MSD	LSD	POT	OUTPUT
1.	000000000000		R_7	0.000V
2.	000000001001		R_3	0.09V
3.	000010011001		R_2	0.99V
4.	100110011001		R_1	9.99V

THREE-DIGIT BCD CONVERTER—Achieves ±0.1% accuracy using three DAC-08 DAC ICs. One DAC is used for each decade, providing 0 to 999 output steps. The first DAC contains the first four significant digits controlling the hundreds digit; the second DAC controls the tens digit; and the third DAC controls steps 0 to 9. The feedback resistor (R_7) sets the zero scale at 0.00 V.—"Signetics Linear LSI Data and Applications Manual," Signetics, Sunnyvale, CA, 1985, p. 4-49.

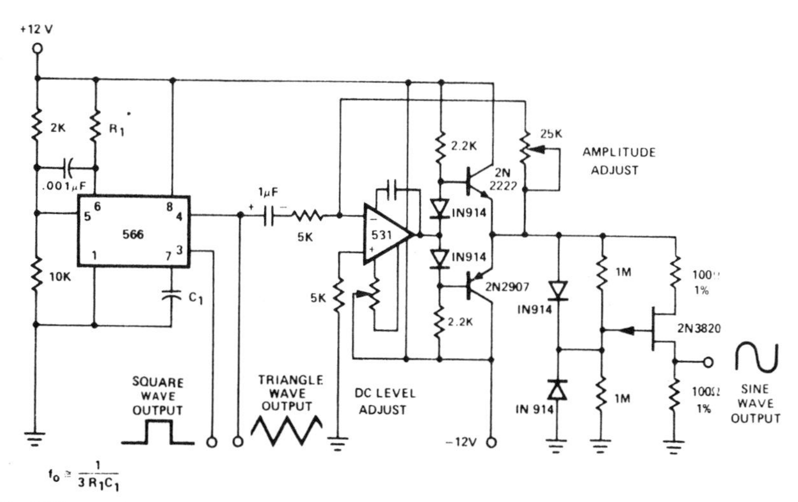

TRIANGLE-TO-SINE CONVERTER— The nonlinear IDS-VDS transfer of a P-channel junction FET (2N3820) is used to shape the triangle waveform. The resulting sinusoid waveform has a distortion of 2% or less. The amplitude of the triangle waveform is critical and must be carefully adjusted for the lowest sinusoidal output.—"Signetics Analog Applications Manual," Signetics, Sunnyvale, CA, 1979, pp. 332–333.

7

Digital Circuits

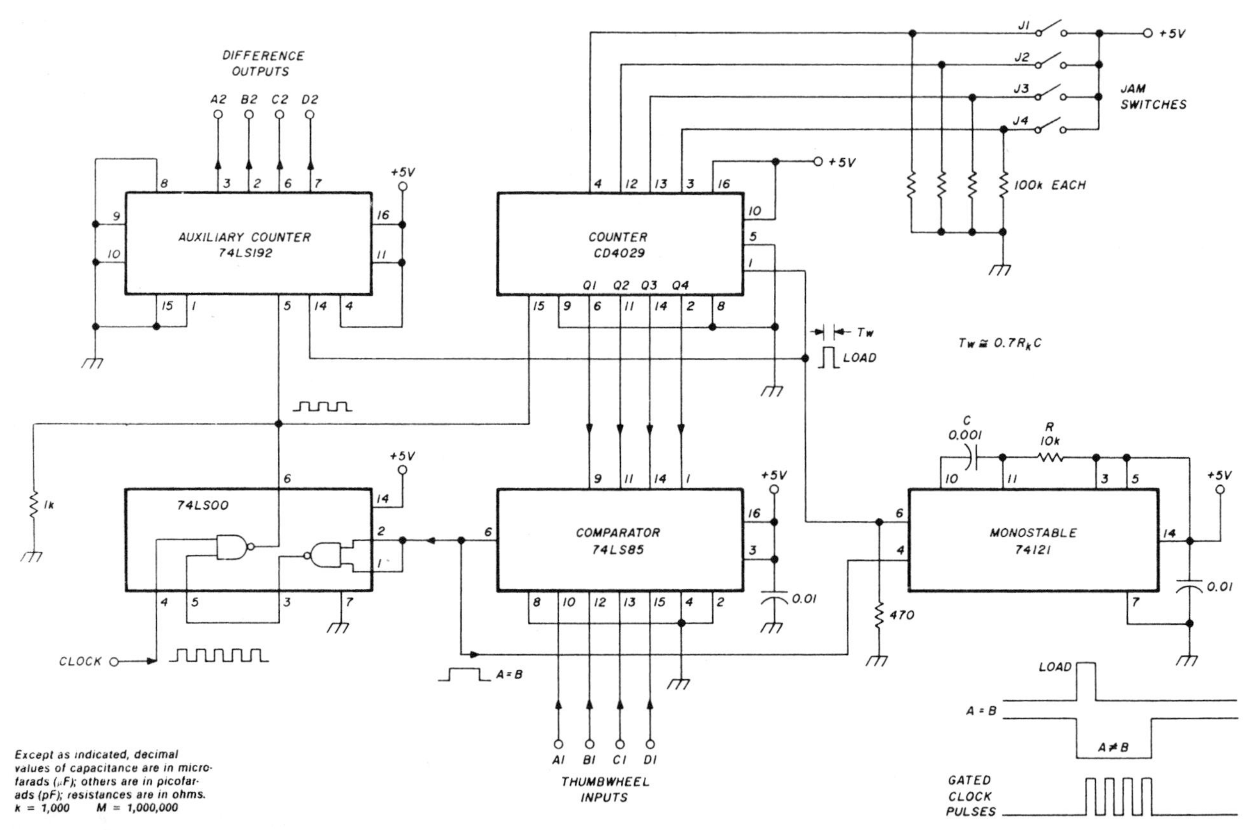

BCD ADDER—Hybrid CMOS/TTL circuit permits addition of BCD. If the Q outputs of the 4029 agree bit for bit with the thumbwheel output bits at the comparator input terminals, the 74LS85 output is set high. This prevents clock pulses from reaching the 4029, and the condition remains until the thumbwheel switch setting is changed. When it is changed, the falling edge of A = B activates the 74121, which in turn generates a single LOAD or PRESET-ENABLE pulse. This pulse presets the 4029 so that its Q outputs agree with the JAM inputs. At the same time, the comparator A = B output goes low and the NAND gate is open, allowing clock pulses to reach the 4029 input terminal. Thereafter, the 4029 Q outputs change starting from their preset values until they again agree with the thumbwheel bits. At this point, the comparator's A = B output goes high, interrupting the clock pulses and "freezing" the contents of the counter. The number of clock pulses necessary to re-establish the A = B state depends on the difference between the thumbwheel number and the JAM number. The pulses are also fed to the 74LS192, which has its DATA inputs "hard-wired" to zero. Each time it receives a LOAD pulse, it resets to zero and accumulates a count numerically equal to the number of clock pulses in the train. The 74LS192 provides a difference output equal to the thumbwheel number minus the JAM number.—N. Foot, Binary Coded Decimal Addition, *Ham Radio,* April 1982, pp. 66-67.

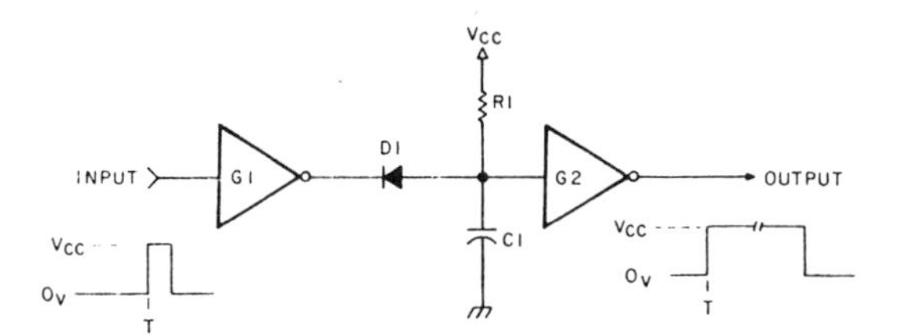

CMOS ONE-SHOT—74C14 Schmitt trigger-type inverter CMOS IC can be used to obtain one-shots. D1 is a 1N4148. Timing depends upon the values of R1 and C1. For a time of 0.1 second, R1 is 1 M and C1 is 0.1 μF. For a time of 0.2 seconds, R1 is 1 M and C1 is 0.2 μF. For 1 second, R1 is 1 M and C1 is 1.0 μF. For 2 seconds, R1 is 2 M and C1 is 1 μF. For 10 seconds, R1 is 5 M and C1 is 2 μF. For 20 seconds, R1 is 10 M and C1 is 2 μF. For 100 seconds, R1 is 5 M and C1 is 10 μF. For 200 seconds, R1 is 3.3 M and C1 is 25 μF. The timing is also dependent upon voltage source stability and capacitor quality.—V. Yingst, Free CMOS Timers, *73 Magazine,* October 1980, p. 113.

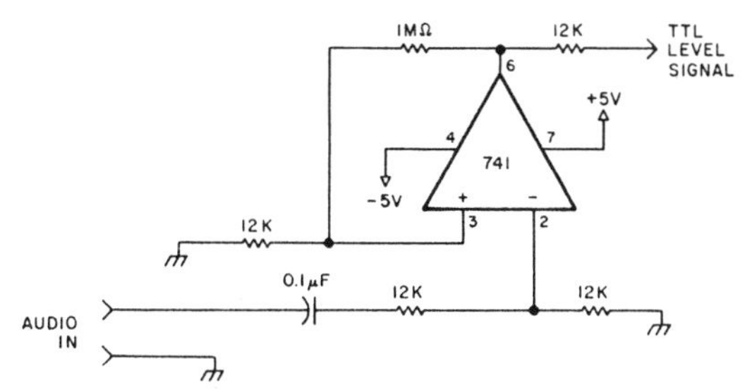

ZERO CROSSING DETECTOR—Produces TTL-level signal from audio input. Each time the input signal crosses the base line (the zero line of the waveform), the TTL signal changes state (0 or 1).—R. Swirsky, Computer Slow Scan, *73 Magazine,* October 1983, pp. 104–106.

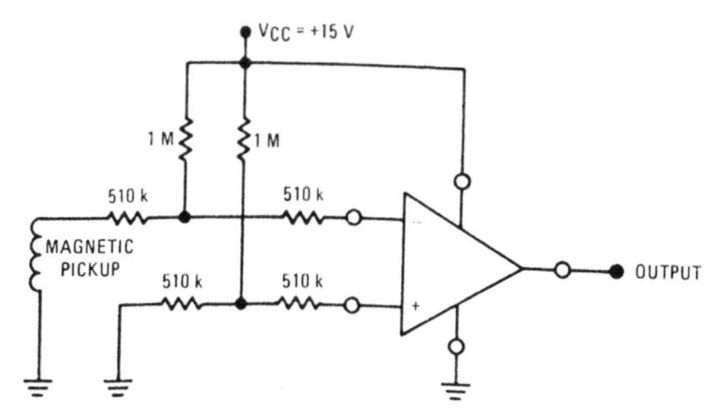

ZERO CROSSING DETECTOR—Produces a square wave output suitable for use with TTL logic from an analog input waveform obtained through the magnetic pickup coil. The circuit uses one amplifier section of an MC3401 quad op amp device.—"Motorola Linear Integrated Circuits Databook," Motorola, Phoenix, AZ, 1979, p. 3-147.

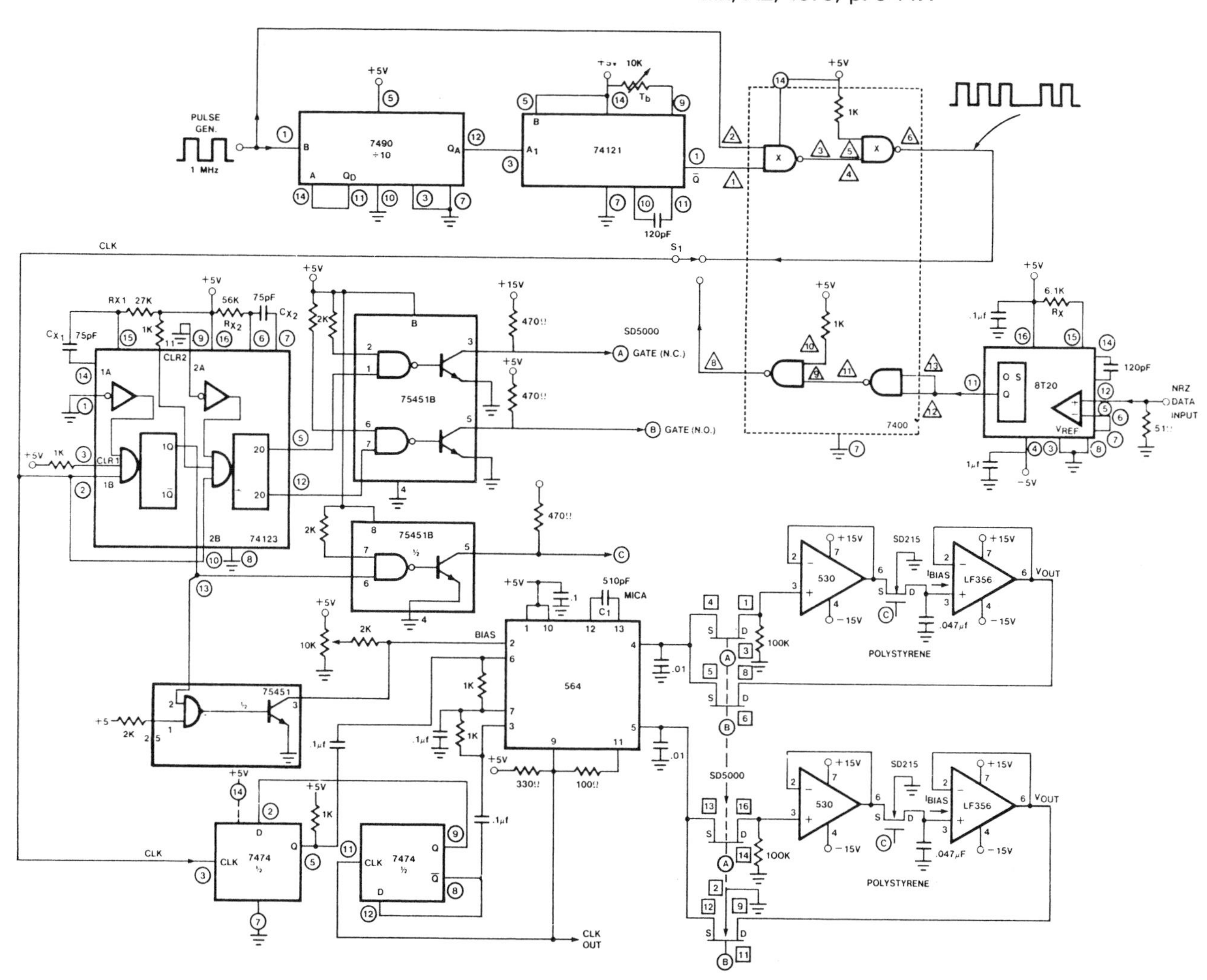

MISSING CLOCK REGENERATOR—Operates at 1000-kHz rate to supply replacement clock signals for those lost on data transmission circuits. 564 PLL device can stay locked for up to 15 missing clock pulses. The circuit is designed to operate in a non-return to zero (NRZ) data transmission system.—"Signetics Analog Applications Manual," Signetics, Sunnyvale, CA, 1979, pp. 322–326.

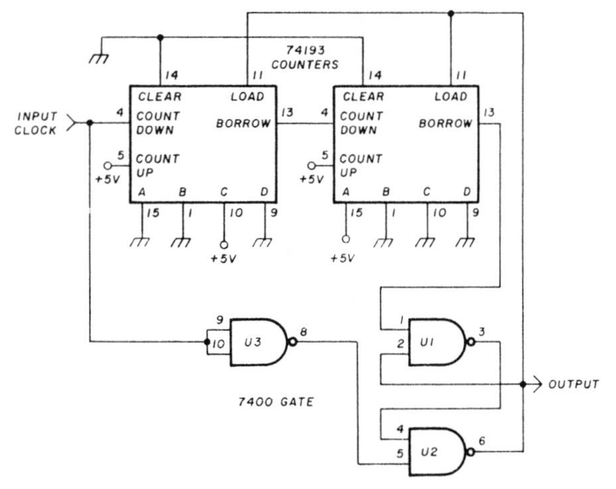

DIVIDE-BY-20 CIRCUIT WITH LENGTHENED OUTPUT PULSE—U1 and U2 form a set-reset latch. Once the latch has been set and the preset load enabled, the latch is reset only when the input clock goes high. The positive edge of the clock would normally toggle the counter, but the 74193 is designed to inhibit counting until the load pin returns high. This means the circuit will skip one input clock.—P. Clower, Digital Techniques, *Ham Radio,* September 1981, pp. 43–46.

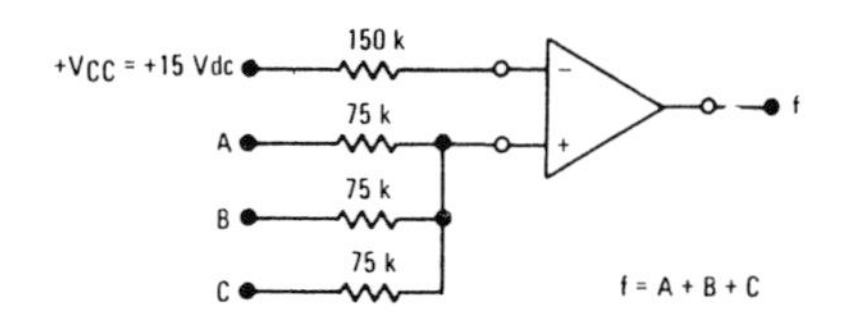

OR GATE—Designed to make use of MC3301 quad op amp device but may be adapted for use with other op amps. Circuit operation follows normal OR gate logic; output will be 1 if one or more inputs A, B, or C are also 1.—"Motorola Linear Integrated Circuits Databook," Motorola, Phoenix, AZ, 1979, p. 3-138.

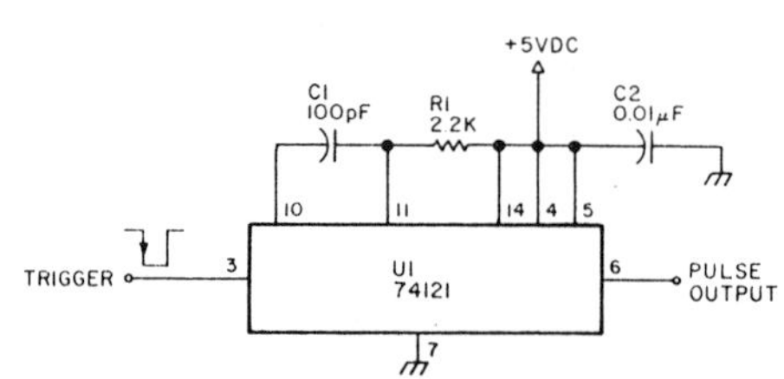

PULSE (ONE-SHOT) GENERATOR—Built around 74121 TTL Schmitt trigger input monostable multivibrator. The drive is provided by a square wave source.—J. Carr, Find Fault with Your Coax, *73 Magazine,* October 1984, pp. 10–14.

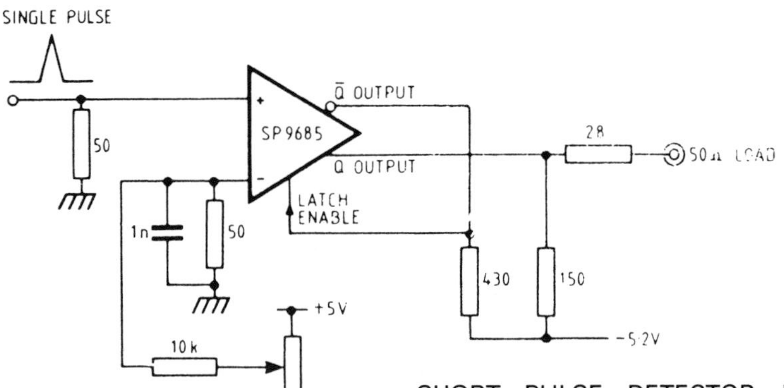

SHORT PULSE DETECTOR—Uses SP9685 comparator IC, and circuit was originally designed for applications in nucleonics and high-energy physics. A positive-going pulse of any width, down to the minimum defined by the propagation delay plus the setup time, will cause the circuit to latch and ignore further inputs. Practical minimum values are a 3-ns pulse and 10-mV amplitude for reliable latching.—"Plessey Applications IC Handbook," Plessey Semiconductor, Irvine, CA, 1982, pp. 174–176.

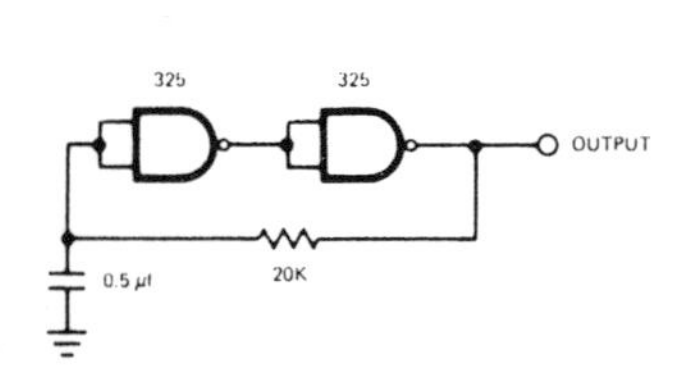

POWER-UP RESET CIRCUIT—Output goes low briefly when power is first applied, then remains normally high. The circuit is designed to generate a flip-flop reset pulse in systems that must start up at zero.—"Data Acquisition Design Handbook," Teledyne Semiconductor, Mountain View, CA, 1984, p. 14-33.

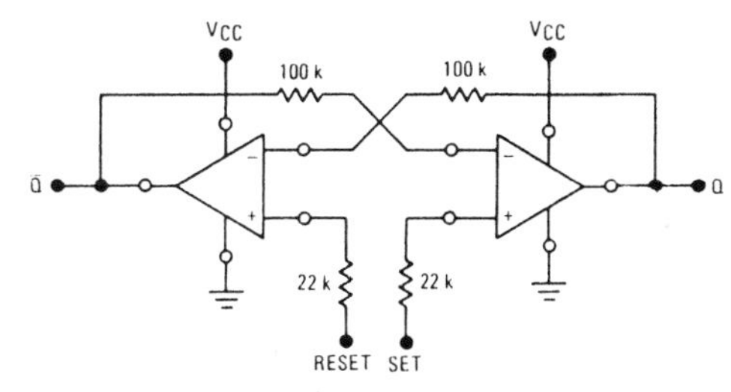

R-S FLIP-FLOP—Designed to make use of MC3301 quad op amp device but may be adapted for use with other op amps. Circuit operation follows normal R-S flip-flop logic. An input at SET (S) will activate output Q; Q will remain the output until an input is made at RESET (R). Then the output will become Q and remain so until an input is made at SET.—"Motorola Linear Integrated Circuits Databook," Motorola, Phoenix, AZ, 1979, p. 3-139.

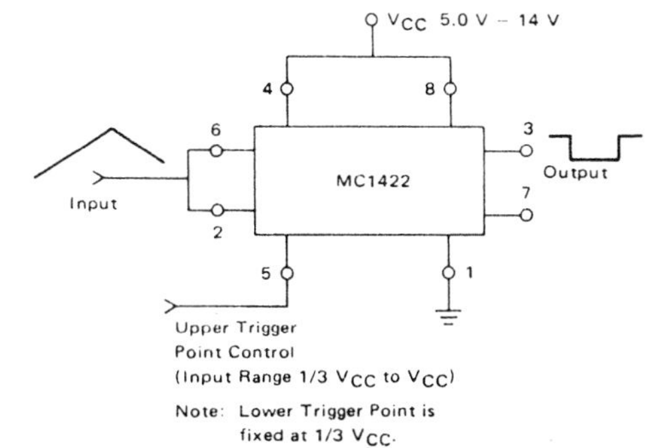

SCHMITT TRIGGER—Uses MC1422 monolithic timing IC. The lower trigger point is fixed at one-third V_{CC}. The upper trigger point is adjustable through pin 5 from one-third V_{CC} to just slightly less than V_{CC}. The circuit will operate with input frequencies up to 50 kHz.—"Motorola Linear Integrated Circuits Databook," Motorola, Phoenix, AZ, 1979, p. 6-17.

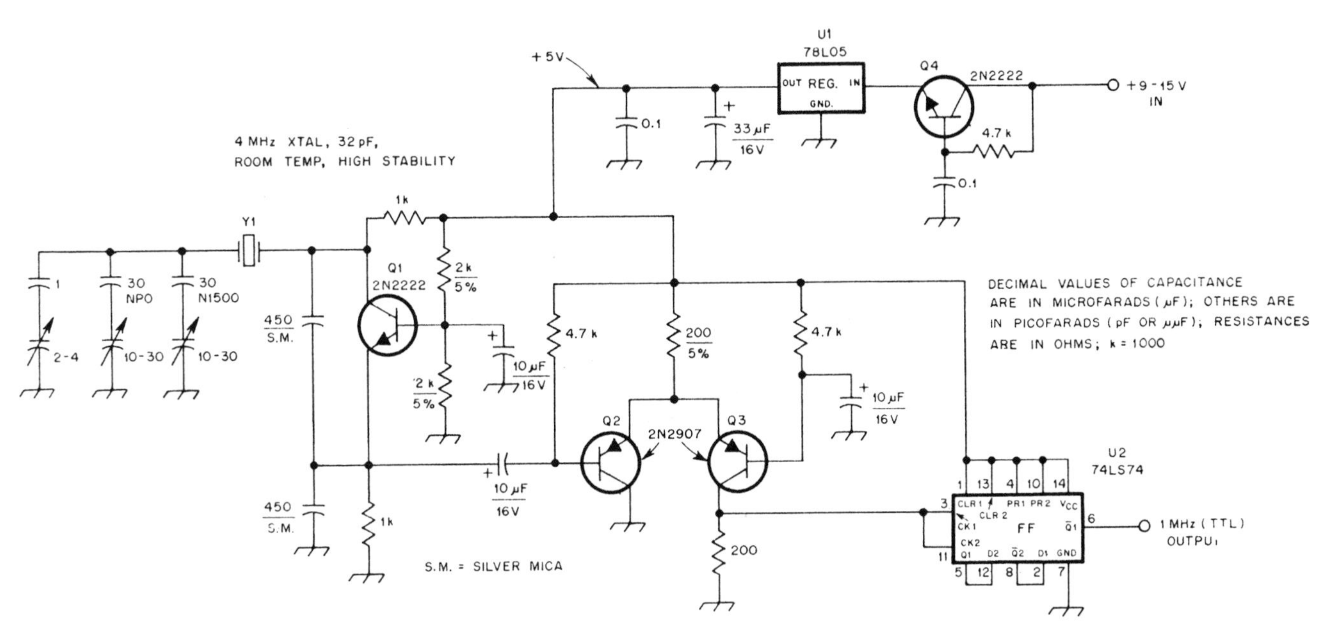

1-MHz FREQUENCY STANDARD—Produces a 1-MHz TTL-level output suitable for driving TTL or CMOS dividers. The stability of the circuit at room temperature is excellent. The circuit was developed to provide "window" timing sequences for coherent CW operation.—C. Woodson, Coherent CW—The Practical Aspects, *QST,* June 1981, pp. 18–23.

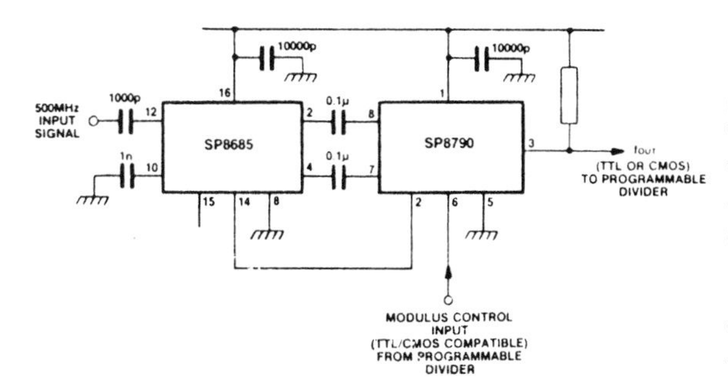

500-MHz TWO-MODULUS DIVIDER—500-MHz divide-by-40/41 divider is obtained by combining an SP8685 (divide-by-10/11) divider and SP8790 (divide-by-4) extender. Total propagation delays are typically 29 ns.—"Plessey Integrated Circuit Databook," Plessey Semiconductor, Irvine, CA, 1983, pp. 1–2.

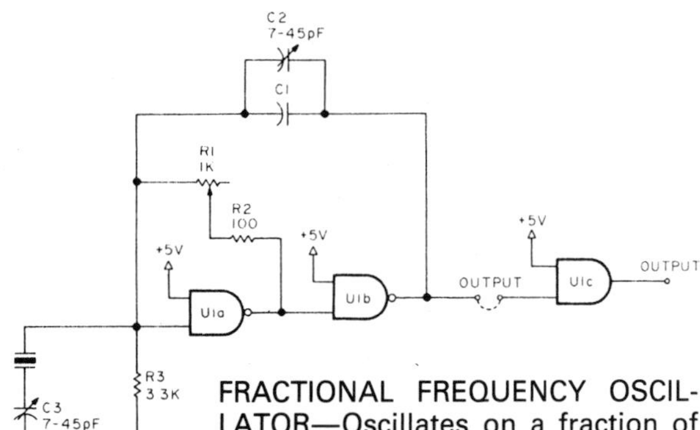

FRACTIONAL FREQUENCY OSCILLATOR—Oscillates on a fraction of the fundamental frequency of the crystal. The division factor range depends upon the value of C1; the final adjustment is obtained by the setting of R1. If C1 is 150 pF, the division factor is 1 to 3. If C1 is 300 pF, the division factor is 2 to 6. If C1 is 560 pF, the division factor is 3 to 13. U1 can be any TTL NAND device such as the 7400.—J. Westenhaver, Undertones, *73 Magazine,* October 1980, p. 56.

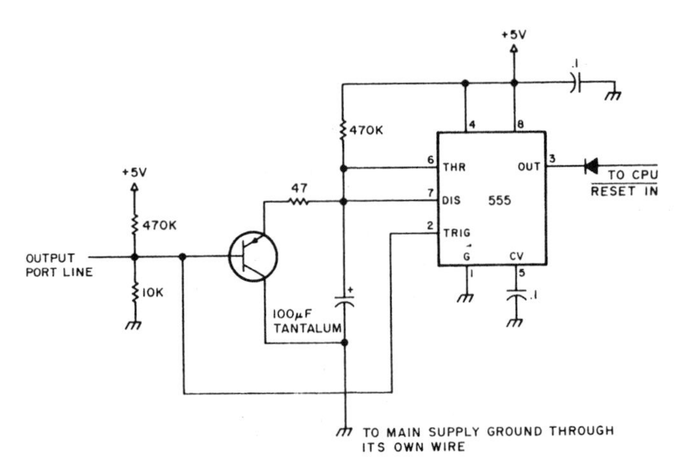

MISSING PULSE DETECTOR—Circuit monitors microcomputer output at output port detector and generates reset signal if a pulse is missed.—E. Ingber, A Computer-Controlled Talking Repeater, *73 Magazine,* November 1980, pp. 132–141.

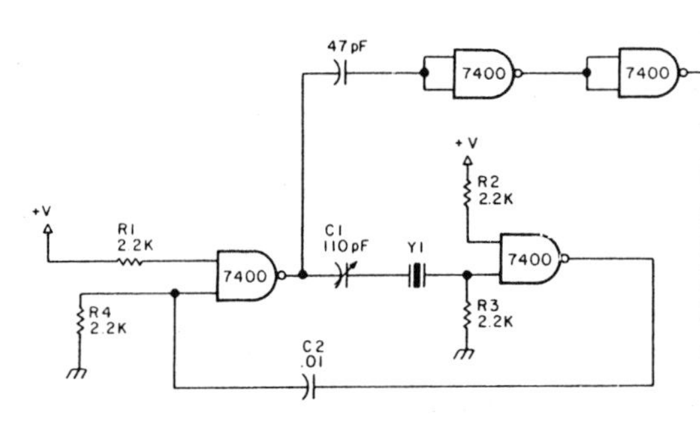

TTL CLOCK—Built around 7400 NAND gates, this oscillator generates square waves over range of 100 kHz to 3 MHz (although it may be balky toward lower end of range). Y1 is cut to the desired frequency, and oscillator operation can be adjusted by the 100-pF variable capacitor.—J. Carr, Clock Blocks, *73 Magazine,* October 1980, pp. 192–194.

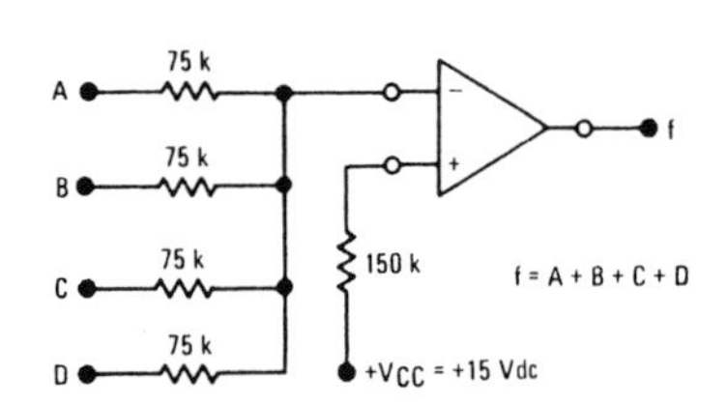

NOR GATE—Designed to make use of MC3301 quad op amp device but may be adapted for use with other op amps. Circuit operation follows normal NOR gate logic; all inputs must be 0 to get a 1 output.—"Motorola Linear Integrated Circuits Databook," Motorola, Phoenix, AZ, 1979, p. 3-139.

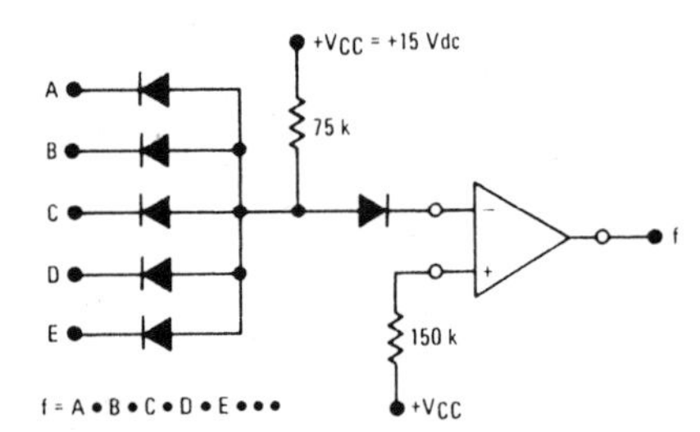

NAND GATE—Designed to make use of MC3301 quad op amp device but may be adapted for use with other op amps. Circuit operation follows normal NAND gate logic; all inputs must be 1 to obtain a 0 output.—"Motorola Linear Integrated Circuits Databook," Motorola, Phoenix, AZ, 1979, p. 3-139.

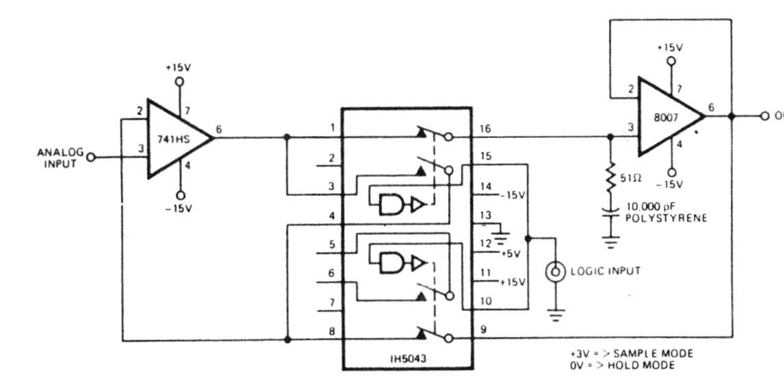

IMPROVED SAMPLE AND HOLD—Uses Intersil IH5043 high-level CMOS analog gate for improved performance. +3 V > sample mode and 0 V > hold mode.—"Intersil Data Book," Intersil, Cupertino, CA, 1981, p. 3-108.

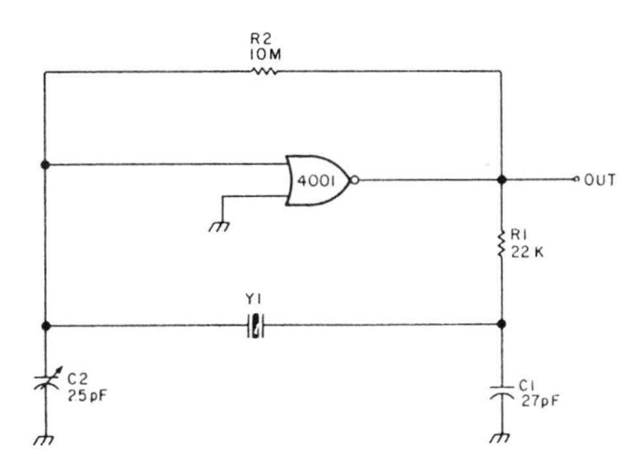

CMOS CRYSTAL OSCILLATOR—Based upon CD4001 CMOS IC, circuit is suitable for frequencies below 100 kHz where TTL oscillators may not function well.—J. Carr, Clock Blocks, *73 Magazine,* October 1980, pp. 192–194.

8

Frequency Synthesis Circuits

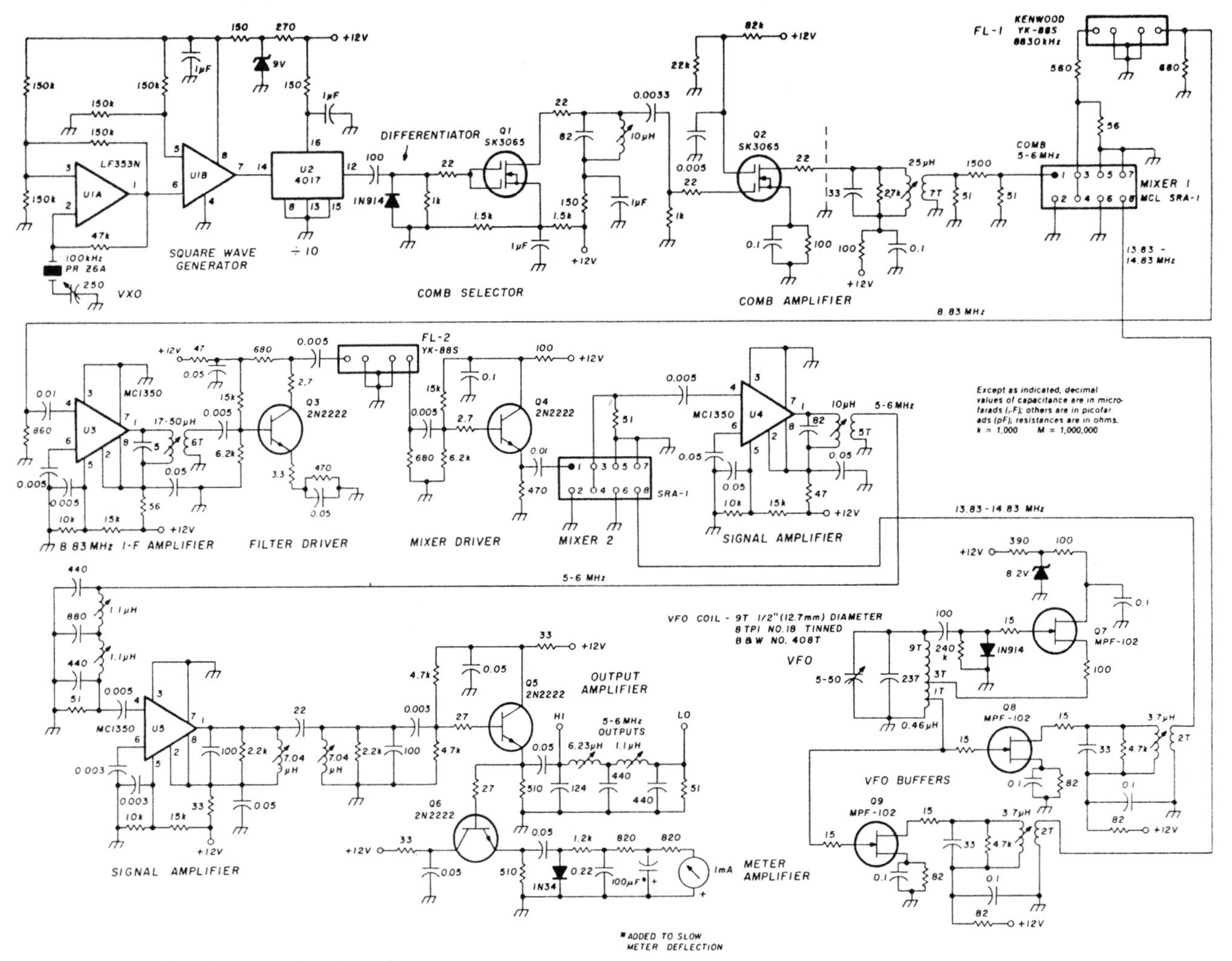

VXO-BASED FREQUENCY SYNTHESIZER—Generates all frequencies from 5-6 MHz with less noise than similar PLL circuits. The heart of the system is a VXO with a range of 10–10.2 MHz. The VXO output frequency is divided by 10 and fed to a pulse generator. There a short pulse is formed with a PRF continuously variable over the 10- to 10.02-MHz range, directly controlled by the VXO. A bandpass filter passes the 500th through 600th harmonics of the pulse, producing a comb with about 10 kHz of spacing which moves up approximately 10 kHz as the VXO frequency is increased. If the 500th harmonic is selected, its frequency will move continuously from 5000 to 5010 kHz. Then if the 501st harmonic is selected, its frequency will move continuously from 5010 to 5020.2 kHz, and so forth. This enables the obtaining of all possible frequencies from 5 to 6 MHz. U1 is a LF353N.—F. Noble, Frequency Synthesis by VXO Harmonic Selection, *Ham Radio*, February 1984, pp. 12–18.

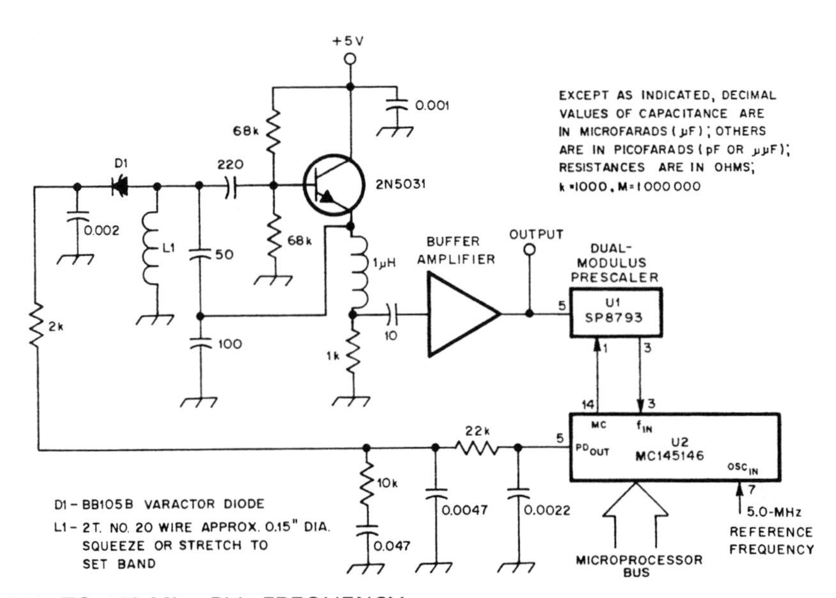

144- TO 148-MHz PLL FREQUENCY SYNTHESIZER—Acquires frequency lock to less than 200 Hz within 50 ms. It operates in 5-kHz steps in a 3-MHz range between 144 and 148 MHz.—R. Simpson, CMOS PLL Notes, *QST*, August 1983, p. 49.

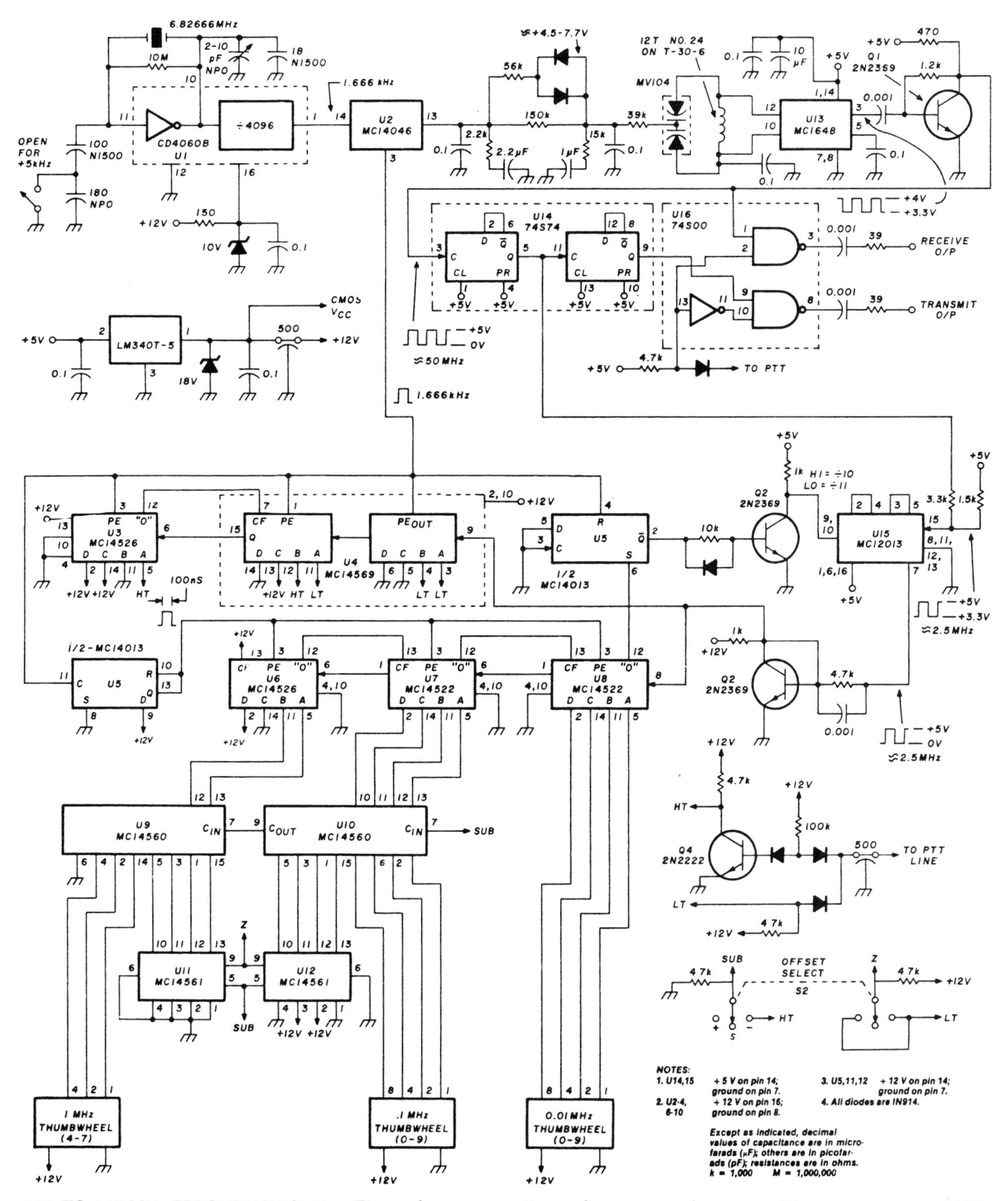

144- TO 148-MHz FREQUENCY SYNTHESIZER—Covers entire 144- to 148-MHz range in 5-kHz increments with output drive at 12 MHz for transmit and 45 MHz for receive. Power consumption is less than 250 mA. The reference oscillator frequency must be in the 5- to 10-MHz range; the circuit uses a 6.82666-MHz crystal oscillator to provide it. The transmit frequency can be offset 600 kHz above or below the desired receive frequency. The temperature stability is good over the entire range (typically less than ±50 Hz at 12-MHz output).—K. Grant, Genesis of a Synthesizer, *Ham Radio,* March 1981, pp. 38–42.

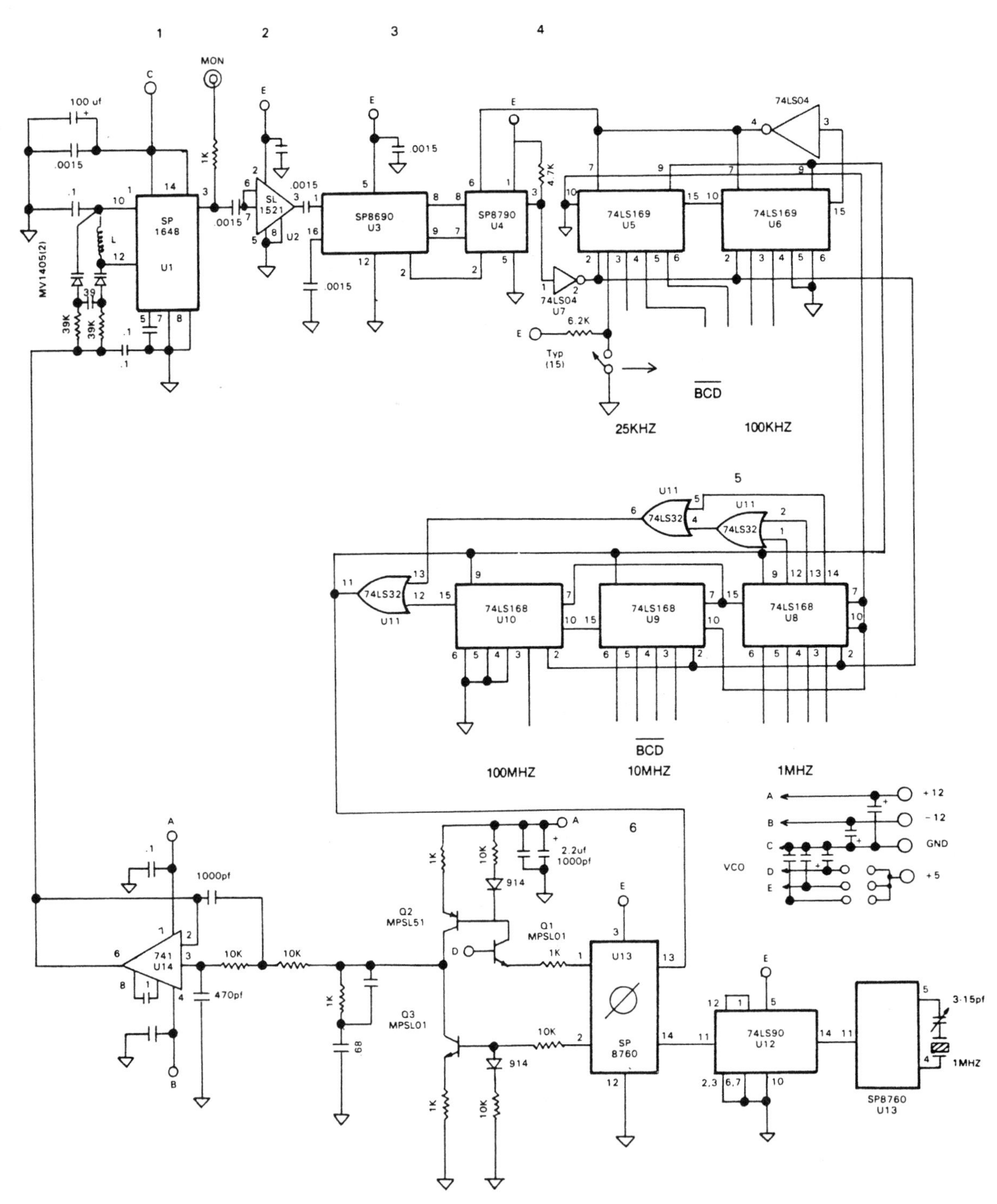

108- TO 174-MHz FREQUENCY SYNTHESIZER—PLL-based circuit covers frequency range in 25-kHz steps. The VCO is a SP1648. An SL1521 wideband amplifier provides isolation between the VCO and the digital dividers. The prescaler is an SP8690 divide-by-10/11 UHF counter. It is combined with an SP8790 to form a divide-by-40/41 counter with an operating frequency over 200 mHz. The original source includes a PC board layout.—"Plessey Applications IC Handbook," Plessey Semiconductor, Irvine, CA, 1982, pp. 237–239.

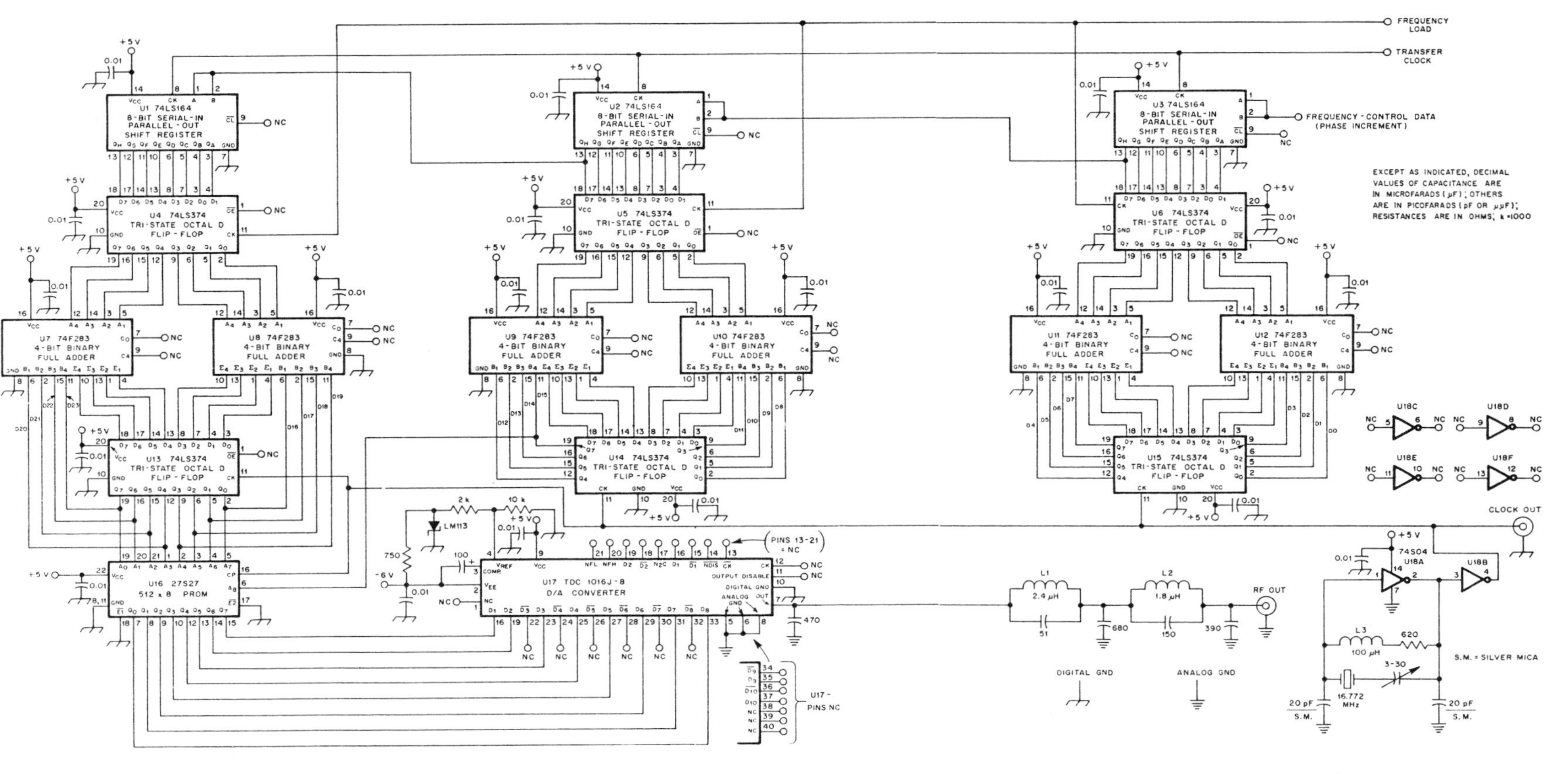

DIGITAL FREQUENCY SYNTHESIZER—Operates from 1 to 7.5 MHz. It is claimed that it provides superior performance compared with PLL circuits having reduced phase noise. The synthesizer may be controlled by a microcomputer or individual switches can be used to enter the desired frequency directly into the accumulator. The desired frequency is entered as a 24-bit binary value, shifted in MSB first.—F. Williams, A Digital Frequency Synthesizer, *QST*, April 1984, pp. 24–30.

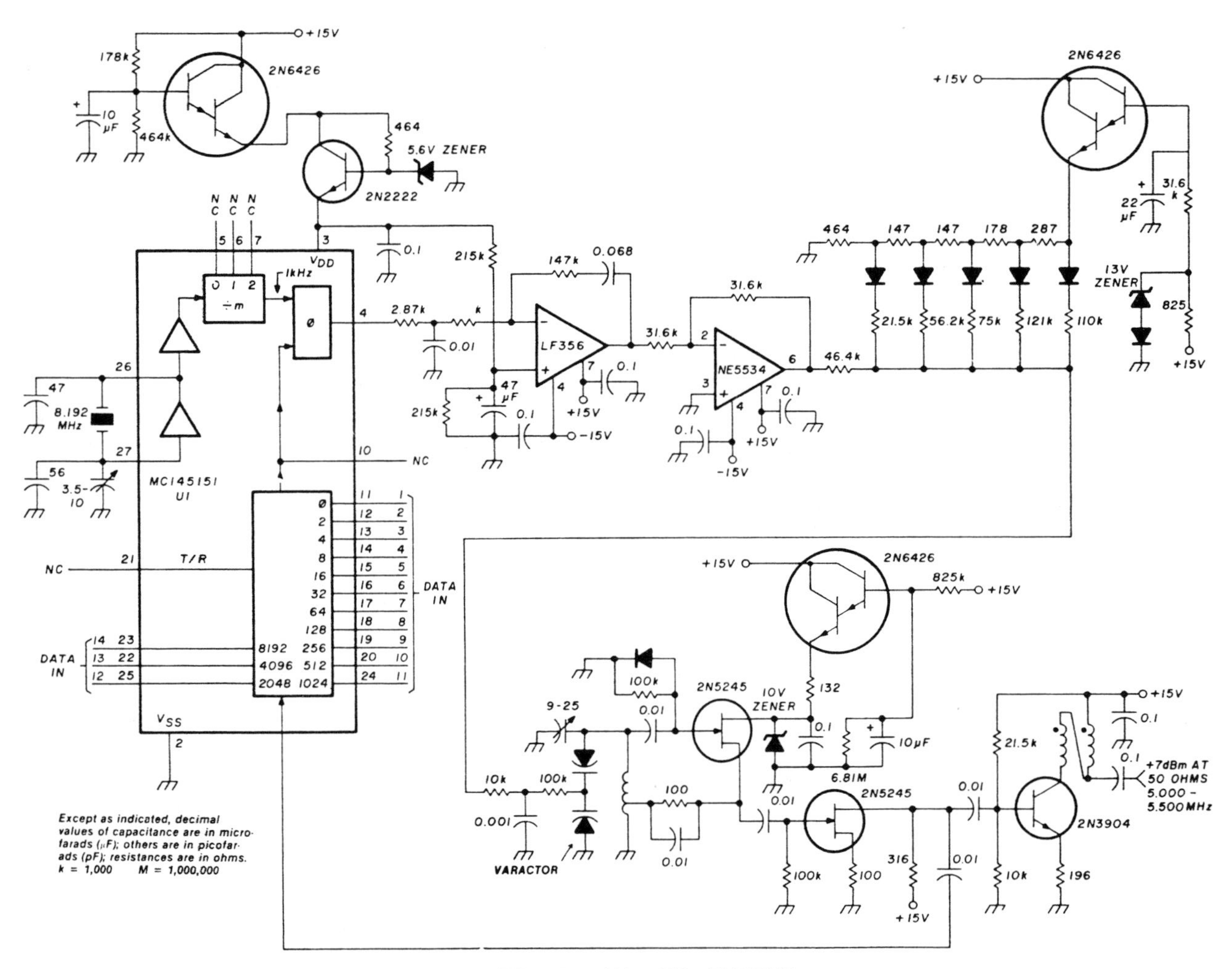

5000- TO 5500-kHz RF SYNTHESIZER—Covers 5000- to 5500-kHz range in 1-kHz increments using single PLL (Motorola MC145151). The circuit works well as a local oscillator for both receivers and transmitters. PLL lock time is about 30 ms. The synthesizer can be constructed in physical space as small as 2 × 4 in. The original article includes extensive design equations for the circuit.—C. Corsetto, RF Synthesizers for HF Communications, *Ham Radio,* October 1983, pp. 17–26.

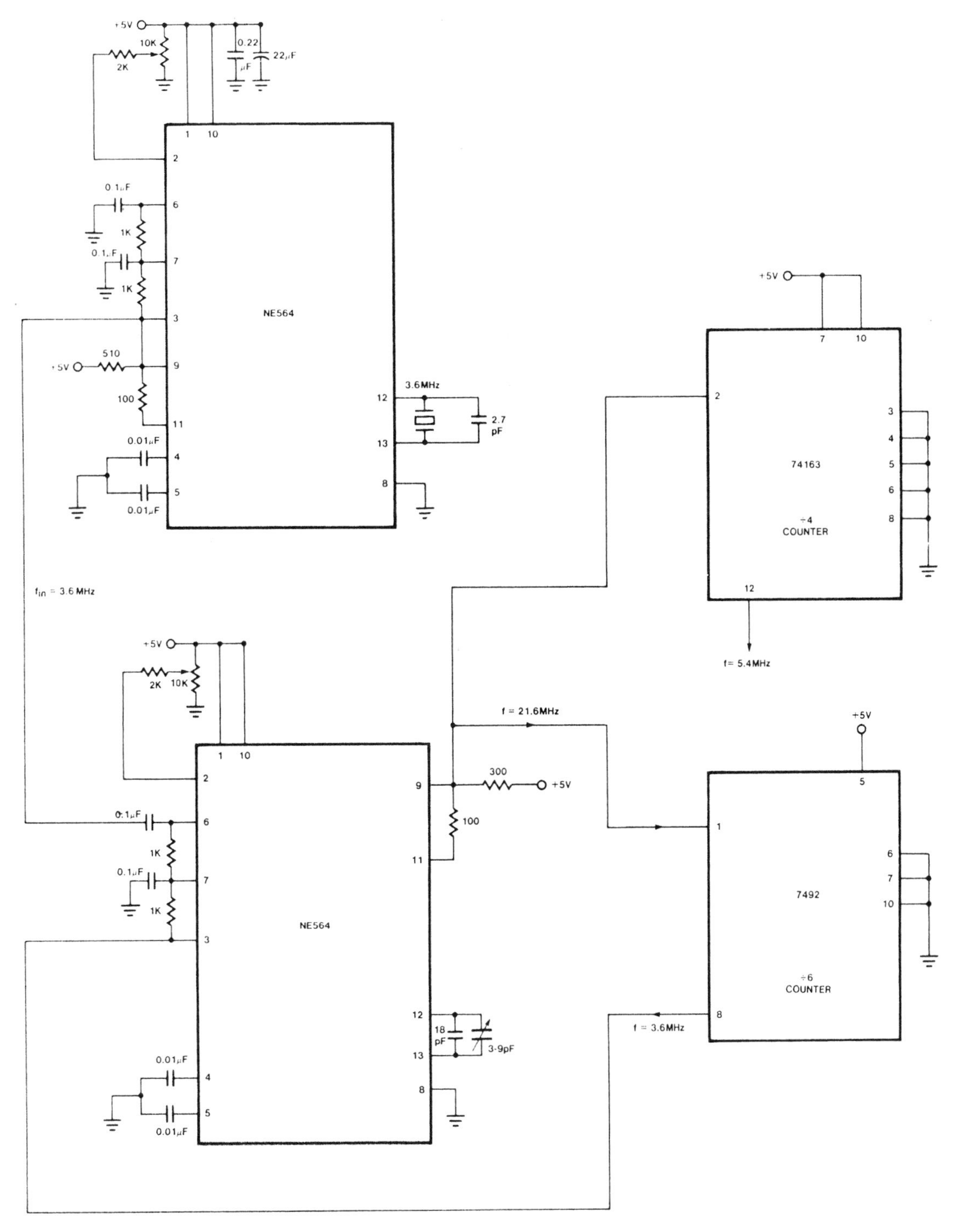

CRYSTAL CONTROLLED FREQUENCY SYNTHESIS—Generates frequencies of 5400 and 21,600 kHz from a 3600-kHz crystal controlled source. Reference signal input is produced by using the crystal as the frequency-determining element in the VCO of a second PLL. The thermal stability of all three frequencies will be the same as the stability afforded by the crystal.—"Signetics Analog Applications Manual," Signetics, Sunnyvale, CA, 1979, pp. 312–314.

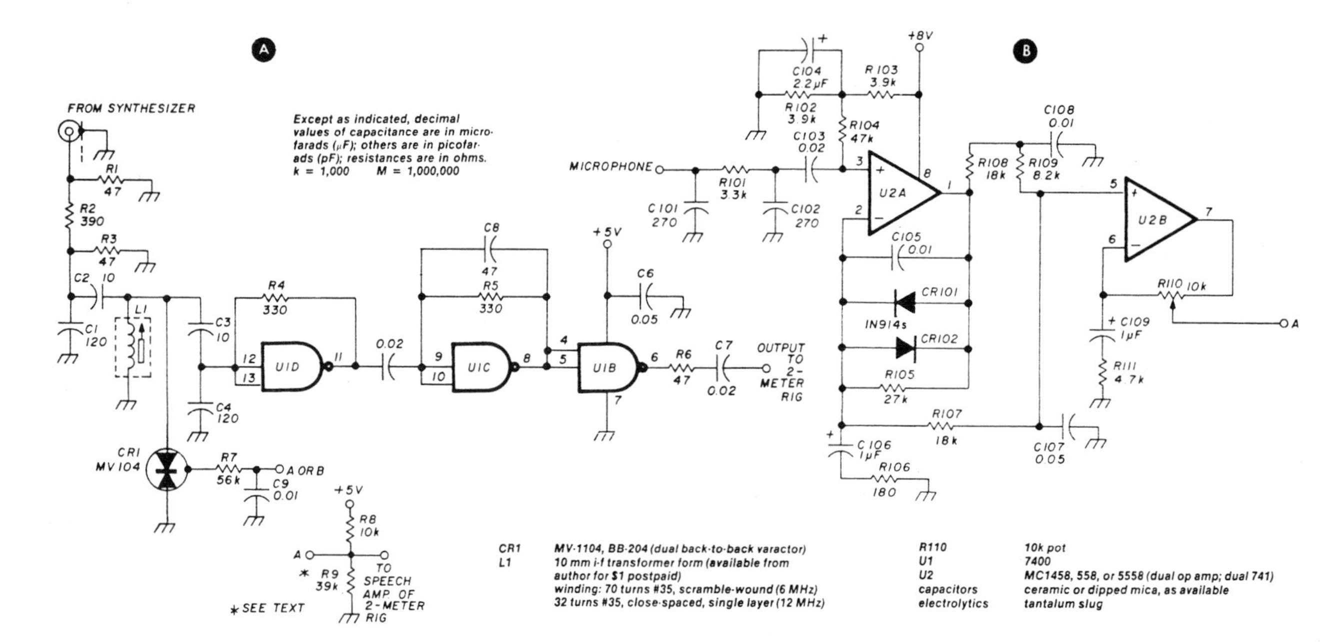

FREQUENCY MODULATOR FOR SYNTHESIZER—Provides frequency-modulated output for synthesizers operating in the 144- to 148-MHz amateur radio band. The modulator uses a varactor-tuned tank circuit. Passing a CW signal through the tank circuit and varying the bias on the varactor results in a variable phase shift and consequently phase modulation. The PM signal is then converted to frequency modulation. Section A of the schematic is the modulator itself while B is an optional speech preamplifer.—T. Cornell, Frequency Modulator for a Two-Meter Synthesizer, *Ham Radio,* April 1981, pp. 68–70.

9

Interfacing Circuits

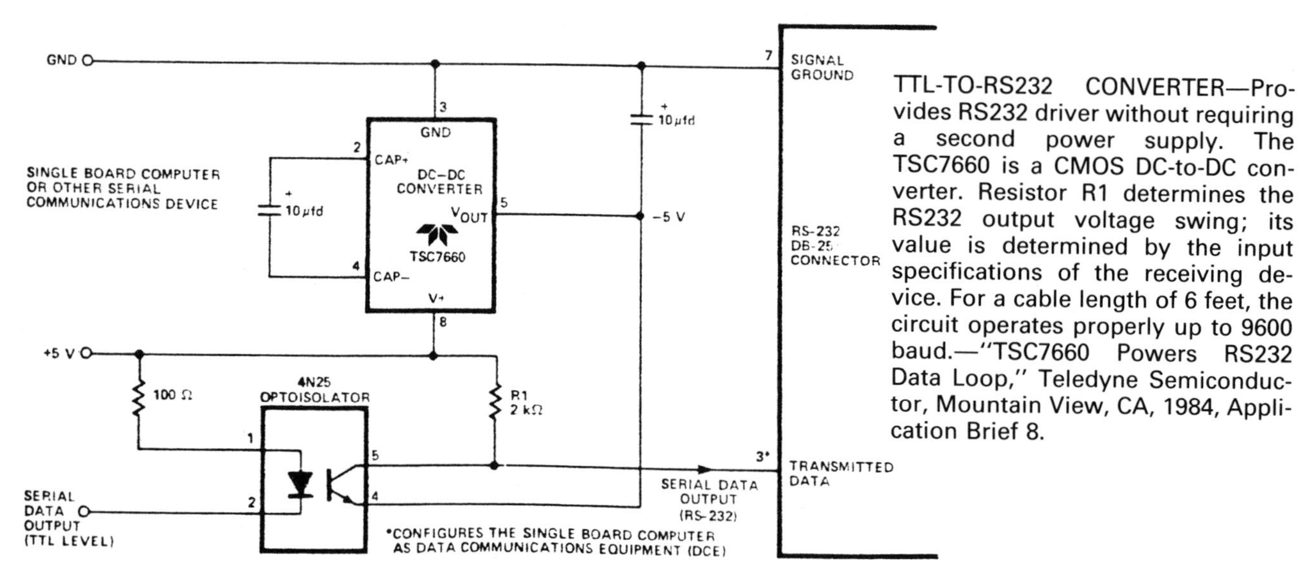

TTL-TO-RS232 CONVERTER—Provides RS232 driver without requiring a second power supply. The TSC7660 is a CMOS DC-to-DC converter. Resistor R1 determines the RS232 output voltage swing; its value is determined by the input specifications of the receiving device. For a cable length of 6 feet, the circuit operates properly up to 9600 baud.—"TSC7660 Powers RS232 Data Loop," Teledyne Semiconductor, Mountain View, CA, 1984, Application Brief 8.

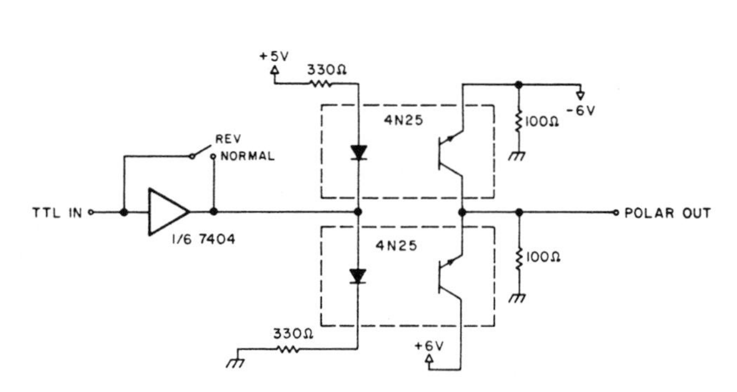

TTL TO "POLAR" CONVERSION—Converts TTL-level signals to ±6 V suitable for driving printing devices using "polar" input.—M. Leavey, RTTY Loop, *73 Magazine,* April 1981, p. 20.

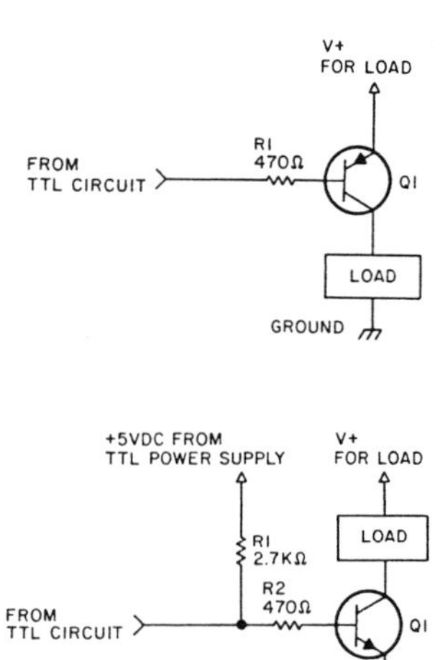

INTERFACING TTL—Circuits using PNP or NPN transistors provide current-carrying ability that TTL devices cannot. The transistors must be selected on the basis of the current-carrying ability necessary to drive a load.—R. Swirsky, Connecting TTL to the Outside World, *73 Magazine,* August 1984, p. 85.

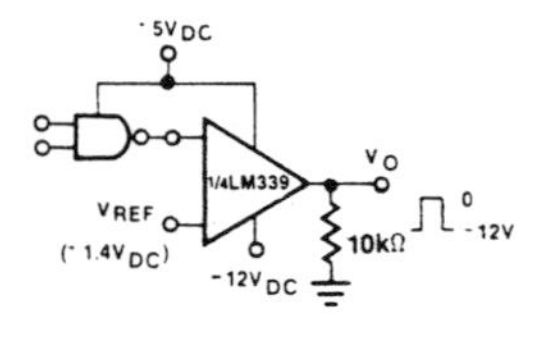

TTL TO MOS INTERFACE—Built around LM339 quad voltage comparator and raises TTL-level signal to −12-V signal for MOS devices. The LM339 requires a +1.4-V reference. All unused inputs on the LM339 should be grounded.—"Signetics Linear LSI Data and Applications Manual," Signetics, Sunnyvale, CA, 1985, p. 14-115.

TTL-TO-RS232 INTERFACE—Allows driving of RS232-level devices with TTL-level signals. R6 protects Q2 in the event the RS232 output is shorted.—S. Freeberg, Simple Transistor TTL-To-RS232 Interface, *73 Magazine,* January 1984, p. 115.

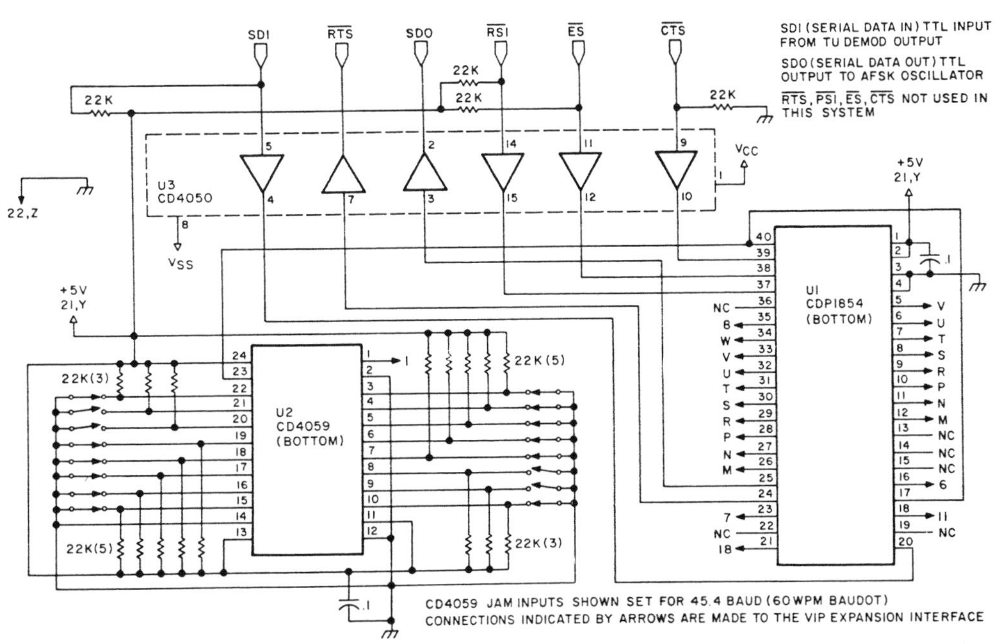

UART INTERFACE FOR RCA VIP MICROCOMPUTER—Allows the 1802-based RCA VIP microcomputer to be used as terminal for ASCII or Baudot with appropriate software.—M. Levy, RTTY Loop, *73 Magazine,* January 1981, p. 18.

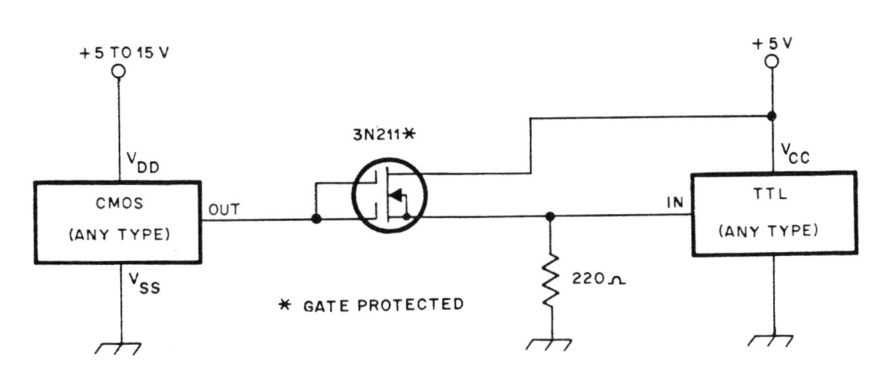

DRIVING TTL FROM CMOS—Dual-gate MOSFET serves as buffer between CMOS and TTL. The circuit can also be used with split supplies provided the positive CMOS output excursion is at least 5 V.—"The Radio Amateur's Handbook," American Radio Relay League, Newington, CT, 1981, pp. 4-56 to 4-57.

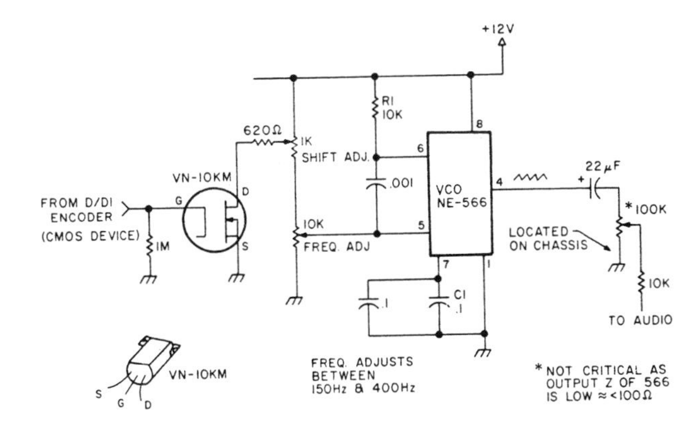

VCO TO DECODER INTERFACE—Matches a garage door opener decoder/encoder chip to a 566 VCO. The 10-K variable resistor adjusts the frequency between 15 and 400 Hz.—W. Desnoes, VCO/Decoder Interface, *73 Magazine,* March 1983, p. 110.

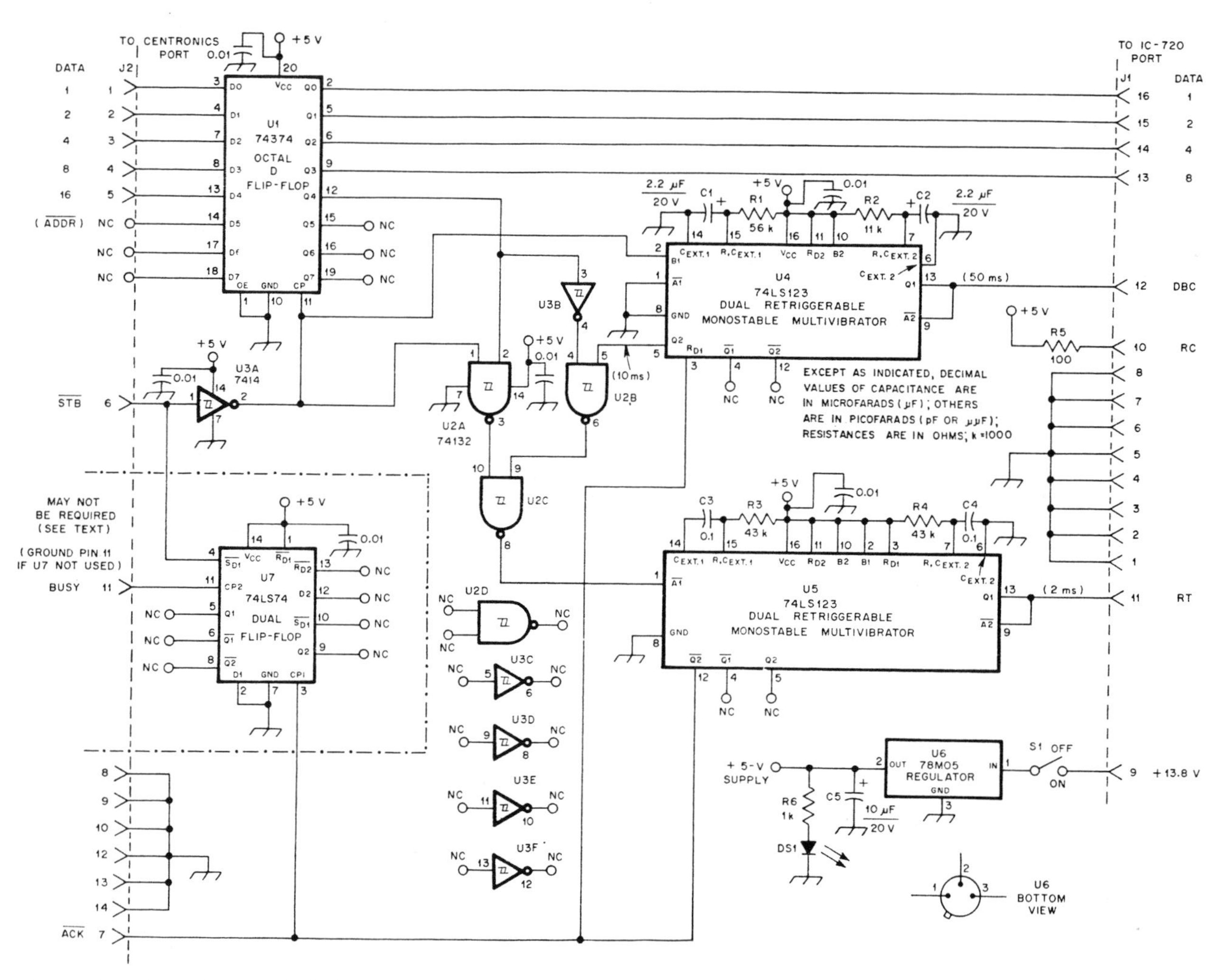

TRANSCEIVER TO MICROCOMPUTER INTERFACE—Allows remote control of an ICOM IC-720 HF transceiver to perform functions such as changing operating frequency and mode of emission. The circuit can be adapted to work with similar transceivers. The original article discusses software for the interface.—L. Studebaker, Simple, Low-Cost Computer Control for the ICOM IC-720, *QST*, July 1984, pp. 34–38.

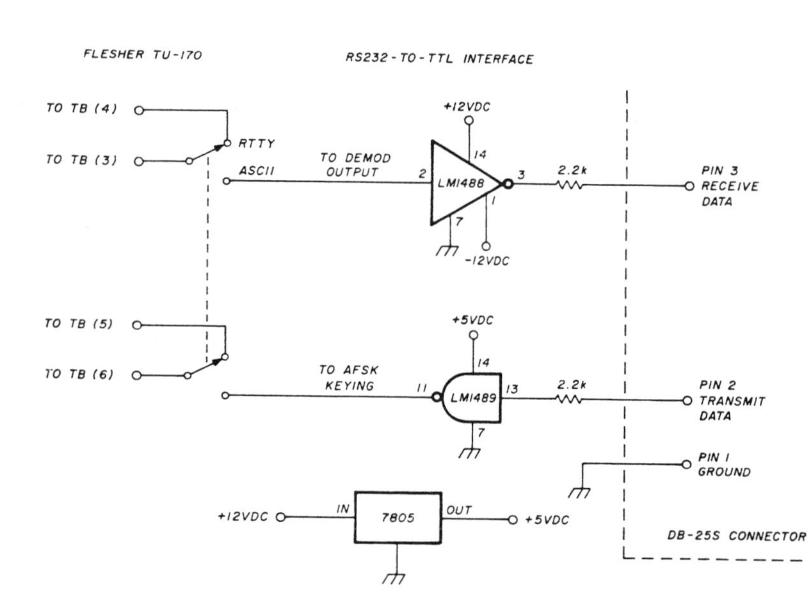

TELEPRINTER RS232 TO TTL INTERFACE—Allows interfacing of teleprinter units with TTL-level I/O to microcomputers with RS232 serial interface port. Toggle switch allows normal teleprinter operation of RS232 operation.—B. Harvey, A RS232 to TTL Interface, *Ham Radio*, November 1982, pp. 70–71.

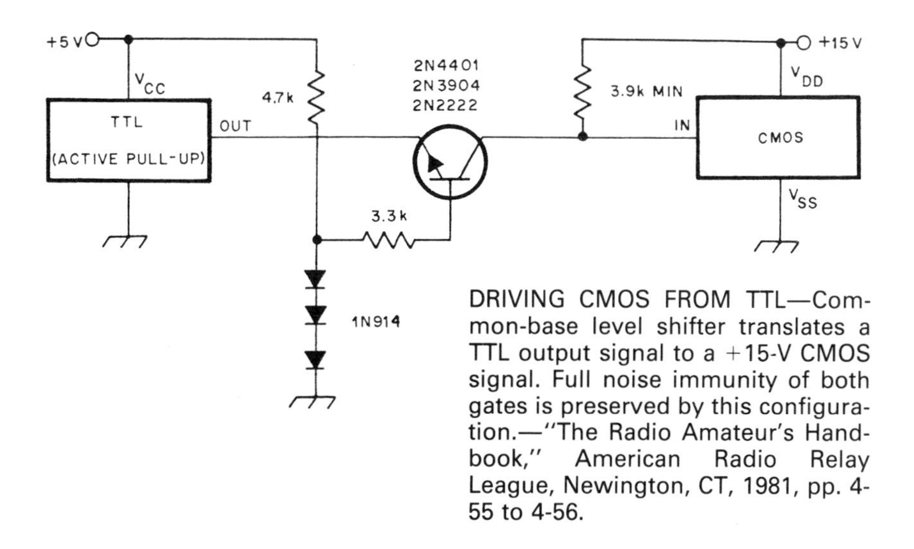

DRIVING CMOS FROM TTL—Common-base level shifter translates a TTL output signal to a +15-V CMOS signal. Full noise immunity of both gates is preserved by this configuration.—"The Radio Amateur's Handbook," American Radio Relay League, Newington, CT, 1981, pp. 4-55 to 4-56.

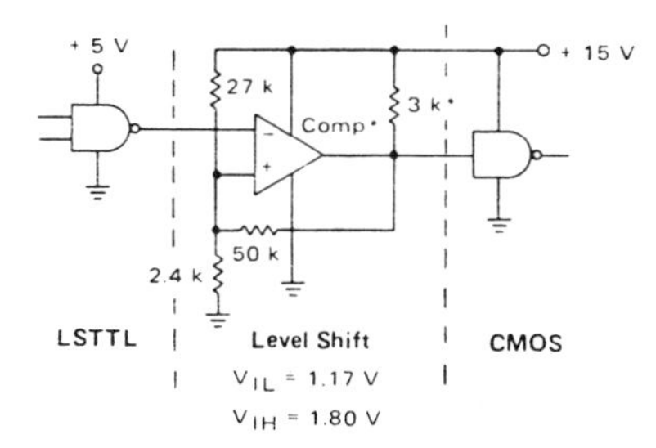

LSTTL TO CMOS INTERFACE—Designed around MC3405 dual op amp and comparator; same configuration may be used with an op amp if the 3-K resistor is omitted.—"Motorola Linear Integrated Circuits Databook," Motorola, Phoenix, AZ, 1979, p. 6-124.

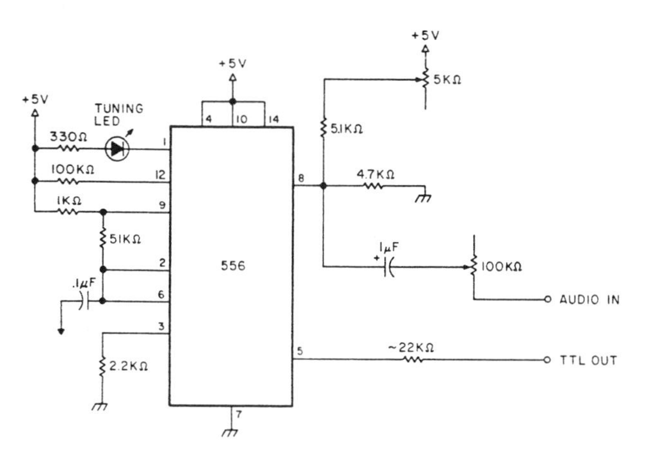

AF TO TTL INTERFACE—Converts audio signals to TTL-level (+5.0 V) output. Originally designed to convert AFSK signals to TTL for driving a microcomputer, the circuit may also be used to convert the output of a cassette program or data storage to TTL.—D. Fait, Ham Over Fist, *73 Magazine,* November, 1984, pp. 64–66.

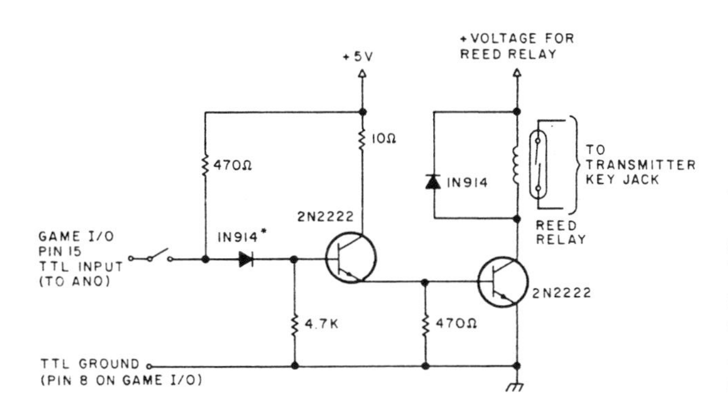

KEYING INTERFACE—Designed to allow keying of a CW transmitter using an Apple II microcomputer and software described in original article. The circuit can also be used for other control functions with different software.—R. Swirsky, Apple, Morse, and You, *73 Magazine,* July 1983, pp. 54–55.

10

LED and Optoelectronic Circuits

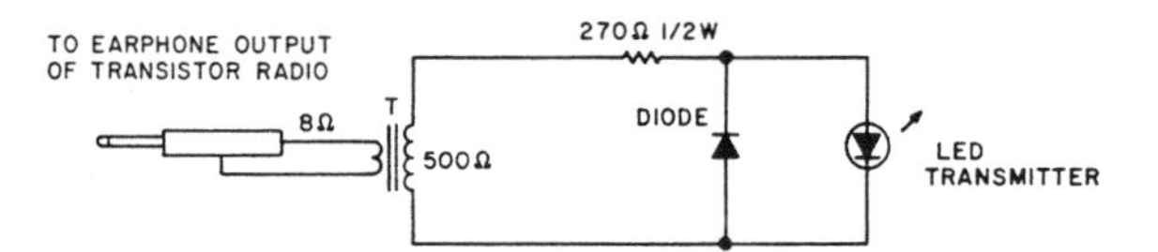

LED TRANSMITTER—Produces energy for light-beam communications. The transmitter is driven from the output of a transistor radio or similar low-power audio source. The receiver can be a silicon or selenium solar cell connected to the input of an audio amplifier.—C. Graf, LEDs You've Never Seen, *73 Magazine,* March 1984, pp. 58–65.

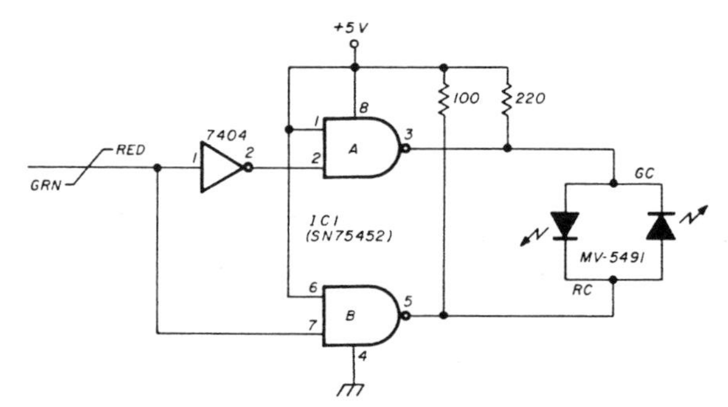

DUAL LED DRIVER—LED changes color according to input signal. The LED can be green or red in color. With the driver input low, the input to IC1A is high and its output is low. In this state, 4.8 V will be dissipated across the 220-Ω resistor, placing the green cathode (GC) at 0.2 V. IC1B, with its low input, will have a high or nonconducting output. This allows the green diode to be forward-biased, and 1.8 V will be dissipated across the 100-Ω resistor. When the circuit input goes high, the IC1A and IC1B outputs will change state, and 4.8 V will be dissipated across the 100-Ω resistor. The red diode goes into conduction, dissipating 3.15 V across the 220-Ω resistor. Rapid transitions between states will produce an amber color.—K. Powell, Light-Emitting Diodes: Theory and Application, *Ham Radio,* August 1980, pp. 12–19.

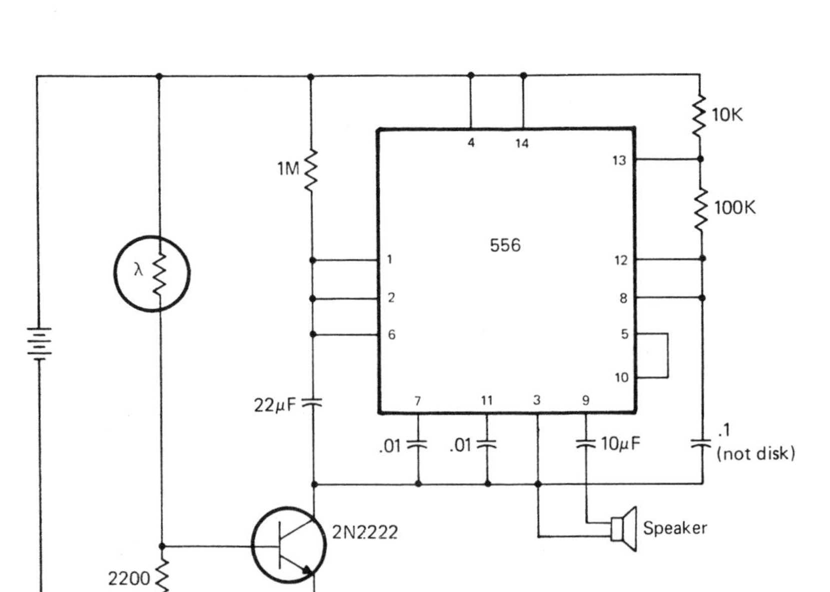

PHOTORESISTOR - ACTIVATED ALARM—Photoresistor pulls up the base of 2N2222 when illuminated, completing negative supply return for 556 dual timer device. The first timer provides the initial delay; then the second timer is turned on and generates the alarm tone.—M. Winston, A Refrigerator Alarm that Saves Energy and Calories, *CQ,* November 1981, p. 85.

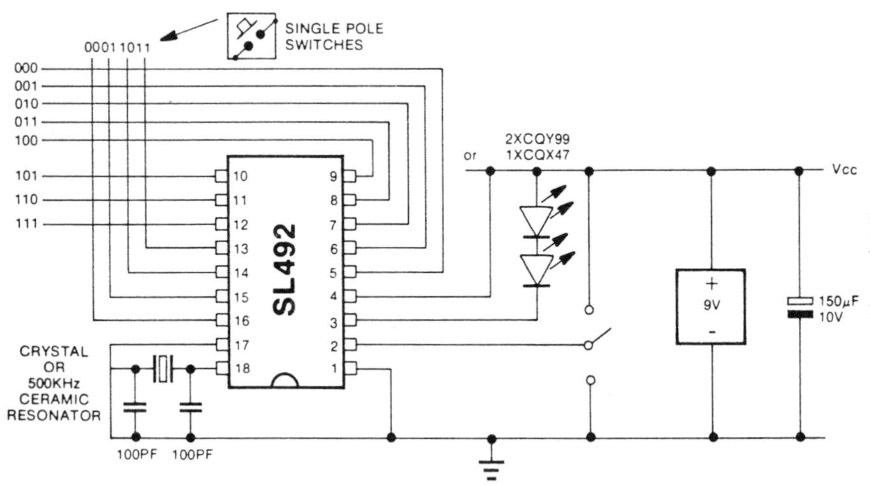

5-BIT INFRARED TRANSMITTER—Operates at 500-kHz oscillation frequency and produces a 5-bit output, including start and stop pulses, in PPM. Supply voltage may range from 3 to 11 V. The original source includes a timing diagram and code table.—"Plessey Integrated Circuits Databook," Plessey Semiconductor, Irvine, CA, 1983, pp. 427–430.

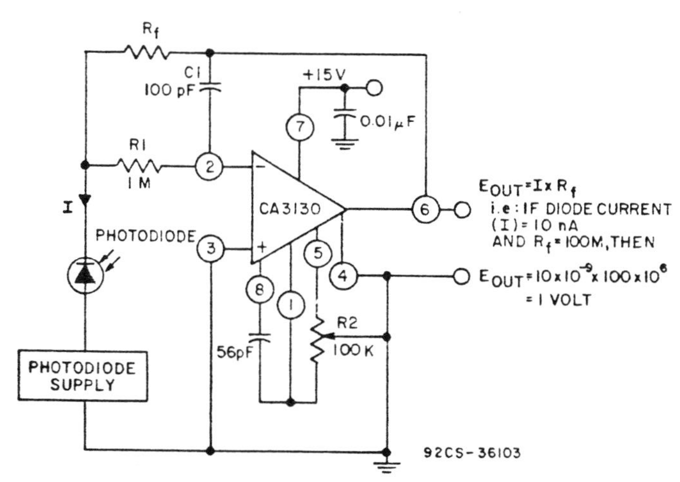

PHOTODIODE CURRENT-TO-VOLTAGE CONVERTER—Circuit provides a ground-referenced output voltage which is proportional to the current flowing through a photodiode in the input circuit. R1 is used to limit the input current to a safe level in case the back-biased photodiode should avalanche and expose the input terminal of the CA3130 to the comparatively high voltages sometimes used in photodiode supplies.—"Understanding and Using the CA3130, CA3130A and CA3130B BiMOS Operational Amplifiers," RCA Solid State Division, Somerville, NJ, 1983, Application Note ICAN-6386.

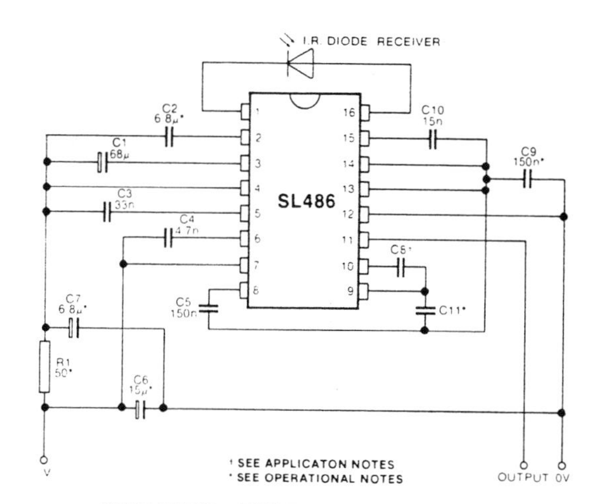

INFRARED RECEIVING PREAMPLIFIER—SL486 IC interfaces between infrared receiving diode and digital input of receiver or microprocessor. Pin 10 is a stretch input, and pin 11 is a stretch output. Output current is 5 mA.—"Plessey Integrated Circuits Databook," Plessey Semiconductor, Irvine, CA, 1983, pp. 417–422.

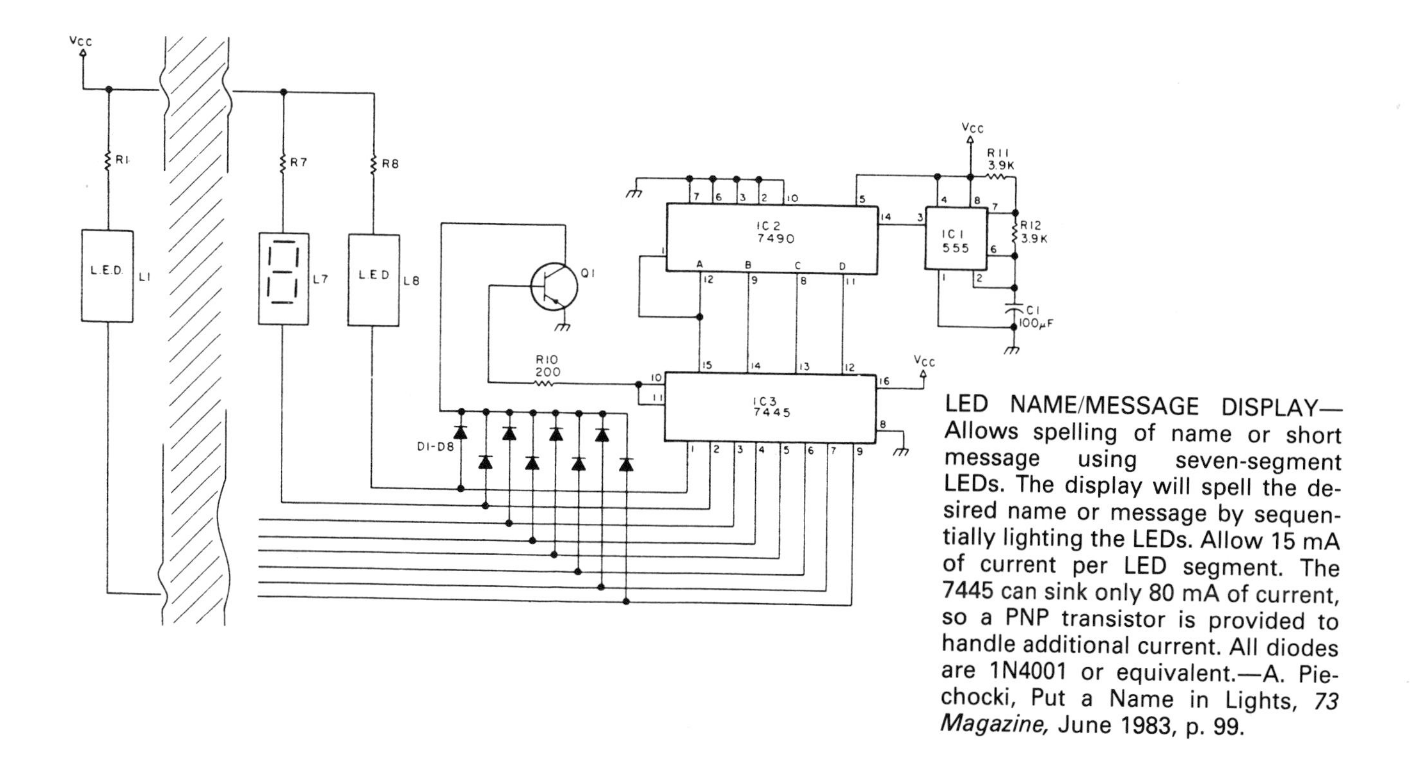

LED NAME/MESSAGE DISPLAY—Allows spelling of name or short message using seven-segment LEDs. The display will spell the desired name or message by sequentially lighting the LEDs. Allow 15 mA of current per LED segment. The 7445 can sink only 80 mA of current, so a PNP transistor is provided to handle additional current. All diodes are 1N4001 or equivalent.—A. Piechocki, Put a Name in Lights, *73 Magazine,* June 1983, p. 99.

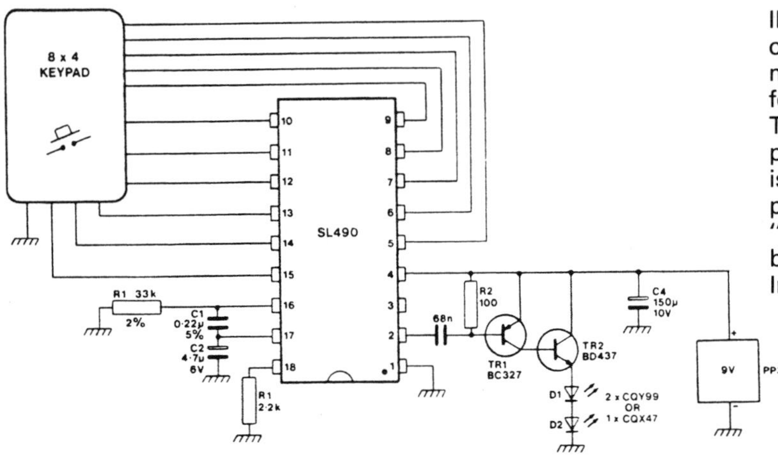

INFRARED TRANSMITTER—Based on SL490 infrared/ultrasonic transmitting IC, PPM output from pin 2 is fed to the base of the PNP transistor TR1, producing an amplified current pulse about 15 µsec wide. This pulse is further amplified by TR2 and applied to infrared diodes D1 and D2.—"Plessey Integrated Circuits Databook," Plessey Semiconductor, Irvine, CA, 1983, pp. 423–426.

INFRARED CONTROL FOR TOYS—Designed for use with SL490 infrared transmitter, ML925 receives/decodes PPM remote control commands. The original source includes a table of command codes for car or truck toys. The circuit is designed to control either a toy vehicle with a two-speed drive motor and a three-position latching steering system, or a vehicle with momentary action steering and a third motor, typically a winch.—"Plessey Integrated Circuits Databook," Plessey Semiconductor, Irvine, CA, 1983, pp. 445–448.

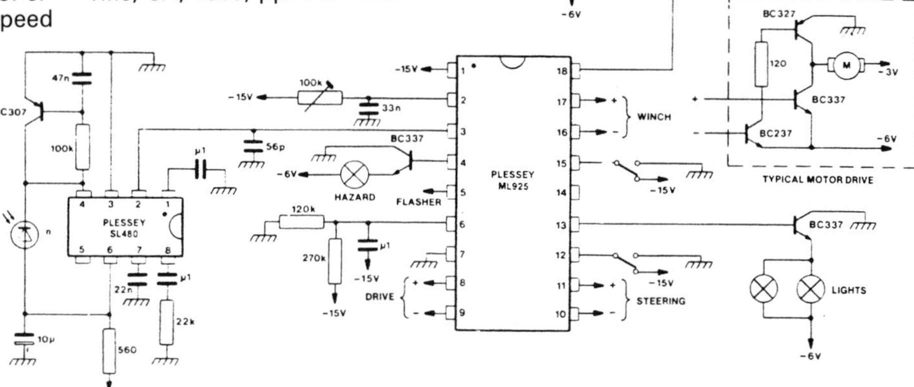

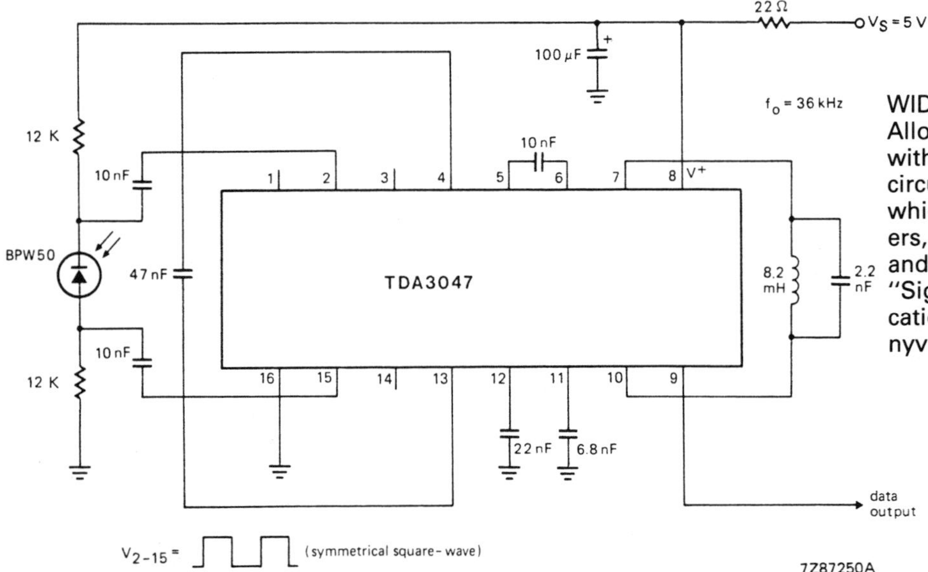

WIDEBAND INFRARED RECEIVER—Allows reception of infrared signals with low power consumption. The circuit is built around a TDA3047 IC, which incorporates a limiter, amplifiers, an AGC detector, a pulse shaper, and an output buffer on the chip.—"Signetics Linear LSI Data and Applications Manual," Signetics, Sunnyvale, CA, 1985, p. 5-126.

11

Measurement and Indication Circuits

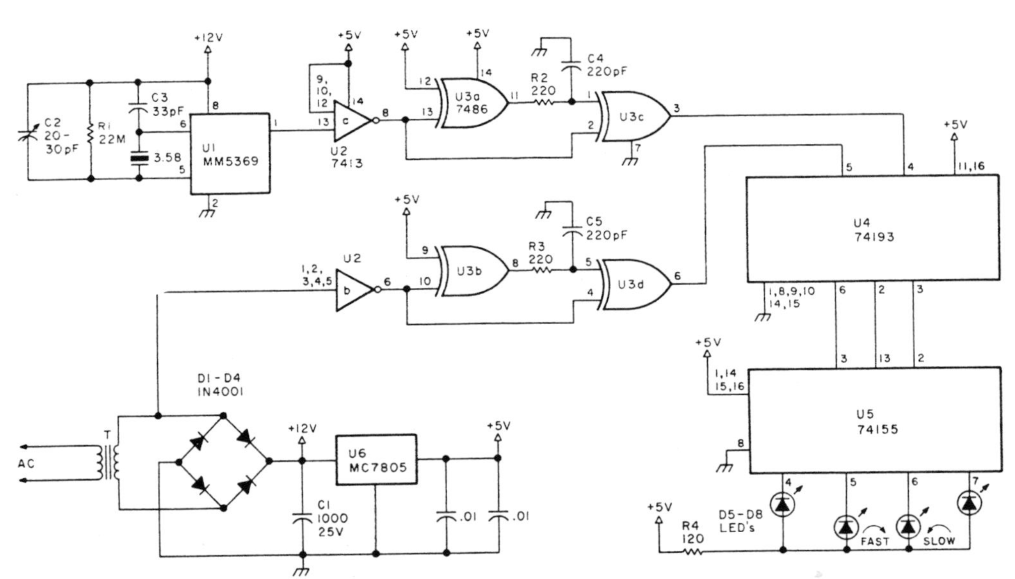

60-Hz FREQUENCY INDICATOR—Basically a frequency comparator, rotation of LEDs indicates whether AC line frequency is above or below standard 60 Hz. U6 should be mounted on a small heatsink.—J. Lunacek, Build a 60-Hz Frequency Monitor, *73 Magazine,* February 1981, pp. 72–73.

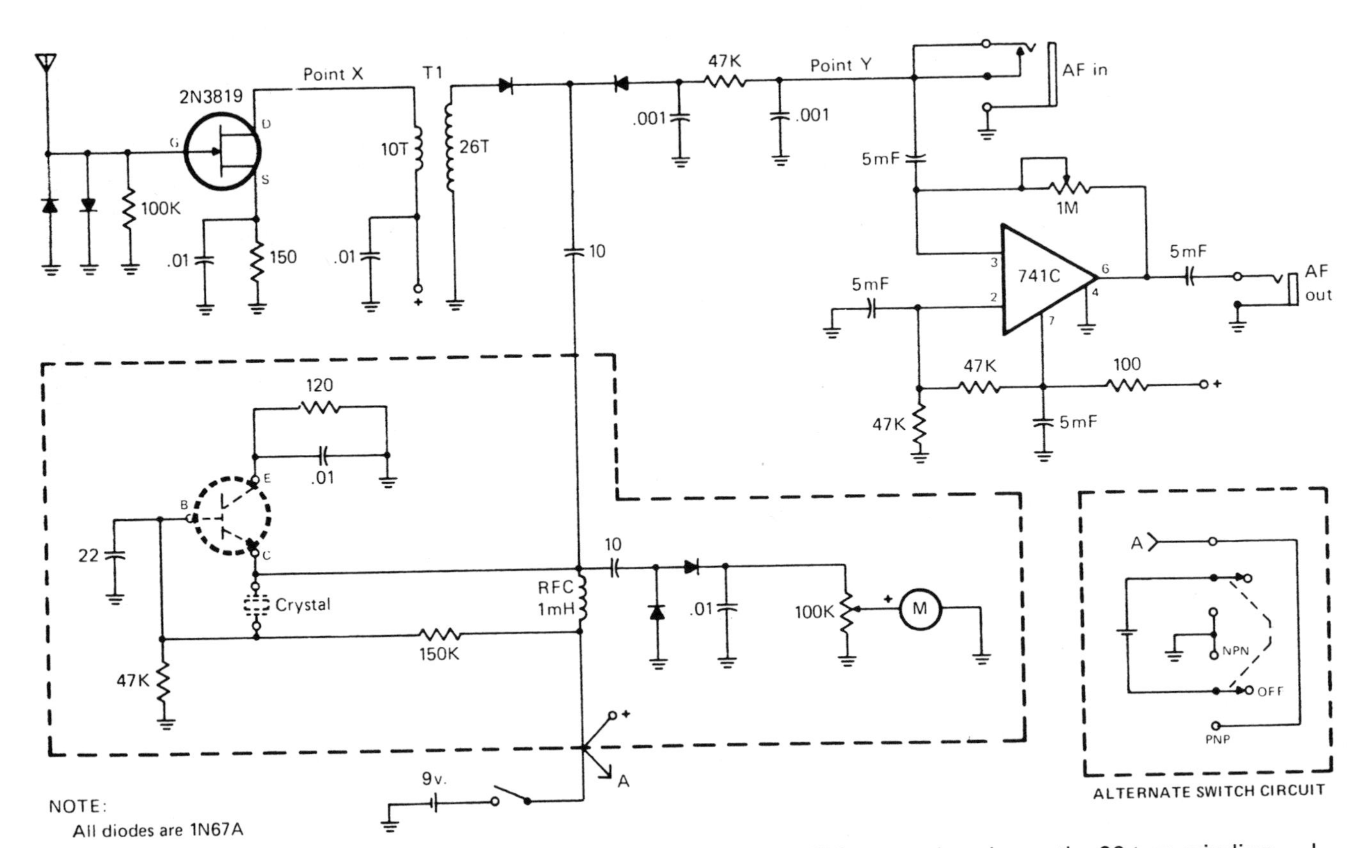

SSB MODULATION MONITOR—Allows monitoring of "off the air" audio from an SSB transmitter. The circuitry shown enclosed in dashed lines is a universal crystal oscillator and output level monitor. It will function with almost any fundamental mode crystal from 3 to 20 MHz. T1 is wound with any fine gauge wire on a T372 core with the 10-turn winding placed over the 26-turn winding.—J. Schultz, Build Your Own Combined SSB Modulation Monitor/Multi-Function Test Instrument, *CQ,* February 1984, pp. 48–52.

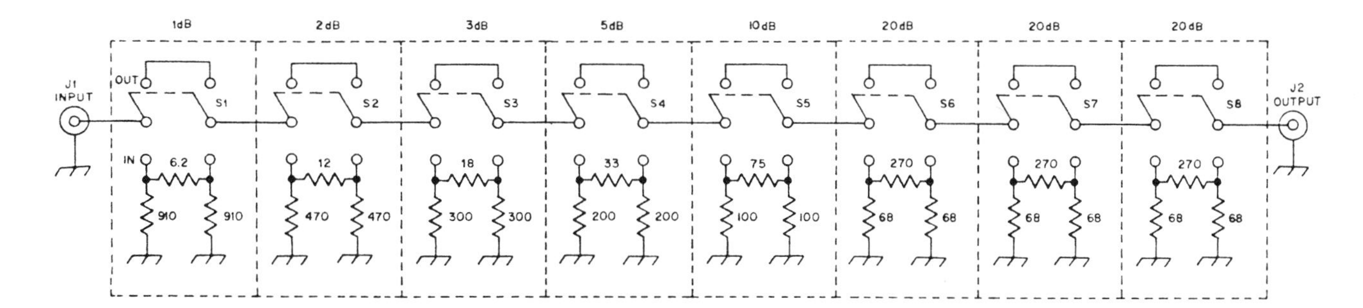

LOW-POWER STEP ATTENUATOR—Eight pi network resistive sections provide variable attenuation in 1-dB steps. The total attenuation available is 81 dB. The maximum attenuation in any section is 20 dB. All resistors are rated at 0.25 W and 5% tolerance.—B. Shriner and P. Pagel, A Step Attenuator You Can Build, *QST*, September 1982, pp. 11–13.

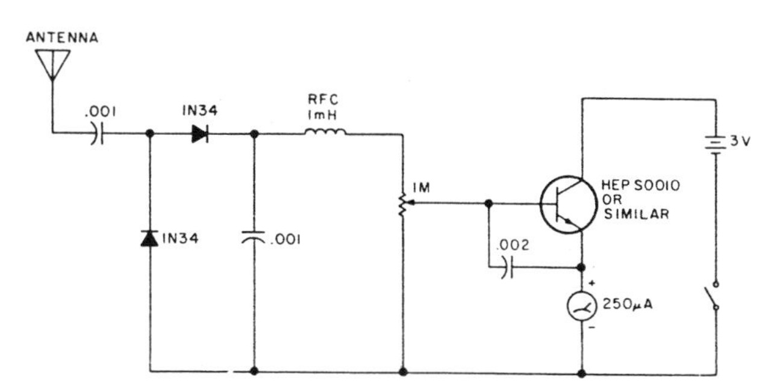

FIELD STRENGTH METER—Originally designed to use meter and case of discarded VOM, this circuit is capable of measuring RF energy up to 144 MHz. Performance on HF frequencies can be improved by adding an amplifier stage following the antenna.—*73 Magazine* Staff, Field Strength for Free, *73 Magazine*, October 1980, pp. 82–84.

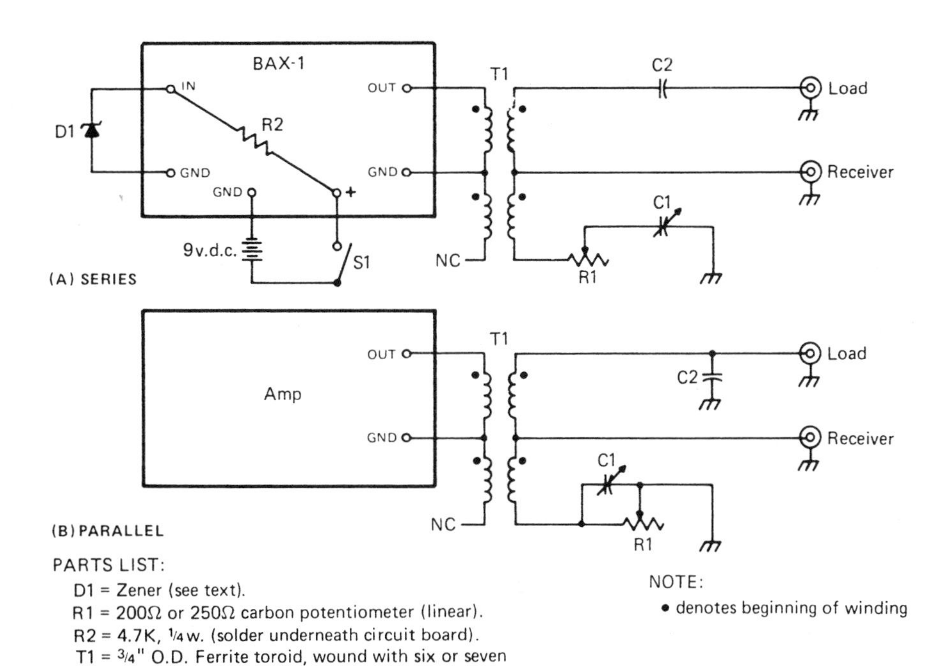

PARTS LIST:
D1 = Zener (see text).
R1 = 200Ω or 250Ω carbon potentiometer (linear).
R2 = 4.7K, 1/4 w. (solder underneath circuit board).
T1 = 3/4" O.D. Ferrite toroid, wound with six or seven quadrifilar turns of #26 or 28 enamel wire.
C1 = 360pF miniature variable (straight line capacity) or similar.
C2 = 180pF mica or polystyrene (exactly half of variable).

RF NOISE BRIDGE—Used to measure impedance of antenna system. The device produces a wideband random noise applied to the bridge transformer; a receiver tuned to the appropriate frequency is used as a null detector. Resistive and reactive components of an antenna system may then be measured. The bridge configuration can be connected in either series or parallel. BAX-1 is an International Crystals broadband amplifier; a similar broadband RF amplifier circuit may be used. D1 should be back-biased to its avalanche voltage.—C. Nouel, How to Build a Cheap and Easy RF Noise Bridge, *CQ*, May 1984, pp. 56–57.

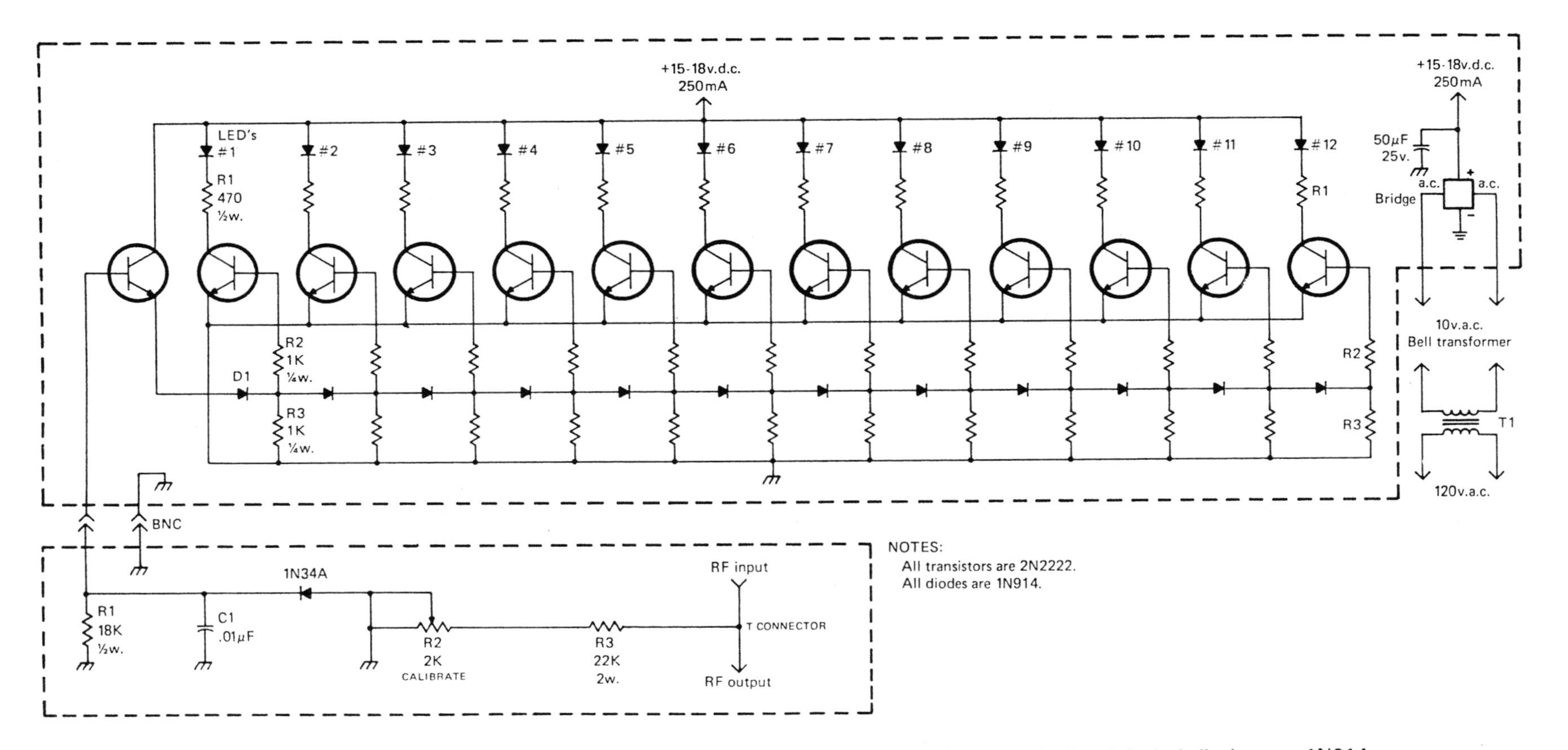

PEAK READING LED RF WATTMETER—Indicates peak RF output power from a transmitter/transceiver by means of LEDs. The circuit is actually a voltmeter calibrated to display watts. The transmission line voltage is sampled and RF energy is rectified via the smaller circuit (shown enclosed in dashed lines). The resulting voltage is used to drive the display unit (larger circuit shown in dashed lines). All transistors are 2N2222, and all unlabeled diodes are 1N914. If one LED is lit, the RF output is 14 W; if all LEDs are lit, the RF output is 2000 W.—R. Beard, Construct a Peak-Reading LED RF Wattmeter, *CQ,* August 1983, pp. 48–49.

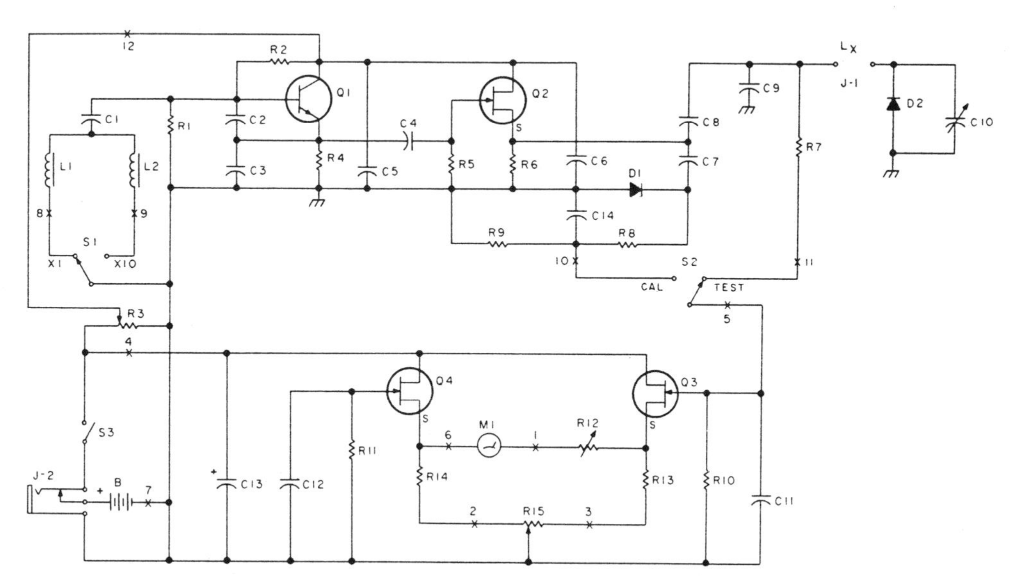

Q METER—Measures inductances from 0.5 to 50 μH and Q to 200. The circuit consists of a dual-frequency RF oscillator, FET voltmeter, power supply, and a tank circuit that indicates the inductor of unknown value (L_x). Q1 is 2N2222, while Q2, Q3, and Q4 are all MPF102.D1 is 1N270, and D2 is a MBD101 hot-carrier diode. L1 is 21 turns of No. 24 wire on a T-37-2 toroid form giving 1.97 μH. L2 is 70 turns of No. 32 wire on a T-37-2 toroid form giving 19.7 μH. C1 is 620-pF and C2 and C3 are 68-pF NPO ceramics. C4 is a 56-pF NPO ceramic. C5, C6, C11, and C12 are 0.01 ceramics. C7 is a 100-pF ceramic. C8 is a 22-pF NPO ceramic. C9 is 1500 pF. C10 is a 138-pF variable, and C13 is a 10-μF 25-V electrolytic. R1 is 47 K, and R2 is 100 K. R3 is a 10-K linear potentiometer. R4 is 1.5 K. R5 and R9 are 1 M. R6 is 390 Ω. R7, R10, and R11 are 2.2 M. R8 is 100 K. R12 is a 100-K trimmer. R13 and R14 are 150 Ω. R15 is a 2-K potentiometer. M1 is a 200-μA meter.—E. Miller, Around and Around and Around, *73 Magazine,* January 1984, pp. 70–72.

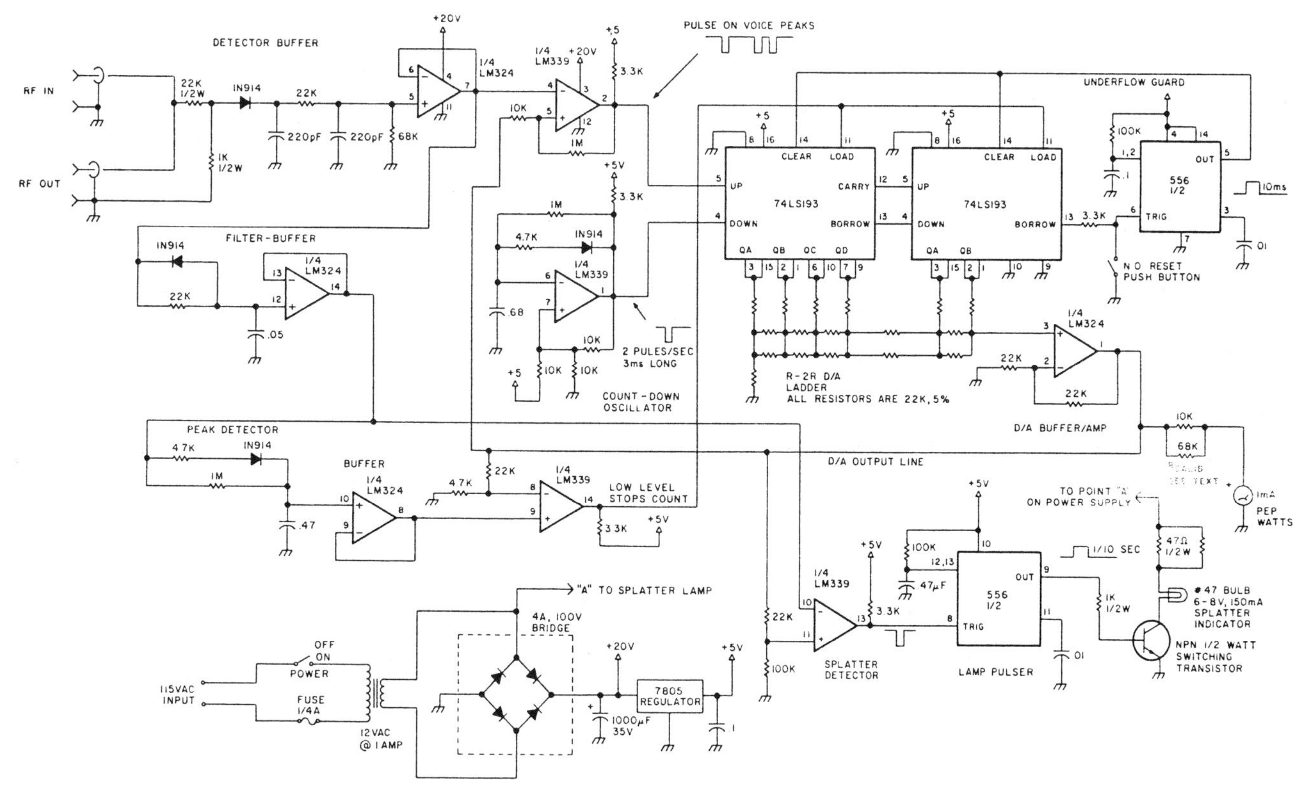

OVERMODULATION INDICATOR—Monitors output of a SSB transmitter and flashes a warning light whenever it detects the signal "flat-topping," which results in distorted audio and spurious signals up to 50 kHz from transmitted frequency. The meter indicates the PEP of the SSB signal—P. Clower, The Splattometer, *73 Magazine,* October 1982, pp. 10–18.

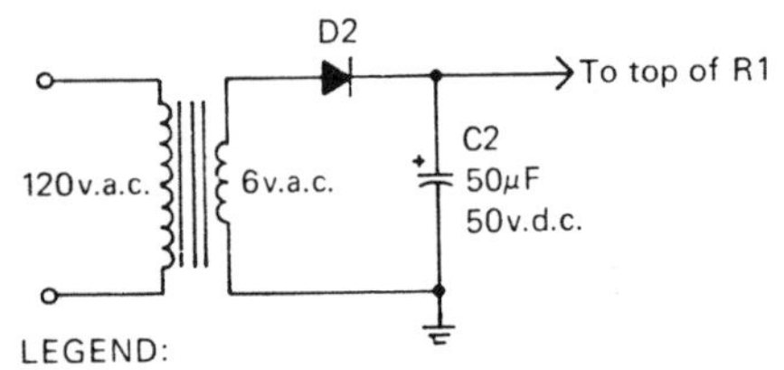

LEGEND:
D2 = 1 amp rectifier diode – Radio Shack 276-1101
T1 = 120v.a.c./6v.a.c. transformer – Radio Shack 273-1381

OVERVOLTAGE WARNING—Buzzer sounds when monitored voltage exceeds desired level. IC1 and IC2 form a conventional LED voltmeter with a range of 0–4.5 V. R1 forms a voltage divider to set the scale of the voltmeter. Two modes of operation are possible. Point B may be connected to pin 10 of IC1. A test voltage is then applied to the top of R1, and R1 is adjusted so that all LEDs are lit. Or point B may be connected to pin 15 of IC1, and R1 is adjusted so that only the fifth LED is lit. Line voltage monitoring requires the use of the AC/DC converter.—P. Danzer, Warning! Over-Voltage Ahead, *CQ,* April 1982, pp. 28–29.

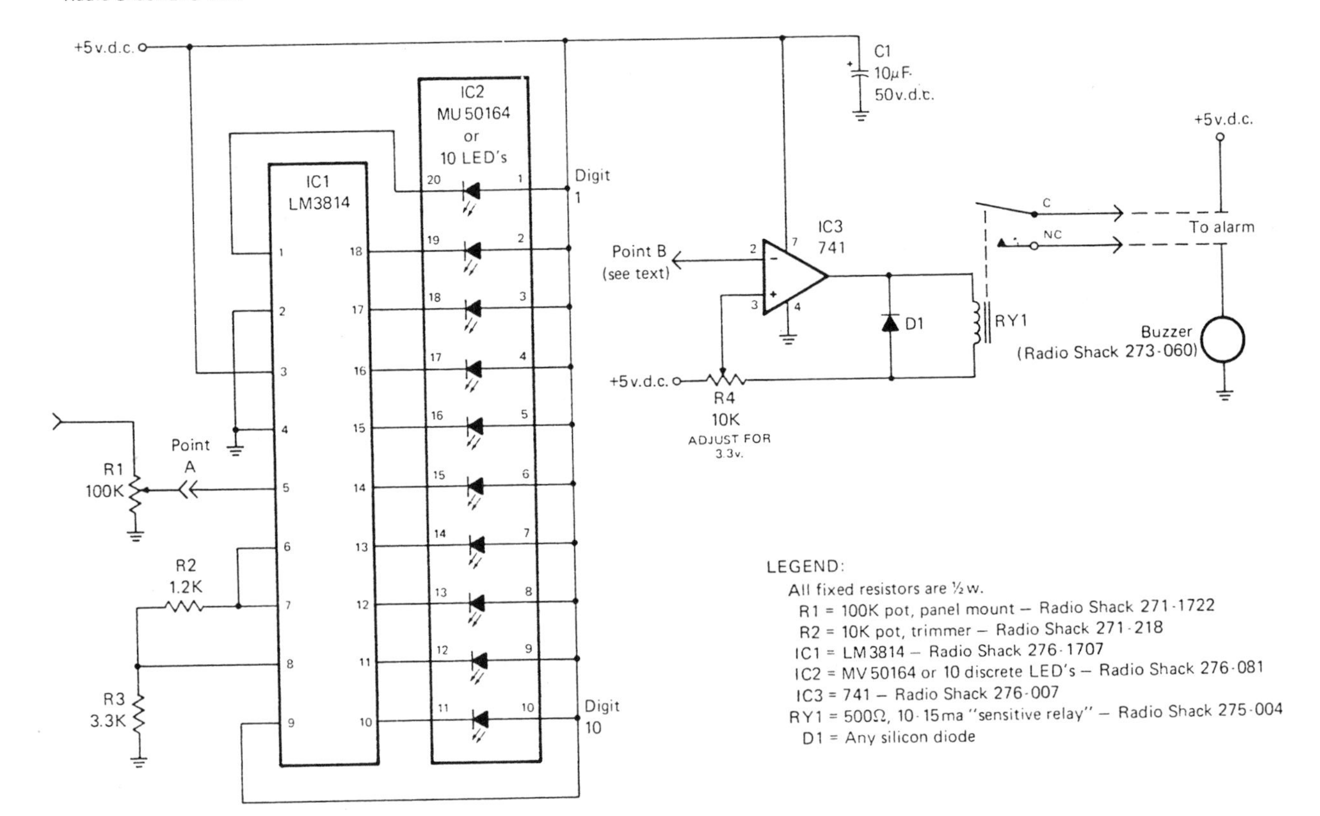

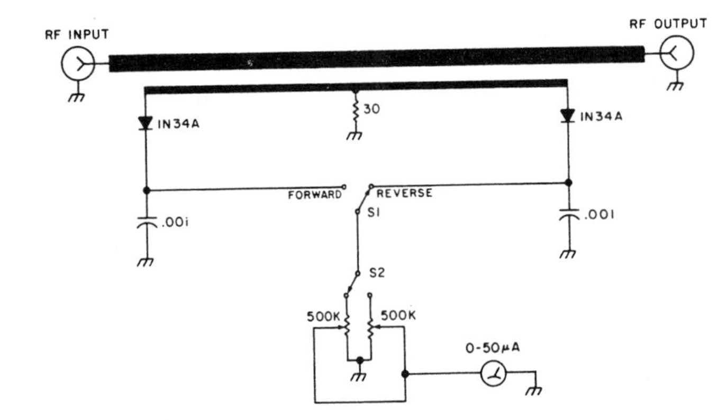

50- TO 500-MHz WATTMETER—Coupling between measuring circuit and RF path is accomplished by stripline etched onto G-10 epoxy PC board. The circuit design can also be used to measure SWR. Calibration is achieved with another wattmeter or bridge of known accuracy.—P. Putman, Elementary, My Dear: Watts 'n' SWR, *73 Magazine,* September 1984, pp. 14–16.

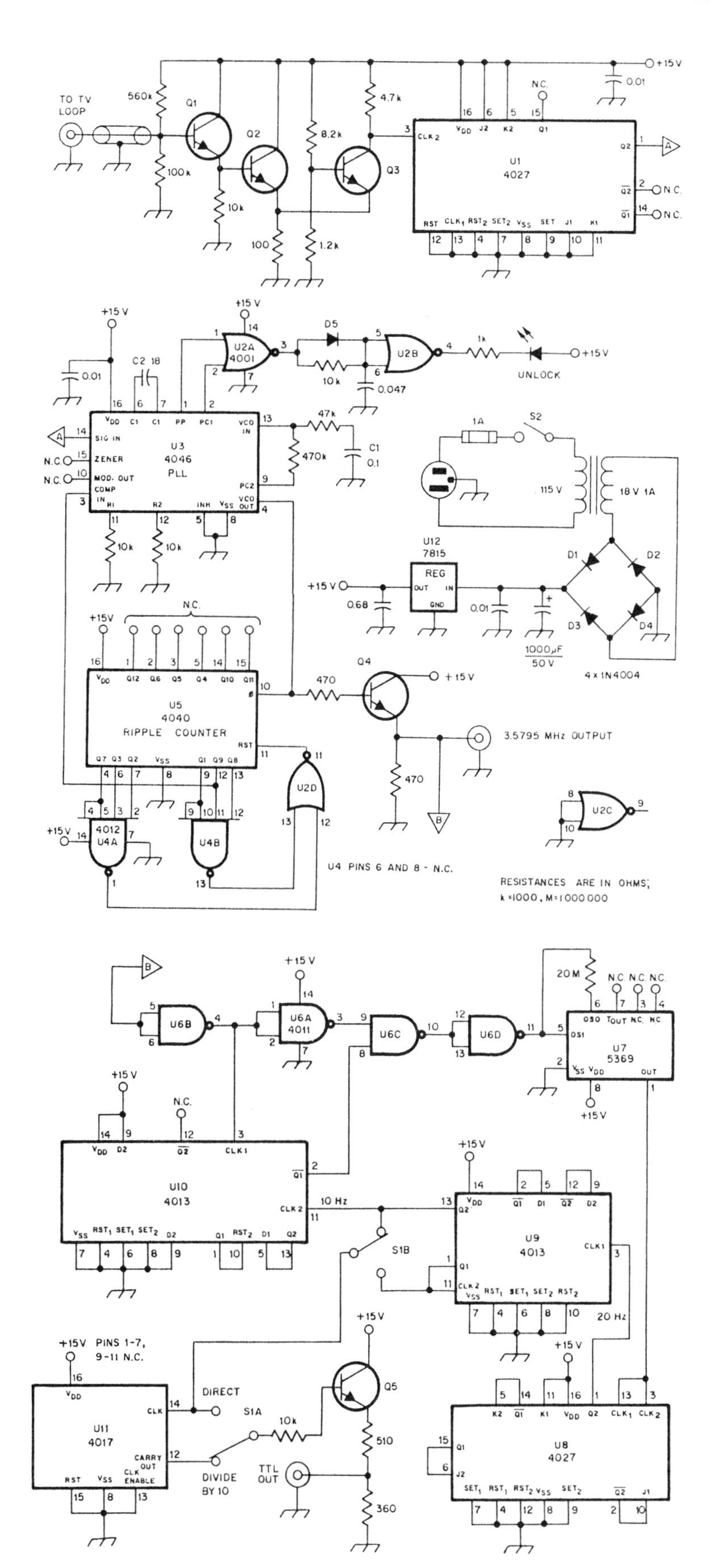

FREQUENCY STANDARD AND REFERENCE—Uses television color burst frequency, 3.579545 MHz, as reference frequency for frequency counters. All transistors are 2N2222A, and D5 is 1N4154.—D. Bissen, The N0AJY CB Standard, *QST*, August 1983, pp. 30–32.

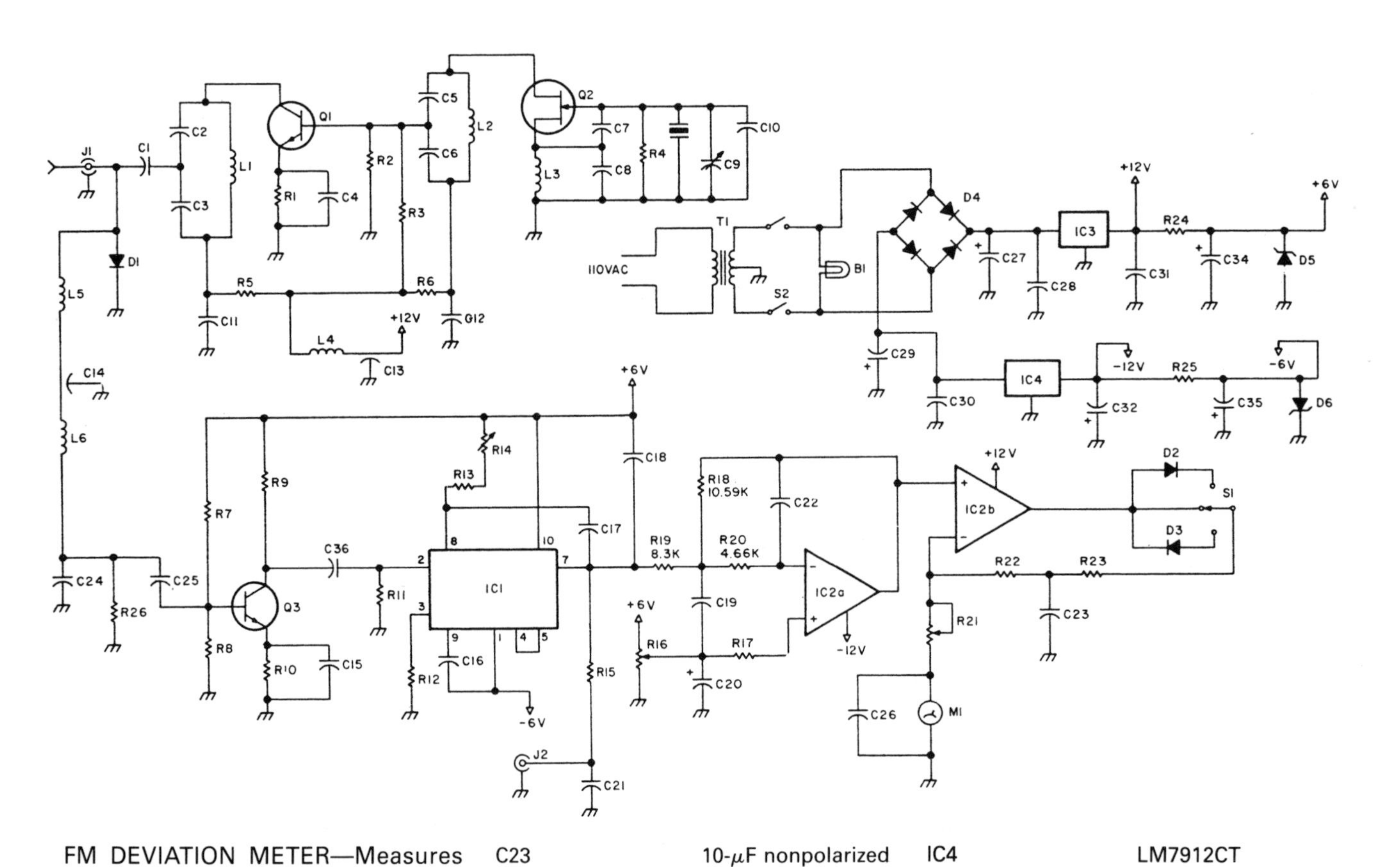

FM DEVIATION METER—Measures deviation ±10 kHz of center frequency. Audio output with 750-μsec de-emphasis is available for oscilloscope monitoring. The parts indicated on schematic are as follows:

B1	24-V power-on indicator
C1	3.3-pF disc
C2	5 pF
C3	50 pF
C4, C11, C12	500-pF disc
C5	15 pF
C6	33 pF
C7	25 pF
C8	100 pF
C9	1.8-8.7-pF air variable
C10	5-pF disc, NPO, or silver mica
C13, C14	100-pF feedthrough
C15, C25, C36	0.02 μF
C16, C22	0.0022 μF
C17	1000 pF
C18, C19	0.01 μF
C20	1-μF, 16-V DC tantalum or electrolytic
C21, C28, C30, C31	0.05-μF disc
C23	10-μF nonpolarized electrolytic
C24	0.003 μF
C26	0.1 μF disc
C27, C29	220-μF 25-V DC electrolytic
C32, C34, C35	4.7-μF 16-V DC tantalum or electrolytic
D1	1N84
D2, D3	1N914
D4	1-A, 50 PIV bridge rectifier
D5, D6	6-V zener (1N4735 or equivalent)
L1	4.5 turns of No. 22 wire on Miller 20A000-4 form
L2	7 turns of No. 22 wire on Miller 20A000-4 form
L3	100-μH RF choke
L4, L5	1.72-μH RF choke
L6	2-mH RF choke
Q1	MPS918
Q2	MPF102
Q3	2N3904
IC1	LM565
IC2	1458
IC3	LM3405T12
IC4	LM7912CT
R1	470 Ω
R2, R10	1 K
R3, R17	10 K
R4	150 K
R5, R6	100 Ω
R7	33 K
R8	8.2 K
R9	6.8 K
R11, R12, R13	4.7 K
R15	15 K
R22	47 K
R23	100 Ω
R24, R25	390 Ω
R14, R16	5-K 10-turn variable resistor
R21	20-K 10-turn variable resistor
R18	10 K
R19	8.2 K
R20	4.7 K
T1	24- to 30-V AC CT transformer
Xtal	Crystal cut to resonance at desired test frequency

R. Six, Strictly for FM Deviates, *73 Magazine*, February 1984, pp. 36–41.

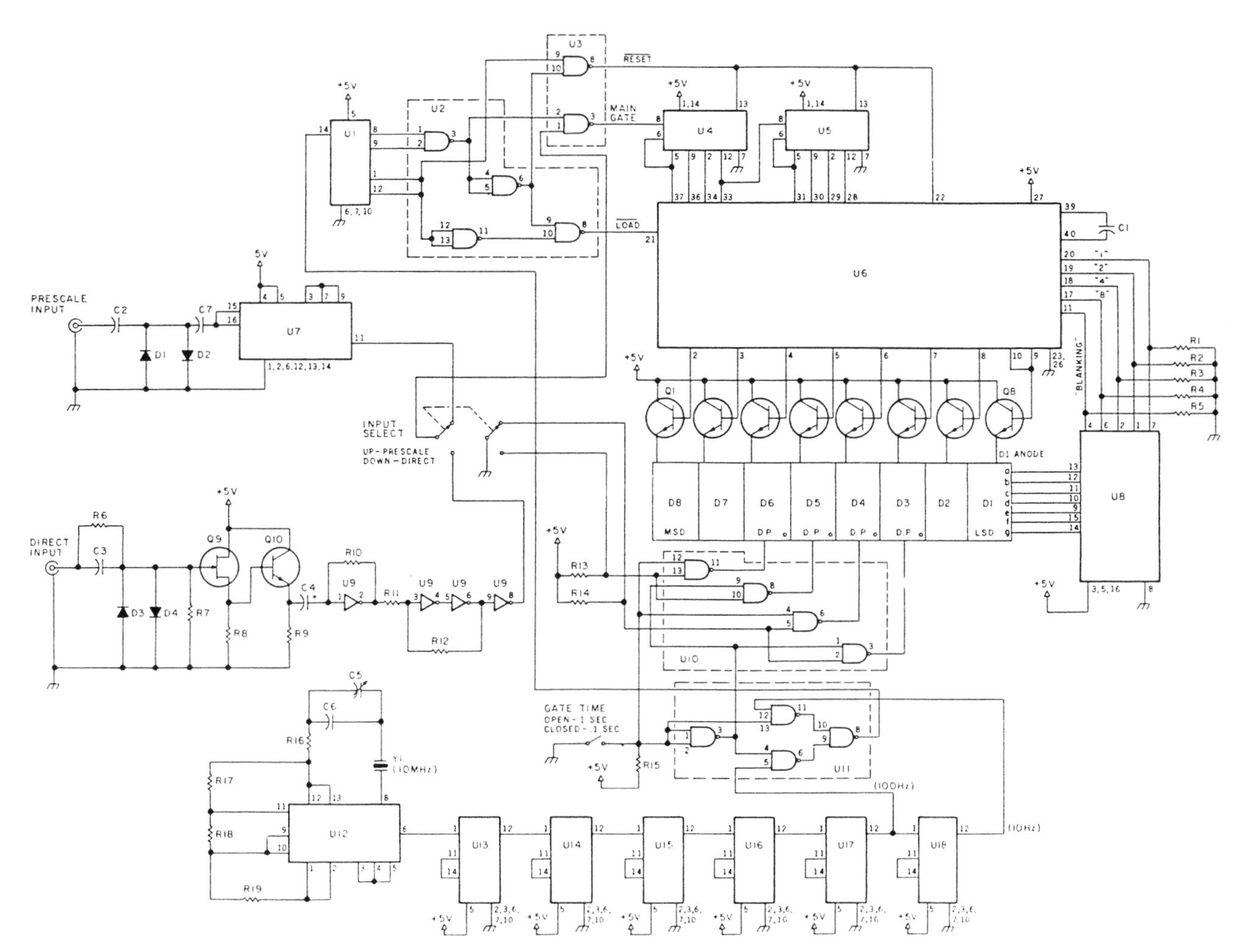

500-MHz 8-DIGIT FREQUENCY COUNTER—Requires only a single 5-V power supply and offers resolution to 1 Hz direct or 10 Hz on prescale. Direct input is good to 50 MHz and prescaled input to 500 MHz. The display reads kHz in direct mode and MHz in prescaled mode. Gate timing is selectable between 1 second and 0.1 second. U1 is a 7492, while U2, U10, and U11 are all 7400. U3 and U12 are both 74LS00. U4 and U5 are both 74196. U6 is LS7031, and U7 is 11C90. U8 is 7447. U9 is 74LS04. U13 through U18 are all 7490. Q1 through Q8 are all 2N3704. Q9 is MPF 102, and Q10 is 2N708. D1 through D4 are 1N914, and D5 through D8 are all 1N4001. Displays D1 through D8 are FND507 or equivalent common-anode display. R1 through R5 and R6 through R15 are all 1 K. R5, R10, and R19 are all 560 Ω. R6 is 100 K. R7 is 1 M. R8 is 4.7 K. R9, R16, and R18 are all 220 Ω. R11 is 470 Ω. R12 is 15 K. R17 is 1.8 K. C1 is 500 pF. C2 and C7 are 0.01 F. C3 is 68 pF. C4 is 47 μF 10 V. C5 is a 20-pF trimmer capacitor. C6 is 15 pF. Y1 is 10 MHz.—K. Erendson, Counting with Class, *73 Magazine,* October 1980, pp. 134–139.

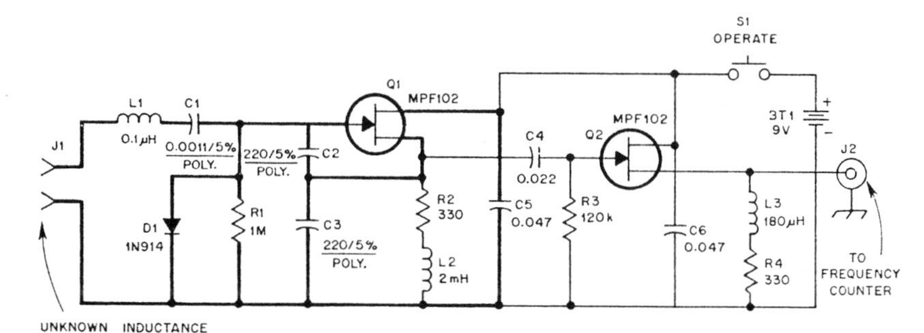

INDUCTANCE MEASURING AID—Used in conjunction with frequency counter to measure unknown inductances. The circuit works as desired if L2 is larger than the largest inductance one wishes to measure and the effective series resistance of the capacitor path (C1, C2, C3) is kept low. L1 is seven turns of No. 24 wire on a T-37-12 core. L2 is a pie-wound phenolic core RF choke.—A. Reinertsen, The L-Meter, *QST,* January 1981, pp. 28–29.

PRECISION PROGRAMMABLE LABORATORY TIMER—Accurately measures preselected time intervals of 0 to 99 seconds or 0 to 99 minutes. The 5-V buzzer sounds when the interval expires. The desired interval is selected by thumbwheel switches S4 and S5. Switch S2 starts the timer, and S3 turns off the buzzer.—"Intersil Data Book," Intersil, Cupertino, CA, 1981, p. 6-131.

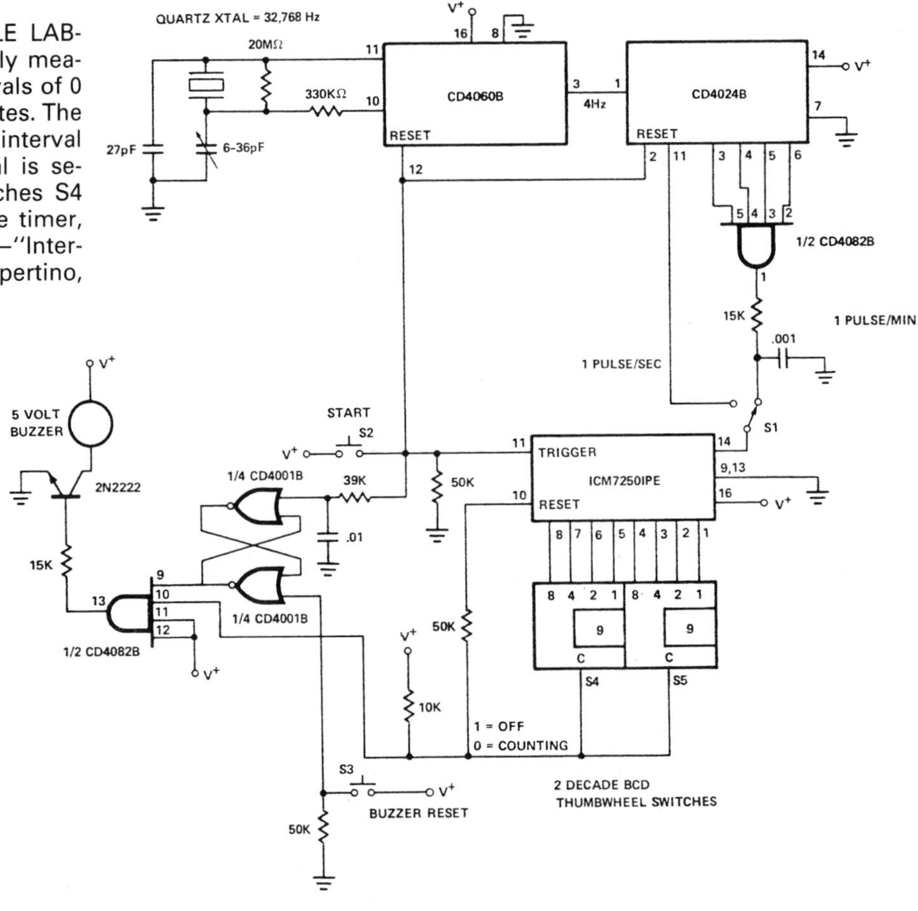

RESONANCE DETECTOR—Detects resonance of tuned circuits (such as quarter wave stubs) and transmission lines over range of 1-2250 MHz. The circuit is used in conjunction with an RF signal generator and a VOM in a manner similar to a dip meter.—R. Six, Build a Resonance Detector, *CQ,* August 1983, pp. 40–41.

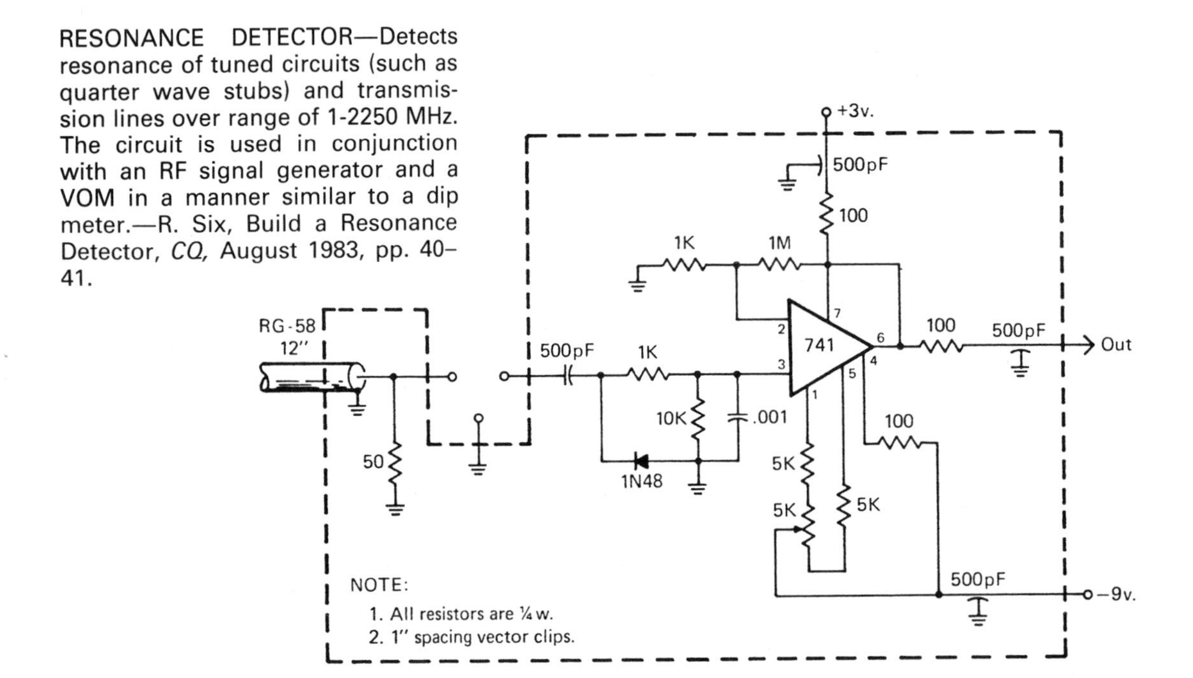

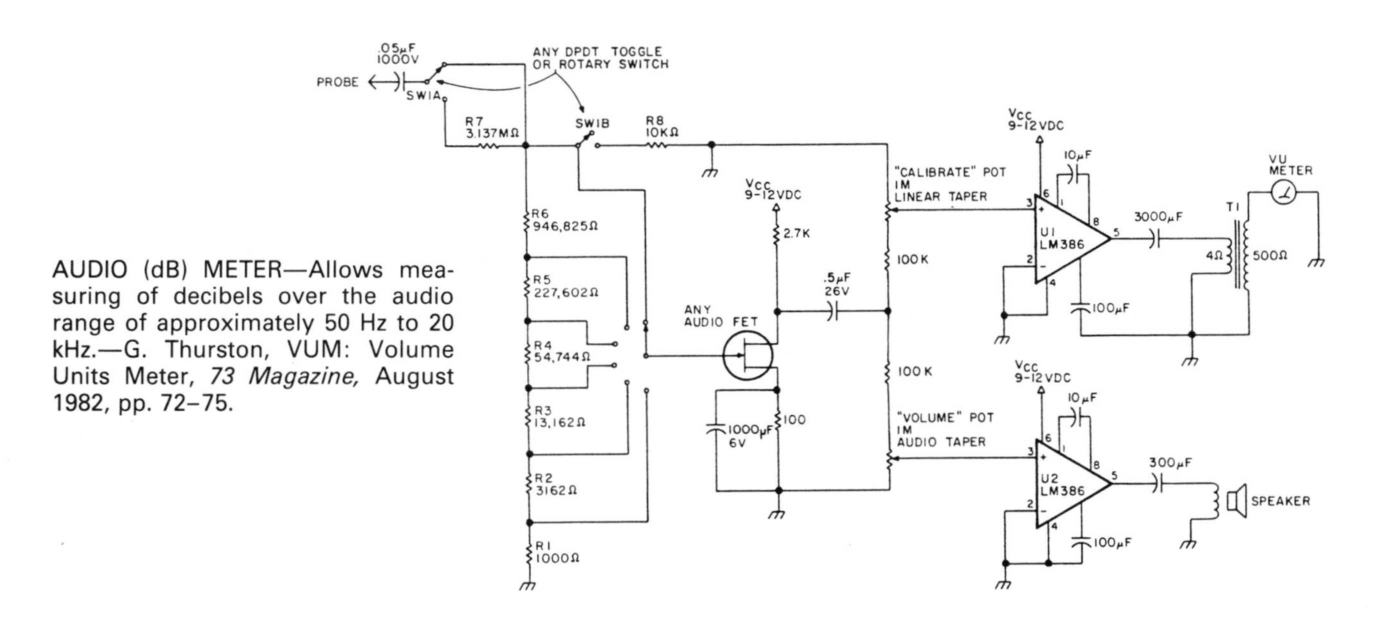

AUDIO (dB) METER—Allows measuring of decibels over the audio range of approximately 50 Hz to 20 kHz.—G. Thurston, VUM: Volume Units Meter, *73 Magazine,* August 1982, pp. 72–75.

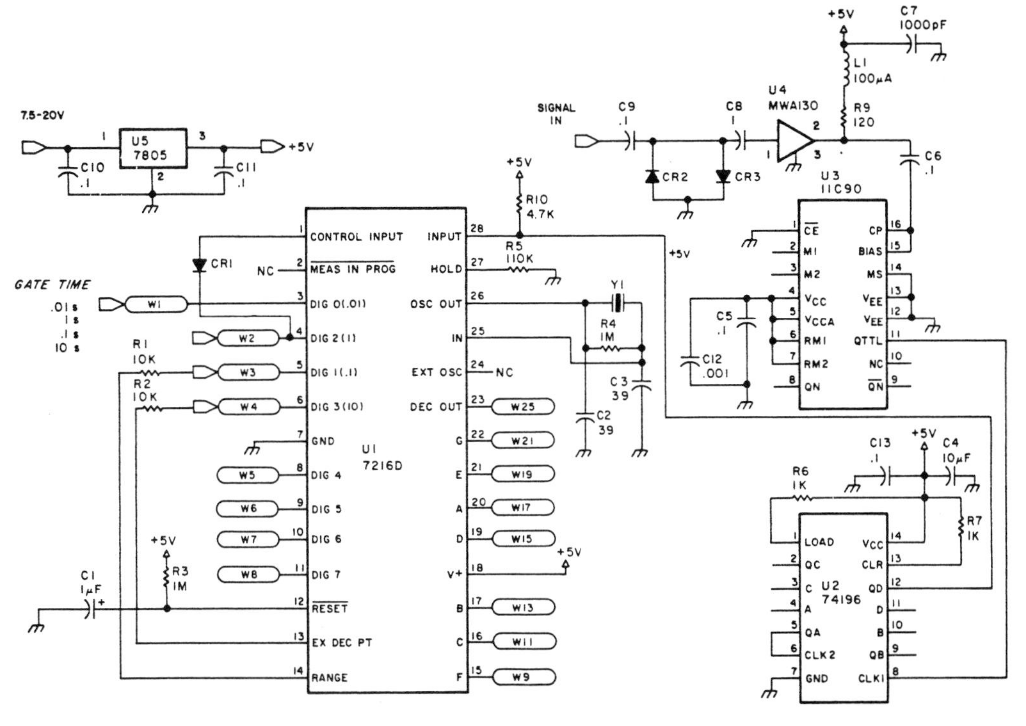

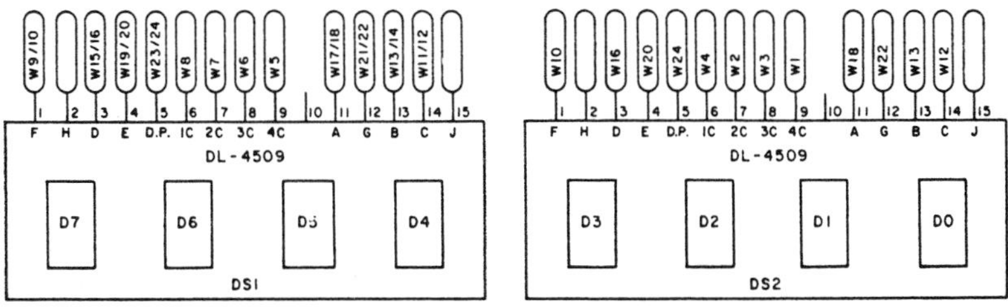

500- TO 600-MHz FREQUENCY COUNTER—Accurately displays frequencies from 500 to 600 MHz. The 11C90 prescaler is programmed for a divide-by-10 mode, and only the TTL outputs are used. Y1 is a 10-MHz crystal. The counter has four possible gate times, with the 0.1-second time used for simplicity. A 1-second gate time will increase resolution to 100 Hz but the display will update at a much slower rate.—J. Berardi, East Berardi Building, *73 Magazine,* June 1984, pp. 10–14.

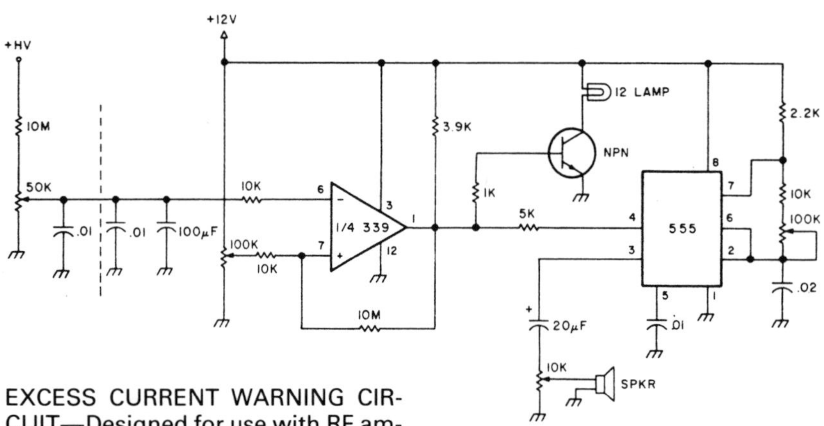

EXCESS CURRENT WARNING CIRCUIT—Designed for use with RF amplifier circuits. It operates on the principle that the voltage in circuits drops when the current increases; the resistor divider samples the high voltage. When it drops, the 339 triggers the oscillator and a warning light and sound are produced.—L. Cebik, Smart Meters: The New Movement, *73 Magazine,* June 1983, pp. 42–46.

12

Miscellaneous Circuits

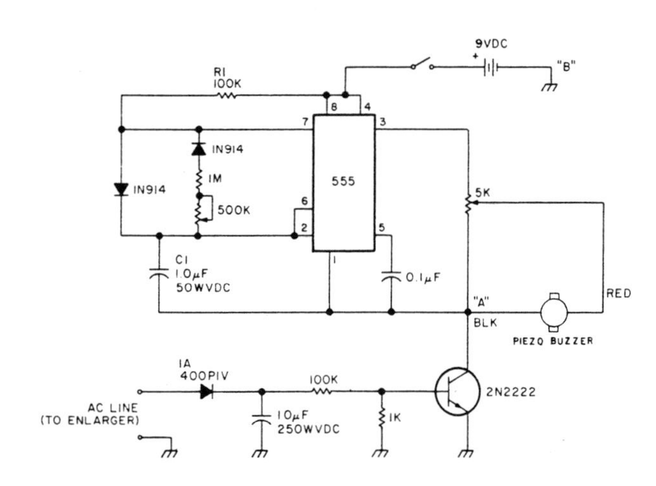

ADJUSTABLE BEEPER—Produces "beeps" of approximately 0.1-second duration at a rate adjustable by the 500-K variable resistor. Volume is adjusted using the 5-K variable resistor.—T. Bowman and B. Long, Here's the Split Second Timer, *73 Magazine,* February 1984, pp. 54–56.

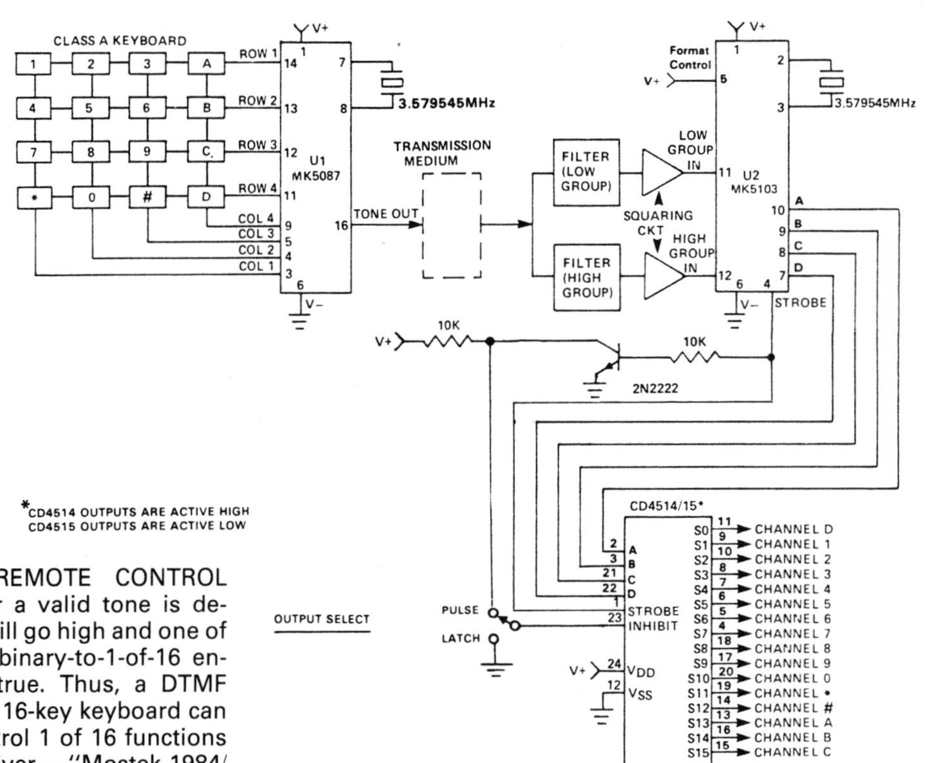

16-CHANNEL REMOTE CONTROL SYSTEM—After a valid tone is detected, strobe will go high and one of 16 outputs on binary-to-1-of-16 encoder will go true. Thus, a DTMF transmitter and 16-key keyboard can be used to control 1 of 16 functions in a DTMF receiver.—"Mostek 1984/85 Microelectronics Data Book," Mostek Corporation, Carrollton, TX, 1984, pp. XV-8 to XV-9.

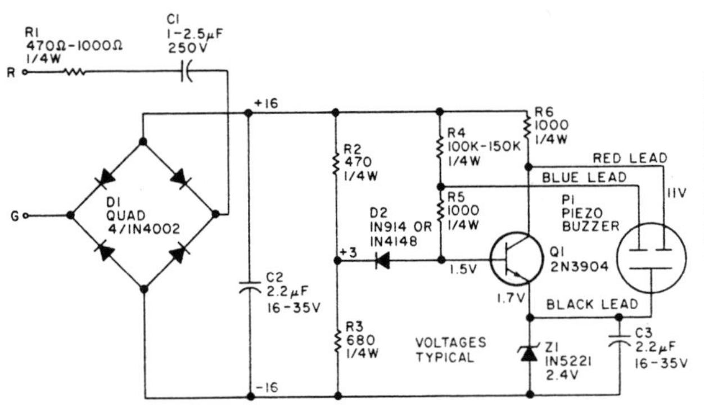

ELECTRONIC BELL—Requires no power supply. The buzzer will be louder if R1 is eliminated. If the piezo buzzer has a third (typically blue) lead, it should be connected to the base of Q1.—E. Sherrill, Ding-a-Ling, *73 Magazine,* April 1983, p. 106.

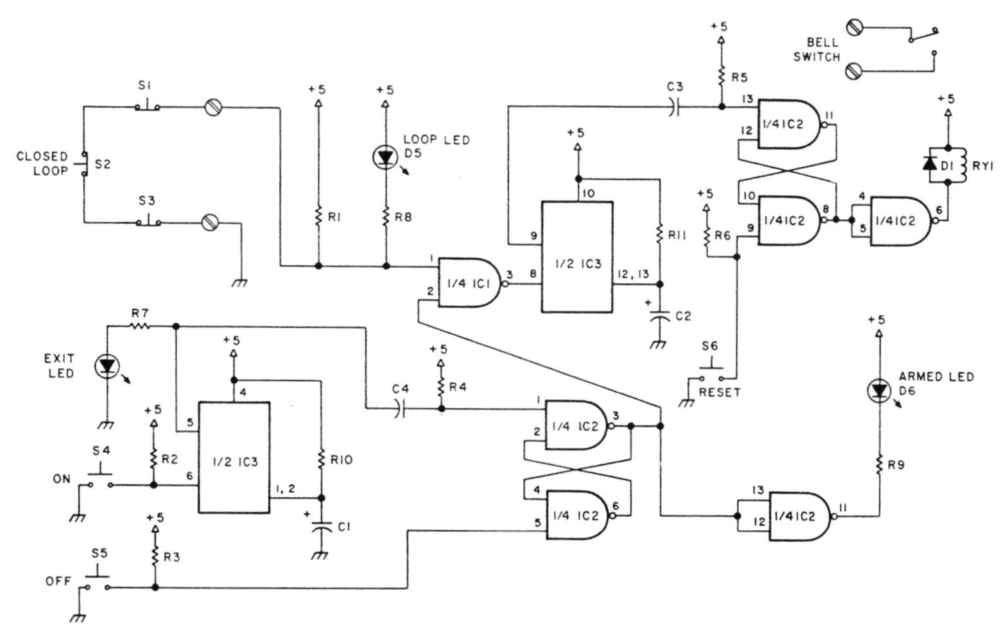

ALARM CONTROL PANEL—Provides entry and exit delay times controlled by C1, C2, R10, and R11. Delay times of approximately 14 seconds can be obtained if C1 and C2 are 10-μF 16-V electrolytics and R10 and R11 are 1 M. IC1 and IC2 are both 7400 NAND devices, while IC3 is a 556 dual timer. RY1 is a 5-V low-current relay. R1 through R6 are 2.2 K and R7 through R9 are 330 Ω. C5 and C6 are 4.7-μF tantalums. C3 and C4 are 0.01-μF ceramic discs. D1 through D3 are all 1N4001 and D4 to D6 are general-purpose LEDs.—J. Willis, Some Alarming Techniques, *73 Magazine,* January 1984, pp. 32–34.

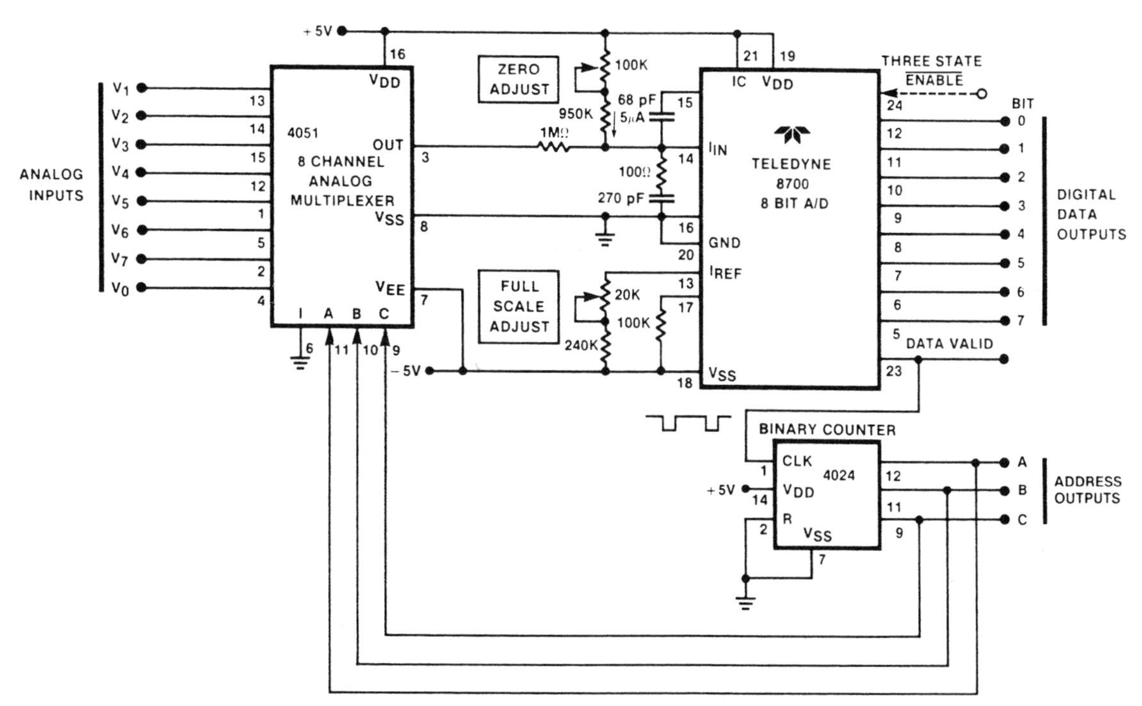

EIGHT-CHANNEL DATA ACQUISITION SYSTEM—Uses 8700 CMOS analog-to-digital converter, 4051 eight-channel CMOS multiplexer, and 4024 binary counter. The input voltage range is limited by the 4051 to ±5 V. Each input is measured for 1 ms, and the digital value is placed in the output latch and remains for 1 ms while the next input is measured. After each 1-ms measurement (conversion), the data valid lane goes low for 5 μs to indicate that the output latch is being updated (the data must not be read during this period).—"Data Acquisition Design Handbook," Teledyne Semiconductor, Mountain View, CA, 1984, p. 15-56.

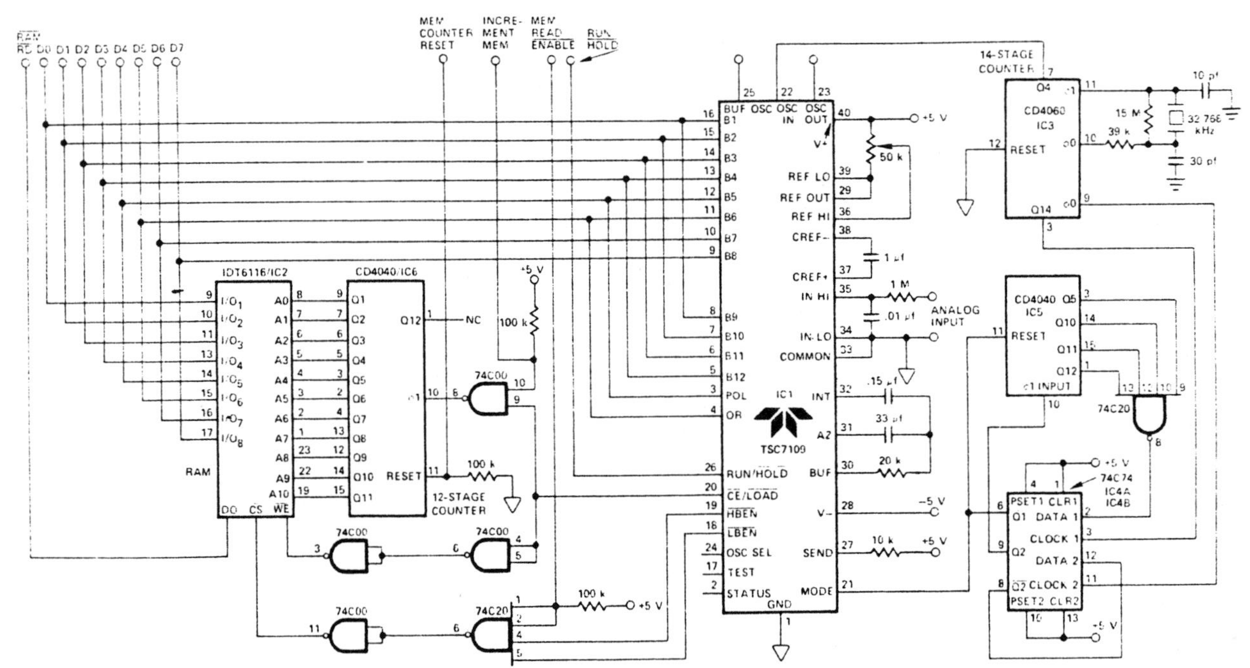

13-BIT REMOTE DATA LOGGER—Circuit is built around a TSC7109 12-bit plus sign CMOS A/D converter. The TSC7109 has a handshake mode in which the result of the latest conversion is output in two consecutive bytes each time the MODE input is strobed high. The data logger stores each byte in sequential RAM locations for later processing by a host computer. The IDT6116 CMOS static RAM stores 2048 bytes and therefore can store 1024 readings. This permits one 13-bit measurement per hour for 7 weeks. Timing for the circuit is provided by IC3 (CD4060) operating with a 32-kHz crystal.—"TSC7109 Records Remote Data Automatically," Teledyne Semiconductor, Mountain View, CA, 1984, Application Note 24.

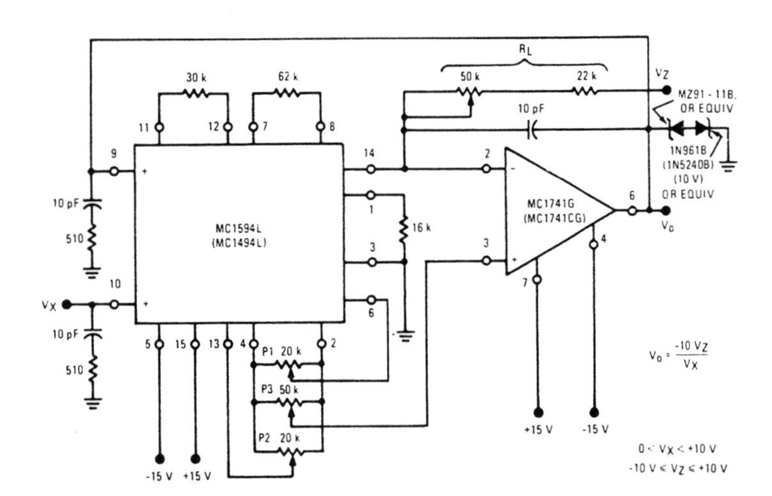

DIVIDE CIRCUIT—Produces output voltage that is the quotient of division of two input voltages. Output voltage V_O is determined by dividing $-10\ V_Z$ by V_X. The original source contains details on the adjustment of the division.—"Motorola Linear Integrated Circuits Databook," Motorola, Phoenix, AZ, 1979, p. 6-69.

HEADPHONE AMPLIFIER—SL630C audio amplifier IC delivers 40-dB gain and has an output capability of 200 mW into a 40-Ω load. The circuit shown requires a 6- to 12-V power supply.—"Plessey Applications IC Handbook," Plessey Semiconductors, Irvine, CA, 1982, pp. 60–61.

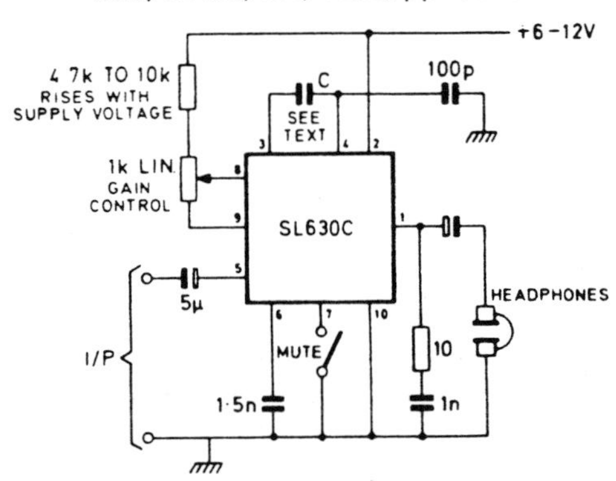

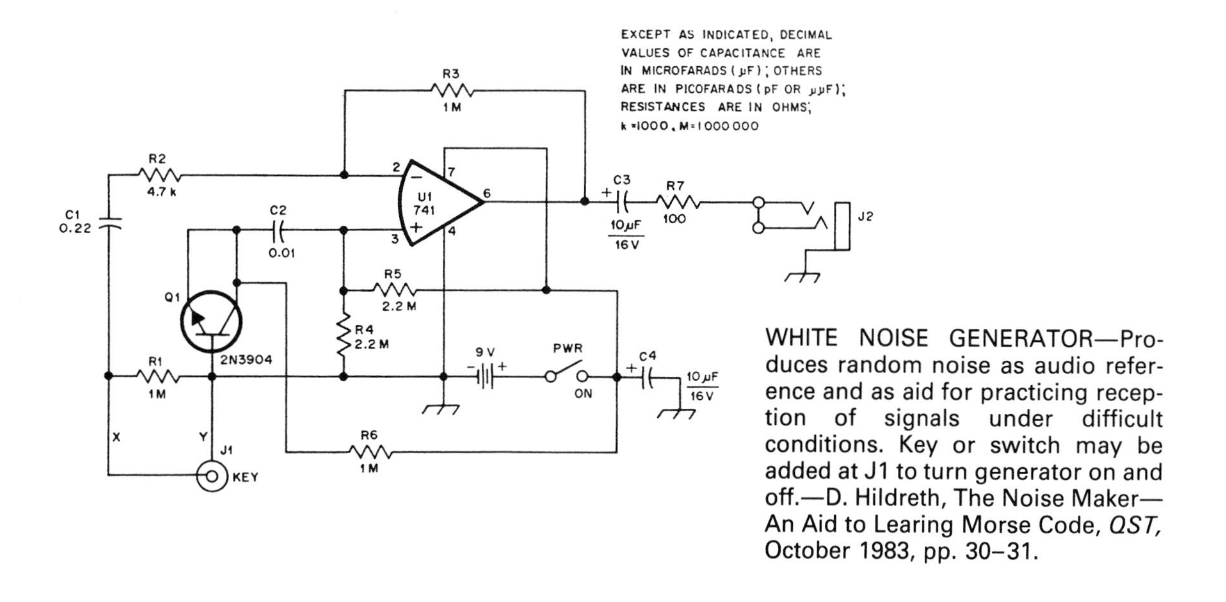

WHITE NOISE GENERATOR—Produces random noise as audio reference and as aid for practicing reception of signals under difficult conditions. Key or switch may be added at J1 to turn generator on and off.—D. Hildreth, The Noise Maker—An Aid to Learing Morse Code, *QST,* October 1983, pp. 30–31.

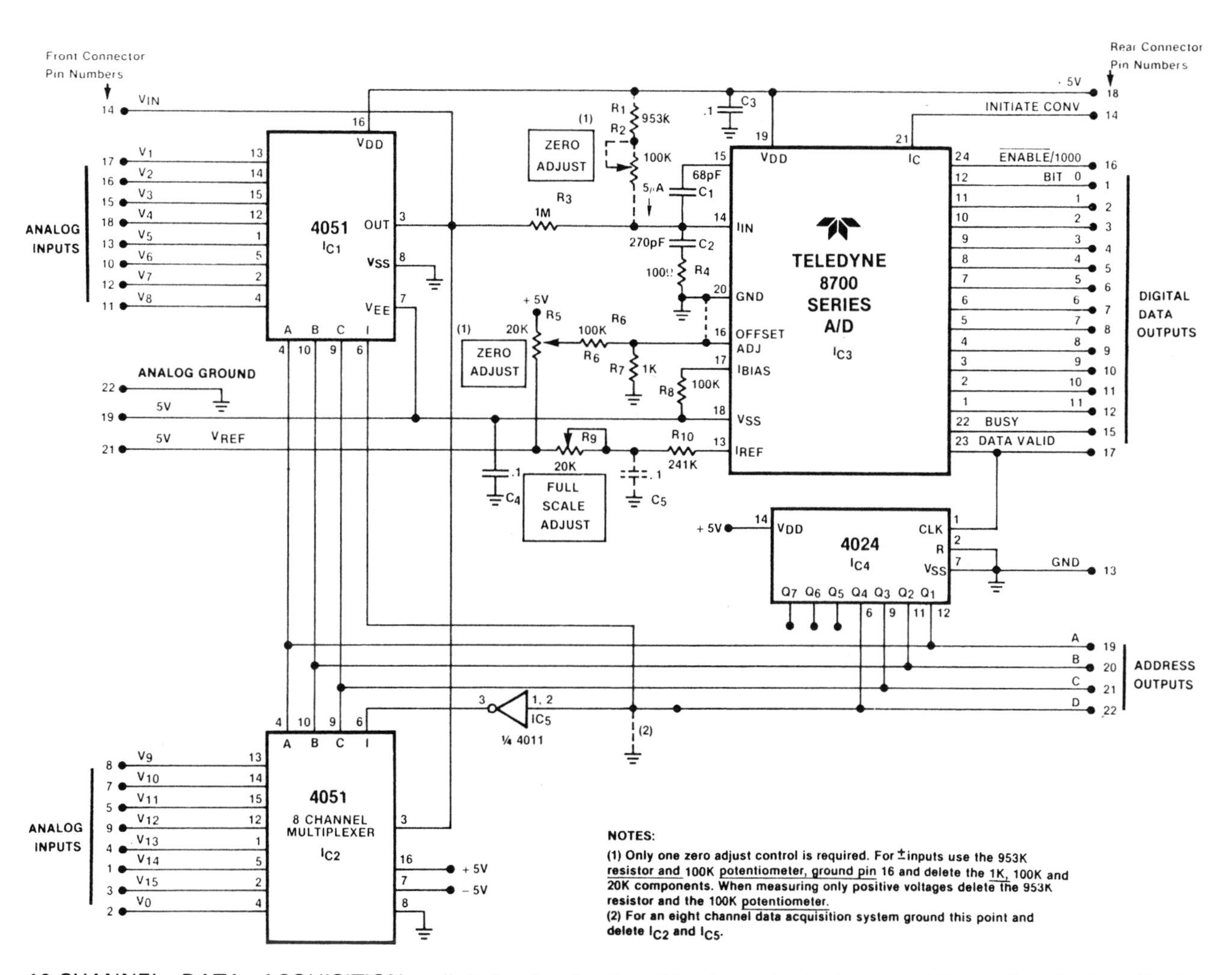

16-CHANNEL DATA ACQUISITION SYSTEM—Designed to operate with a ±5-V input voltage range. Each input is measured for 1 ms, and the digital value is placed in the output latch and remains for 1 ms while the next input is measured. The original source contains the circuit board layout.—"Data Acquisition Design Handbook," Teledyne Semiconductor, Mountain View, CA, 1984, pp. 15-59 to 15-63.

MOTOR SPEED CONTROL—Uses TDA2085 phase control IC. In the diagram, M represents the motor and TG is the tacho generator. Power for the circuit may be from a DC line or AC supply if current limiting components are used. If there is an open circuit in the tacho generator section, the TDA2085 will demand full speed and power.—"Plessey Integrated Circuits Databook," Plessey Semiconductor, Irvine, CA, 1983, pp. 197–201.

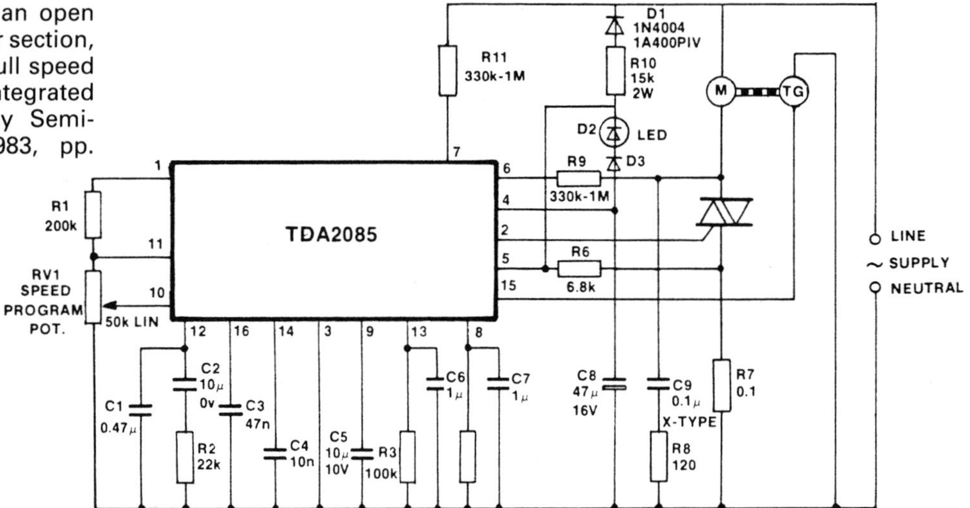

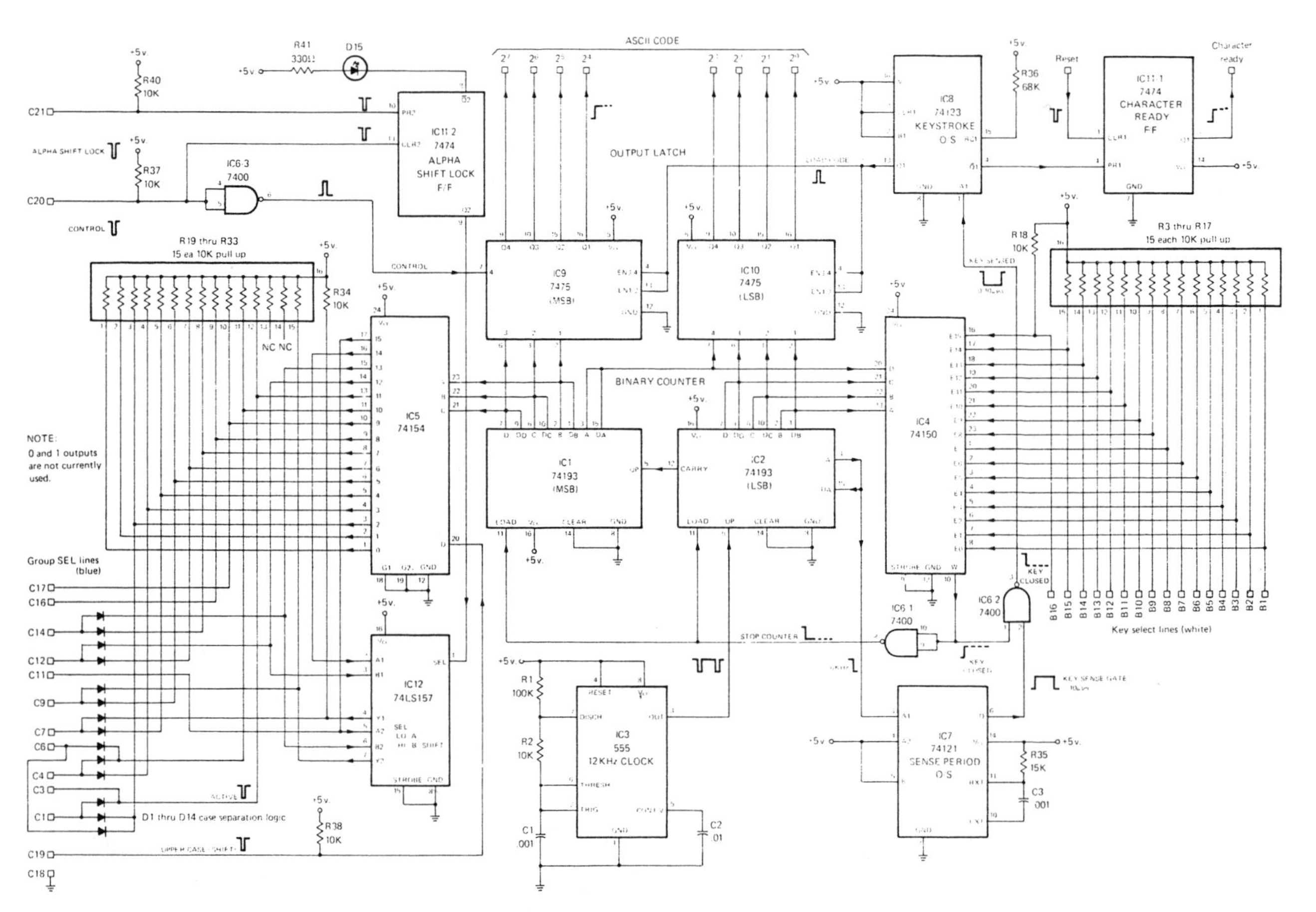

ASCII KEYBOARD CONTROLLER—Provides upper- and lower-case ASCII codes and 8-bit parallel data. The circuit is adaptable to either polled or interrupt operation with CHARACTER READY and RESET signals provided. All interfaces are TTL-compatible. It requires +5 V and approximately 380 mA. The keyswitch debounce is standard.—E. Worner, A Homebrew ASCII Keyboard Controller, *CQ,* December 1981, pp. 10–16.

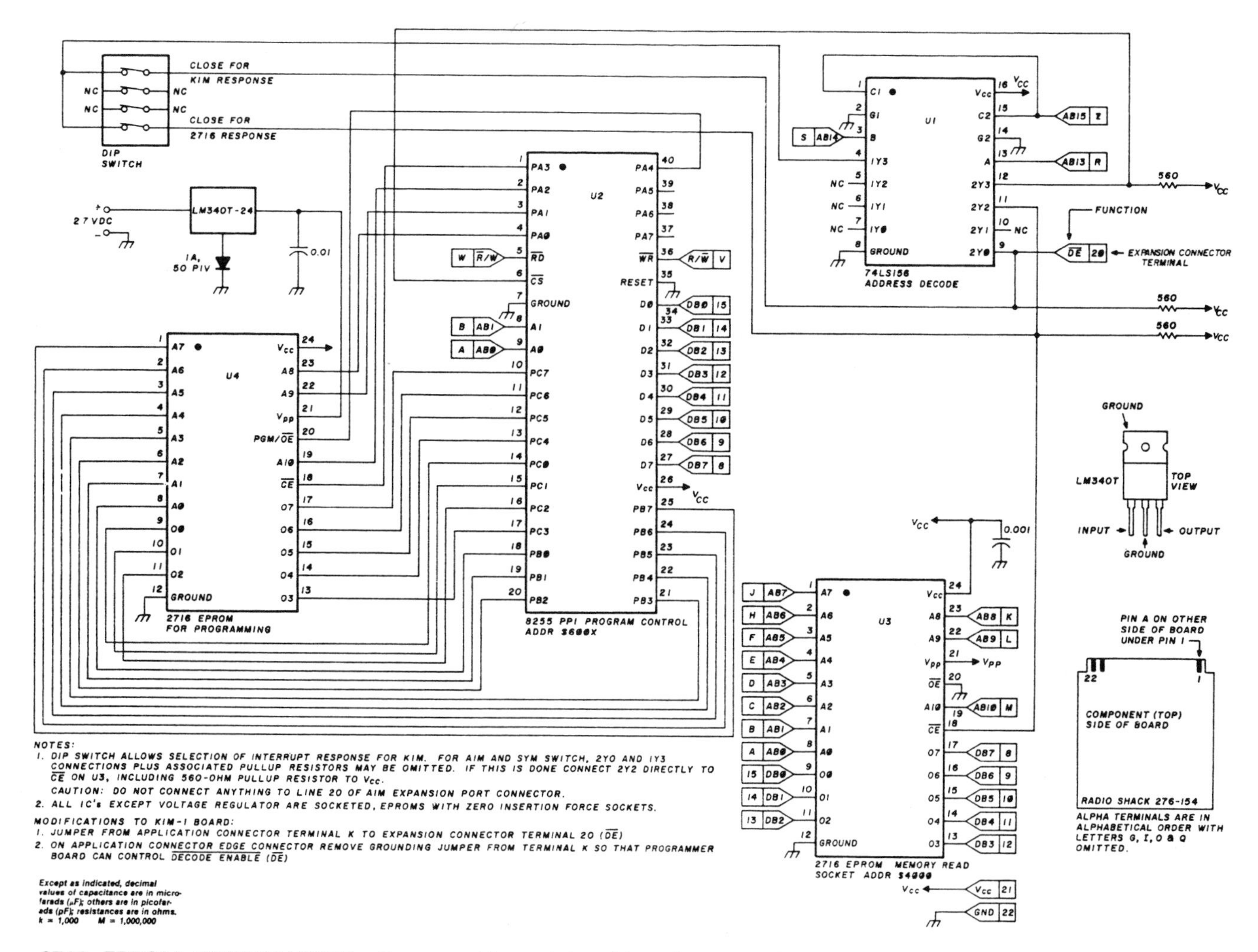

2716 EPROM PROGRAMMER—Designed to program 2716 2-K EPROM devices. The basic programming circuit is built around an 8255 programmable peripheral interface chip. Programming the 2716 requires a +25-V signal, which is derived from an LM340T monolithic regulator. The original article includes a programming algorithm and details.—C. Eubanks, 2716 EPROM Programmer, *Ham Radio*, April 1982, pp. 32–36.

AUDIO OUTPUT STAGE FOR SPEECH SYNTHESIS—Designed for use with MEA8000 speech synthesizer device. The output from the synthesizer is filtered and amplified to drive a 4- or 8-Ω speaker.—"Signetics Linear LSI Data and Applications Manual," Signetics, Sunnyvale, CA, 1985, p. 5-171.

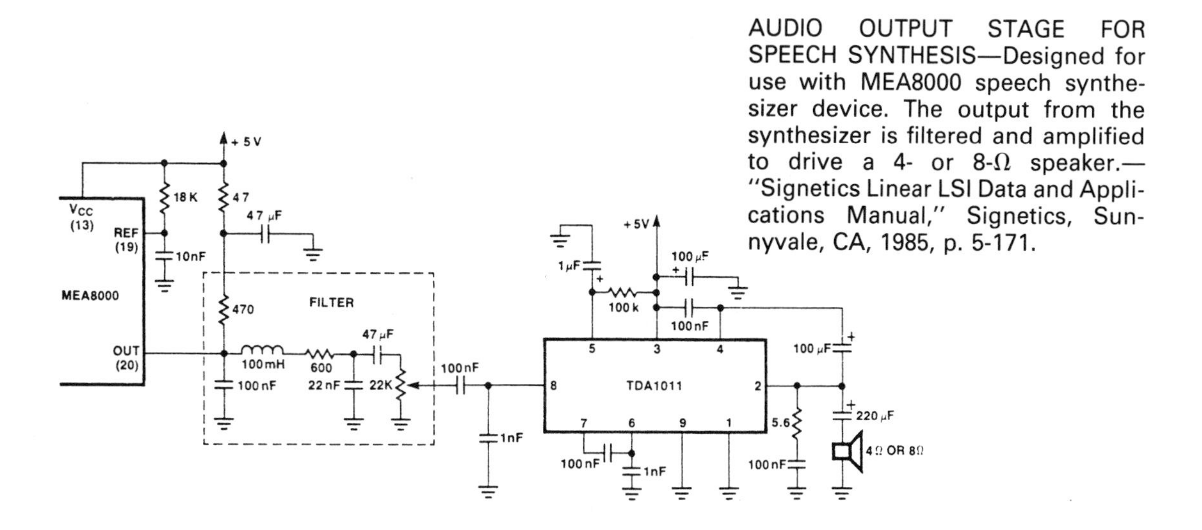

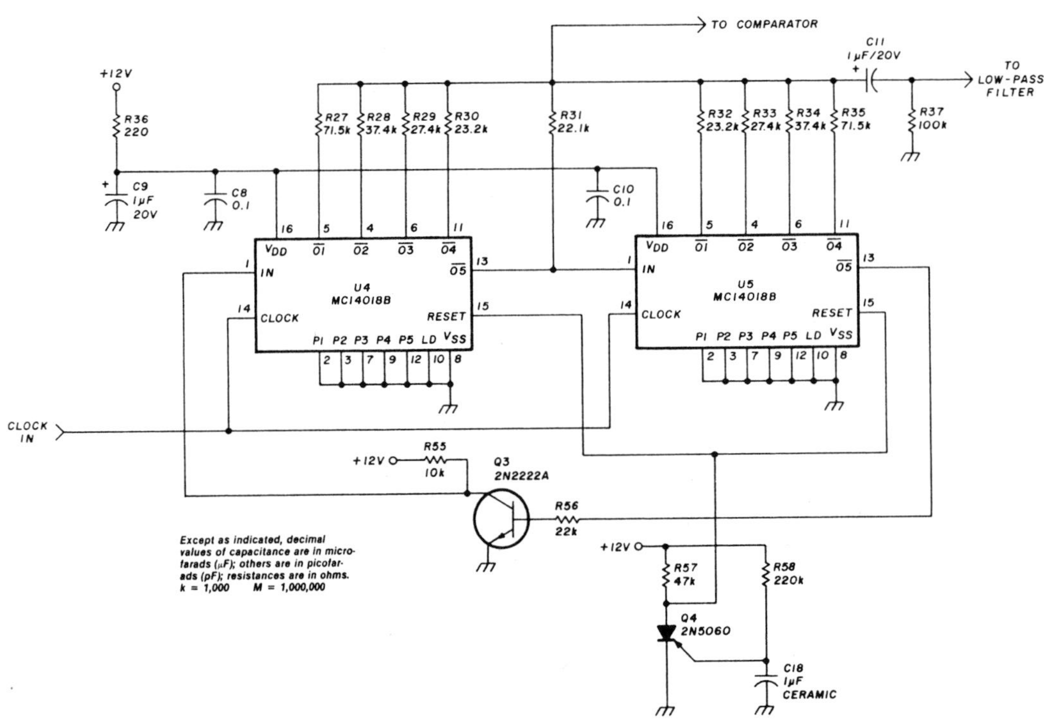

PROGRAMMABLE TONE GENERATOR—Produces any frequency from 67 to 250.3 Hz with accuracy of better than 0.25%. The circuit is designed around three modules: a crystal oscillator, a divide-by-N counter, and a digital sine wave generator. The crystal oscillator drives the divide-by-N counter, and the output of the counter feeds the digital sine wave generator. The digital sine wave generator lowers the frequency still further and produces the desired output waveshape. The seven binary mode control lines of the HCTR0320 should be grounded. U1 is MC14011B while U2, U4, and U5 are all MC14018B. A low-pass filter may be added to the digital sine wave generator to condition the output. The divider control lines can be wired to three BCD thumbwheel switches; a 12-bit latch controlled by a computer could be used instead. Power requirements are 12 V and 20 mA. The original article includes the circuit board layout.—C. Winter, A Programmable PL Tone Generator, *Ham Radio,* April 1984, pp. 51–54.

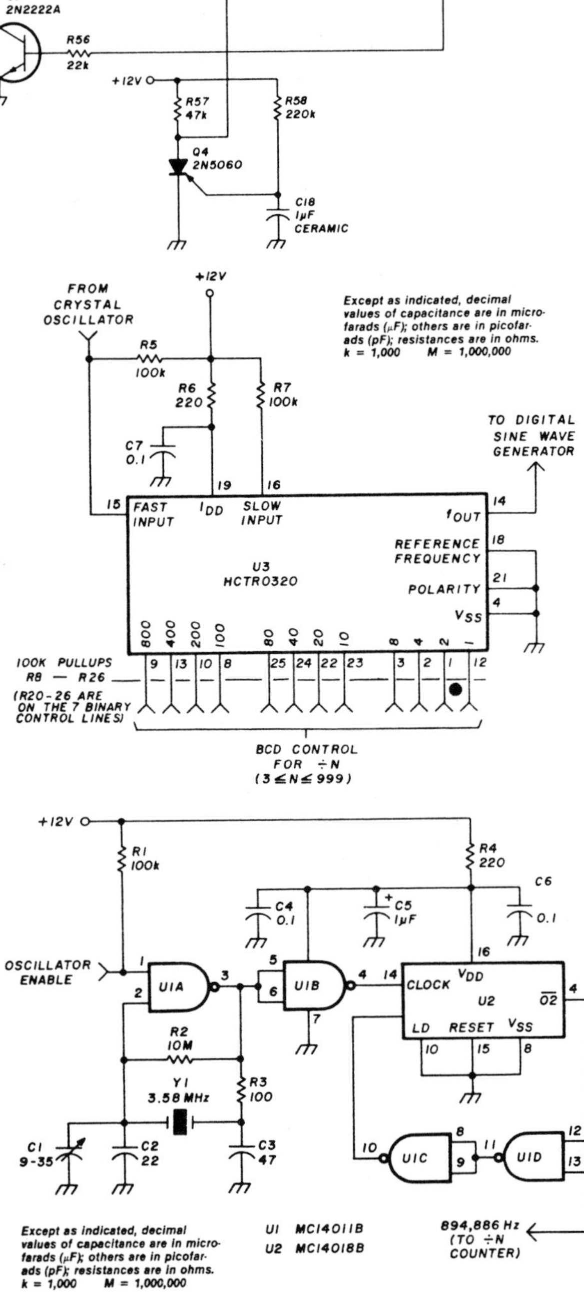

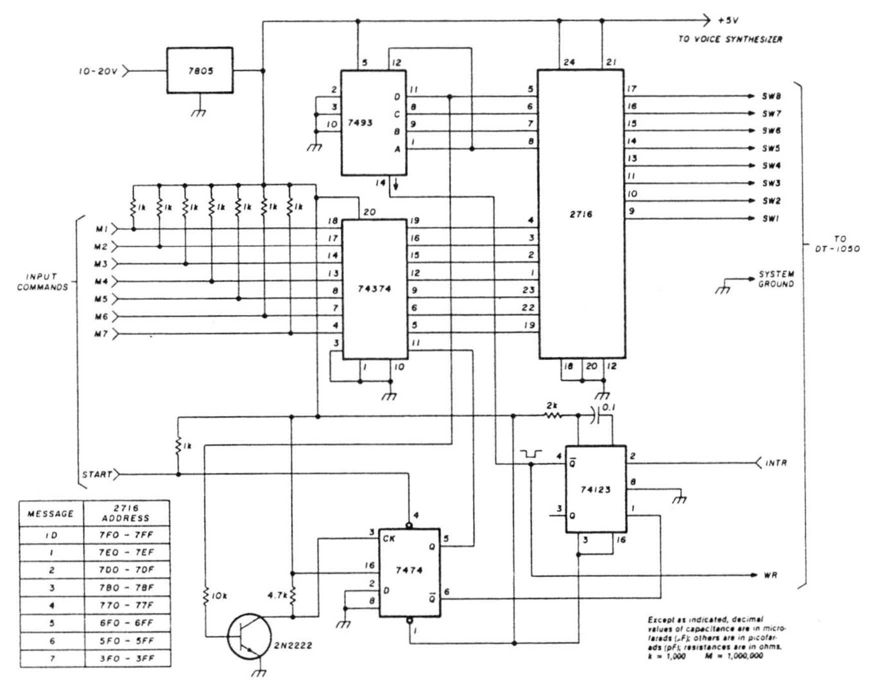

MESSAGE	2716 ADDRESS
ID	7F0 - 7FF
1	7E0 - 7EF
2	7D0 - 7DF
3	7B0 - 7BF
4	770 - 77F
5	6F0 - 6FF
6	5F0 - 5FF
7	3F0 - 3FF

VOICE SYNTHESIZER—Built around National Semiconductor DT-1050 "Digitalker" circuit and matching hardware controller. The unit is capable of generating each discrete letter of the alphabet and over 100 additional words and phrases. The DT-1050 itself consists of an MM5144 and two 8-K ROMs. An audio filter and amplifier have been added to the DT-1050. To control the DT-1050, a character command is placed on the 8-bit latched inputs SW1 (LSB) through SW8 and produces a low to high TTL signal on the WRITE line (pin 4) of the processor. The processor then latches the 8 bits and starts the character speech sequence. It continues until finished or until a new command is entered. When the processor is busy, it produces a TTL low signal on the introduction (pin 6). When the character is complete, the line goes high. This line can be used for "handshaking." The original article includes a vocabulary list and 8-bit binary addresses for words and phrases.—R. Wright, Speech Synthesis for Repeaters, *Ham Radio,* March 1984, pp. 78–81.

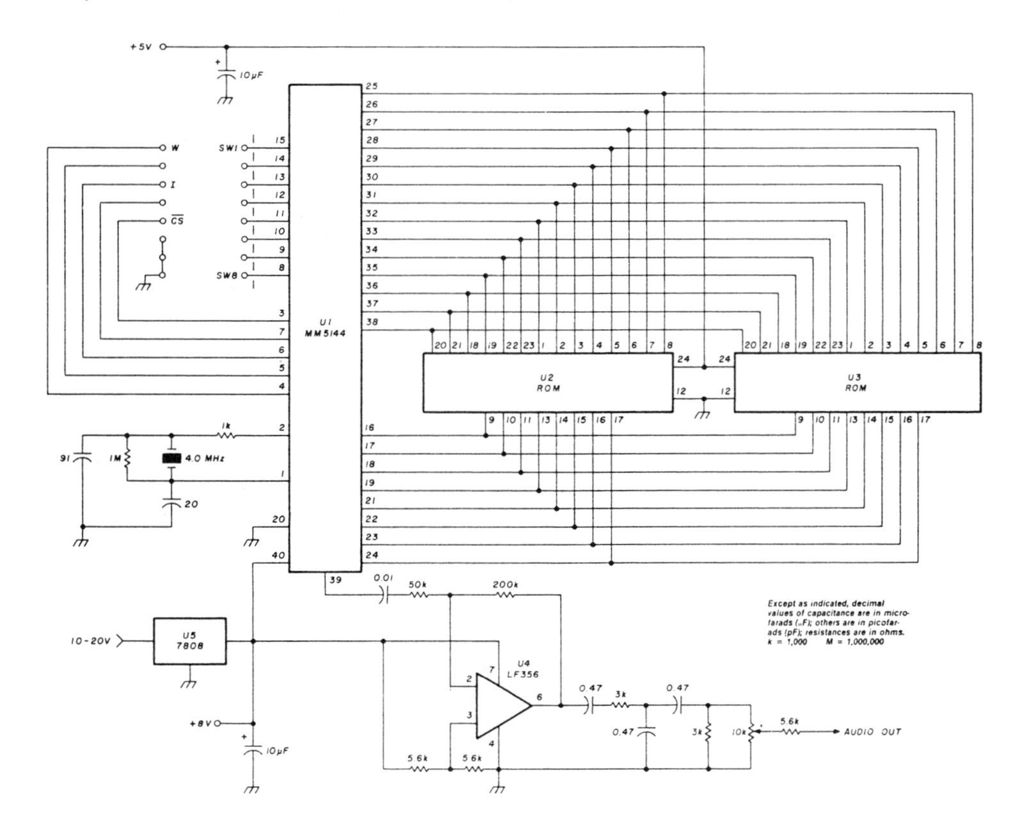

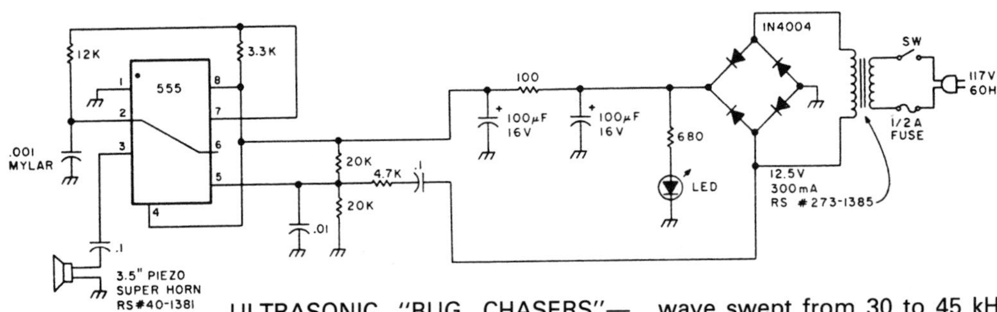

ULTRASONIC "BUG CHASERS"—Generates high-intensity ultrasonic sound which is claimed to be effective in repelling certain types of insects. The circuit generates a square wave swept from 30 to 45 kHz.—L. Hutton, Electronics vs. Creepy Crawlers, *73 Magazine,* February 1983, pp. 90–91.

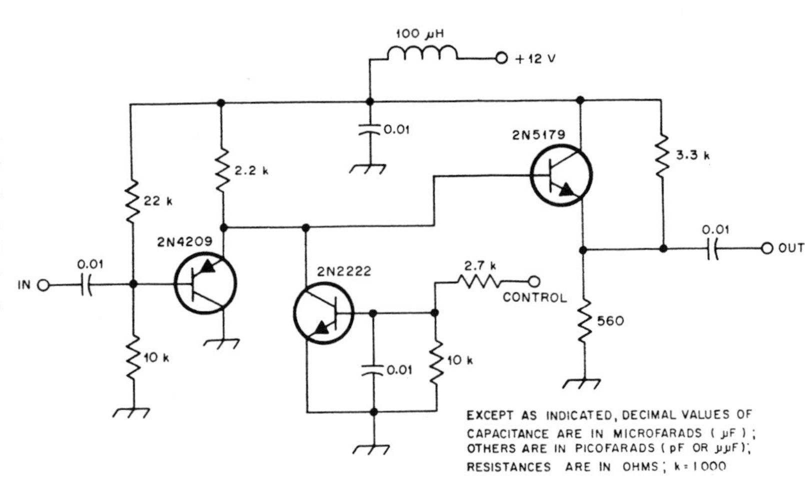

DC-CONTROLLED HF SWITCH—Active switch provides excellent buffer action and reduced losses. Its isolation properties are also excellent. This circuit was designed for use in a 9-MHz IF strip.—D. Monticelli, An Active HF Switch, *QST,* August 1984, pp. 48–49.

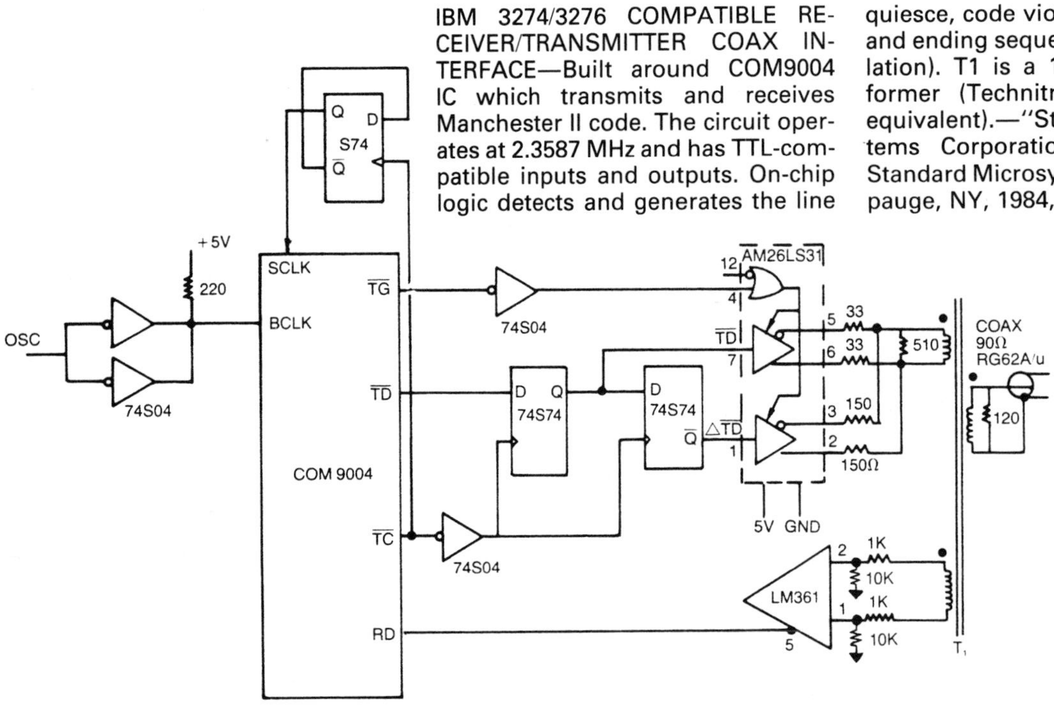

IBM 3274/3276 COMPATIBLE RECEIVER/TRANSMITTER COAX INTERFACE—Built around COM9004 IC which transmits and receives Manchester II code. The circuit operates at 2.3587 MHz and has TTL-compatible inputs and outputs. On-chip logic detects and generates the line quiesce, code violation, sync, parity, and ending sequence (mini code violation). T1 is a 1 : 1 : 1 pulse transformer (Technitrol No. 11LHA or equivalent).—"Standard Microsystems Corporation Data Catalog," Standard Microsystems Corp., Hauppauge, NY, 1984, p. 200.

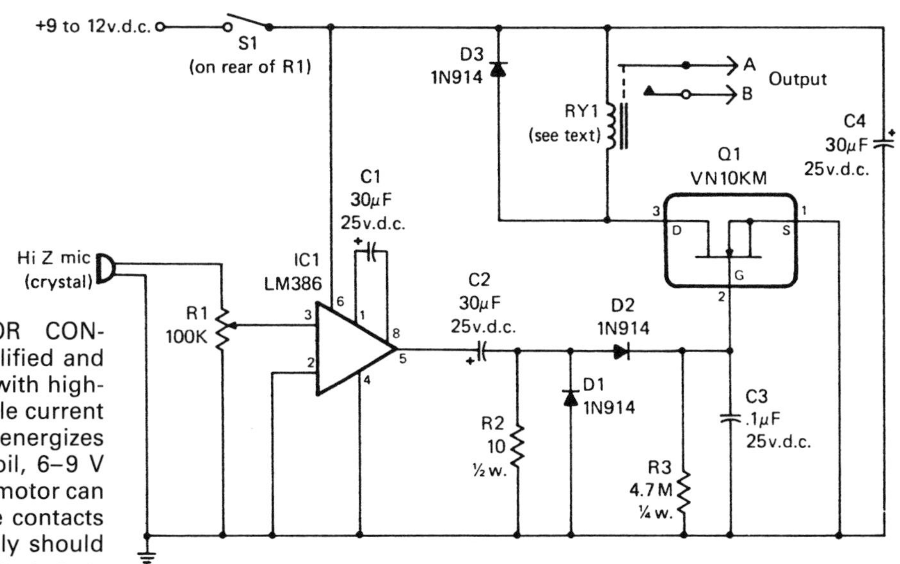

VOICE-ACTIVATED MOTOR CONTROL—Voice input is amplified and turns on Q1, a power FET with high-impedance and considerable current drive. When Q1 is on, it energizes RY1 (SPDT relay, 500-Ω coil, 6–9 V DC, 10–12 mA). A 12-V DC motor can be wired in series with the contacts of RY1. Motor power supply should be isolated from control power supply.—P. Danzer, The Ultimate Contest Accessory, *CQ,* October 1983, pp. 34–36.

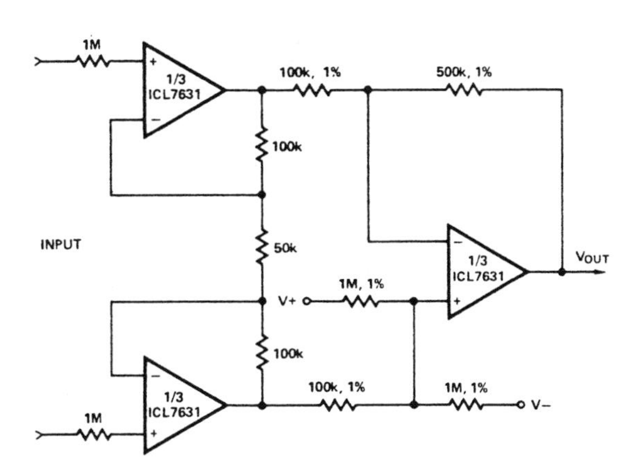

MEDICAL INSTRUMENT PREAMPLIFIER—Uses ICL7631 triple low-power op amp device. The input current is limited to <5 μA under fault conditions and is obtained through sensors attached to the patient. Power is supplied through a single ni-cad battery; the operating voltage can range from 0.5 to 8 V.—"Intersil Data Book," Intersil, Cupertino, CA, 1981, p. 5-151.

TOUCH-ACTIVATED ALARM—Operation depends upon a source of 60-Hz signals (such as a standard 110-V AC power line). The sensitivity of the alarm is variable. The alarm is hung from a doorknob by a hook made from heavy-gauge copper. When the doorknob is touched, the alarm sounds.—D. Pharr, Stop That Heist!, *73 Magazine,* pp. 72–73.

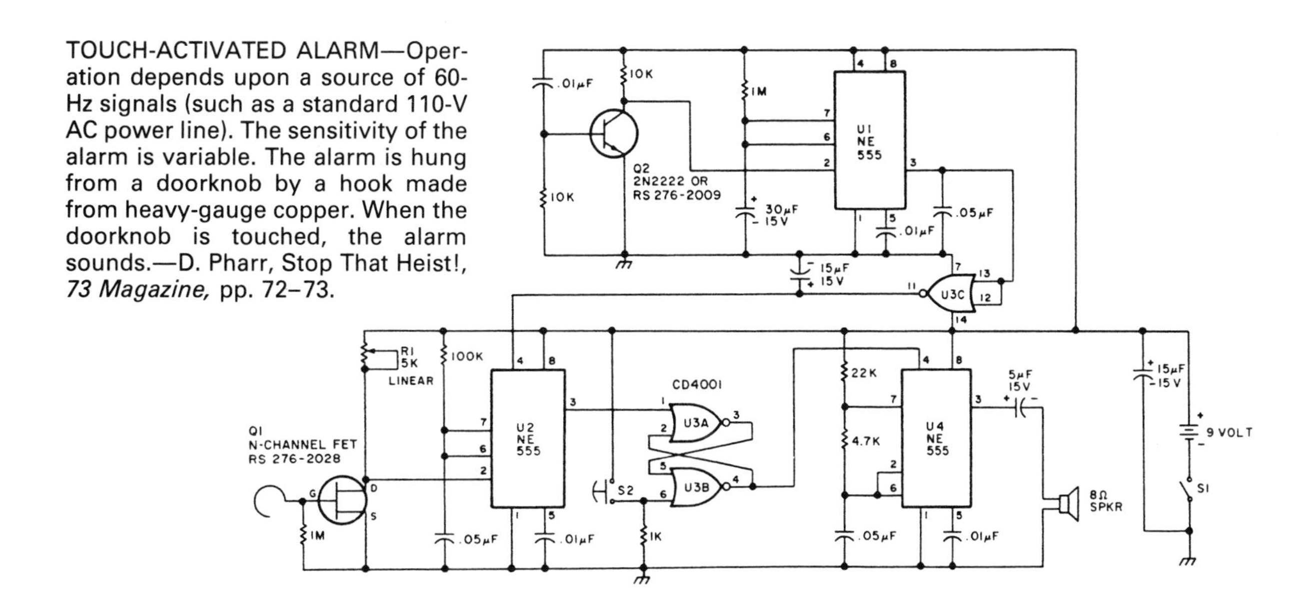

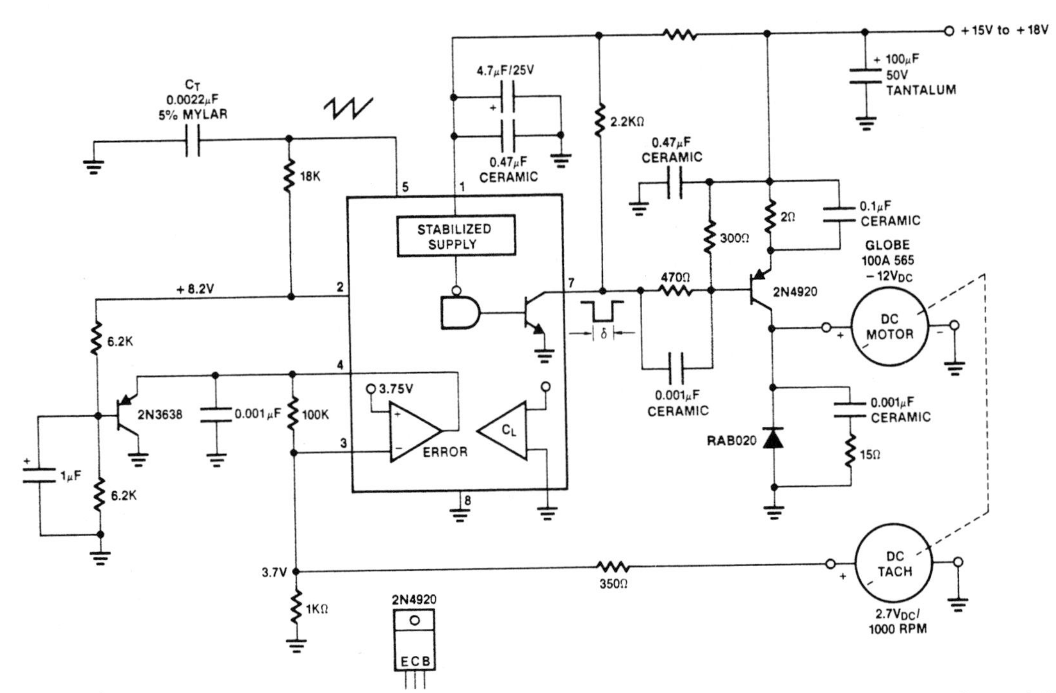

DC MOTOR DRIVE WITH FIXED SPEED CONTROL—Incorporates a switch mode approach to DC motor control which is free of dissipation problems inherent in linear drives. The efficiency is also high. The IC used is a NE5561 which provides pulse proportional drive and speed control based on DC tachometer feedback. The 2N4920-based switching circuit is used to deliver programmed pulse energy to the motor.—"Signetics Linear LSI Data and Applications Manual," Signetics, Sunnyvale, CA, 1985, p. 9-248.

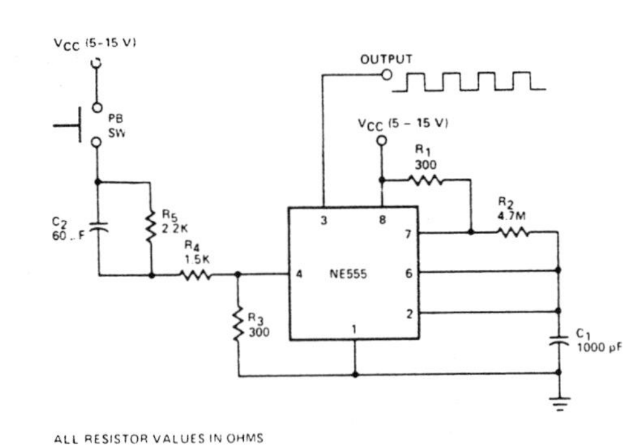

SQUARE WAVE TONE BURST GENERATOR—Pressing pushbutton switch produces square wave tone bursts whose duration depends on time period for which voltage at pin 4 exceeds threshold. R1, R2, and C1 cause the astable action of the timer.—"Signetics Analog Applications Manual," Signetics, Sunnyvale, CA, 1979, p. 158.

STEPPER MOTOR DRIVE—TDA3717 controls and drives current in one winding of stepper motor using switch mode technique. Logic control inputs can accept TTL or CMOS signals. Output current control may range from 20 to 500 mA. Typical component values are 10 mH for L, 1 Ω for RE, 1 K for RA, 820 pF for CA and CT, and 56 K for RT.—"Plessey Integrated Circuits Databook," Plessey Semiconductor, Irvine, CA, 1983, pp. 203–208.

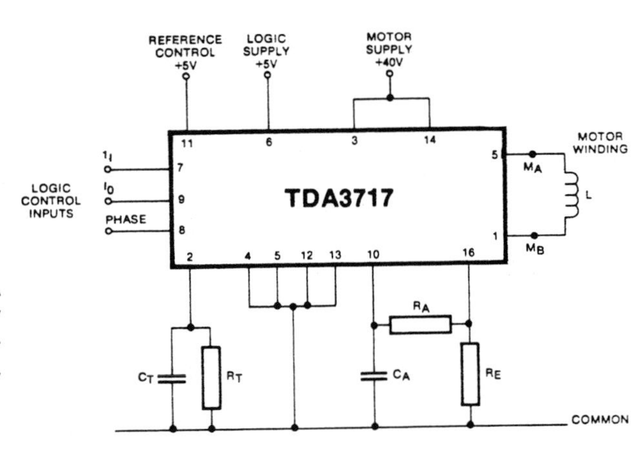

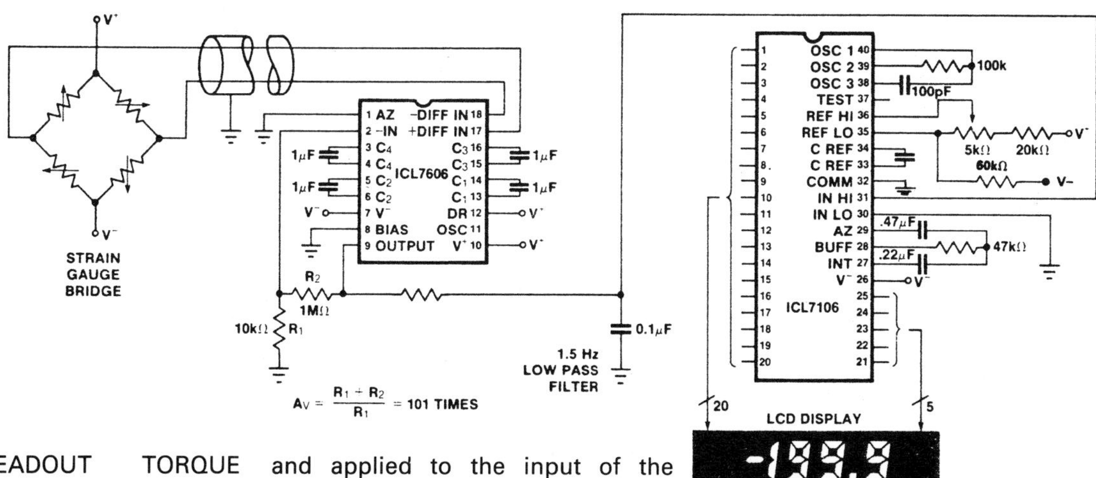

$A_V = \frac{R_1 + R_2}{R_1} = 101$ TIMES

DIGITAL READOUT TORQUE WRENCH—The ICL7606 instrumentation amplifier is used as a preamplifier, taking the differential voltage of the bridge and converting it to a single-ended voltage reference to ground. The signal is then amplified and applied to the input of the ICL7106, which drives the LED display. The circuit may be adapted to several strain gauge system applications.—"Intersil Data Book," Intersil, Cupertino, CA, 1981, p. 5-137.

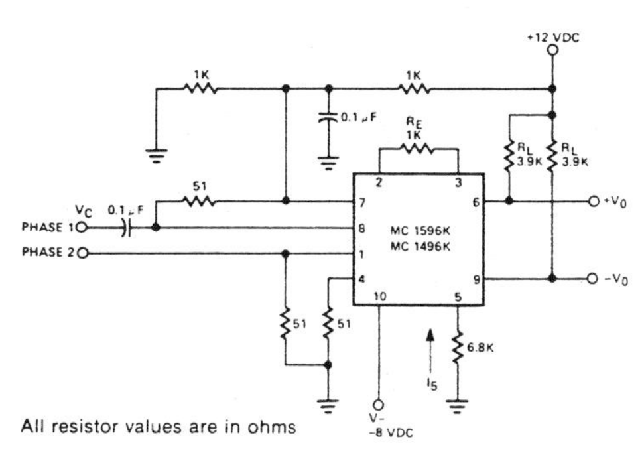

All resistor values are in ohms

PHASE DETECTOR/COMPARATOR—Two signals of equal frequency are applied to the inputs. The frequencies are multiplied together, producing sum and difference frequencies. Equal frequencies cause the difference component to become DC, while the undesired sum component is filtered out. The DC component is related to the phase angle. At 90° the cosine becomes zero while being at maximum positive or maximum negative at 0° and 180°, respectively.—"Signetics Analog Applications Manual," Signetics, Sunnyvale, CA, 1979, pp. 189–190.

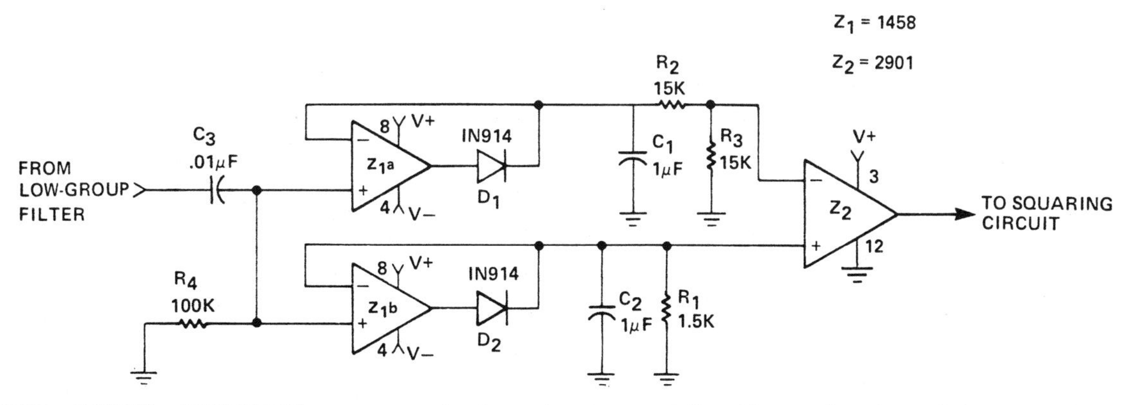

ENVELOPE DECAY DETECTOR—Consists of two precision rectifiers, two sample and hold circuits, and a comparator. C3 couples the low-group filter output to the envelope detector. Z_{1a}, D1, C1, R2, and R3 then rectify the incoming signal and store a peak value. The R2/R3/C1 time constant is set for 20 ms so that the voltage at the inverting input of Z_2 will represent 50% of the peak value of the incoming signal. Z_{1b}, D2, R1, and C2 also rectify the incoming signal and store a peak value, but the time constant is set for 1.4 ms so that the voltage at the noninverting input of Z_2 will represent the instantaneous peak value of the incoming waveform. As long as the instantaneous value is greater than 50% of the peak value, the comparator output will be high.—"Mostek 1984/85 Microelectronics Data Book," Mostek Corporation, Carrollton, TX, 1984, pp. XV-17 to XV-18.

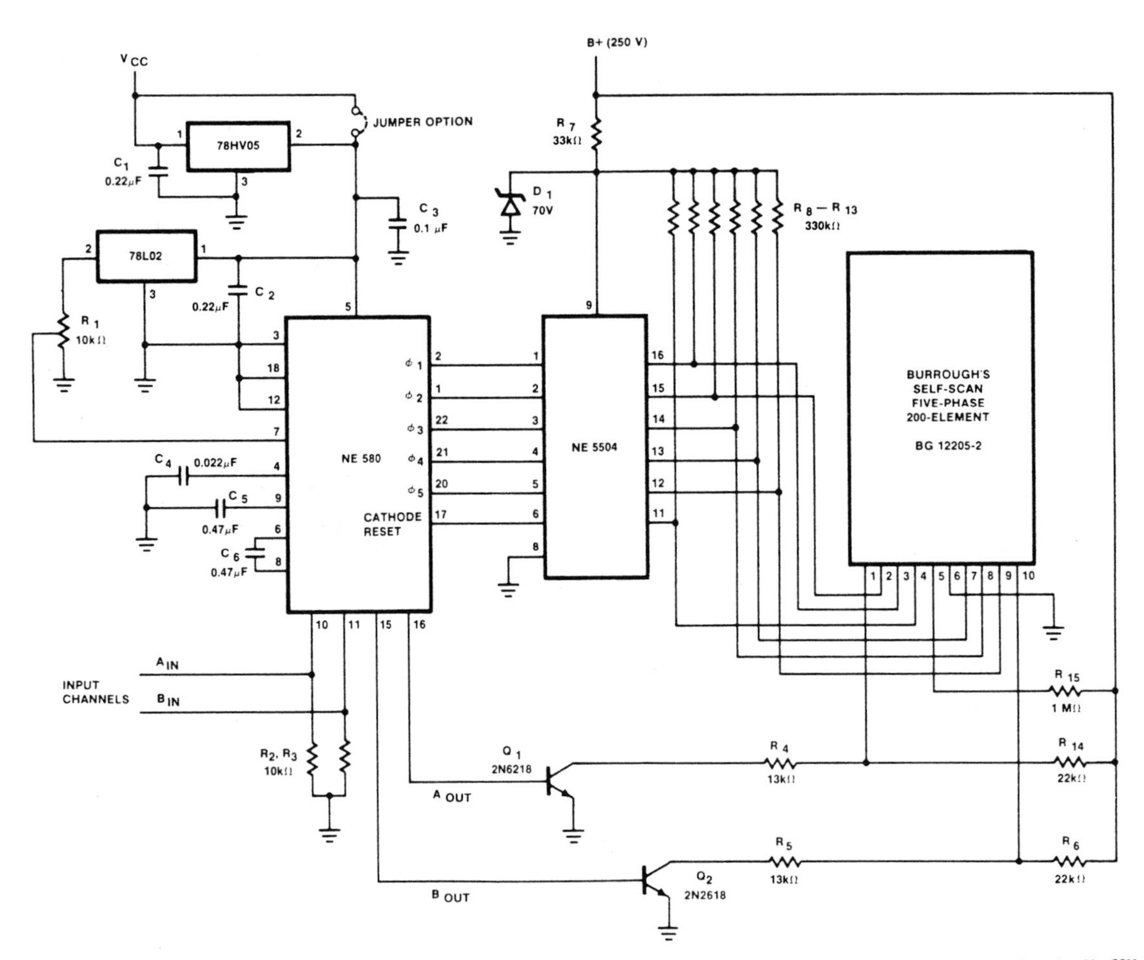

GAS DISCHARGE BARGRAPH DRIVER—Anode outputs A and B drive the bases of the 2N6218 transistors. When the anode is on, the anode voltage is 250 V. With the anode off, Q_1 and Q_2 will drop the anode to about 95 V. The cathode output phases of the NE580 and the reset pulse are used to drive a high-voltage Darlington array (NE5504). It is necessary to clamp the cathode "off" bias of the gas discharge display to 72 V. The NE5504 has a breakdown voltage of 100 V and can accommodate the bias. The 70-V zener sets the cathode "off" bias for the display and prevents the maximum voltage potential of the NE5504 from being exceeded.—"Signetics Analog Applications Manual," Signetics, Sunnyvale, CA, 1979, pp. 111–114.

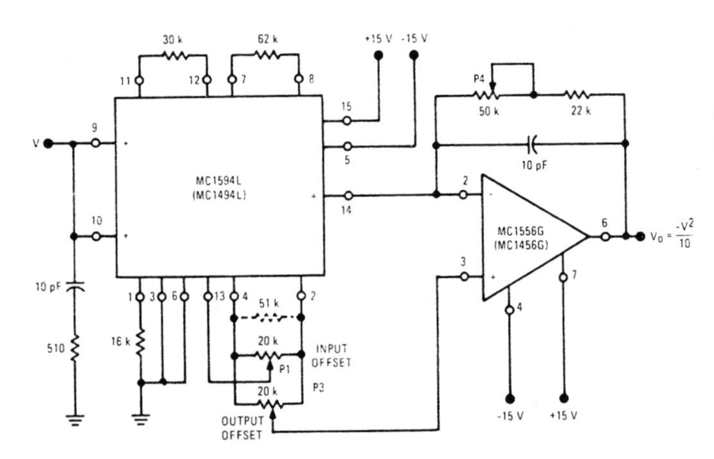

SQUARING CIRCUIT—Produces output voltage proportional to square of input voltage. The circuit uses a MC1494L four-quadrant multiplier device, which has an input voltage range of ±10 V. Linearity is excellent. The original source contains details on AC and DC adjustment procedures.—"Motorola Linear Integrated Circuits Databook," Motorola, Phoenix, AZ, 1979, p. 6-68.

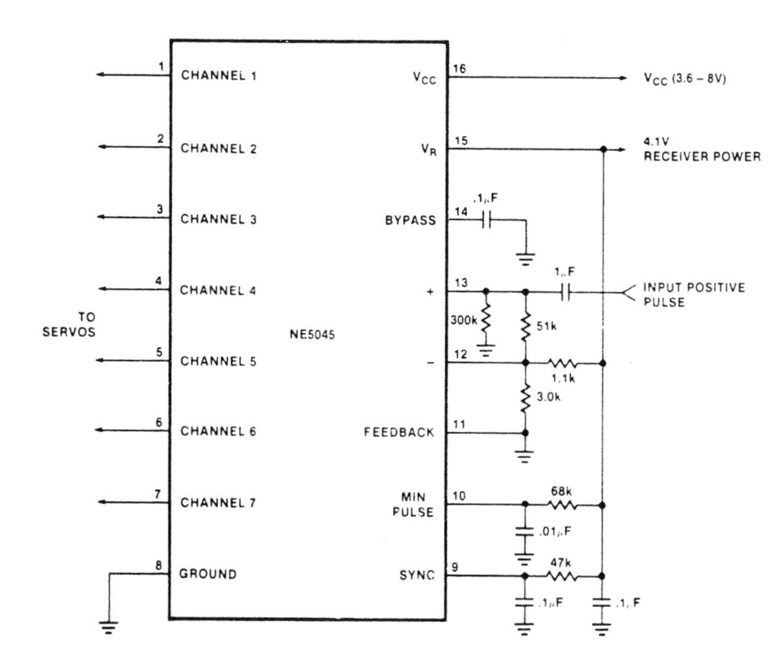

SEVEN-CHANNEL RADIO CONTROL DECODER—Based on NE5045 IC, circuit is a serial input, parallel output system providing up to seven channels of information in PWM form. Either positive or negative input pulses may be used. All outputs are reset to zero when no input signal is present. The circuit is suitable for radio control, robotics, telemetry, or industrial control.—"Signetics Linear LSI Data and Applications Manual," Signetics, Sunnyvale, CA, 1985, p. 7-13.

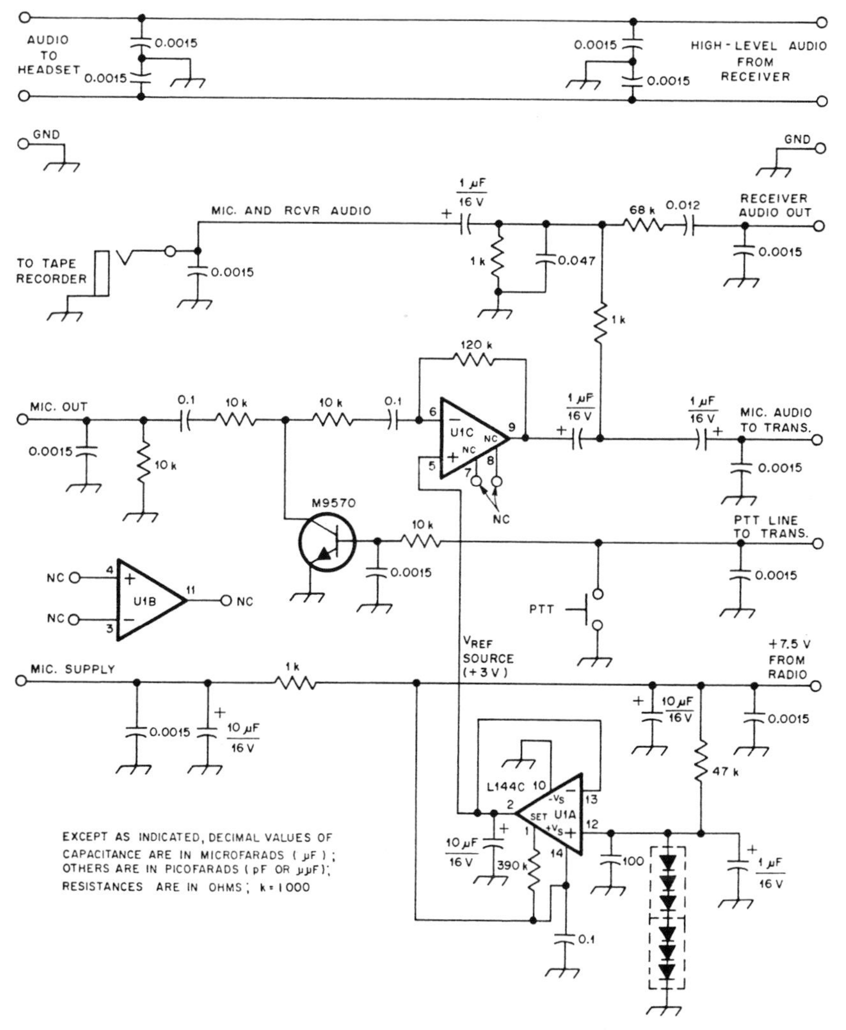

TAPE RECORDER/TRANSCEIVER INTERFACE—Used to interface modified Motorola MX-300-S VHF-FM transceiver to tape recorder and headset microphone. The circuit was used by astronaut Owen Garriott, W5LFL, to record his communications with earthbound amateur radio operators during a Space Shuttle mission. U1 is a Siliconix L144CJ.—T. McMulle, J. Worsham, and H. Sanderson, Amateur Radio's Hand-Held in Space, *QST,* August 1984, pp. 14–17.

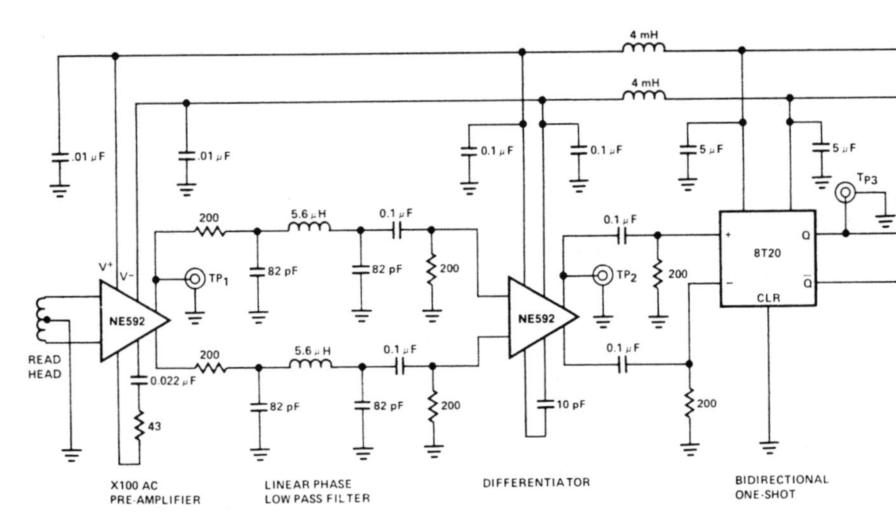

All resistor values are in ohms

PHASE ENCODED DATA SIGNAL CONDITIONER—Preconditions linear data recovered from magnetic disc or drum storage. Read-back data is applied directly to the input of the first NE592. This amplifier functions as a wideband AC-coupled amplifier with a gain of 100. The output of the first-stage amplifier is routed to a linear phase shift low-pass filter, a single-stage constant-K type. The second NE592 is used as a differentiator/amplifier stage. The output of this stage is connected to the 8T20 bidirectional monostable unit to provide the proper pulses at zero crossing points of the differentiator.—"Signetics Analog Applications Manual," Signetics, Sunnyvale, CA, 1979, p. 14-141.

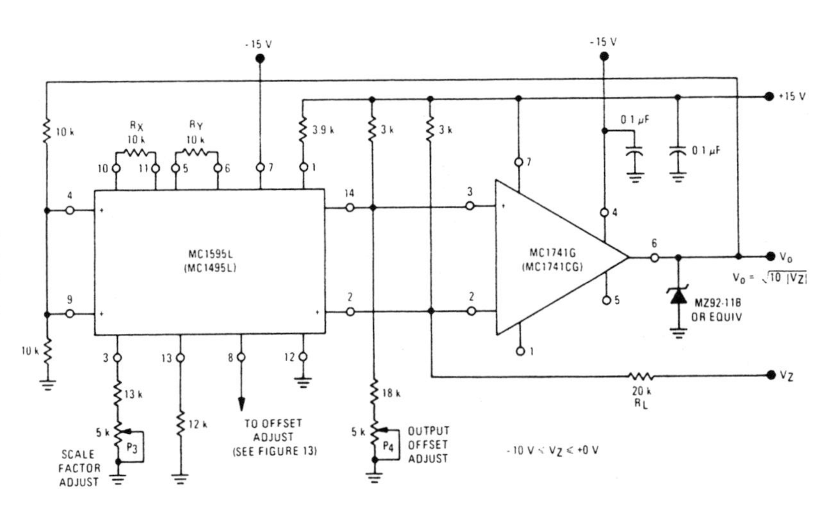

SQUARE ROOT CIRCUIT—Delivers output voltage equal to square root of 10 V_Z, where V_Z is input voltage. The original source contains details on the precise adjustment of the circuit's operation.—"Motorola Linear Integrated Circuits Databook," Motorola, Phoenix, AZ, 1979, p. 6-86.

13

Modem Circuits

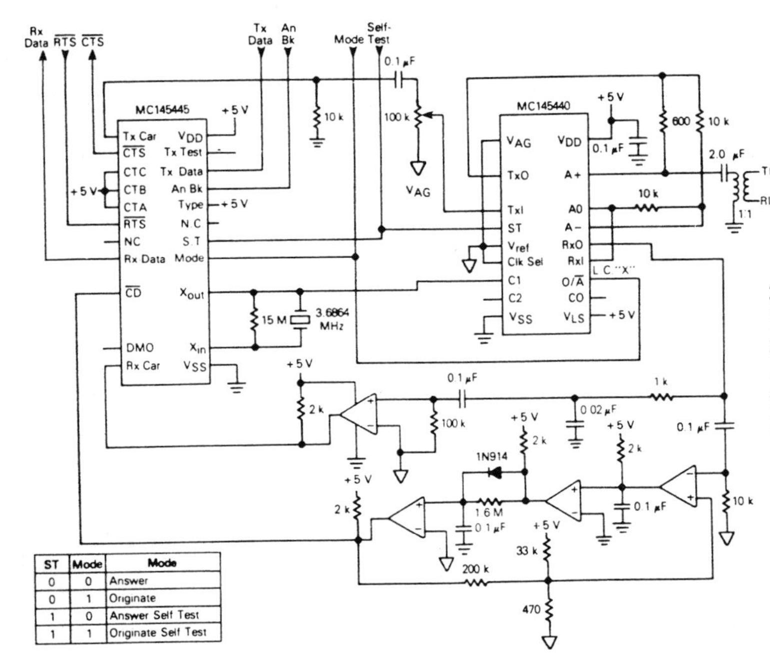

ST	Mode	Mode
0	0	Answer
0	1	Originate
1	0	Answer Self Test
1	1	Originate Self Test

300-BAUD MODEM—Based upon MC145445 modem IC. Operates as a 300-baud Bell® 103 standard modem. The MC145440 IC acts as a filter.—Motorola, Inc., Semiconductor Div., MC145445 Data Sheet.

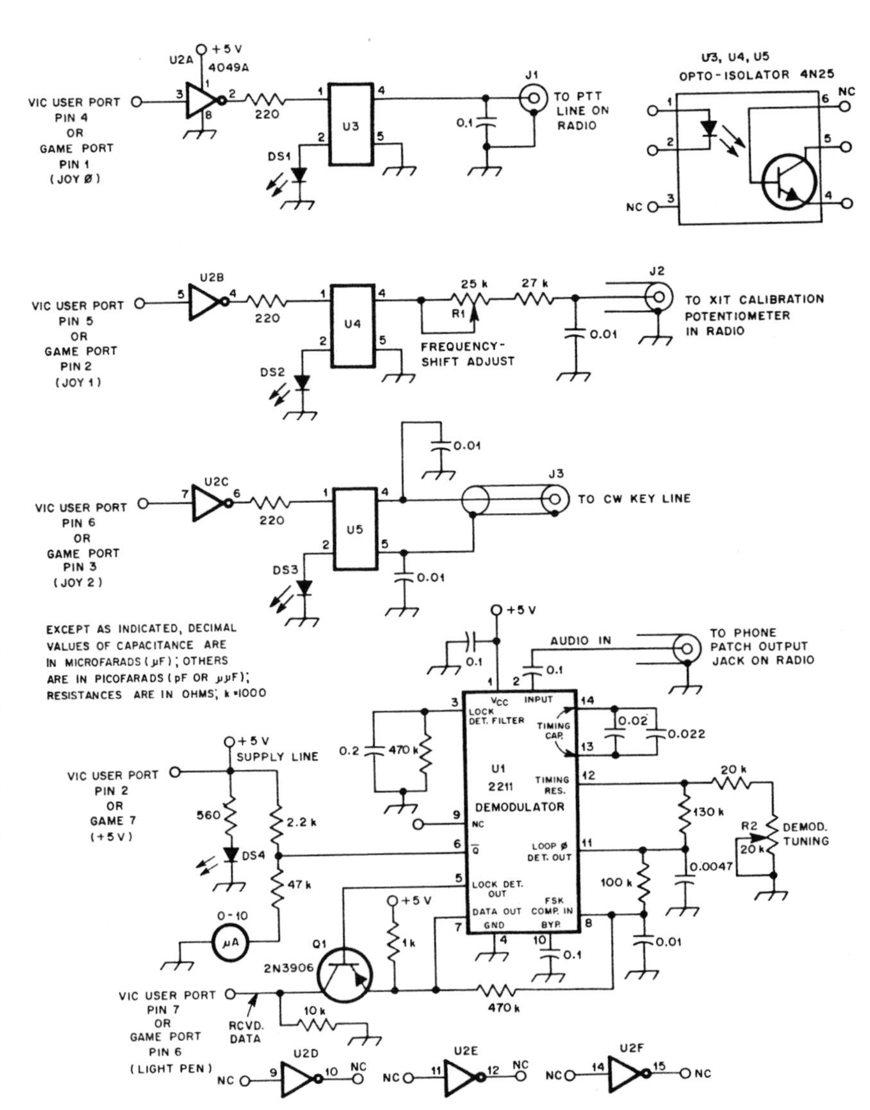

RADIO MODEM WITH FSK—Designed for use with a Commodore VIC-20® microcomputer and appropriate software for transmission and reception of FSK signals. The frequency shift is fixed at 170 Hz, and the value of R1 will vary depending on the transmitter or transceiver used.—J. Dewey, Radio Modem With FSK RTTY For HF Rigs, *QST*, September 1984, pp. 38–39.

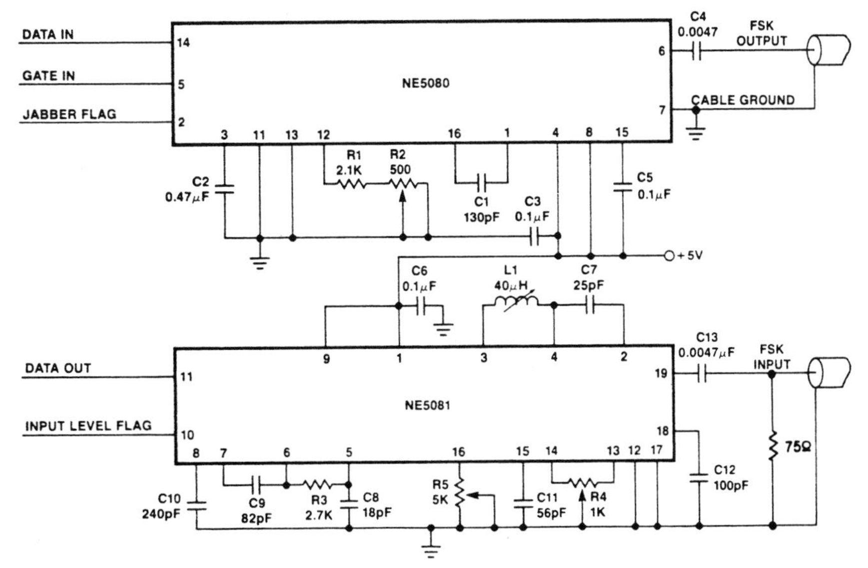

ASYNCHRONOUS FULL DUPLEX FSK MODEM—Carrier frequency can range from 50 kHz to over 20 MHz (circuit shown is for 5-MHz carrier). The dynamic range is 30 dB, and rejection of EMI/RFI is excellent. The distance that can be driven varies with the type of cable used, the number of modems attached to the cable, and the carrier frequency.—"Signetics Linear LSI Data and Applications Manual," Signetics, Sunnyvale, CA, 1985, pp. 9-174 to 9-175.

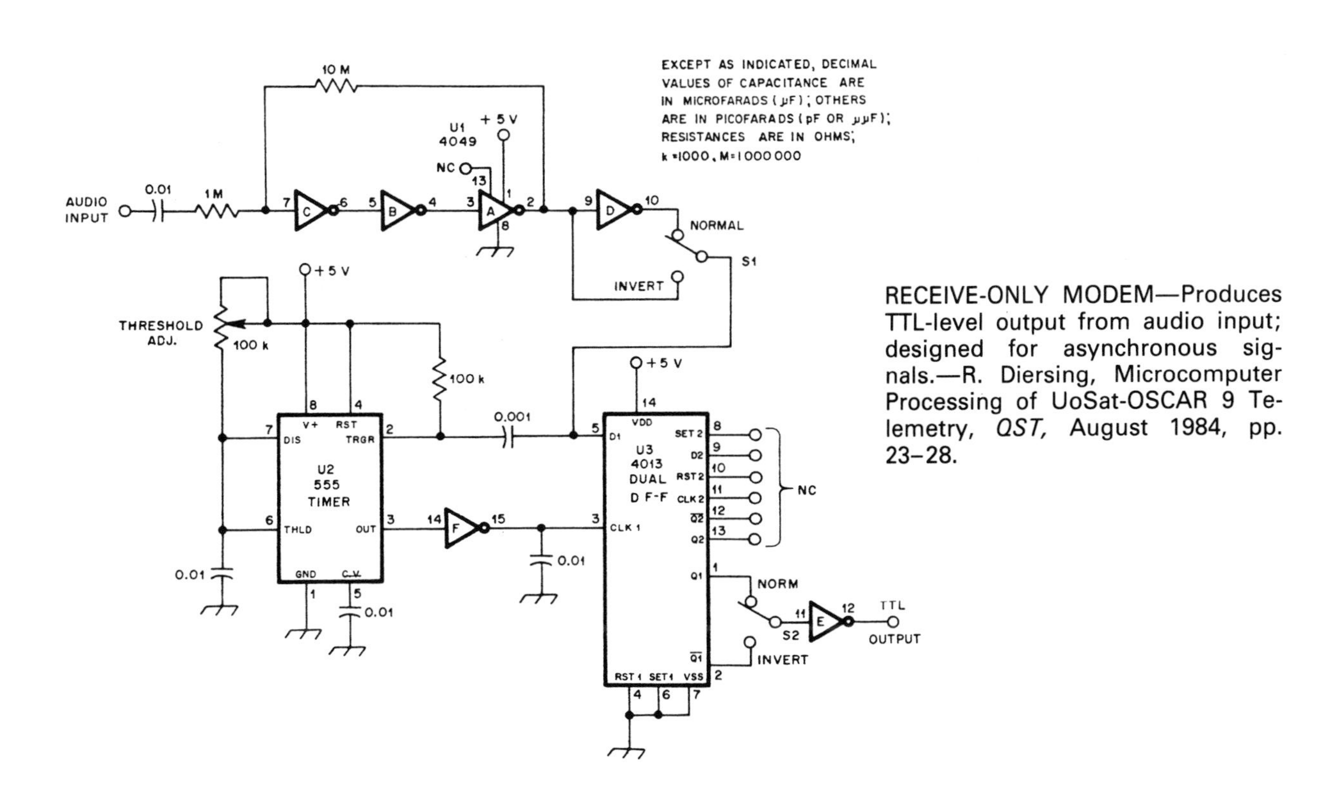

RECEIVE-ONLY MODEM—Produces TTL-level output from audio input; designed for asynchronous signals.—R. Diersing, Microcomputer Processing of UoSat-OSCAR 9 Telemetry, *QST,* August 1984, pp. 23–28.

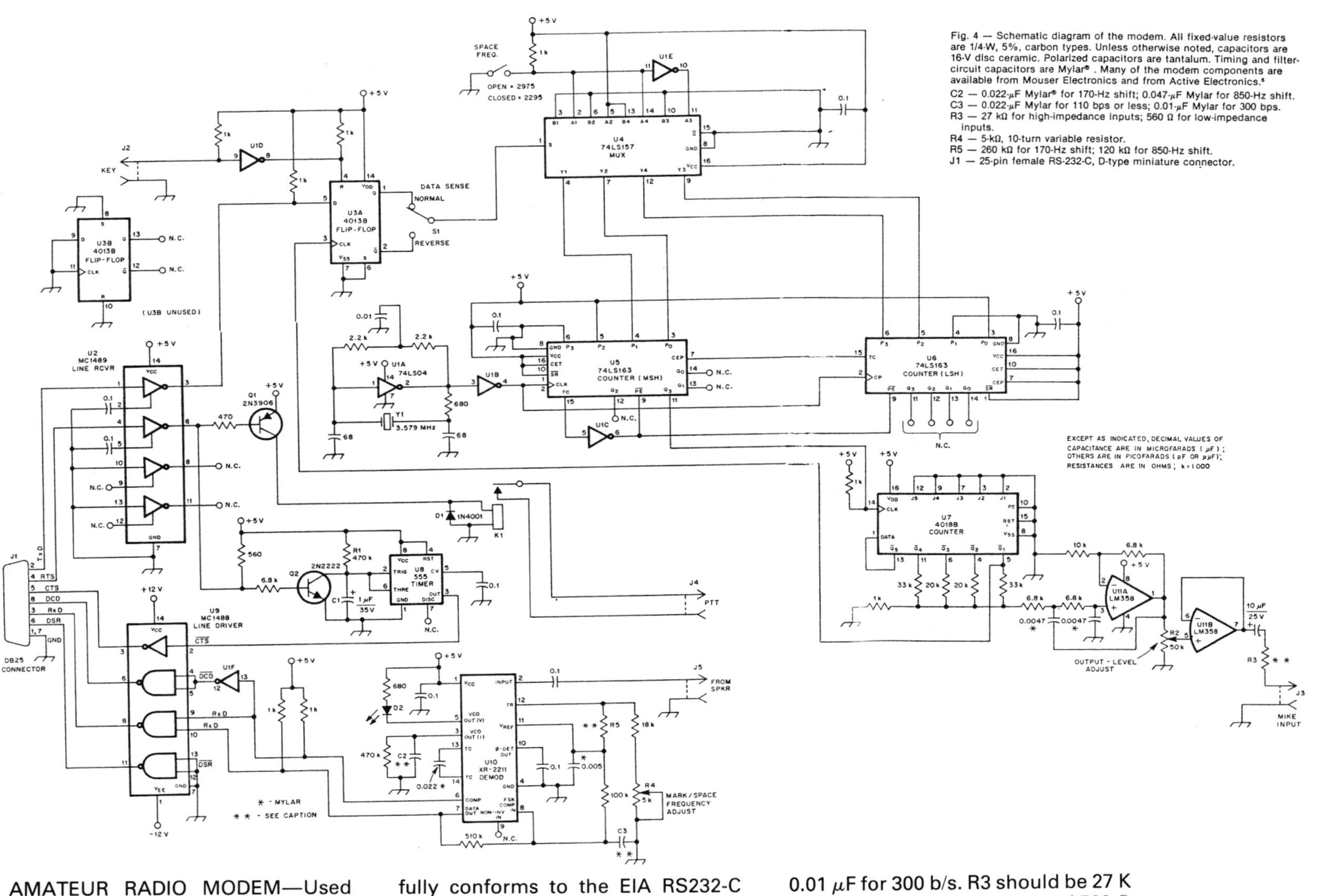

Fig. 4 — Schematic diagram of the modem. All fixed-value resistors are 1/4-W, 5%, carbon types. Unless otherwise noted, capacitors are 16-V disc ceramic. Polarized capacitors are tantalum. Timing and filter-circuit capacitors are Mylar®. Many of the modem components are available from Mouser Electronics and from Active Electronics.[5]
C2 — 0.022-μF Mylar® for 170-Hz shift; 0.047-μF Mylar for 850-Hz shift.
C3 — 0.022-μF Mylar for 110 bps or less; 0.01-μF Mylar for 300 bps.
R3 — 27 kΩ for high-impedance inputs; 560 Ω for low-impedance inputs.
R4 — 5-kΩ, 10-turn variable resistor.
R5 — 260 kΩ for 170-Hz shift; 120 kΩ for 850-Hz shift.
J1 — 25-pin female RS-232-C, D-type miniature connector.

AMATEUR RADIO MODEM—Used for transmission of ASCII via amateur radio. It can also be used to transmit Baudot with the appropriate microcomputer software. The circuit can also be used to perform ASCII-to-Baudot conversion, protocol implementation and error detection, and packet switching. This circuit fully conforms to the EIA RS232-C standard for half-duplex communication over a dedicated line and reliably operates up to 300 baud. The values of certain components depend upon the parameters desired in amateur radio operation. C2 is 0.022 μF for a 170-Hz shift and 0.047 μF for 850 Hz. C3 is 0.022 μF for 110 b/s and 0.01 μF for 300 b/s. R3 should be 27 K for high-impedance inputs and 560 Ω for low-impedance inputs. R5 is 260 K for a 170-Hz shift and 120 K for 850 Hz. J1 is a 25-pin female RS232 connector.—R. Valleau, Build an Amateur Radio Modem, *QST,* October 1983, pp. 32–36.

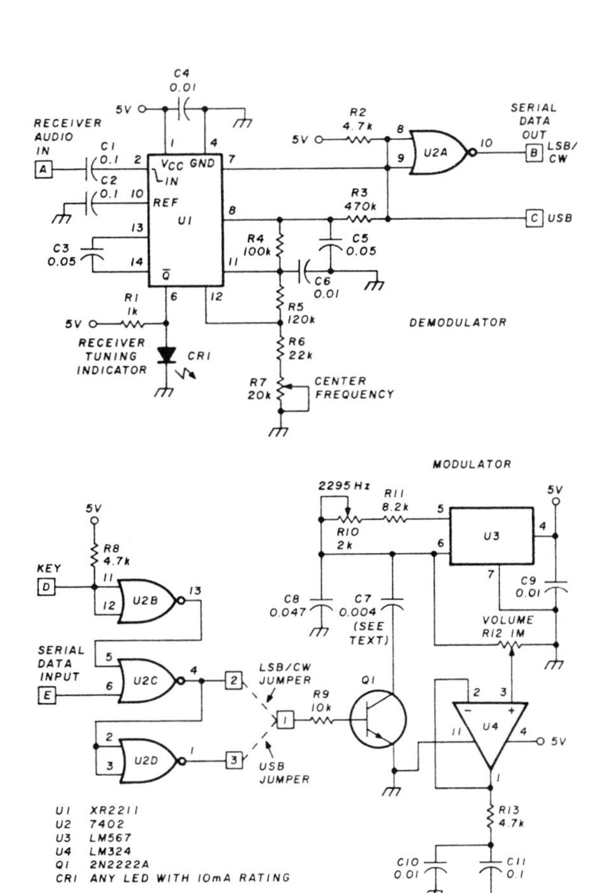

SSB TO FSK MODEM—Designed to allow SSB/CW-only transceivers to be used for AFSK operation. All interfaces are TTL-compatible. U1 is an XR2211 demodulator/tone decoder. U3 (LM567) generates mark and space tones of 2125 and 2295 Hz, respectively. Calibration is accomplished by adjusting R10 until a 2295-Hz tone is obtained with Q1 cut off.—T. Zeltwanger, FSK Adapter for SSB Transmitters, *Ham Radio,* July 1981, pp. 12–15.

14

Morse Code Circuits

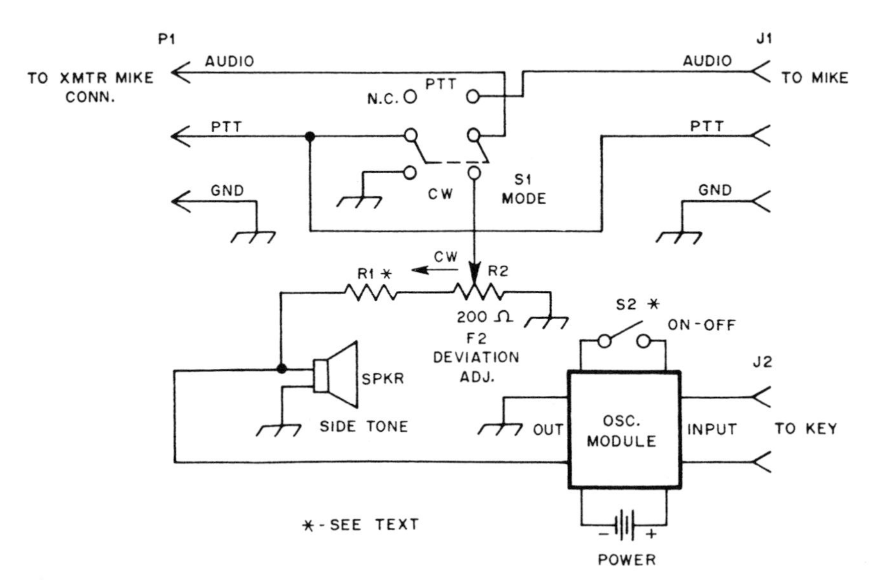

AUDIO TONE-MODULATED MORSE CODE KEYING—Circuit allows transmission of Morse code via audio tones on an AM, FM, or SSB transmitter. A microphone connection allows the operator to add verbal comments to Morse messages. The OSC MODULE may be a code practice oscillator or sidetone oscillator of a keying device.—D. Blum, Use Your 2-Meter Rig for F2 Operation, *QST,* November 1983, pp. 55–56.

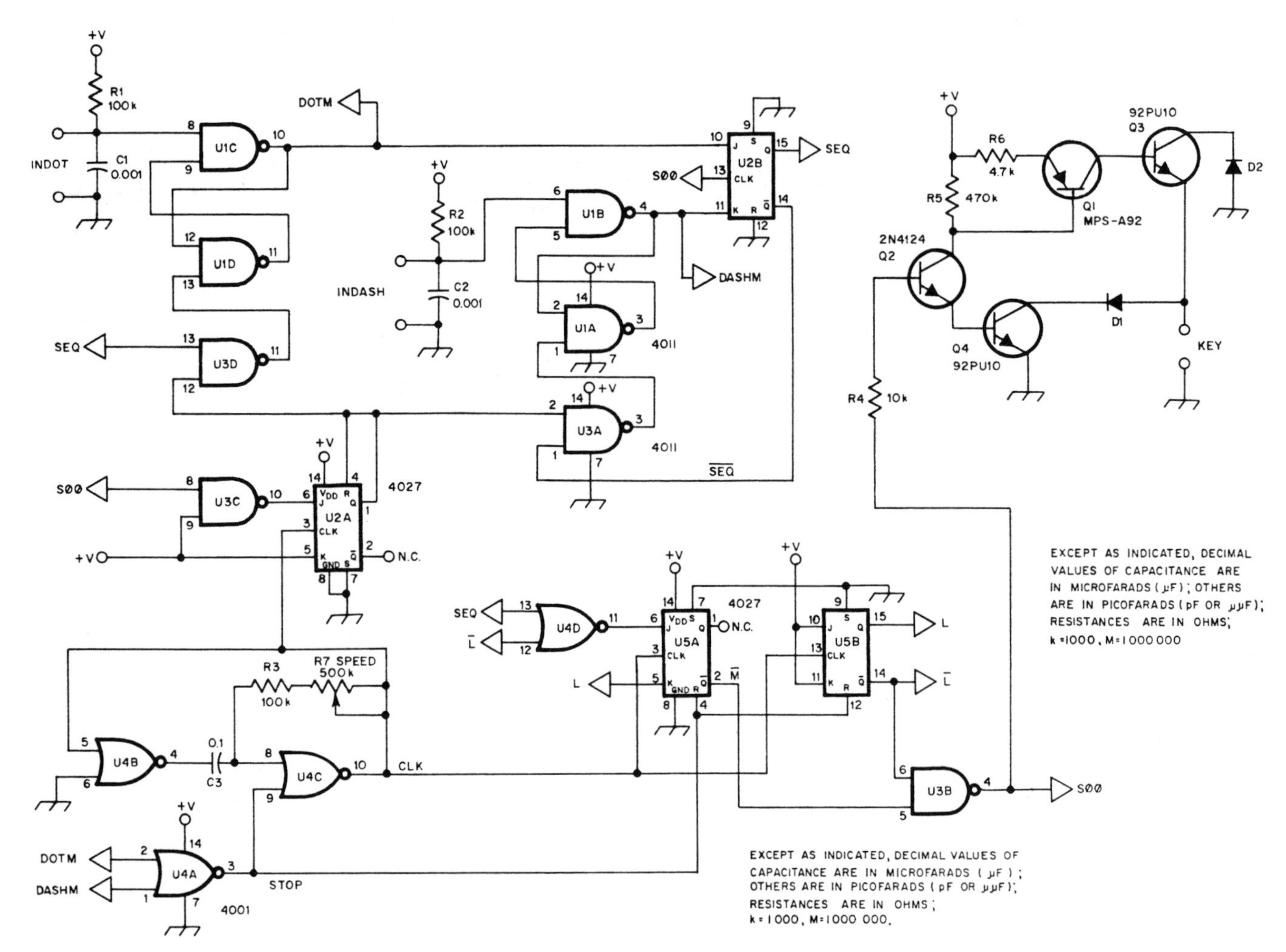

CMOS IAMBIC KEYER—Circuit may be used with iambic (squeeze) keyer or ordinary keyer paddle. The keyer is built around a 2-bit binary counter. The counter has four possible states—0, 1, 2, and 3. During 0, the transmitter is off. When a dot is sent, the counter goes to 1 and back to 0. During a dash, it goes through all the states and then returns to 0. The keyer is capable of switching voltages of +300 to −300 V at currents to 55 mA. U1 and U3 are 4011, U2 and U5 are 4027, and U4 is 4001. D1 and D2 are 1N4004.—T. Theroux, A Digital CMOS Iambic Keyer, *QST,* June 1982, pp. 26–28.

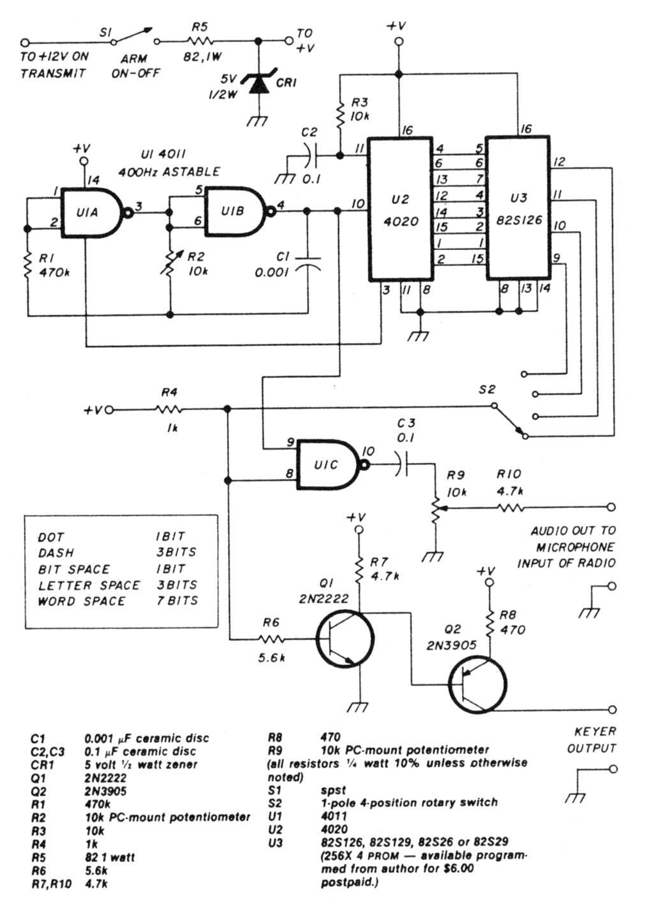

CW IDENTIFIER—82S126 PROM stores station call sign; it is addressed with eight address lines from the 4020 binary counter driven by an astable oscillator formed by one-half of the 4011. R2 controls the clock rate, which affects the CW speed and the tone of transmitted signal. The tone level is controlled by R9. The circuit may be used for repeater, SSTV, or RTTY work. The original article includes details on PROM programming.—M. Di Julio, Versatile CW Identifier, *Ham Radio,* October 1980, pp. 22–25.

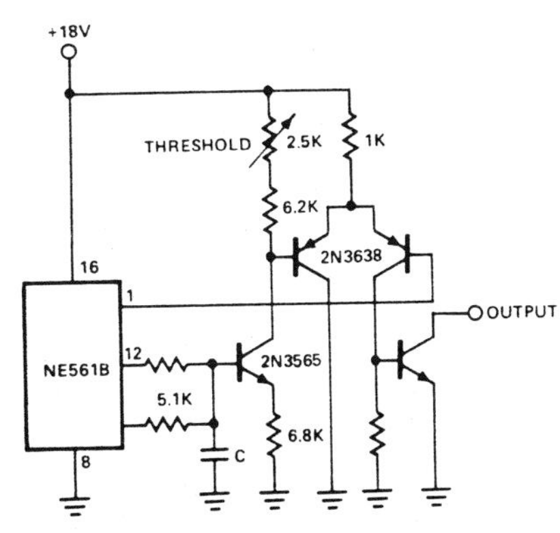

CW DETECTOR—Uses 561 PLL IC, which functions as an AM detector. When CW input is applied, there is a DC shift at the output of the AM detector (pin 1). The shift is small compared with the no-signal DC level, and a reference is necessary. The 2N3565 is used as a constant-current source. The differential amplifier, using the two 2N3638 transistors, amplifies the DC output and allows it to drive an NPN transistor referenced to ground.—"Signetics Analog Applications Manual," Signetics, Sunnyvale, CA, 1979, pp. 291–294.

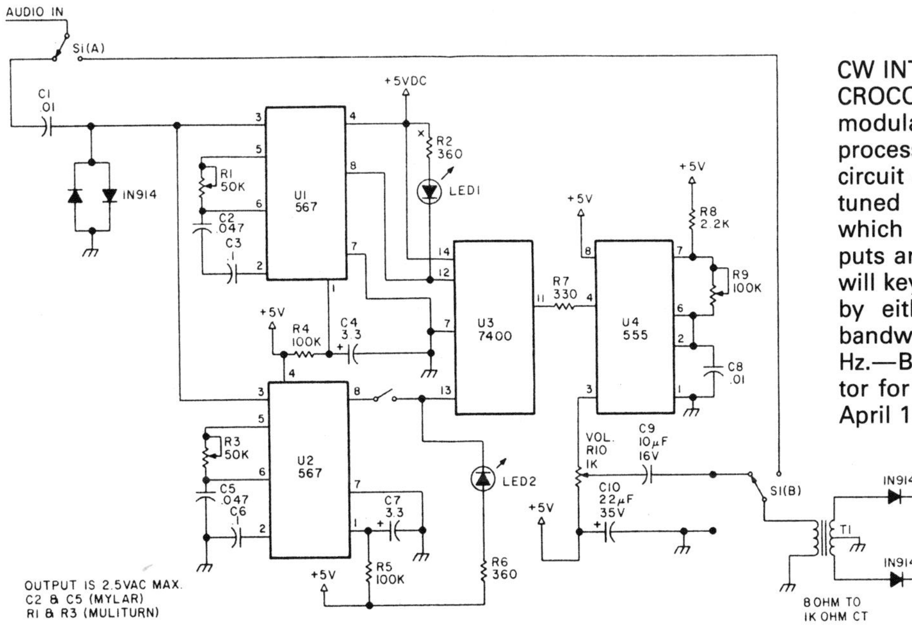

CW INTERFERENCE FILTER FOR MICROCOMPUTERS—Processes demodulated CW signals to ease their processing by microcomputers. The circuit consists of a pair of decoders tuned to slightly different tones which overlap each other. The outputs are fed to a NAND gate, which will key whenever a tone is detected by either detector. The operating bandwidth is approximately 200 Hz.—B. Buckingham, QRM Eliminator for Computer CW, *73 Magazine,* April 1983, p. 106.

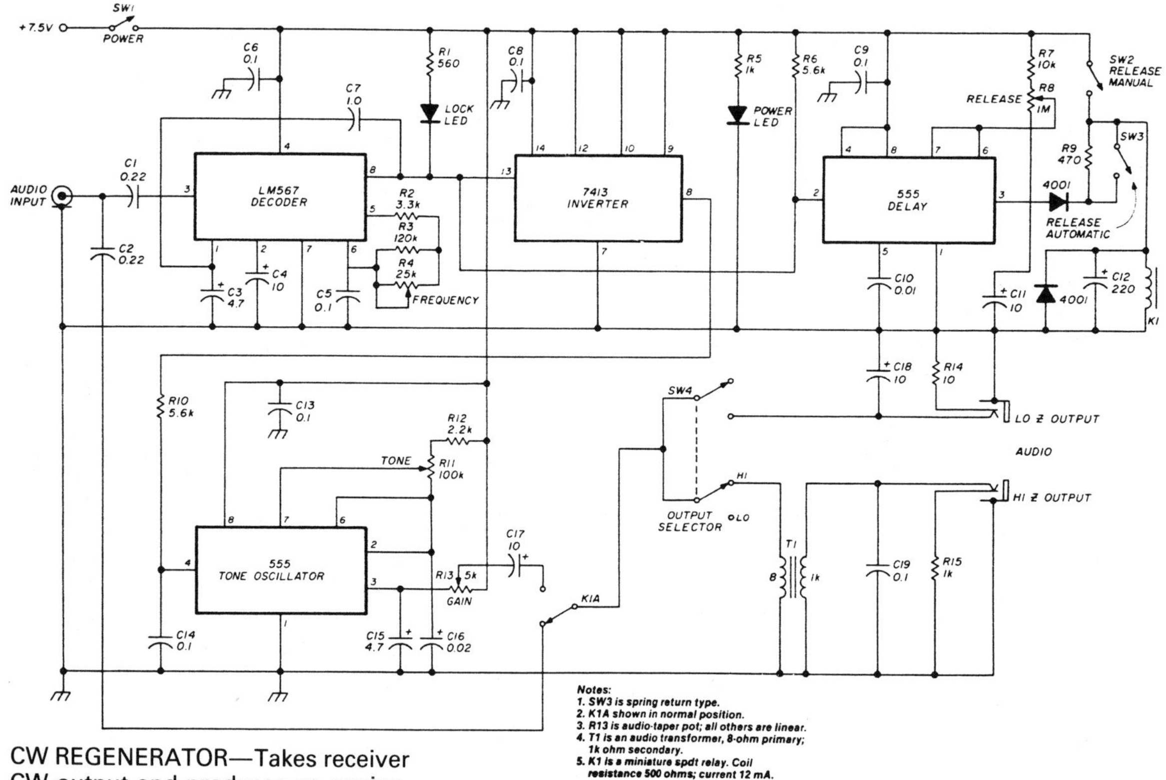

Notes:
1. SW3 is spring return type.
2. K1A shown in normal position.
3. R13 is audio-taper pot; all others are linear.
4. T1 is an audio transformer, 8-ohm primary; 1k ohm secondary.
5. K1 is a miniature spdt relay. Coil resistance 500 ohms; current 12 mA.

CW REGENERATOR—Takes receiver CW output and produces an equivalent "clean" output. The frequency acquisition adjustment range is approximately 400–1800 Hz. The release tie time can vary from 1 to 20 seconds to compensate for various incoming code characteristics. The pitch and gain of the internal tone oscillator are variable; the tone may range from 280 to 800 Hz. When the power is switched on, C1 feeds audio to the LM567. Simultaneously, the audio is connected to the output device through C2 and K1A. Varying R2 allows selection of the particular CW signal to regenerate. When the PLL is locked, the LOCK LED will light in step with the chosen CW input signal.—F. Marcellino, CW Regenerator for CW Receivers, *Ham Radio,* October 1980, pp. 64–67.

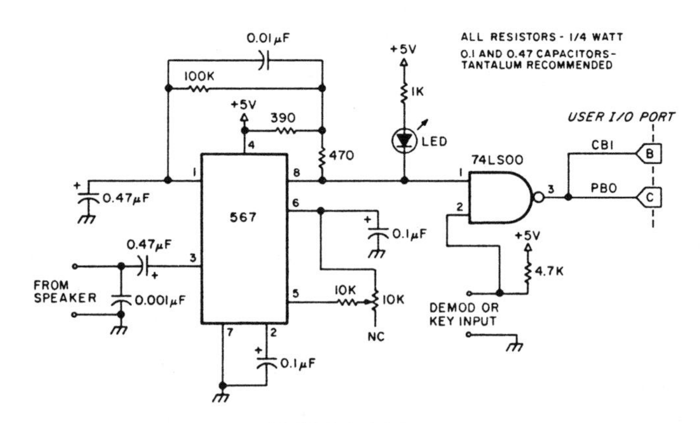

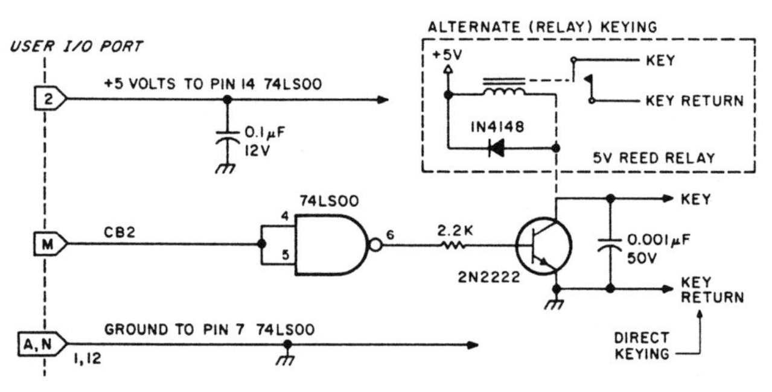

MORSE CODE TRANSMISSION AND RECEPTION WITH MICROCOMPUTER—Circuits are designed for transmission and reception of Morse code signals using Commodore VIC-20® or C64® microcomputers in an amateur radio system. A BASIC langauge program controls the generation and decoding of Morse signals in the microcomputer.—M. Apsey and R. Myers, Sounds Good to Me, *73 Magazine,* June 1984, pp. 38–40.

CW SHAPER—Allows shaping of the waveform of transmitted CW signal to overcome shortcomings in transmitter keying circuitry. U1 is an inverting comparator. It isolates the key from following circuitry and provides a +12-V source with the key closed and a ground with the key up. When the key is closed, D1 is forward-biased. C1 is then charged through R1 (variable attack control). When the key is up, C1 discharges through R2 (decay time control). U2 is a voltage follower that "reads" the voltage of C1 and applies it to the keying transistors.—E. Nichols, Try this Versatile CW Shaper, *QST,* December 1984, pp. 29–30.

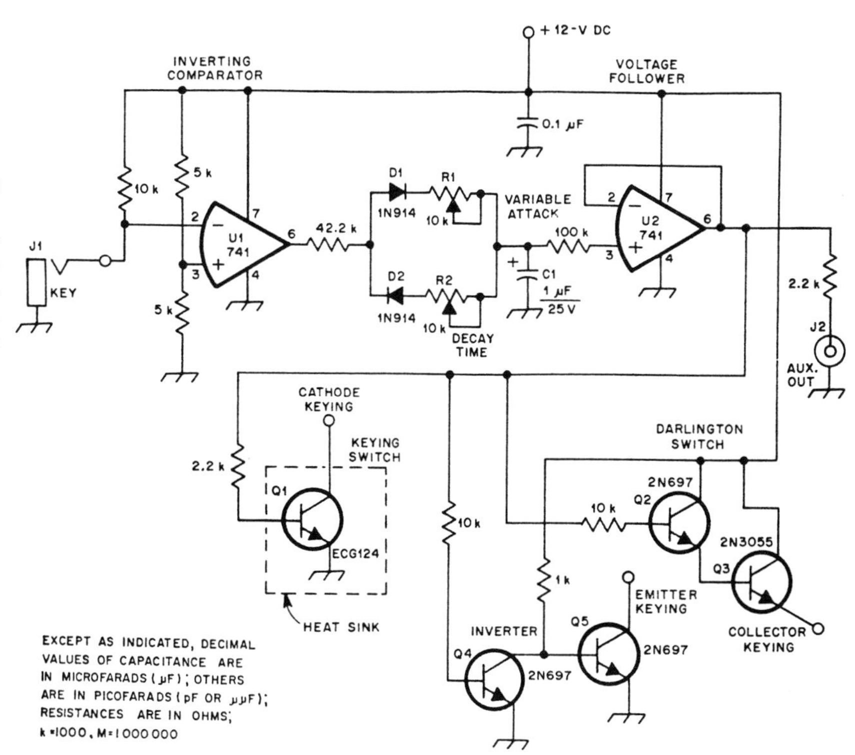

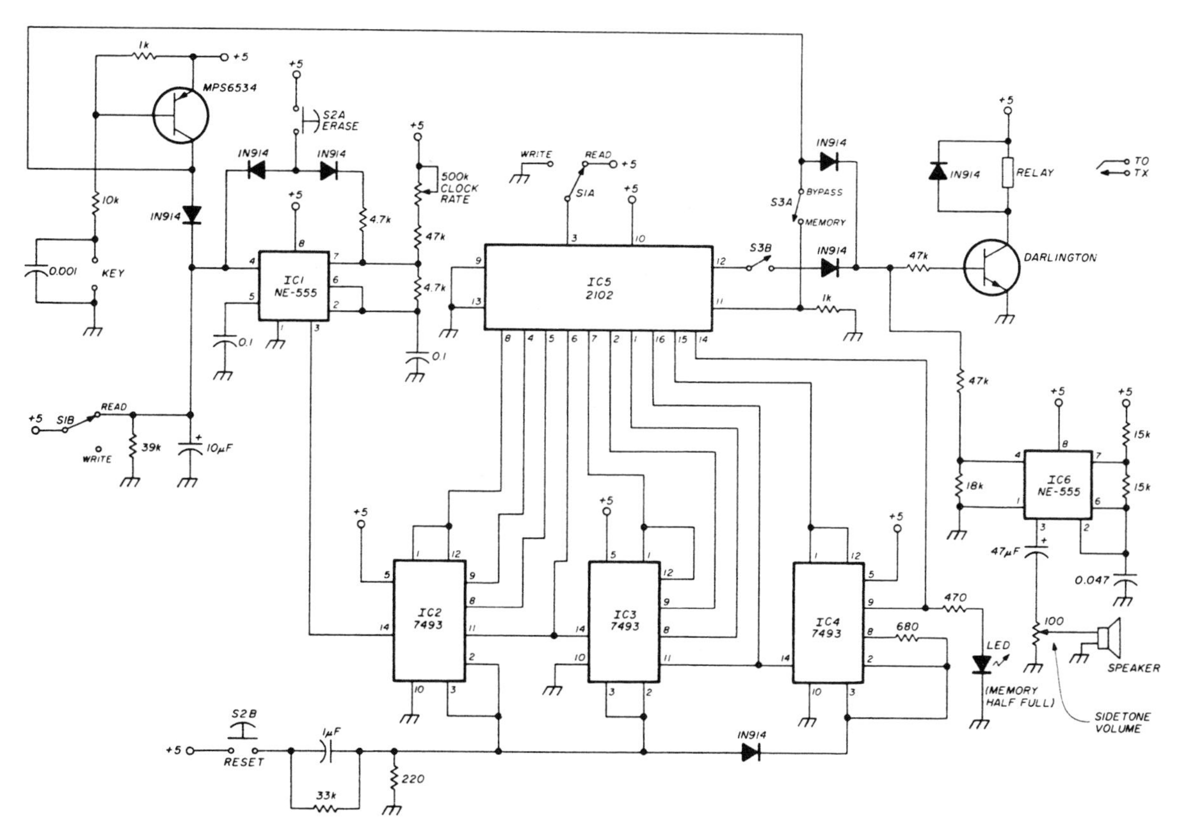

CW MEMORY—Stores desired CW message in 2102 1-K RAM and keys transmitter via relay when S1B is set to READ and S2B is set to RESET. A 555 timer is used for the clock oscillator and drives the three 7493 binary counters. These provide the 10-bit address codes to the RAM. Power is applied to the clock oscillator only when the key is closed. When the unit is first turned on, the memory contains random data and must be cleared by setting S1B to WRITE and S2A to ERASE.—R. Merigirian, Simple CW Memory, *Ham Radio,* November 1980, pp. 46–47.

MORSE CODE TRAINER/KEYER—CPP1 single-chip microcomputer contains Morse code practice and keying software on chip. Closing the appropriate switch (S2A through S2F) selects a different group of practice characters. The speed may be adjusted from 2 to over 40 WPM. The circuit can be used as an electronic keyer by adding an iambic key. IC1 is the CPP1 (manufactured by Micro Digital Technology, Box 1139, Mesa, AZ, 85201). IC2 and IC3 are 555 timer ICs. C1 and C2 are 22-pF ceramics; C3 is 1 μF 10 V DC; C4 and C6 are 0.01 μF; C5 and C8 are 0.1 μF; and C7 and C9 are 100 μF 10 V DC. Q1 is a 2N3904 or equivalent. R1 through R5, R8, and R9 are all 15 K. R6 is 220 K, and R12 is 4.7 K. R7 is a 2-M potentiometer. D1 and D2 are 1N914 or equivalent. The crystal may be in range of 1 to 6 MHz but 3.57 MHz is the nominal frequency.—K7ZOV, You Can Build this Code Trainer, *73 Magazine,* July 1983, pp. 12–16.

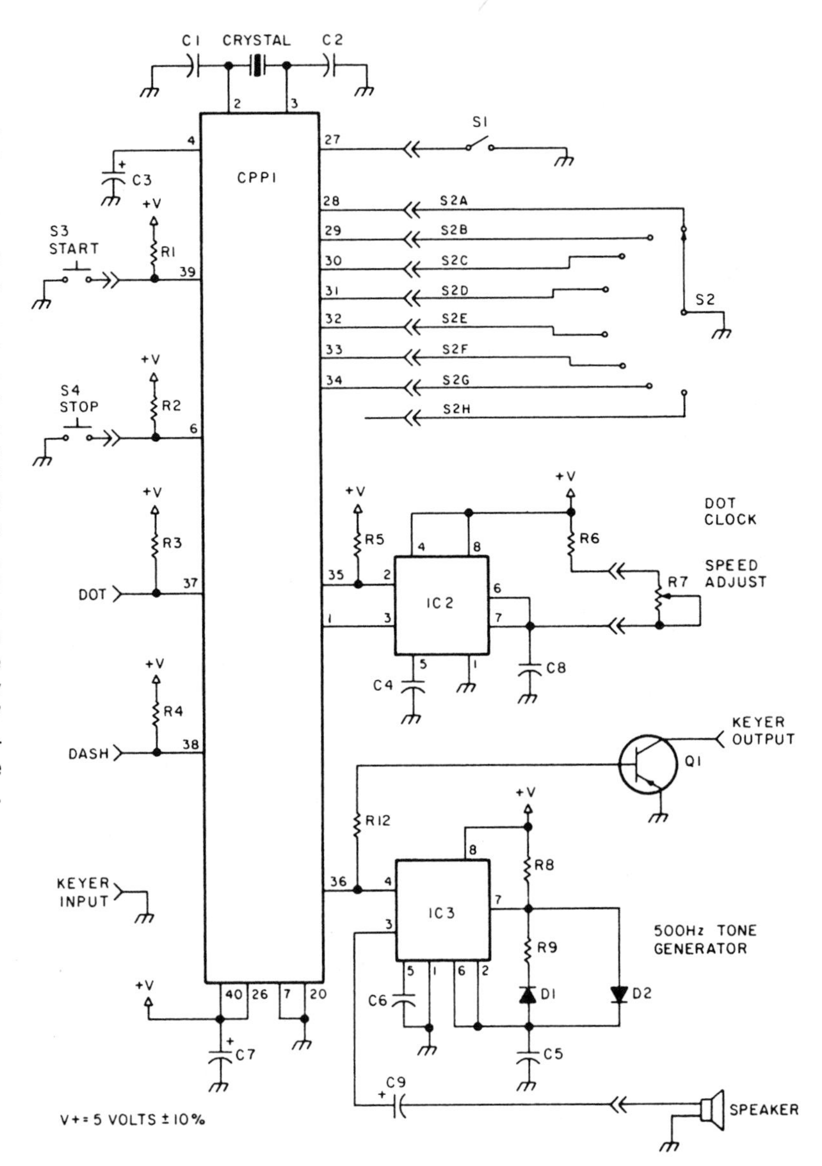

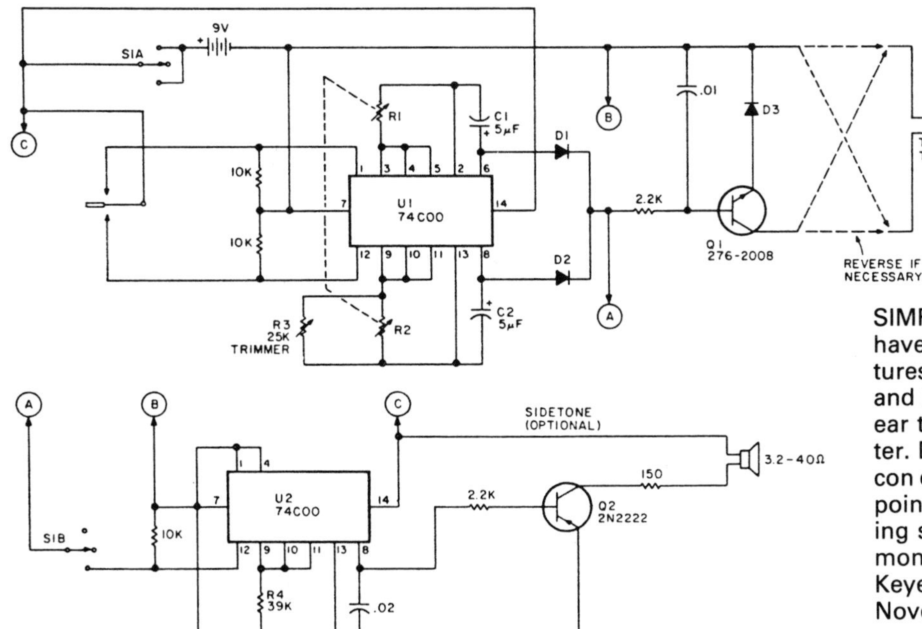

SIMPLE KEYER—Keyer does not have iambic or self-completing features but requires only one IC. R1 and R2 are together a 30- or 50-K linear taper dual "stereo" potentiometer. DS1, D2, and D3 can be any silicon diode. Points A, B, and C indicate points in the circuit where a matching sidetone circuit can be added to monitor the keyer.—J. Donaldson, Keyer on a Shoestring, *73 Magazine,* November 1982, pp. 104–105.

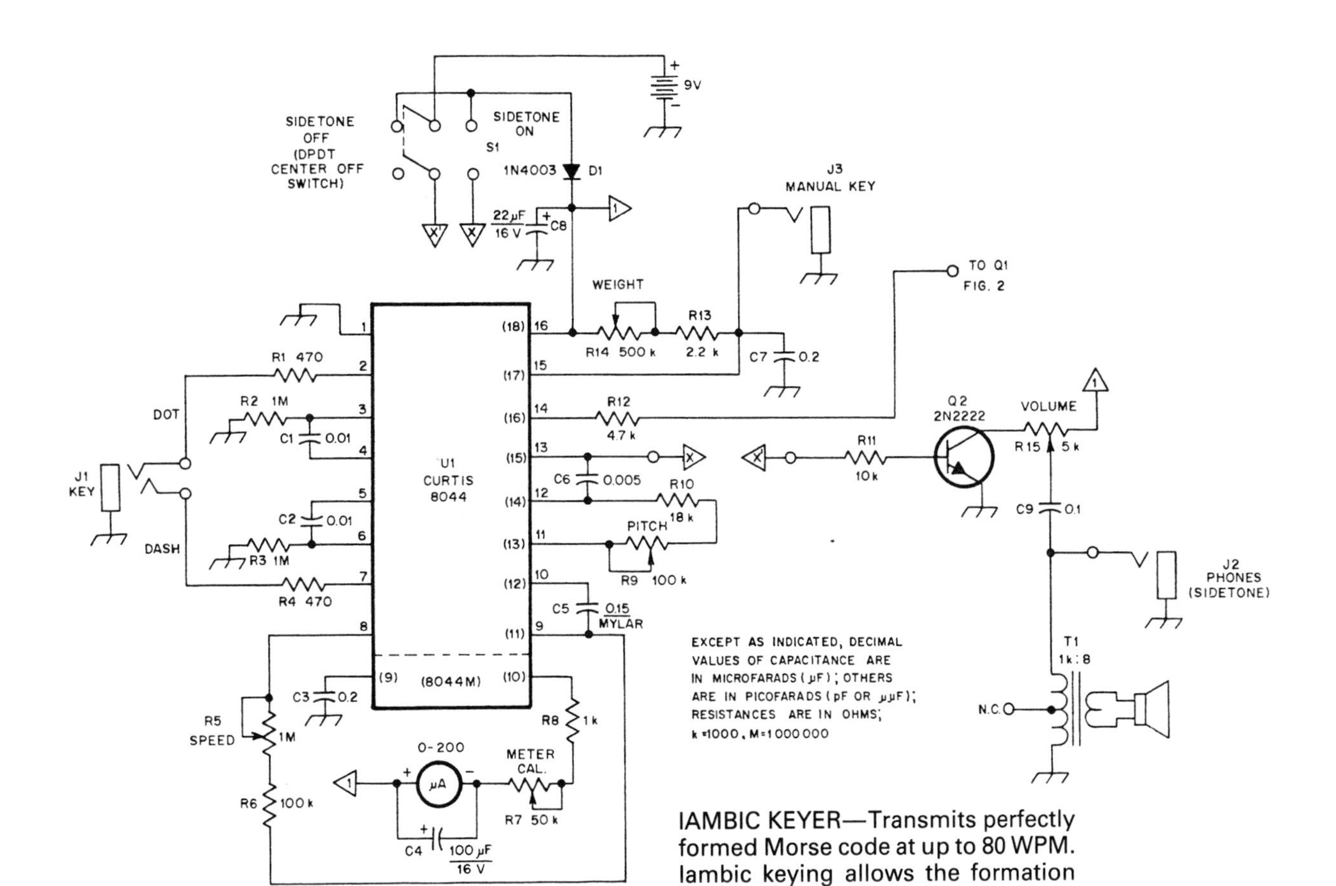

IAMBIC KEYER—Transmits perfectly formed Morse code at up to 80 WPM. Iambic keying allows the formation of characters by squeezing the key paddle rather than moving it from side to side or up and down.—B. Shriner and P. Pagel, CW on a Chip, *QST,* December 1983, pp. 16–19.

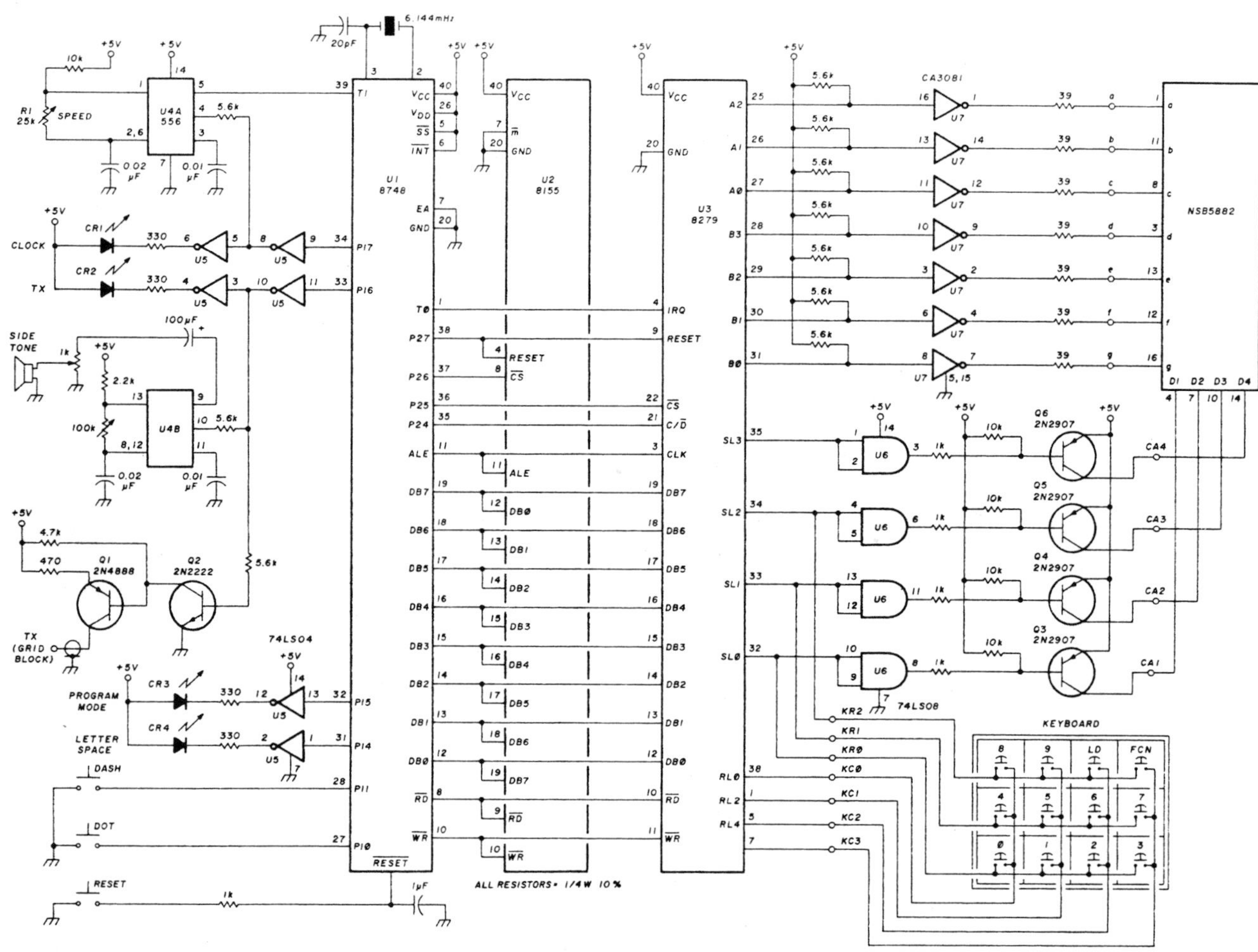

MICROCOMPUTER-BASED KEYER—Based on Intel 8748 programmable microcomputer, keyer offers 10 programmable messages and a four-digit contact counter. A 12-button keypad is used for control. The 8748 contains 1 K of EPROM, 64 bytes of RAM, a programmable counter, and three 8-bit I/O ports. RAM is expanded by 256 bytes through the 8155. The 8279 is a programmable keyboard/display interface. Both the 8155 and the 8279 are controlled by the 8748. The original article includes a flowchart for system software, and a listing of the code for the system is available. Keying speed range is 5–60 WPM. Four separate seven-segment, common-anode displays may be substituted for the NBS7882 display.—A. White, Microcomputer-Based Contest Keyer, *Ham Radio,* January 1981, pp. 36–42.

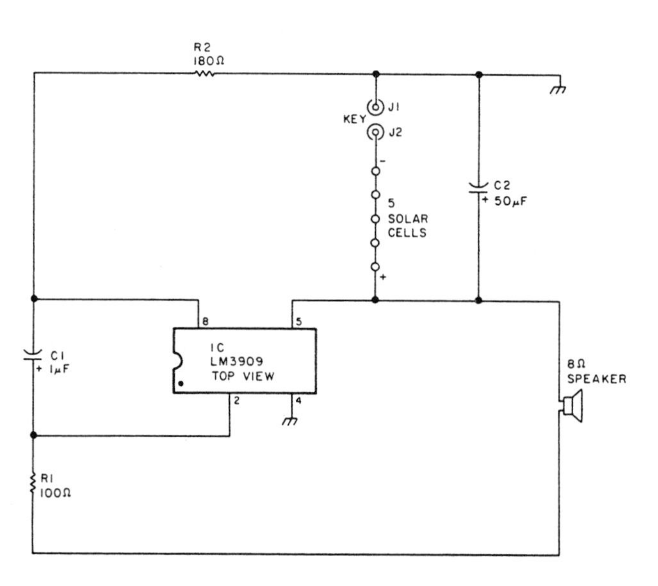

SOLAR-POWERED CODE PRACTICE OSCILLATOR—Unit for practicing Morse code derives its power solely from solar cells. The solar cells used in the original design were taken from the Solar Cell Experimenter's Assortment from Solar Amp, Inc.—H. Davidson, Code Practice: Have You Seen the Light?, *73 Magazine,* December 1983, pp. 10–12.

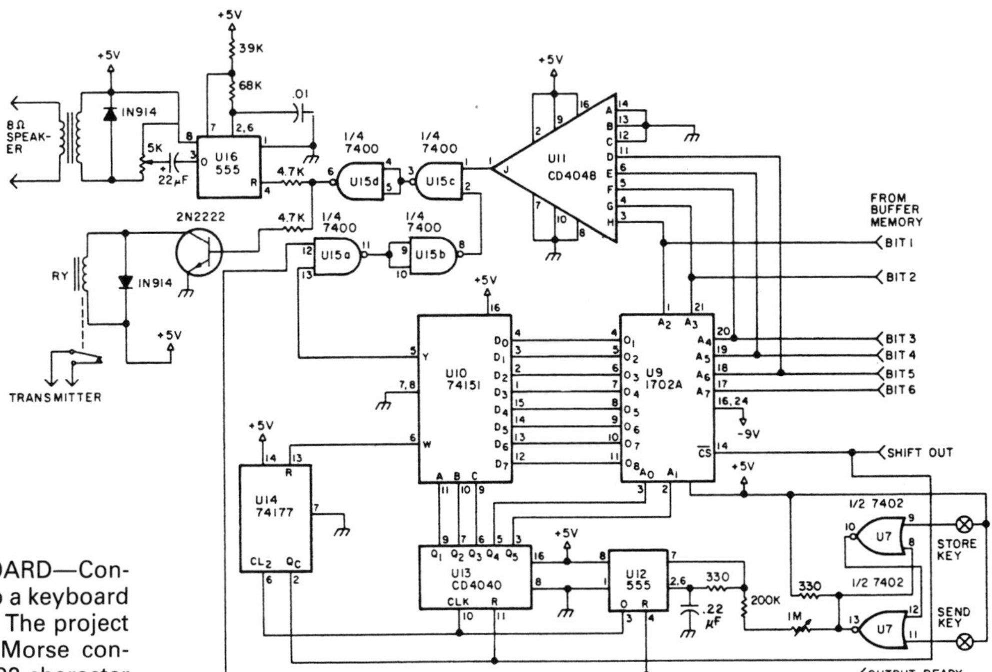

MORSE CODE KEYBOARD—Converts ASCII keyboard into a keyboard for sending Morse code. The project consists of an ASCII-to-Morse converter board and a 192-character buffer memory. The keyboard will send Morse with perfect character formation and spacing at speeds up to 80 WPM.—C. Dahlberg, Keyboard Your Way to Happiness, *73 Magazine,* November 1983, pp. 20–33.

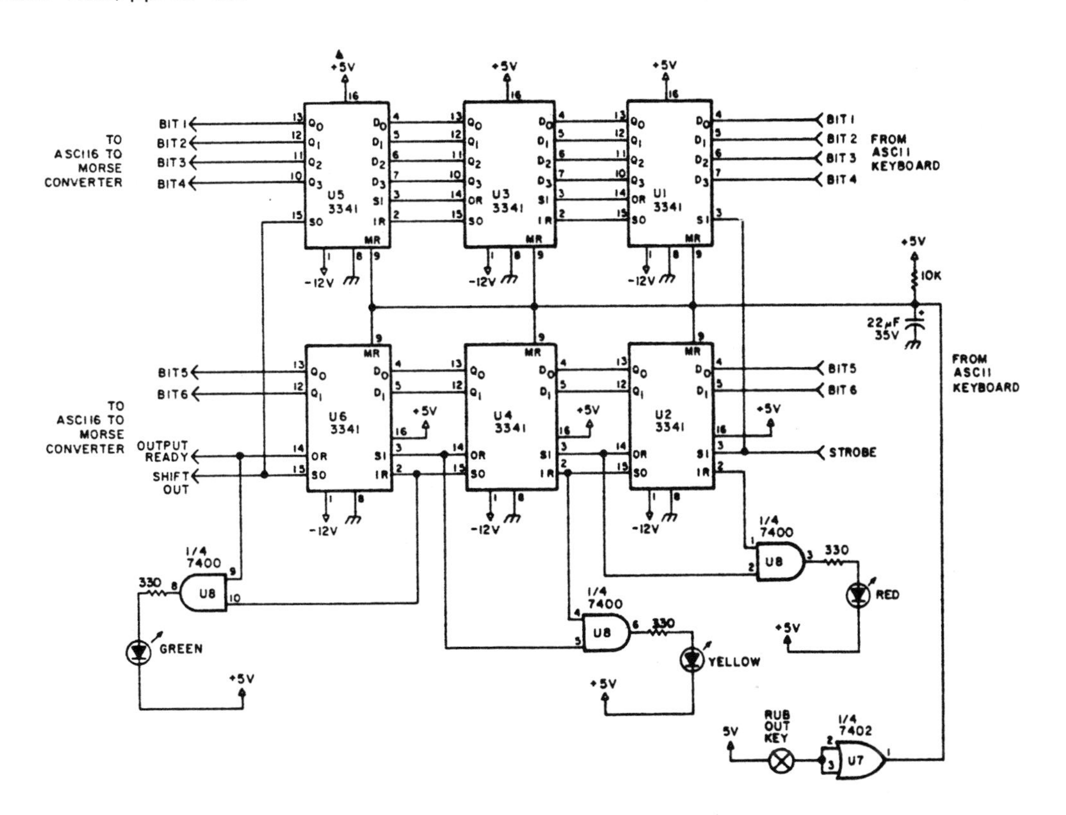

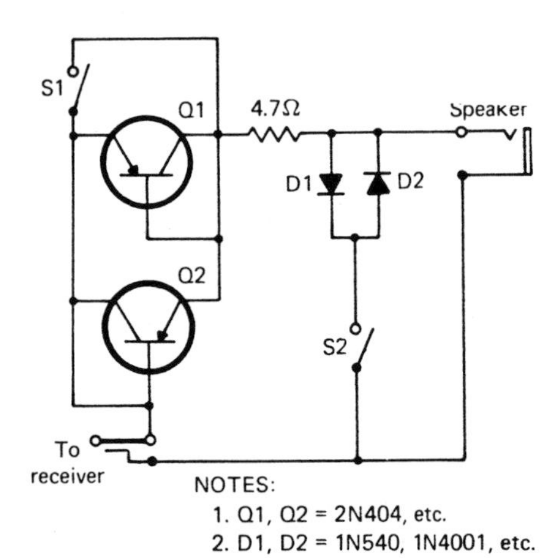

CW THRESHOLD GATE—No output results unless input signal exceeds about 0.2 V, resulting in silencing of background noise between characters of a CW signal. The circuit enhances the signal-to-noise ratio by as much as 40 dB. The circuit goes between receiver audio output and speaker/headphones. Q1 and Q2 are 2N404 or equivalent, and D1 and D2 are 1N540 or 1N4001.—F. Brown, The Threshold Gate: A C.W. Operator's Accessory, *CQ,* June 1983, p. 54.

PORTABLE MORSE CODE KEYBOARD—Designed for portable operation or with battery-powered CW transmitters. The circuit includes a type-ahead buffer of up to 48 characters. The speed is variable from 5 to over 35 WPM. The original article includes details on selecting and adapting an ASCII keyboard for use with the circuit.—T. Davies, QRP Key for Misers, *73 Magazine,* July 1983, pp. 58–63.

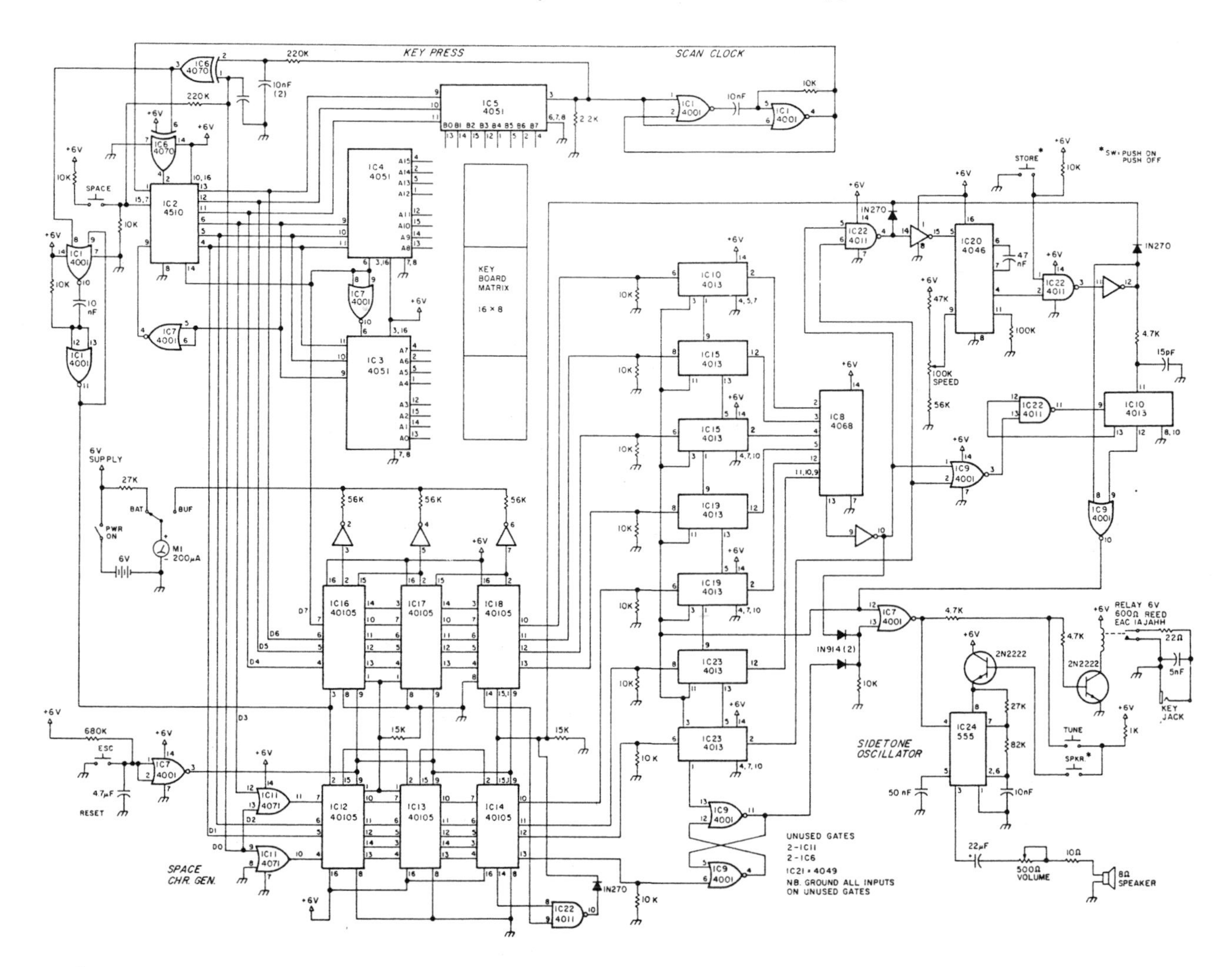

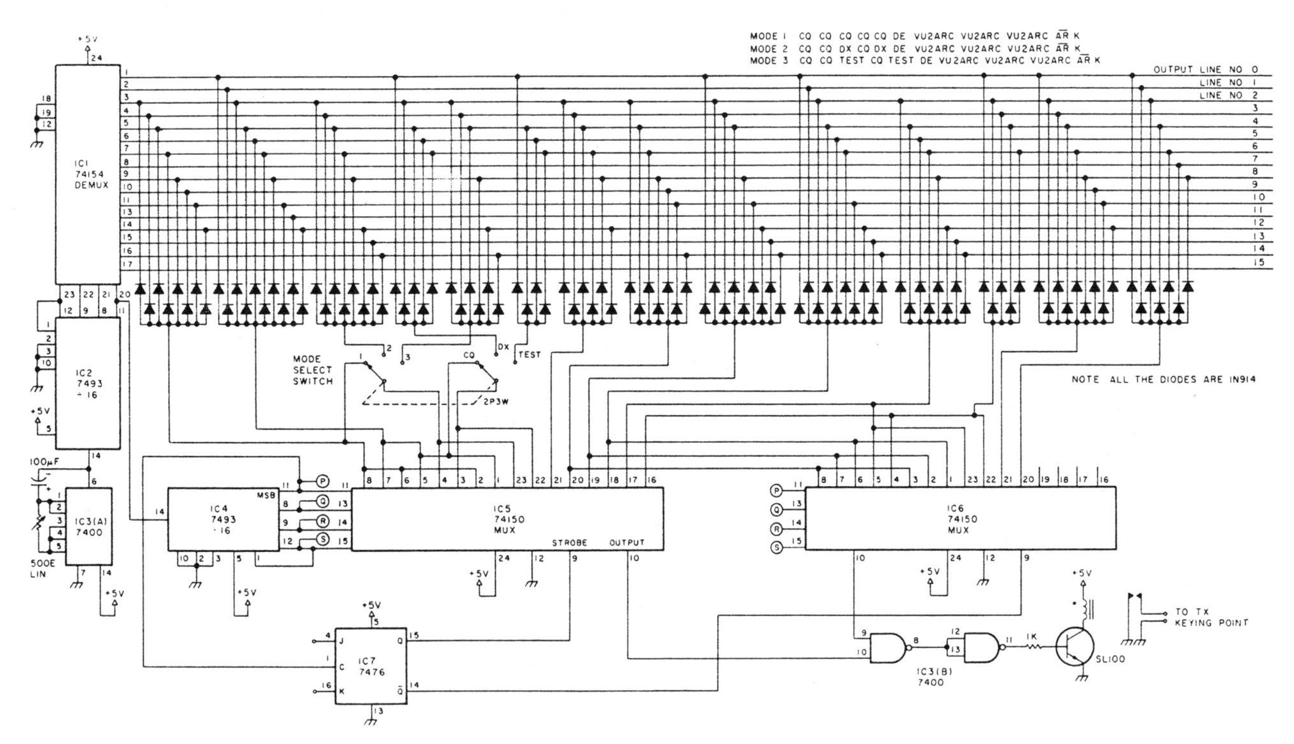

PROGRAMMABLE MORSE CODE KEYER—Diode matrix allows three short messages to be transmitted at speeds from 8 to 15 WPM. The original article gives details on programming the diode matrix.—C. Gururaj, The ROM-less, REAM-less, CQ Sender, *73 Magazine,* July 1983, pp. 90–93.

KEYBOARD KEYER—Allows transmission of Morse code from a keyboard at up to 60 WPM. The design includes all letters and numbers plus punctuation and Morse code procedure characters. Automatic letter spacing is also included, although no character buffer is in the design. The original article includes details on designing and installing the diode matrix for the keyboard itself.—W. Jones, Build the Billboard Keyboard Keyer, *73 Magazine,* July 1983, pp. 44–46.

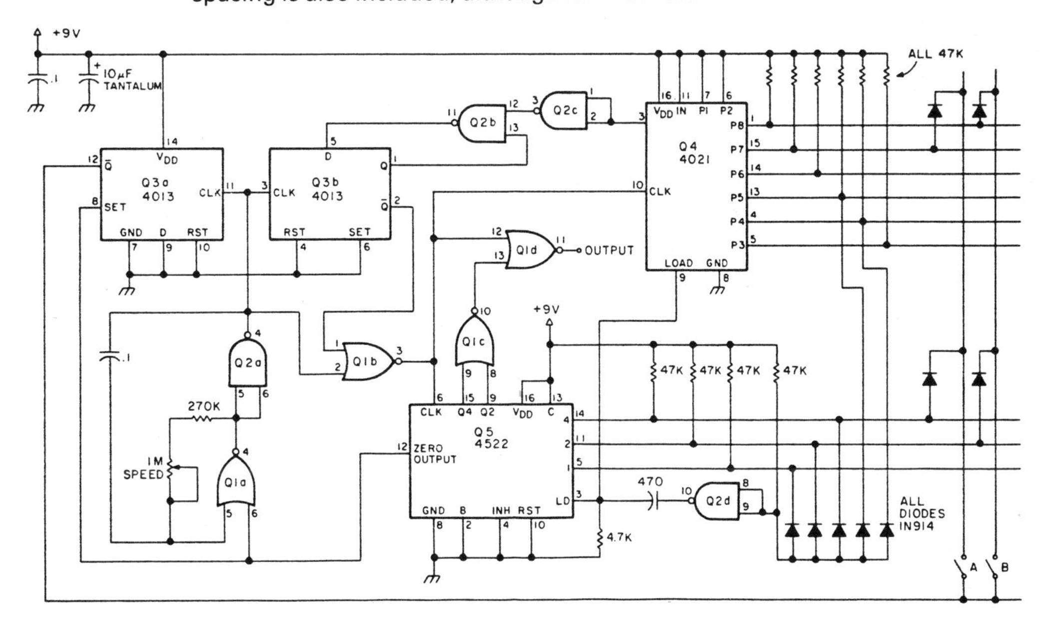

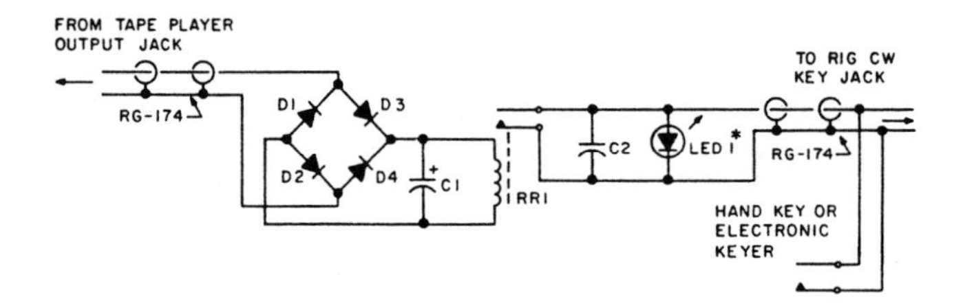

AUDIO KEYER—Accepts Morse code signals as on/off audio tones and uses them to key a CW transmitter. D1 through D4 are rectifier diodes and RR1 is a reed relay. C1 is a 200-μF 15-V electrolytic capacitor, while C2 is a 0.01-μF disc capacitor.—W. Edwards, Inexpensive Memory Keyer, *73 Magazine,* December 1983, p. 99.

15

Oscillator Circuits

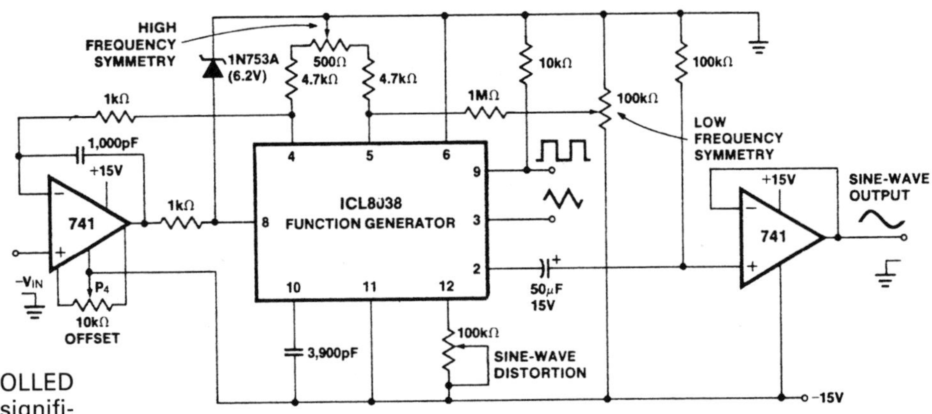

LINEAR VOLTAGE CONTROLLED OSCILLATOR—741 op amp significantly improves the linearity of input sweep voltage versus output frequency. The ICL8038 is a precision waveform generator/voltage controlled oscillator device.—"Intersil Data Book," Intersil, Cupertino, CA, 1981, p. 5-196.

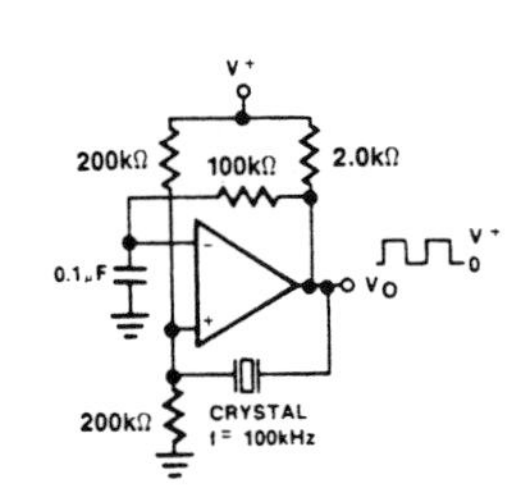

100-kHz CRYSTAL CONTROLLED OSCILLATOR—Uses one section of LM139A quad voltage comparator IC. The supply voltage can range from 2 to 36 V DC. Inputs of unused comparators on the device should be grounded.—"Signetics Linear LSI Data and Applications Manual," Signetics, Sunnyvale, CA, 1985, p. 4-115.

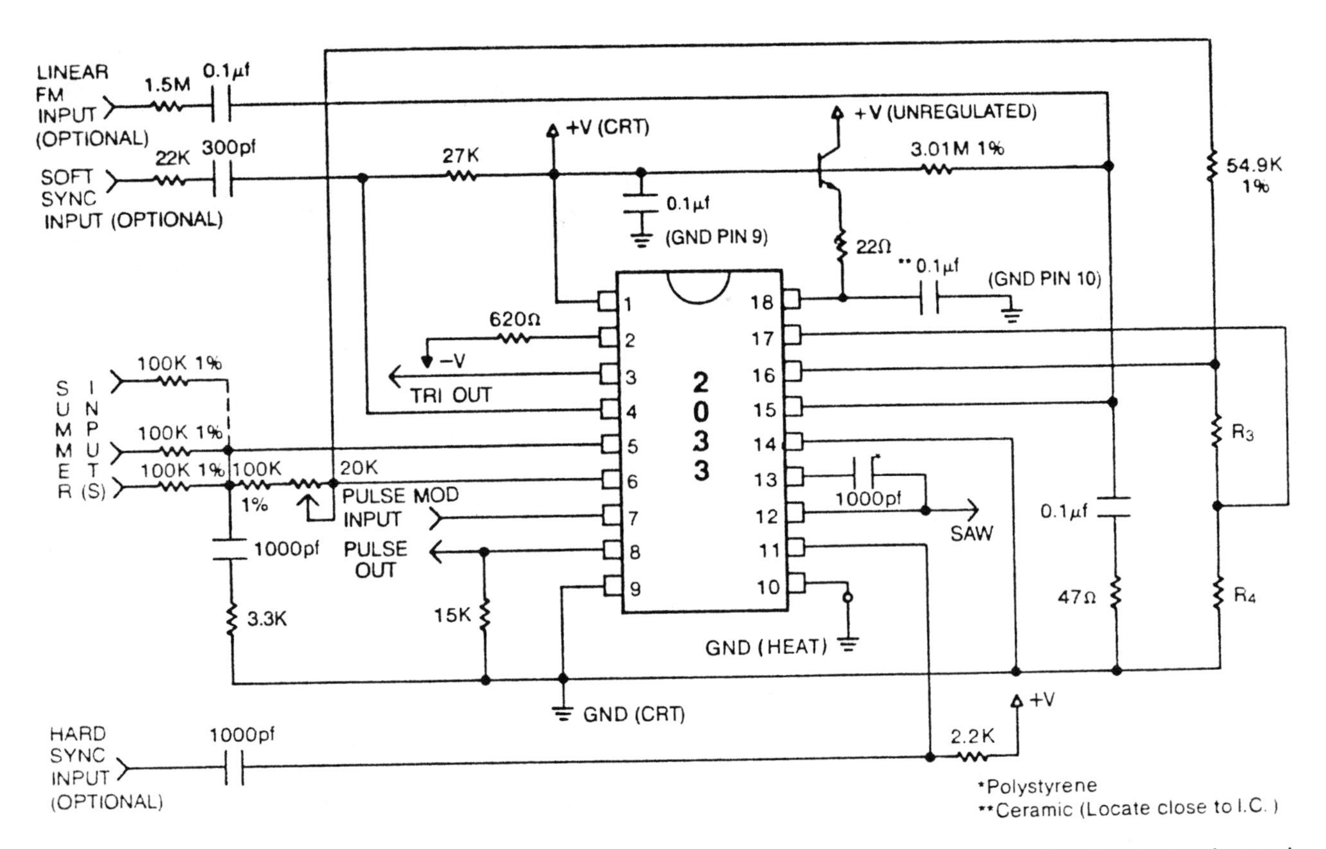

VOLTAGE CONTROLLED OSCILLATOR—SSM 2033 precision voltage controlled oscillator allows operating frequency to be swept over 500,000 : 1 range by simultaneous linear and exponential control inputs. On-chip low-input bias summer and control op amps have been provided. Sawtooth, triangle, and variable width pulse outputs are provided. The circuit was designed for tone generation in electronic music systems.—"Voltage Controlled Oscillator," Solid State Micro Technology for Music, Santa Clara, CA, 1980, SSM 2033.

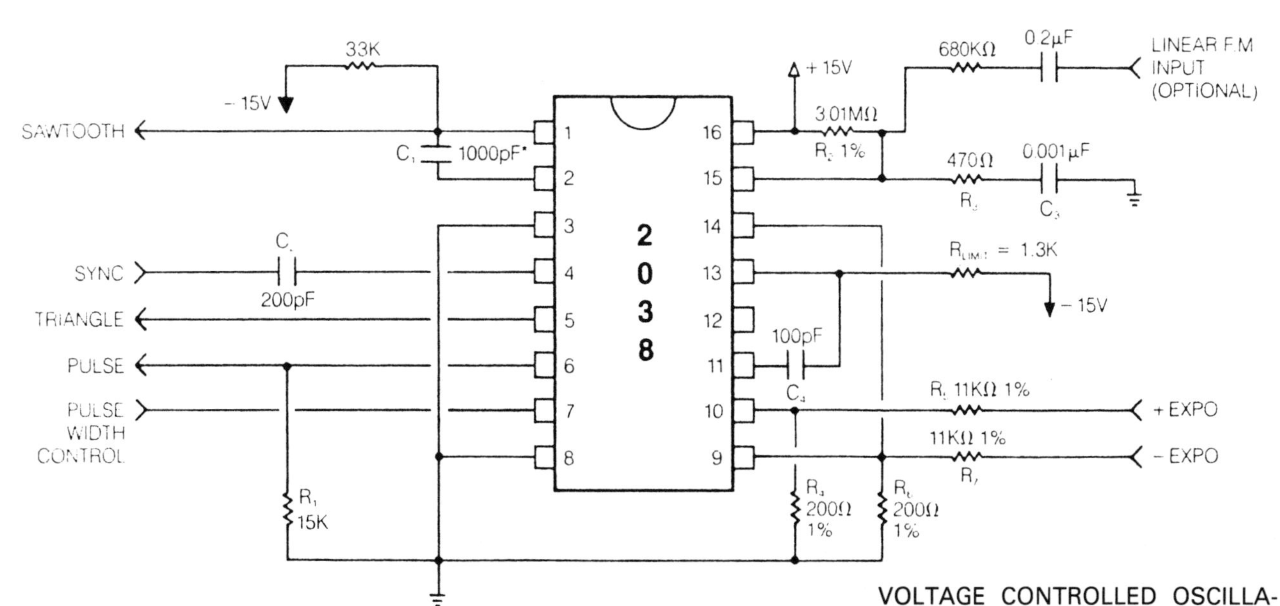

*POLYSTYRENE

VOLTAGE CONTROLLED OSCILLATOR—SSM 2038 precision voltage controlled oscillator allows operating frequency to be swept over 5000 : 1 range by simultaneous linear and exponential control inputs. Sawtooth, triangle, and pulse buffered outputs are provided. The pulse output has a voltage controlled duty cycle from 0 to 100%; clean edges are guaranteed by a comparator with internal hysteresis. The circuit was designed for electronic music systems.—"Voltage Controlled Oscillator," Solid State Micro Technology for Music, Santa Clara, CA, 1983, SSM 2038.

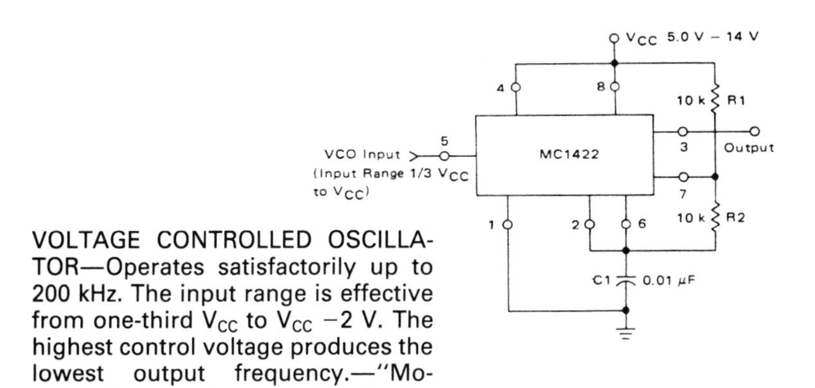

VOLTAGE CONTROLLED OSCILLATOR—Operates satisfactorily up to 200 kHz. The input range is effective from one-third V_{CC} to V_{CC} −2 V. The highest control voltage produces the lowest output frequency.—"Motorola Linear Integrated Circuits Databook," Motorola, Phoenix, AZ, 1979, p. 6-17.

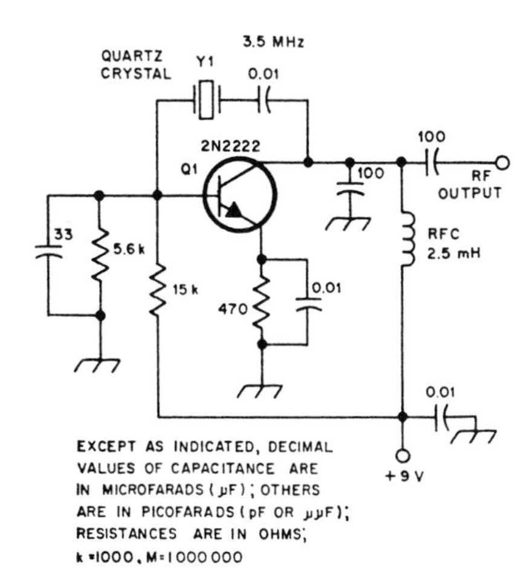

CRYSTAL CONTROLLED PIERCE OSCILLATOR—Oscillates at 3500 kHz and may be used as a lab project, wireless code practice oscillator, or as a very-low-power CW transmitter. The key is added by opening the ground connection for the 470-Ω resistor.—D. DeMaw, The Basics of Transmitters, *QST,* November 1984, pp. 40–44.

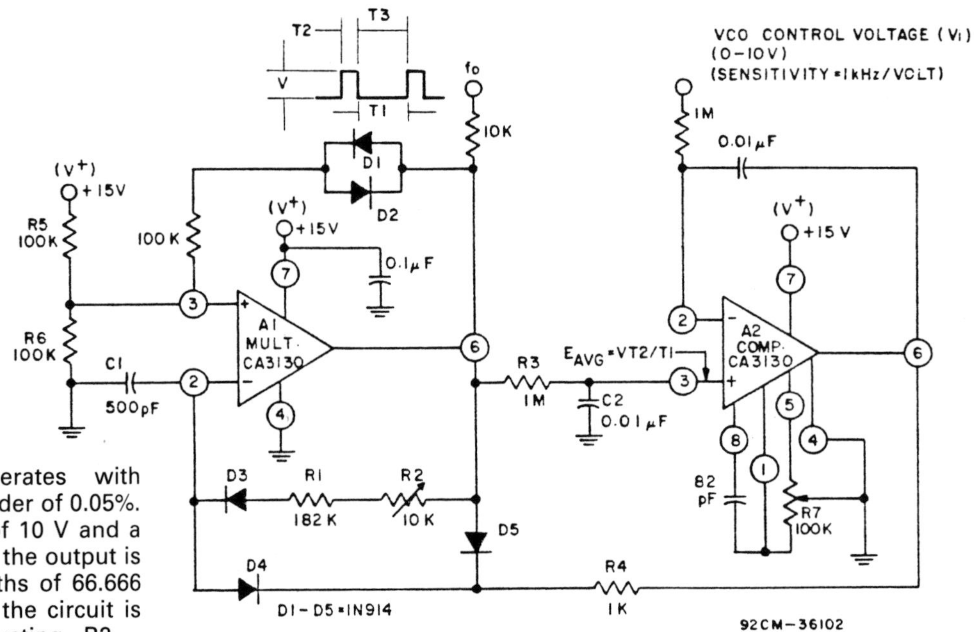

PRECISION VCO—Operates with tracking error on the order of 0.05%. For a control voltage of 10 V and a supply voltage of 15 V, the output is 10 kHz with pulse widths of 66.666 ms. The calibration of the circuit is accomplished by adjusting R2.—"Understanding And Using the CA3130, CA3130A and CA3130B Bi-MOS Operational Amplifiers," RCA Solid State Division, Somerville, NJ, 1983, Application Note ICAN-6386.

THREE-OUTPUT VARIABLE AUDIO OSCILLATOR—Provides square, triangle, and sine wave outputs from 20 Hz to approximately 20 kHz. The circuit uses an ICL8038 precision waveform generator/voltage controlled oscillator device.—"Intersil Data Book," Intersil, Cupertino, CA, 1981, p. 5-196.

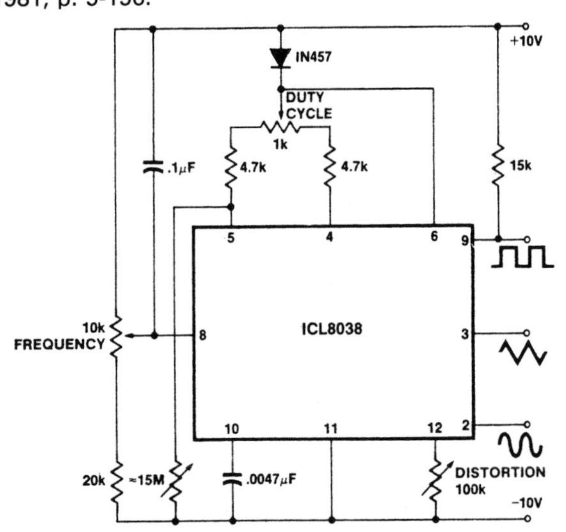

TWIN-T AUDIO OSCILLATOR—Used to generate modulated CW so Morse code may be transmitted via SSB or AM. The circuit was originally designed for use with a microcomputer but it may be adapted whenever a steady sine wave audio signal is needed.—M. Leavey, RTTY Loop, *73 Magazine,* November 1984, pp. 71–72.

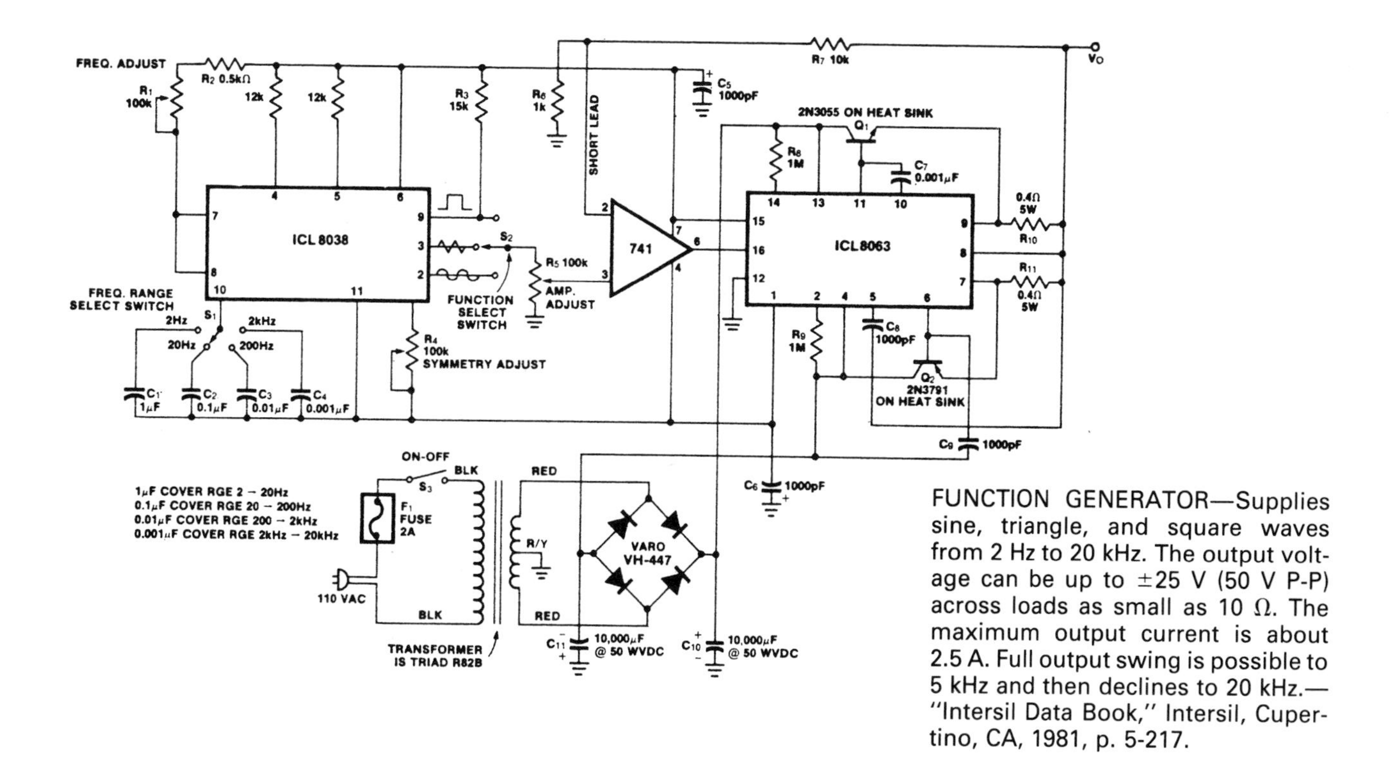

FUNCTION GENERATOR—Supplies sine, triangle, and square waves from 2 Hz to 20 kHz. The output voltage can be up to ±25 V (50 V P-P) across loads as small as 10 Ω. The maximum output current is about 2.5 A. Full output swing is possible to 5 kHz and then declines to 20 kHz.—"Intersil Data Book," Intersil, Cupertino, CA, 1981, p. 5-217.

TRIANGLE/SQUARE WAVE GENERATOR—Frequency and duty cycle are virtually independent of power supply variations since output range swings exactly from rail to rail. The operating voltage can range from ±0.5 to ±8 V.—"Intersil Data Book," Intersil, Cupertino, CA, 1981, p. 5-151.

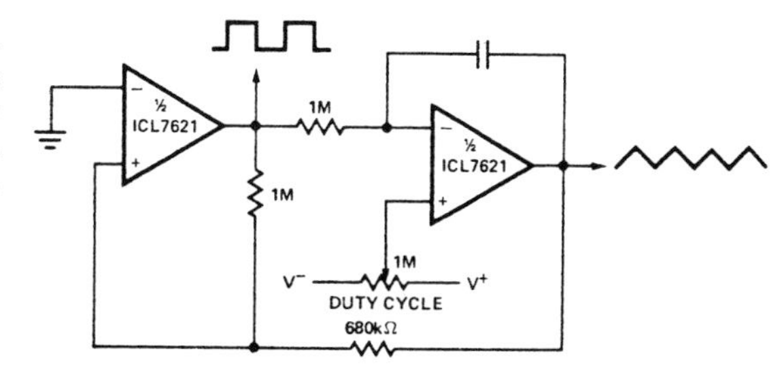

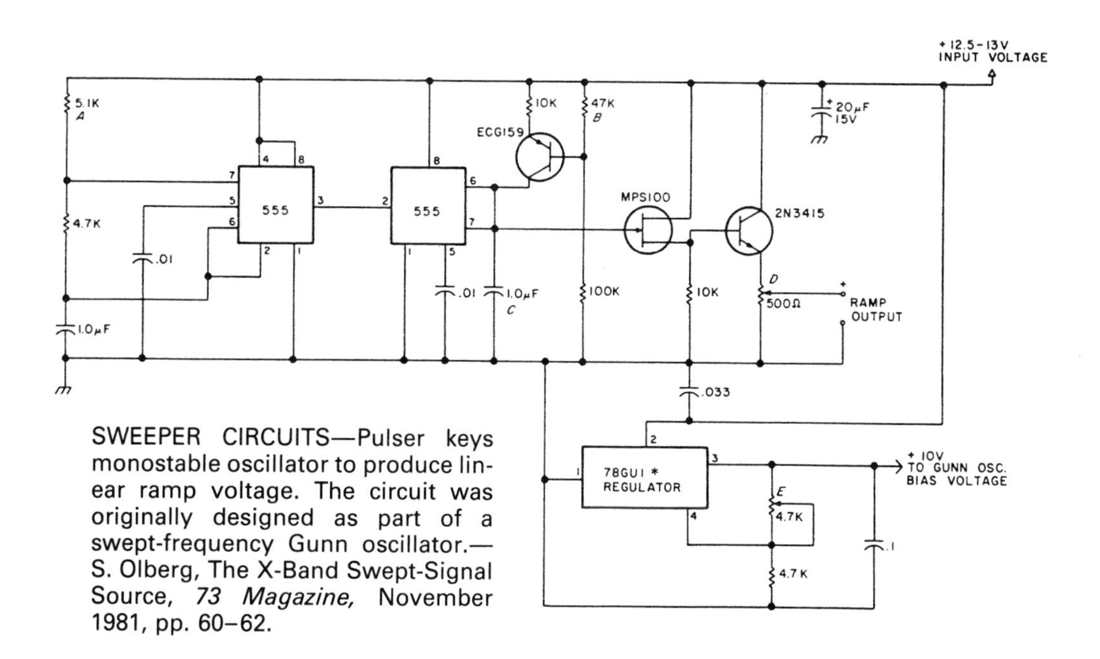

SWEEPER CIRCUITS—Pulser keys monostable oscillator to produce linear ramp voltage. The circuit was originally designed as part of a swept-frequency Gunn oscillator.—S. Olberg, The X-Band Swept-Signal Source, *73 Magazine,* November 1981, pp. 60–62.

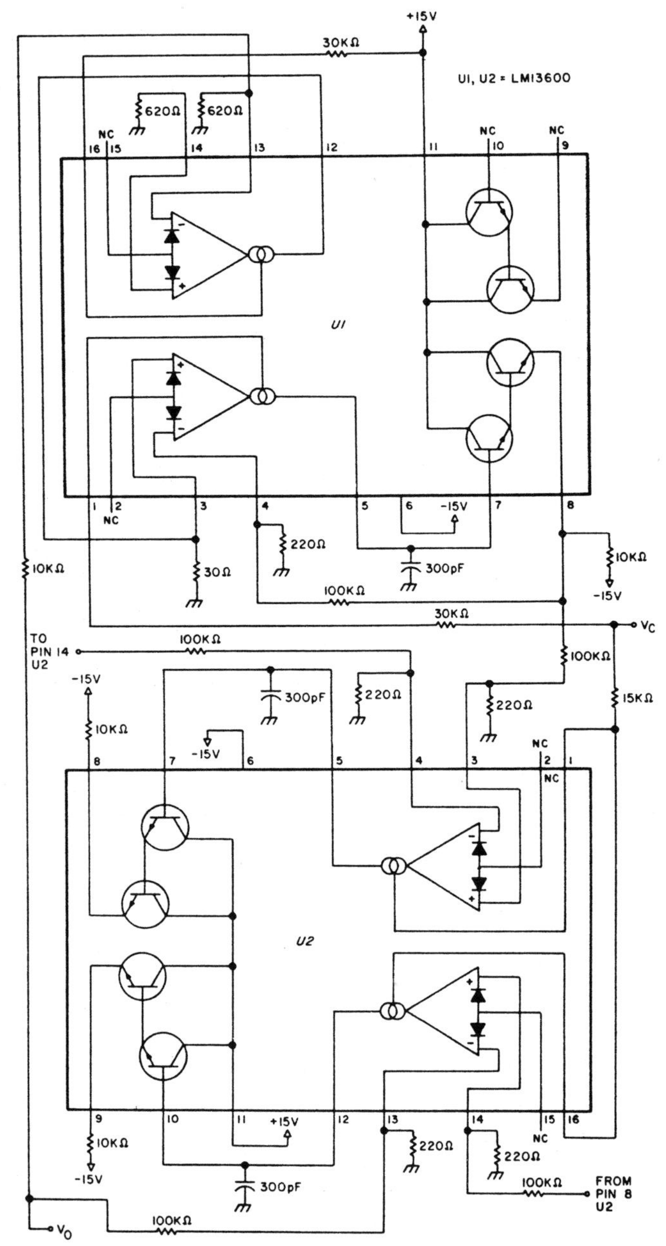

SINUSOIDAL VOLTAGE CONTROLLED OSCILLATOR—Frequency range is from 5 Hz to 50 kHz, depending upon input voltage. The THD is less than 1%.—P. Cavanaugh, Sinusoidal VCO, *73 Magazine,* December 1983, p. 115.

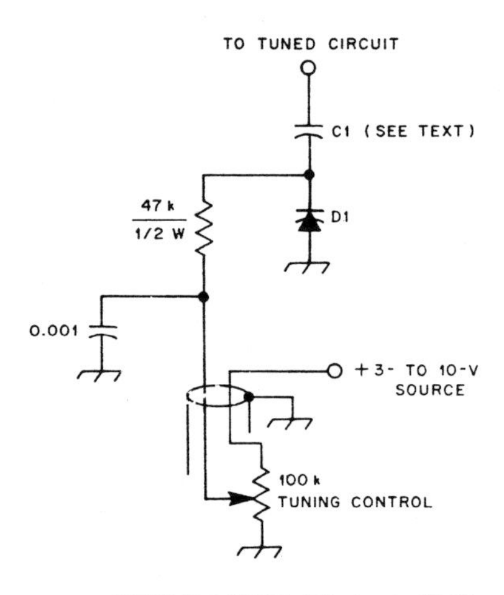

VARACTOR REMOTE TUNING—Varactor acts as variable capacitor if reverse bias is applied. D1 is an SK3327 which has a maximum capacitance of 33 pF and a minimum of 3.2 pF. This capacitance may be adjusted by the value of C1 if necessary. The 100-K potentiometer varies the capacitance. The circuit may be used with VFOs, BFOs, crystal oscillators, etc.—J. Rice, Remote Tuning with a Varactor, *QST,* April 1982, pp. 52–53.

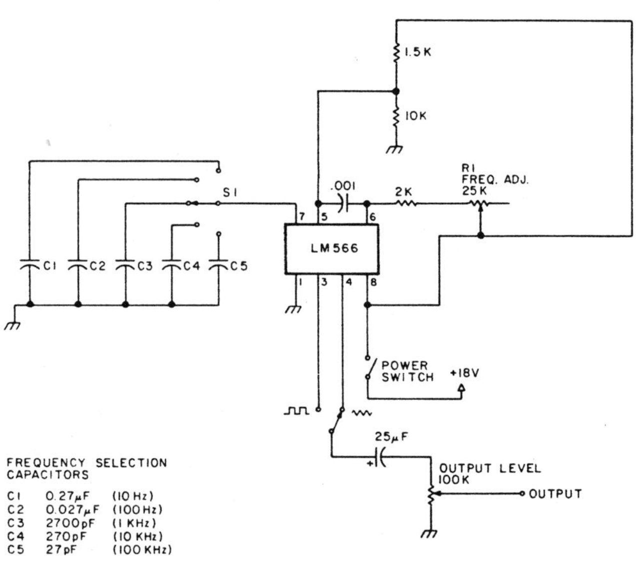

AUDIO OSCILLATOR—Generates square or triangle waves from 10 Hz to 100 kHz, depending on values of C1 through C5.—R. Miara, The Micro-Generator, *73 Magazine,* July 1981, p. 94.

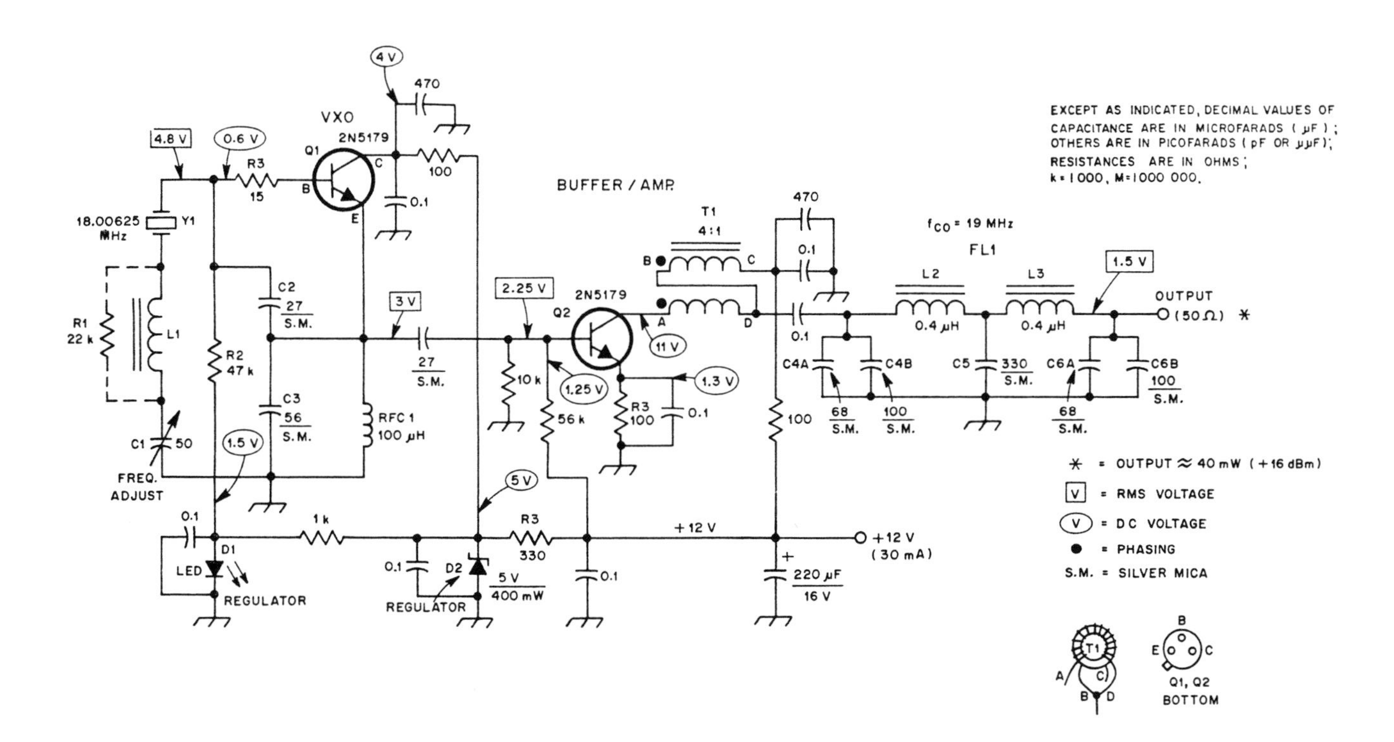

Typical L and C Values for Various Operating Frequencies

Crystal (Y1) Range (MHz)	C2 (pF)	C3 (pF)	C4, C6 (pF)	C5 (pF)	L1	L2, L3	T1
6.000 to 9.000	68	100	390 (for 40-meter use)	820	17 μH max. 55 ts. of no. 26 wire on a T68-2 toroid core.	1 μH. 18 ts. no. 26 wire on a T37-6 toroid core. (40-meter use)	15 bilifar ts. of no. 26 on an FT-37-43 toroid core.
9.000 to 15.000	39	68	287 (for 30-meter use)	560	12 μH max. 49 ts. of no. 28 wire on a T50-2 toroid core	0.72 μH. 15 ts. no. 26 wire on a T37-6 toroid core. (30-meter use)	15 bilifar ts. of no. 26 on an FT-37-61 toroid core.
9.000 to 15.000	39	68	212 (for 20-meter use)	424	12 μH max. Same as above.	0.53 μH. 13 ts. no. 24 wire on a T37-6 toroid core. (20-meter use)	Same as above on FT-37-61 toroid core.
15.000 to 20.000	27	56	168	330	7 μH max. 42 ts. of no. 26 wire on a T50-6 toroid core.	0.42 μH. 12 ts. no. 26 wire on a T37-6 toroid core. (for 18-MHz use)	Same as above on FT-37-61 toroid core.

UNIVERSAL VXO—Capable of operating between 6 to 20 MHz, depending on values of various components. It may be used as the frequency controlling elements in an HF CW transmitter or multiplied to operate on VHF/UHF. D1 is a 1.5-V regulator.—D. DeMaw, A Beginner's Look at Basic Oscillators, *QST,* February 1984, pp. 35–40.

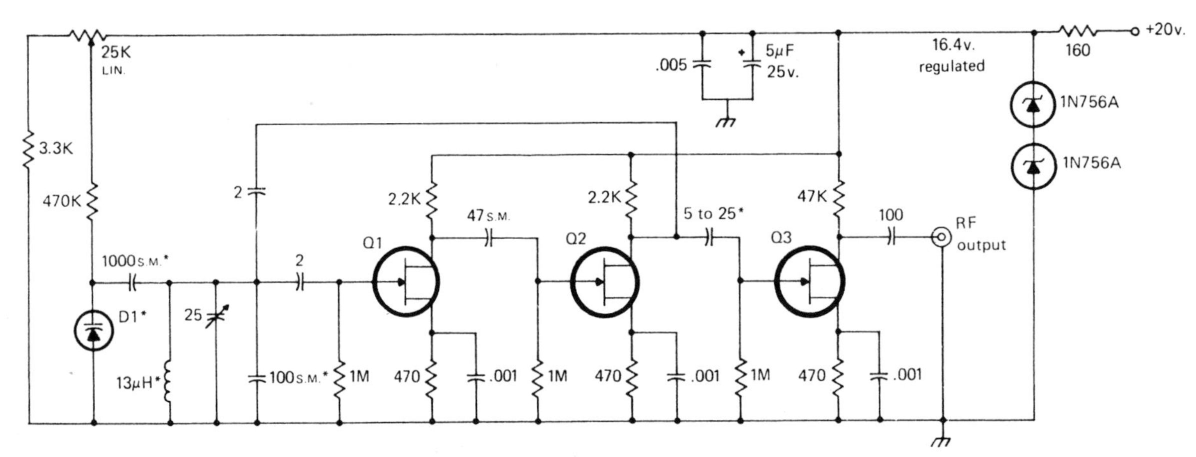

"FRANKLIN" JFET VFO—Capable of covering 80-, 40-, and 20-meter amateur radio bands, depending upon component values and bias applied to D1. The circuit is an adaptation of the "Franklin" oscillator developed with vacuum tubes. N-type JFETs such as the 2N5163 are used for Q1, Q2, and Q3. D1 may be a 1N3182, 1N5450, or equivalent. The 13-μH inductor is 1.6 in in diameter, 2 in long, and formed from 23 turns of No. 16 wire. The circuit illustrated covers 3500–3850 kHz and has a peak output of 5.9–6 V RF. The D1 bias can be 2.2–11.2 V. Stability is good, although the operating environment temperature should be stable. The original article includes details on adapting the circuit for other amateur radio bands.—J. Park, The Franklin VFO—With JFETs, *CQ,* January 1982, pp. 58–60.

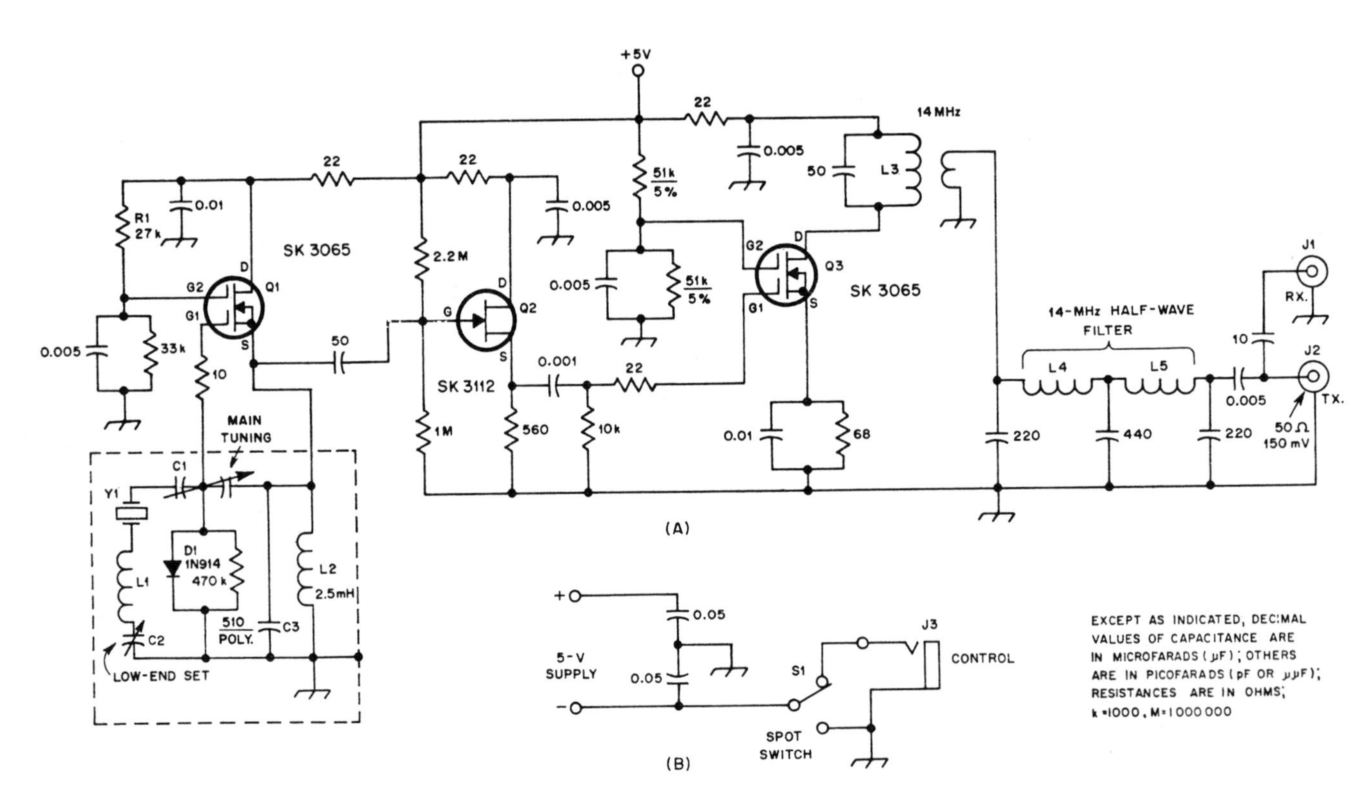

VARIABLE CRYSTAL OSCILLATOR—Covers 14,000–14,025 kHz with greater stability than conventional inductively controlled oscillators. The frequency of crystal Y1 is "pulled" over the operating range by the circuit. C1 is a dual 140-pF tuning capacitor such as a Hammarlund MCD-140-M. C2 is a 140-pF air variable capacitor. L1 is 12 μH and formed from Barker and Williamson "Mininductor" No. 3004, 32 turns per inch of No. 24 wire wound 2 in long and 0.5 in in diameter. L3 is a Miller No. 4404 or equivalent with two turns of hookup wire wound over the cold end of the coil. L4 and L5 are both six turns of hookup wire wound on the same form as L3; the value for each is 0.57 μH. Y1 is a fundamental quartz crystal, type HC6/U, cut for 14,030 kHz with 32 pF.—F. Noble, A Variable Frequency Crystal Oscillator, *QST,* March 1981, pp. 34–37.

TWO-PHASE SINE WAVE OSCILLATOR—Uses two-pole pass Butterworth filter followed by phase shifting single-pole stage fed back through a voltage limiter to achieve sine and cosine outputs. The circuit has oscillation frequency of 2 kHz.—"Signetics Analog Applications Manual," Signetics, Sunnyvale, CA, 1979, pp. 38–40.

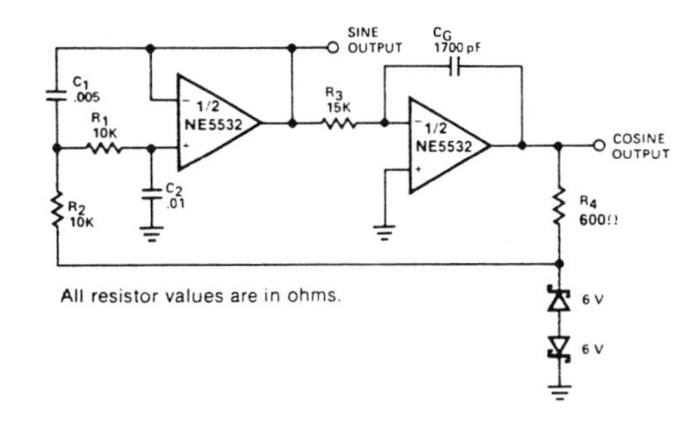

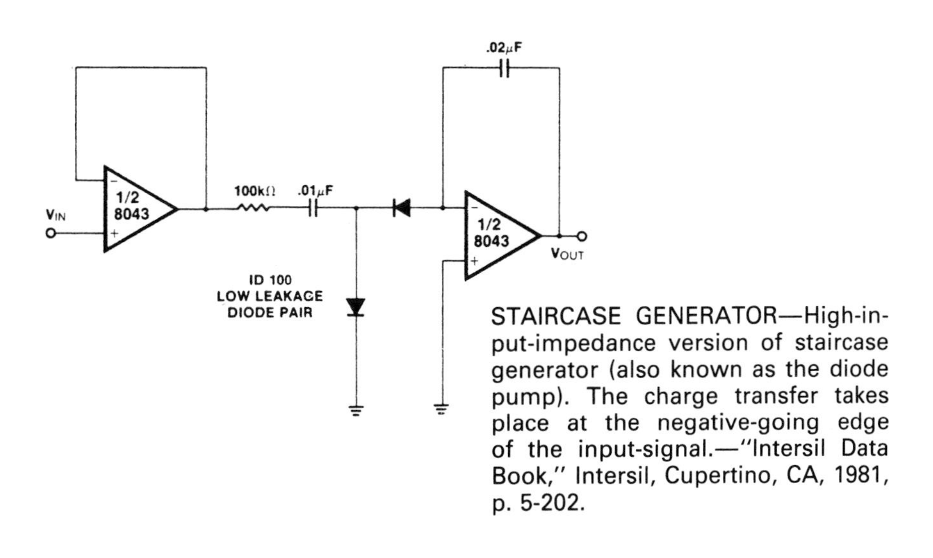

STAIRCASE GENERATOR—High-input-impedance version of staircase generator (also known as the diode pump). The charge transfer takes place at the negative-going edge of the input-signal.—"Intersil Data Book," Intersil, Cupertino, CA, 1981, p. 5-202.

16

Power Supply Circuits

400-Hz POWER SUPPLY—Designed to power radio compass units; operating frequency of 555 timer can be adjusted over wide range through R1. U1 is a 6-A, 50 PIV full wave bridge rectifier. T1 is a 117-V primary, 25-V secondary transformer used to "step up" output voltage. T2 is a 117-V primary, 12- to 18-V secondary transformer with at least a 4-A rating.—G. Wasson, A 400 Hz Power Supply for a Radio Compass, *QST,* July 1983, p. 38.

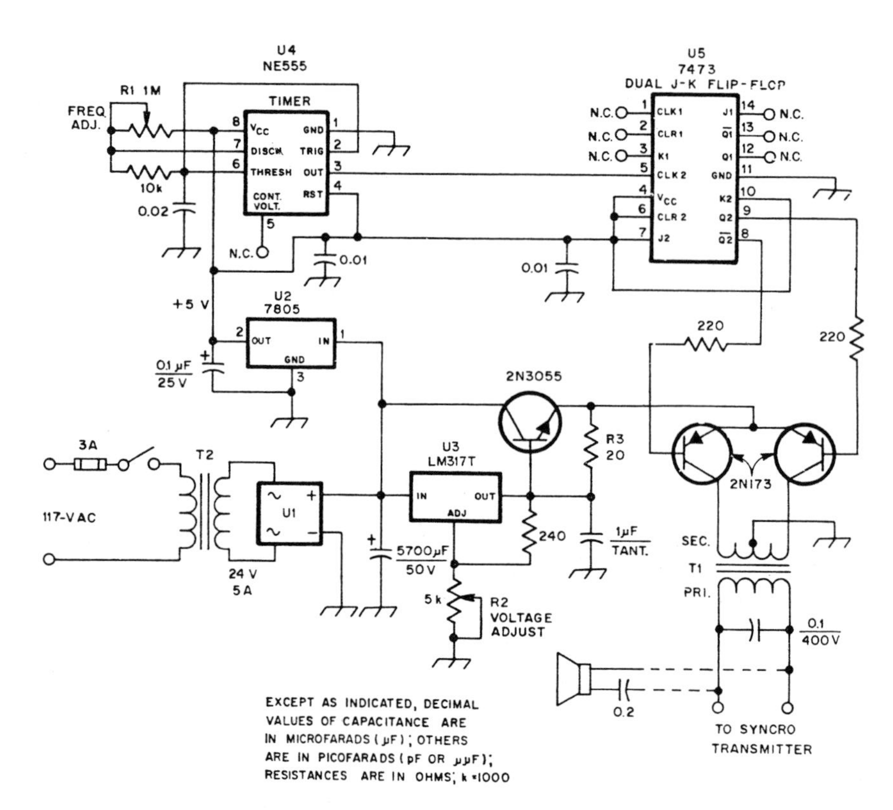

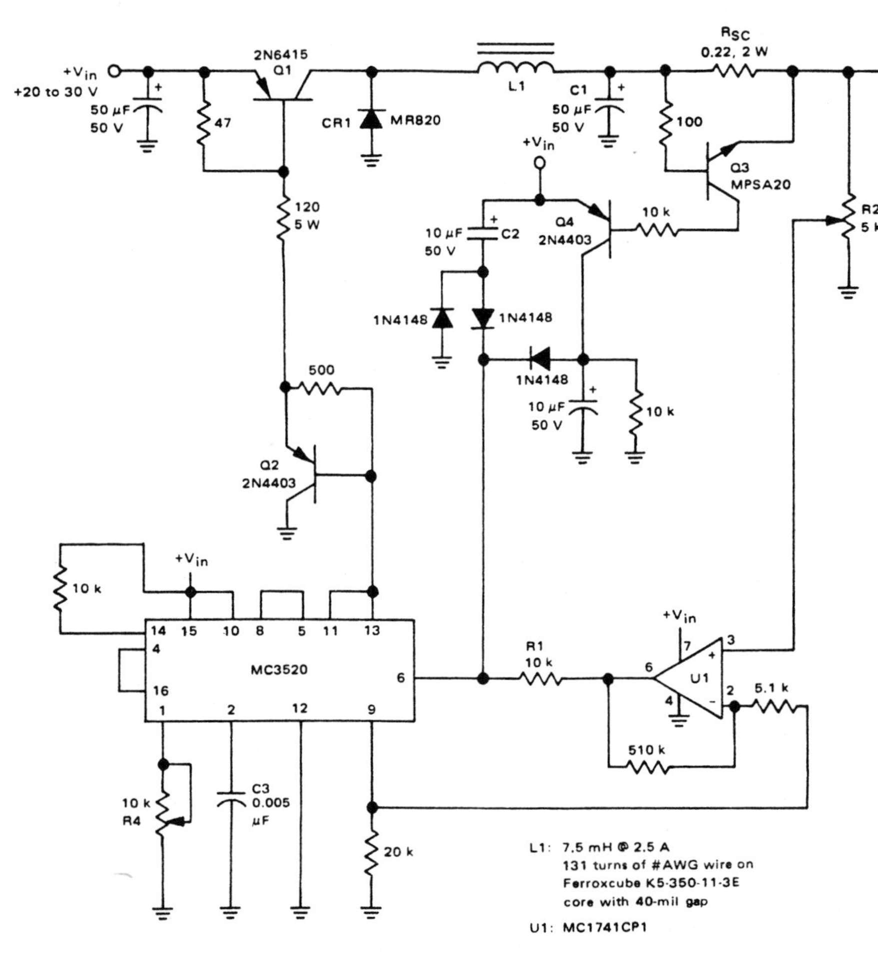

DC-TO-DC CONVERTER—PWM switching power supply uses single series switching element. Q1 chops the DC input voltage at a frequency of approximately 25 kHz, and the resulting waveform is filtered by L1 and C1 to provide the DC output voltage. U1 is an MC1741CP1 or equivalent. L1 is 7.5 mH at 2.5 A and is made from 131 turns of No. AWG wire on a Ferroxcube K5-350-11-3E core with a 40-mil gap.—"Motorola Linear Integrated Circuits Databook," Motorola, Phoenix, AZ, 1979, p. 4-121.

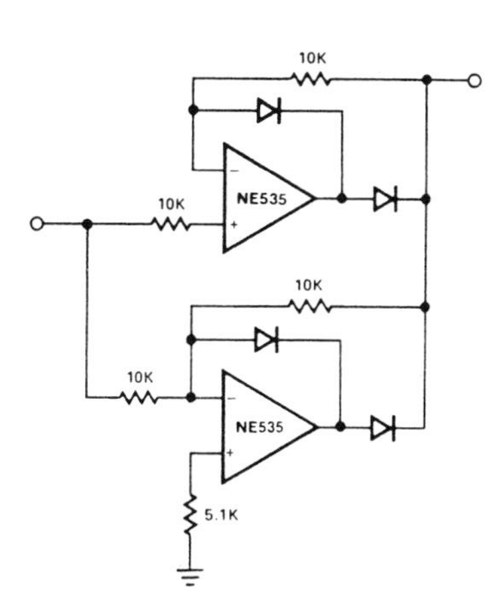

PRECISION FULL WAVE RECTIFIER—Provides accurate full wave rectification with low-output impedances for both input polarities. The output will not sink heavy currents, except a small amount through the 10-K resistors; the load applied should be referenced to ground or to a negative voltage. Reversing all diode polarities will reverse the output polarity.—"Signetics Analog Applications Manual," Signetics, Sunnyvale, CA, 1979, pp. 37–39.

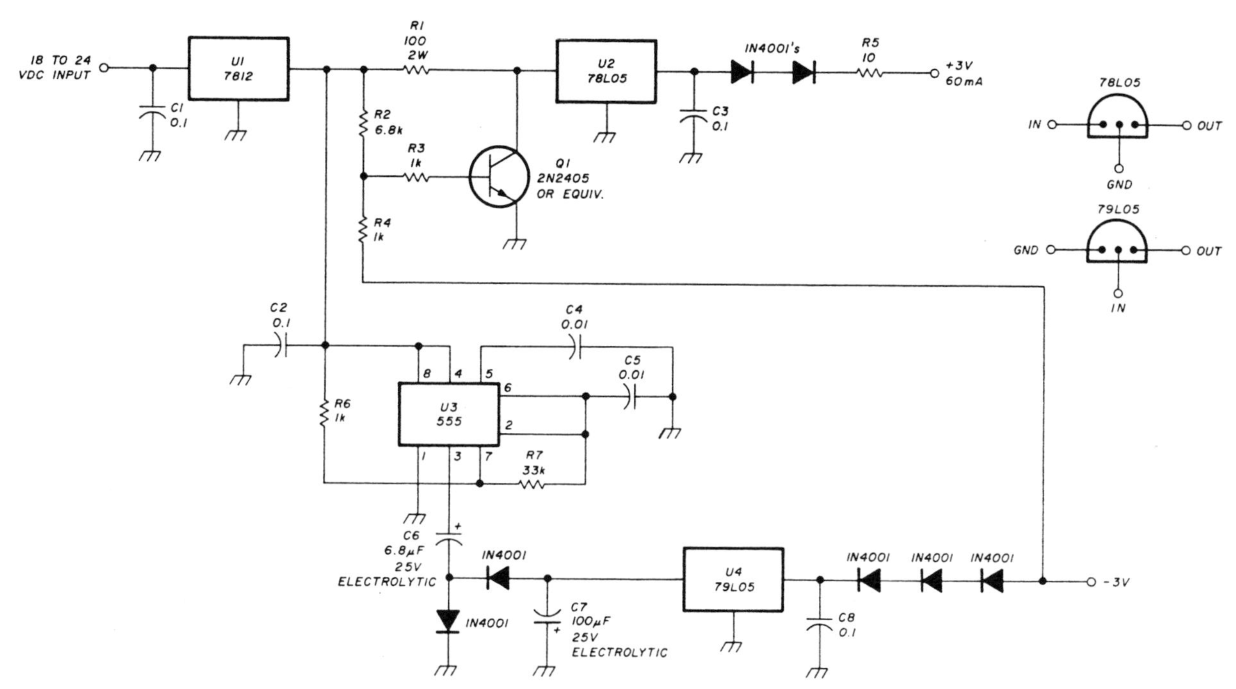

GaAs FET CIRCUIT POWER SUPPLY—Designed for circuits using GaAs FETs (such as receiving preamplifiers). The supply furnishes back (negative) bias first; without back-bias, GaAs FETs may go into saturation and be destroyed. The supply also maintains reverse bias through C3 when the power is switched off until the positive current has run down.—N. Foot, Safe Power for Your Low-Noise GaAs FET Amplifier, *Ham Radio,* November 1982, pp. 18–20.

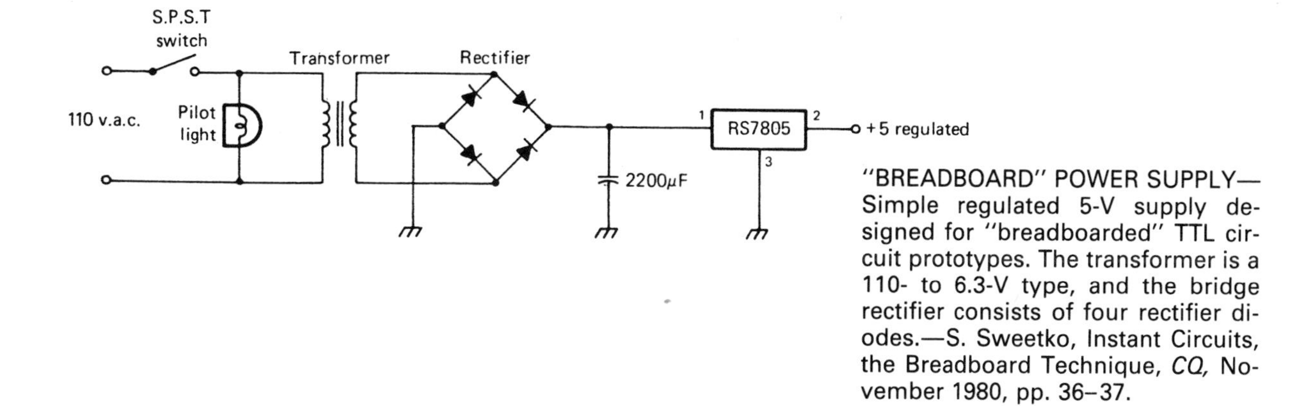

"BREADBOARD" POWER SUPPLY—Simple regulated 5-V supply designed for "breadboarded" TTL circuit prototypes. The transformer is a 110- to 6.3-V type, and the bridge rectifier consists of four rectifier diodes.—S. Sweetko, Instant Circuits, the Breadboard Technique, *CQ,* November 1980, pp. 36–37.

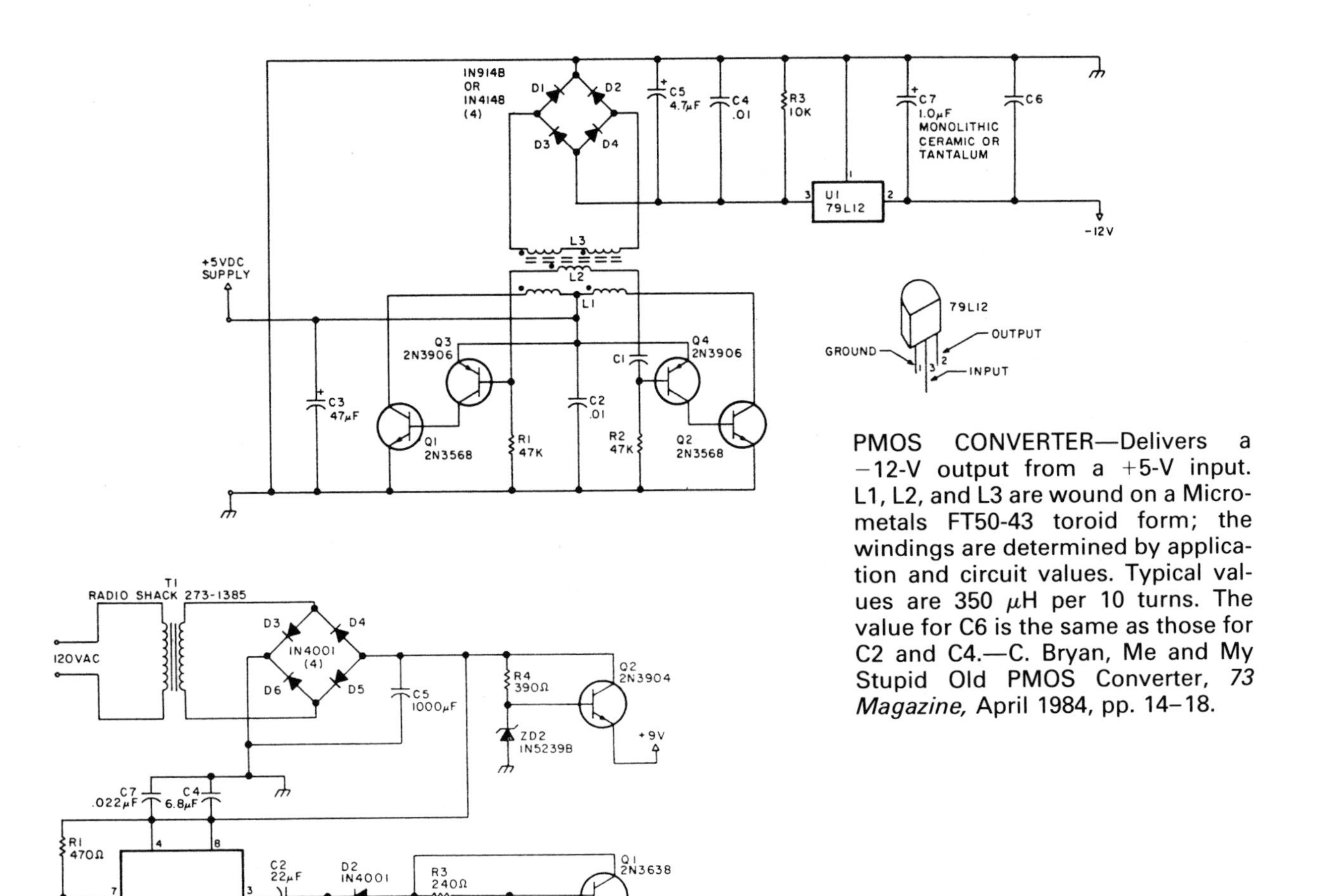

PMOS CONVERTER—Delivers a −12-V output from a +5-V input. L1, L2, and L3 are wound on a Micrometals FT50-43 toroid form; the windings are determined by application and circuit values. Typical values are 350 μH per 10 turns. The value for C6 is the same as those for C2 and C4.—C. Bryan, Me and My Stupid Old PMOS Converter, *73 Magazine,* April 1984, pp. 14–18.

DUAL-POLARITY POWER SUPPLY—Delivers both +9- and −9-V DC voltages at approximately 60–70 mA per terminal.—H. Minchow, When Plus Goes Minus, *73 Magazine,* July 1980, pp. 58–59.

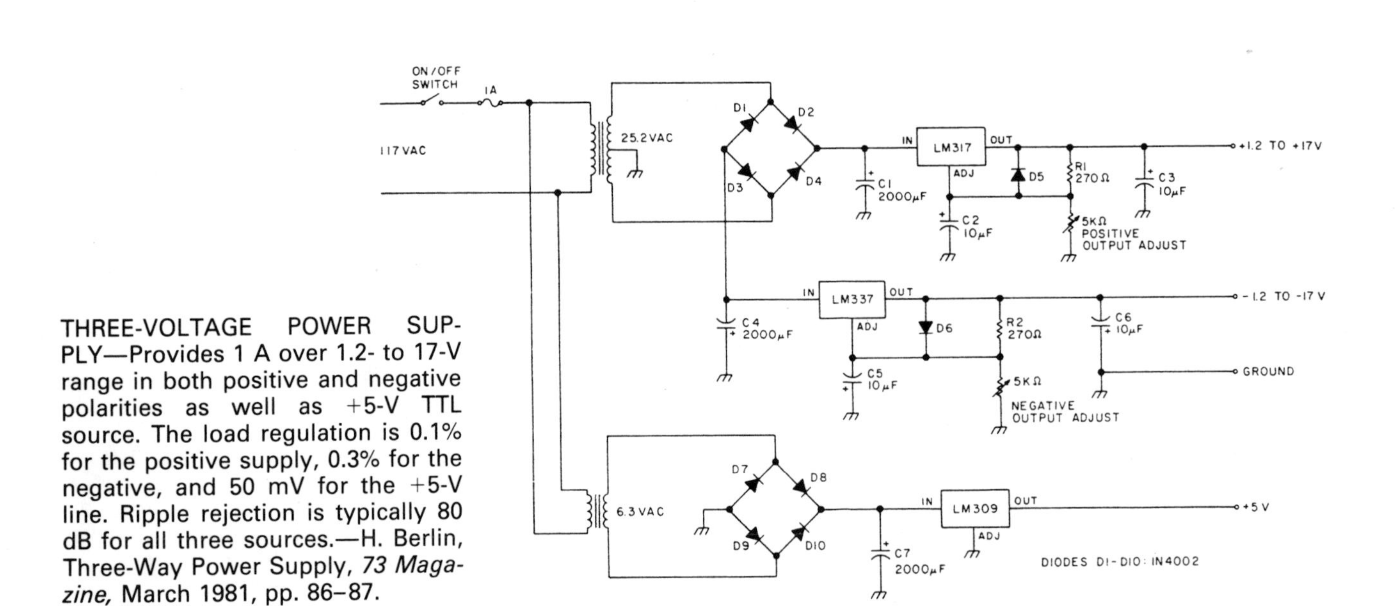

THREE-VOLTAGE POWER SUPPLY—Provides 1 A over 1.2- to 17-V range in both positive and negative polarities as well as +5-V TTL source. The load regulation is 0.1% for the positive supply, 0.3% for the negative, and 50 mV for the +5-V line. Ripple rejection is typically 80 dB for all three sources.—H. Berlin, Three-Way Power Supply, *73 Magazine,* March 1981, pp. 86–87.

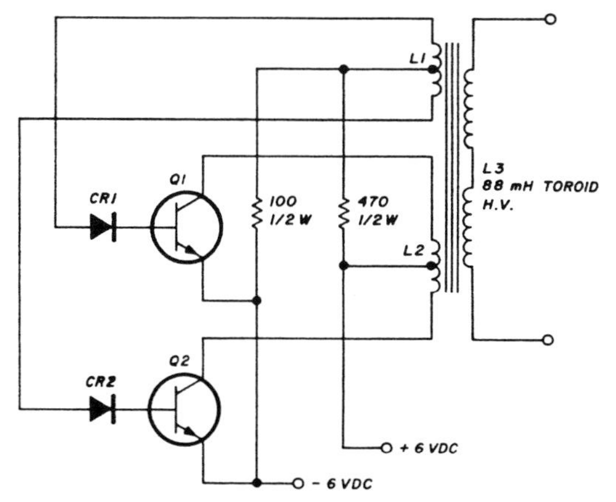

CR1,CR2 silicon diode, 100 PIV
L1 12 turns No. 28 center-tapped (feedback)
L2 24 turns No. 24 center-tapped (primary)
L3 original 88-mH toroid winding
Q1,Q2 2N3053 or similar
Note: L1 and L2 wound over center of 88-mH toroid

transistor	DC input	no load AC output
2N3394	6 V 160 mA	600
2N3415	6 V 280 mA	750
2N2726	6 V 180 mA	600
2N2222	6 V 280 mA	800
2N3053	6 V 280 mA	800

HIGH-VOLTAGE SWITCHING POWER SUPPLY—Uses surplus 88-mH toroid with added primary and feedback windings. L1 and L2 are wound in the center of the toroid (L3) away from the ends of the original winding (which are at high-voltage potential). Input and output voltages depend upon which transistor is used for Q1 and Q2. Q1 and Q2 should be adequately heatsinked.—J. Najork, A Switching High-Voltage Power Supply, *Ham Radio,* April 1984, pp. 48–49.

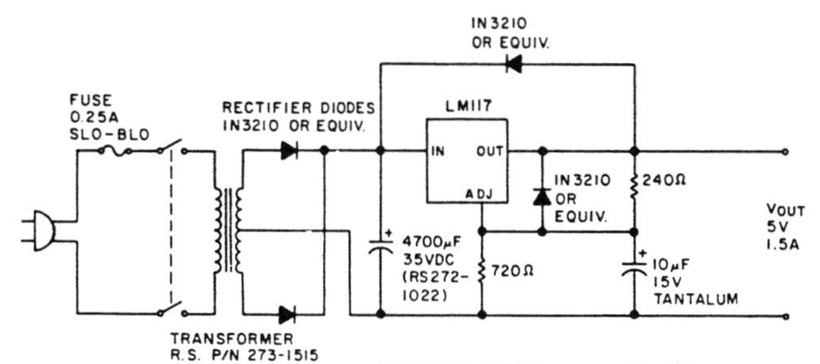

REGULATED 5-V SUPPLY—Delivers 5 V at 1.5 A. The LM117 should be mounted on a heat sink.—B. Myers, Unlimited Guarantee for Power Supplies, *73 Magazine,* December 1983, pp. 20–26.

REGULATED DC-TO-DC CONVERTER—Produces positive and negative 15-V DC outputs from a +5-V DC input. The line and load regulation is 0.1%. T_1 is a Shafer Magnetics SMC3359.—"Signetics Analog Applications Manual," Signetics, Sunnyvale, CA, 1979, pp. 158–159.

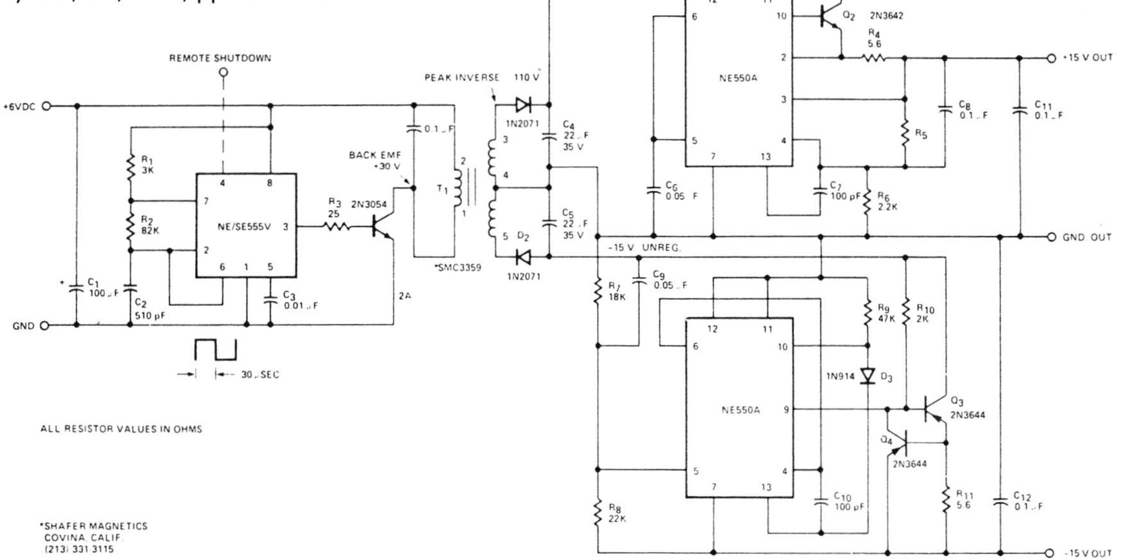

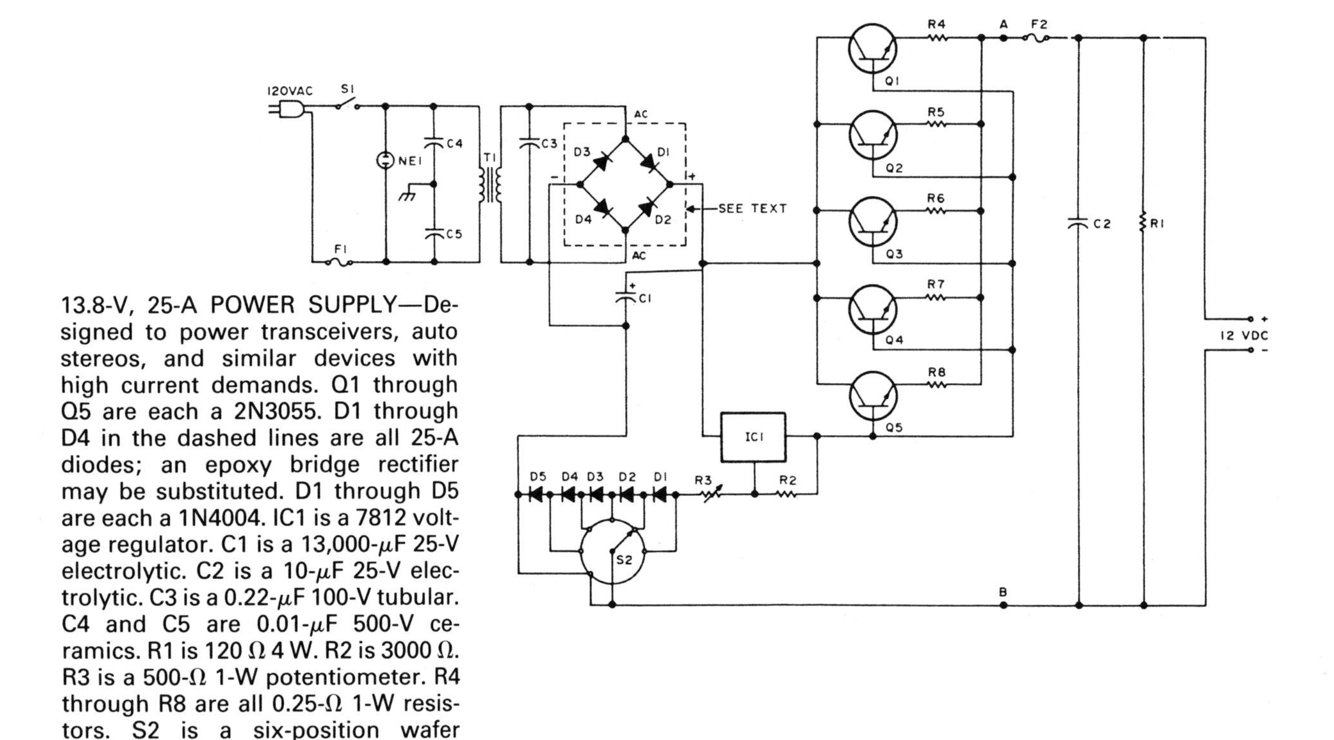

13.8-V, 25-A POWER SUPPLY—Designed to power transceivers, auto stereos, and similar devices with high current demands. Q1 through Q5 are each a 2N3055. D1 through D4 in the dashed lines are all 25-A diodes; an epoxy bridge rectifier may be substituted. D1 through D5 are each a 1N4004. IC1 is a 7812 voltage regulator. C1 is a 13,000-μF 25-V electrolytic. C2 is a 10-μF 25-V electrolytic. C3 is a 0.22-μF 100-V tubular. C4 and C5 are 0.01-μF 500-V ceramics. R1 is 120 Ω 4 W. R2 is 3000 Ω. R3 is a 500-Ω 1-W potentiometer. R4 through R8 are all 0.25-Ω 1-W resistors. S2 is a six-position wafer switch. T1 is a 120/17-24 V AC power transformer. F1 is a 5-A fuse, and F2 is a 30-A fuse.—V. Weiss, Cheap and Simple, *73 Magazine,* January 1981, pp. 50–52.

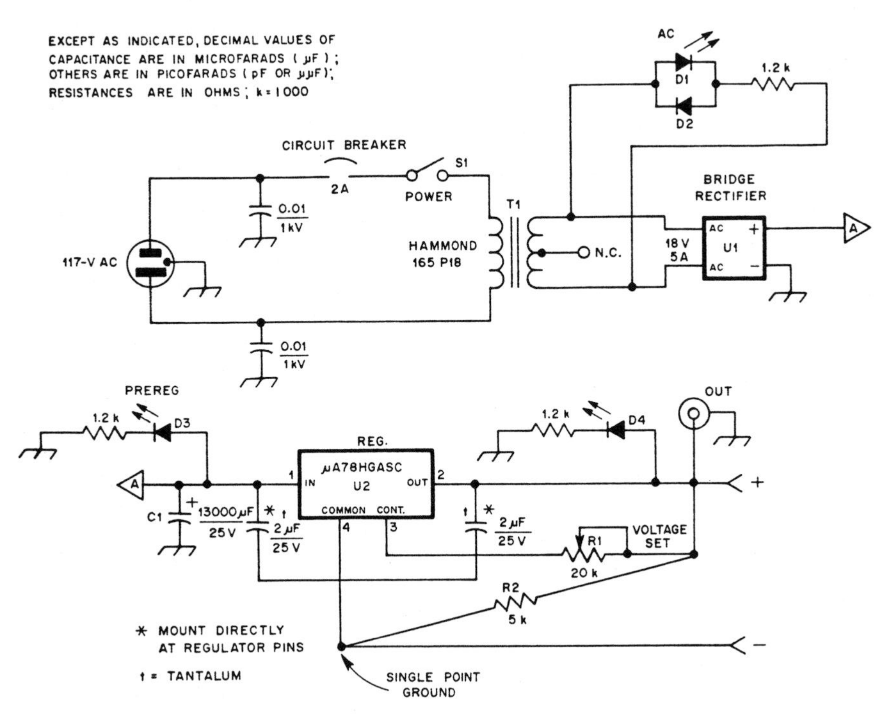

5- TO 20-V, 5-A POWER SUPPLY—Incorporates short-circuit and overvoltage protection and allows adjustment of output voltage through R1. A circuit breaker is used in place of a line fuse. LEDs monitor different sections of the circuit and indicate its operating status. The PREREG LED indicates output from the rectifier filter. The OUTput LED goes out if there is a short circuit at the output or the regulator IC shuts down. The AC LED goes out if the AC input is interrupted or the circuit breaker trips. D1, D2, and D4 are miniature LEDs. D2 is a 1N4001. U1 is a 25-A, 50 PIV silicon full wave bridge rectifier. U2 is a Fairchild A78HGASC adjustable voltage regulator IC. R1 is a 20-K 10-turn potentiometer (Bourns 3006 or equivalent). The circuit breaker is a 2-A television replacement type.—G. Hull, Introducing The PS5—A Dependable, 5-A Portable Power Supply, *QST,* June 1983, pp. 19–20.

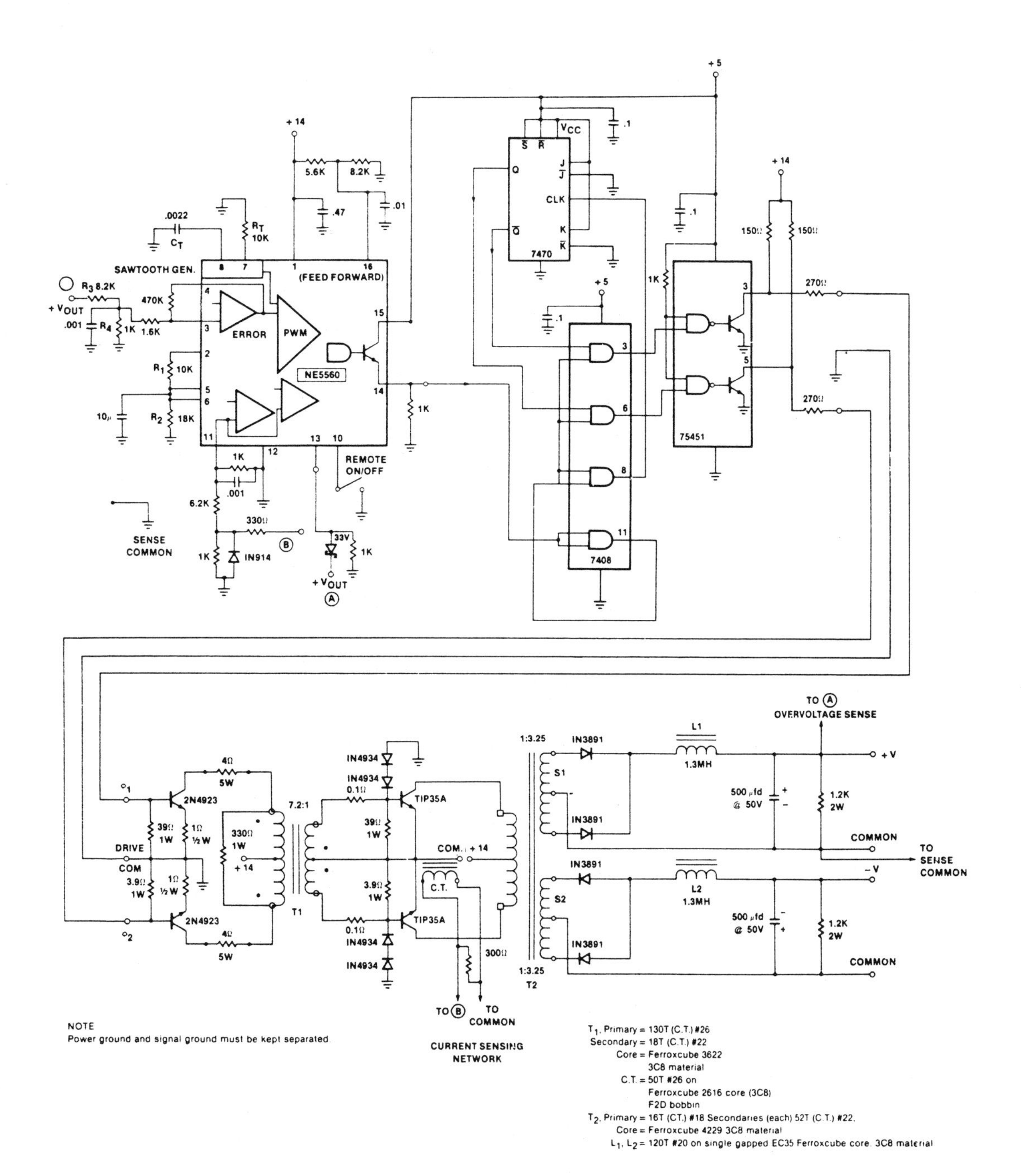

NOTE
Power ground and signal ground must be kept separated.

T_1: Primary = 130T (C.T.) #26
Secondary = 18T (C.T.) #22
Core = Ferroxcube 3622
3C8 material
C.T. = 50T #26 on
Ferroxcube 2616 core (3C8)
F2D bobbin
T_2: Primary = 16T (C.T.) #18 Secondaries (each) 52T (C.T.) #22.
Core = Ferroxcube 4229 3C8 material
L_1, L_2 = 120T #20 on single gapped EC35 Ferroxcube core. 3C8 material

PUSH-PULL SWITCH MODE REGULATED SUPPLY WITH TTL DRIVE CONVERSION LOGIC—Includes feed-forward input compensation and cycle-to-cycle drive current protection. The input voltage range is +12 to +18 V for an output of +30 and −30 V at a maximum load current of 1 A. The average efficiency is 81%. The main pulse width modulator operates to 48 kHz with power switching at 24 kHz.—"Signetics Linear LSI Data and Applications Manual," Signetics, Sunnyvale, CA, 1985, p. 7-37.

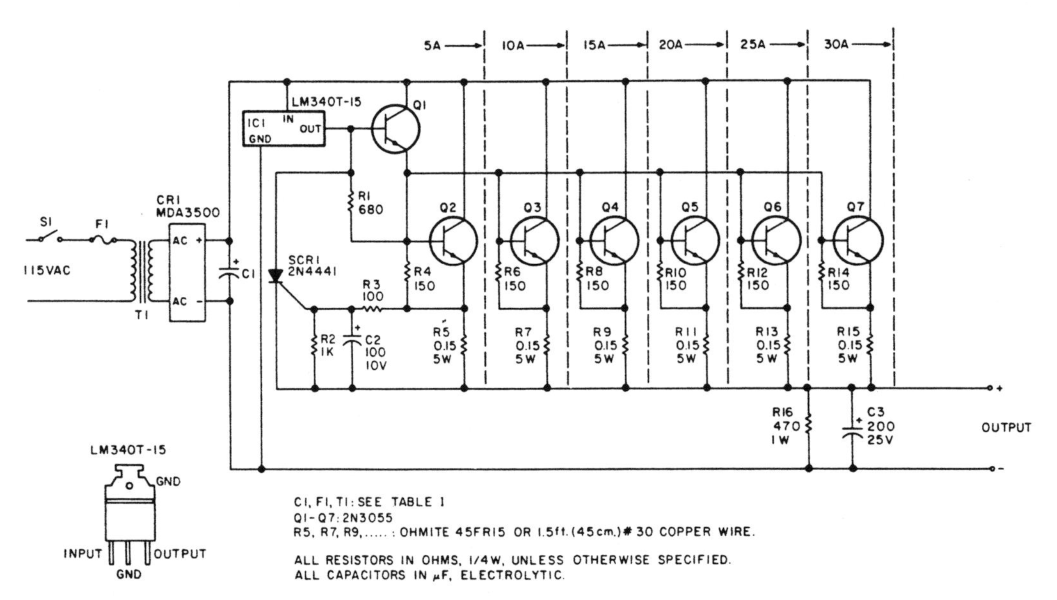

VARIABLE CAPACITY POWER SUPPLY—Capable of delivering 13 V DC at 5 to 30 A, depending upon C1, F1, and T1. Each additional stage (indicated by dashed lines) delivers the amperage indicated. All transistors are 2N3055. For 5A, C1 is 7500 μF 25 V, F1 is 2 A, and T1 is a Triad F-242u. For 10 A, C1 is 15,000 μF 25 V, F1 is 4 A, and T2 is a Triad F-234u. For 15 A, C1 is 22,000 μF, F1 is 6 A, and T1 is a Triad F-244u. For 20 A, C1 is 30,000 μF 25 V, F1 is 8 A, and T1 is a Triad F-244u. For 25 A, C1 is 40,000 μF 25 V, F1 is 10 A, and T1 is a Triad F-245u. For 30 A, C1 is 50,000 μF 25 V, F1 is 12A, and T1 is a Triad F-245u. All fuses are the "slow blow" type.—D. Oliver, Construct this Customized Power Supply, *73 Magazine,* August 1983, pp. 8–10.

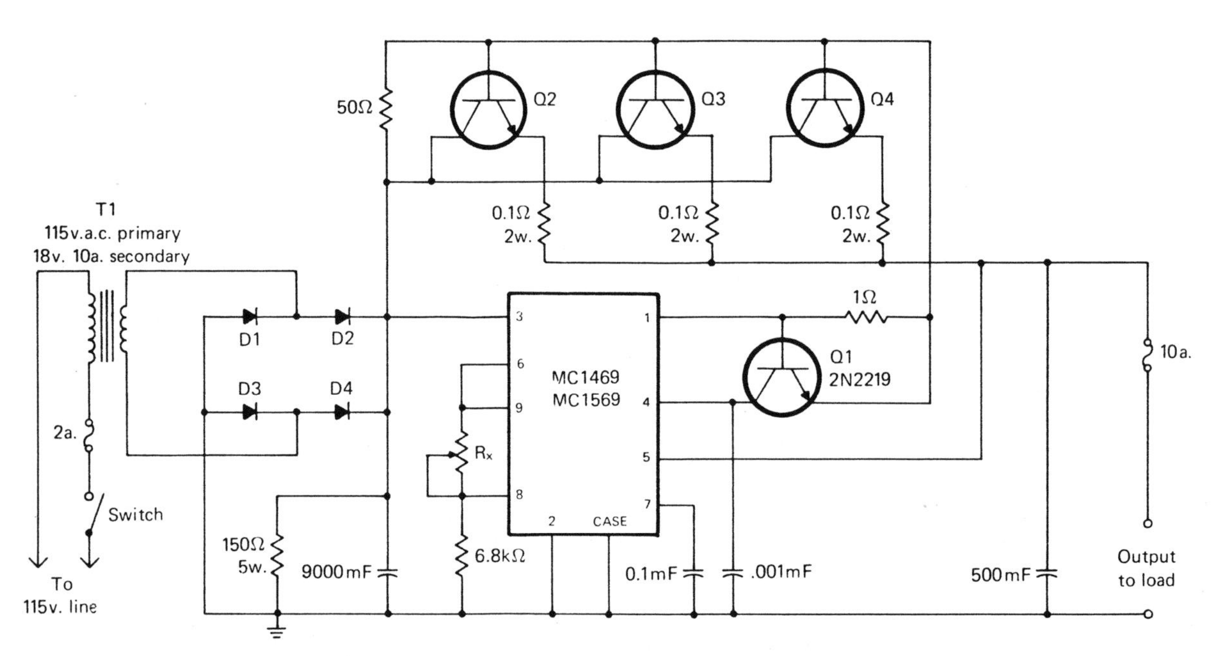

13.8-V, 10-A POWER SUPPLY—Designed for use with transceivers with large current requirements. The MC1469/1569 should be mounted on a No. 20 gauge copper heatsink formed into a U shape. Q2 through Q4 should be heavy-duty silicon NPN power transistors such as 2N3902, 2N3055, or equivalent rated at 3.5 A. D1 through D4 are 10-A diodes. Rx adjusts the desired output voltage; its value should be 16 K.—G. Moynahan, A Voltage-Regulated, High-Current 13.8 Volt Power Supply, *CQ,* February 1981, pp. 11–15.

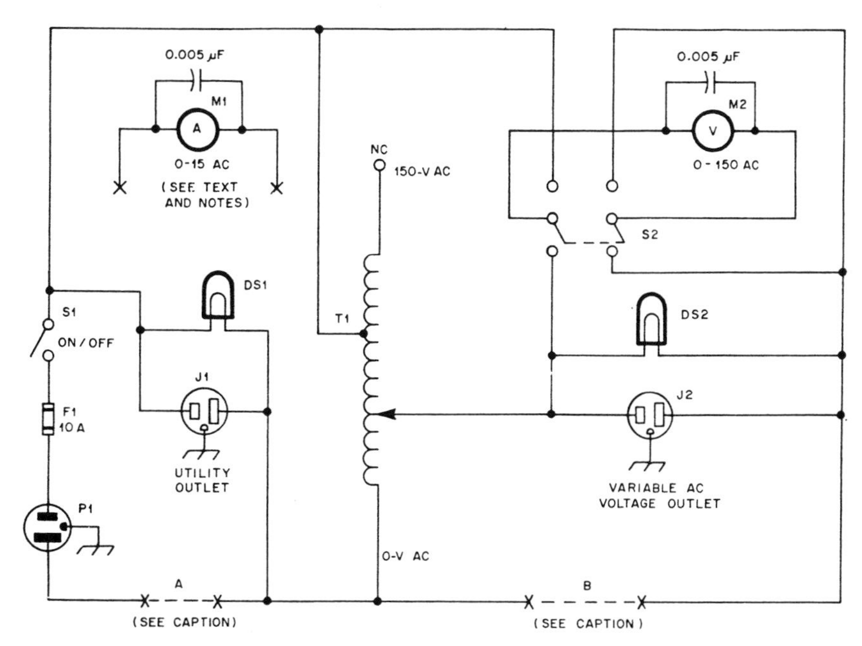

VARIABLE AC VOLTAGE SOURCE—Provides AC voltages from 0 to 150 V at 0 to 15 A.—J. Magnusson, A Variable AC-Voltage Source, *QST*, August 1984, pp. 29–30.

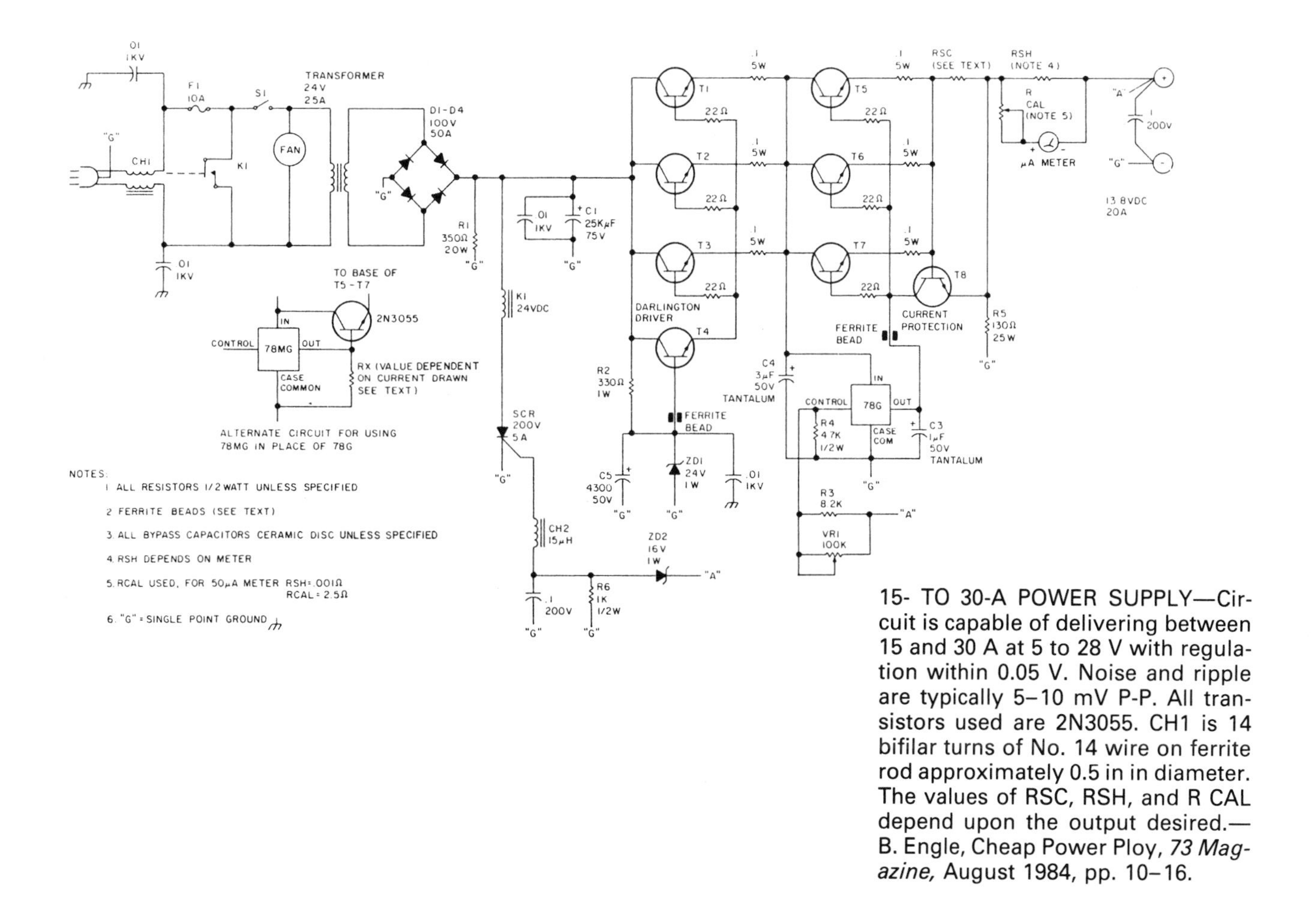

15- TO 30-A POWER SUPPLY—Circuit is capable of delivering between 15 and 30 A at 5 to 28 V with regulation within 0.05 V. Noise and ripple are typically 5–10 mV P-P. All transistors used are 2N3055. CH1 is 14 bifilar turns of No. 14 wire on ferrite rod approximately 0.5 in in diameter. The values of RSC, RSH, and R CAL depend upon the output desired.—B. Engle, Cheap Power Ploy, *73 Magazine*, August 1984, pp. 10–16.

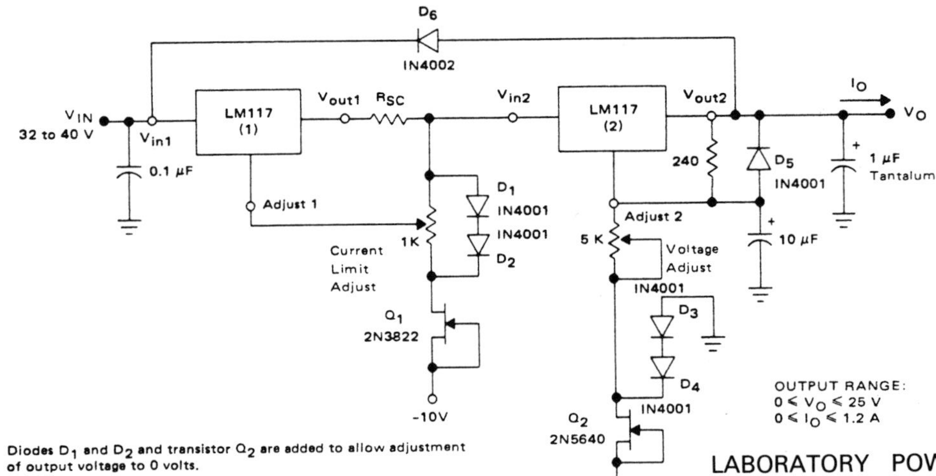

LABORATORY POWER SUPPLY—Features adjustable current limit and output voltage from 0 to 25 V at 0 to 1.2 A. D_1, D_2, and Q_2 are added to allow the adjustment of the output voltage to 0 V. D_6 protects both LM117 devices in the event of an input short circuit.—"Motorola Linear Integrated Circuits Databook," Motorola, Phoenix, AZ, 1979, p. 4-23.

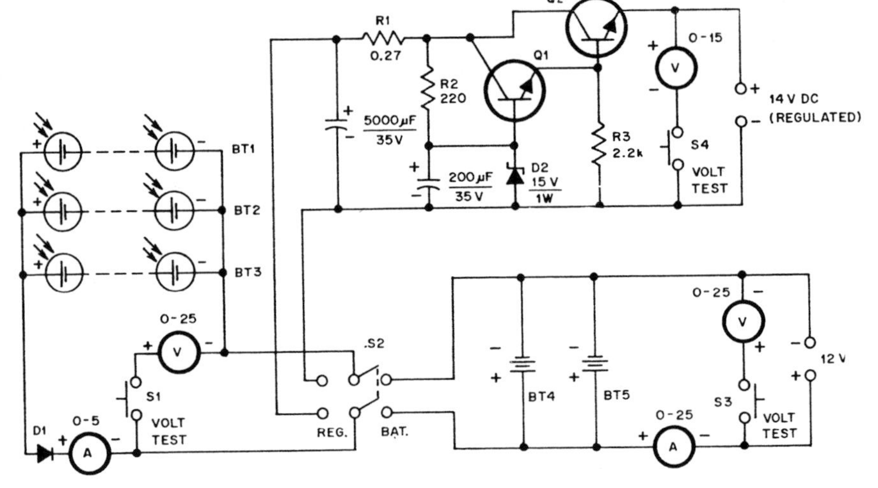

SOLAR POWER SUPPLY—Delivers 12–14 V DC at up to 2.5 A in sunlight. BT1 through BT3 are 20-V 0.5-A solar panels by Spectrolab. BT4 and BT5 are both 12-V lead-acid automobile storage batteries. D1 is a Motorola MR752/7414 or equivalent. Q1 is a TIP31 or equivalent, while Q2 is a 2N3055 or equivalent.—J. Halliday, Solar Powering a Ham Station, *QST*, August 1980, pp. 11–12.

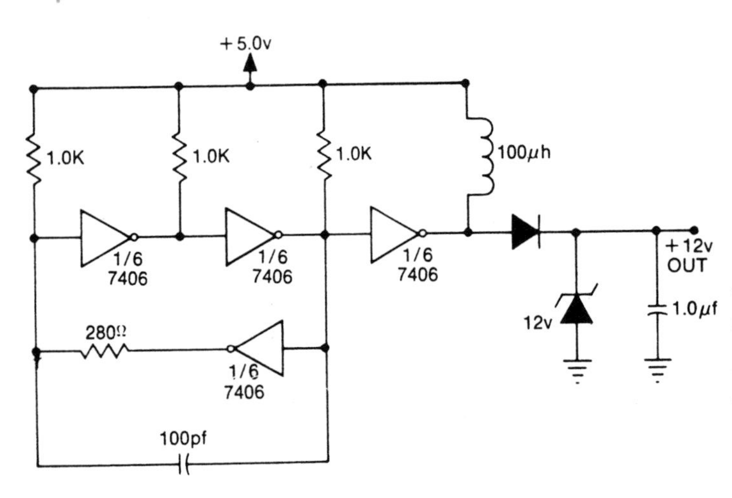

VOLTAGE CHARGE PUMP—Produces a +12-V output from a +5-V input; current is typically 22 mA.—"Standard Microsystems Corporation Data Catalog," Standard Microsystems Corp., Hauppauge, NY, 1984, p. 235.

17

Receiving Circuits

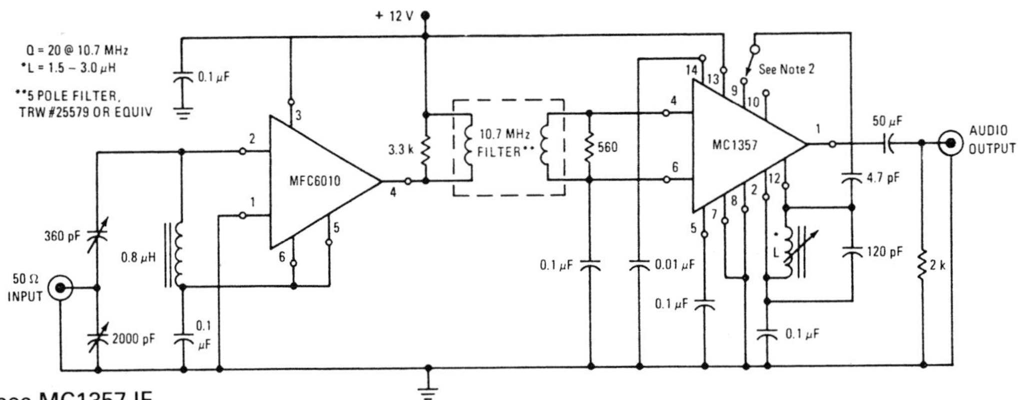

10.7-MHz IF STRIP—Uses MC1357 IF amplifier and quadrature detector IC for IF amplifier section for television sound or FM receiver. There may be an optional input to the quadrature coil from either pin 9 or pin 10. Pin 9 is commonly used to avoid overload with various tuning techniques. However, a significant improvement in limiting sensitivity can be obtained by using pin 10.—"Motorola Linear Integrated Circuits Databook," Motorola, Phoenix, AZ, 1979, p. 5-77.

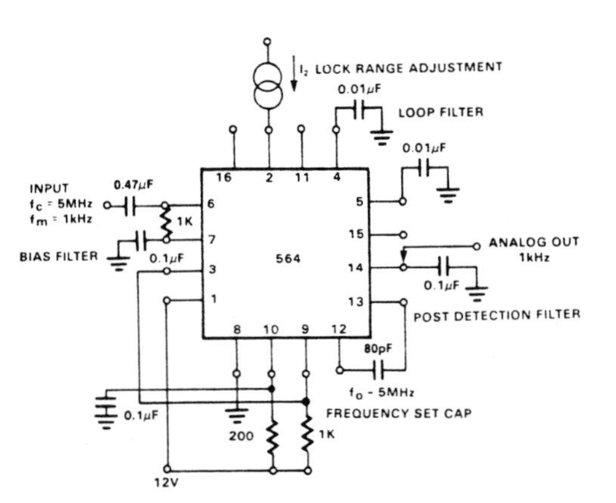

FM DEMODULATOR—Input signal is AC coupled with output signal being extracted at pin 14. Loop filtering is provided by the capacitors at pins 4 and 5. To obtain a sufficient demodulated output signal, the frequency deviation in the input signal should be 1% or higher. The supply voltage is 12 V.—"Signetics Linear LSI Data and Applications Manual," Signetics, Sunnyvale, CA, 1985, p. 5-138.

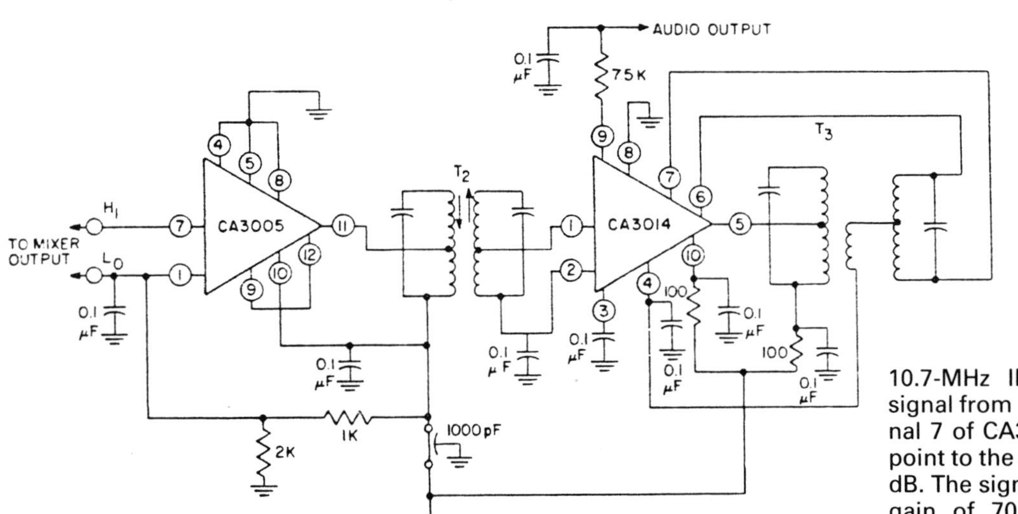

T2 = TRW No. 22468 or equiv.

T3 = TRW No. 21590 or equiv.

10.7-MHz IF STRIP—10.7-MHz FM signal from mixer is applied to terminal 7 of CA3005. The gain from this point to the input of the CA3014 is 25 dB. The signal receives an additional gain of 70 dB and limiting from terminal 1 to terminal 5 in the CA3014.—"Integrated Circuits for FM Broadcast Receivers," RCA Solid State Division, Somerville, NJ, 1970, Application Note ICAN-5269.

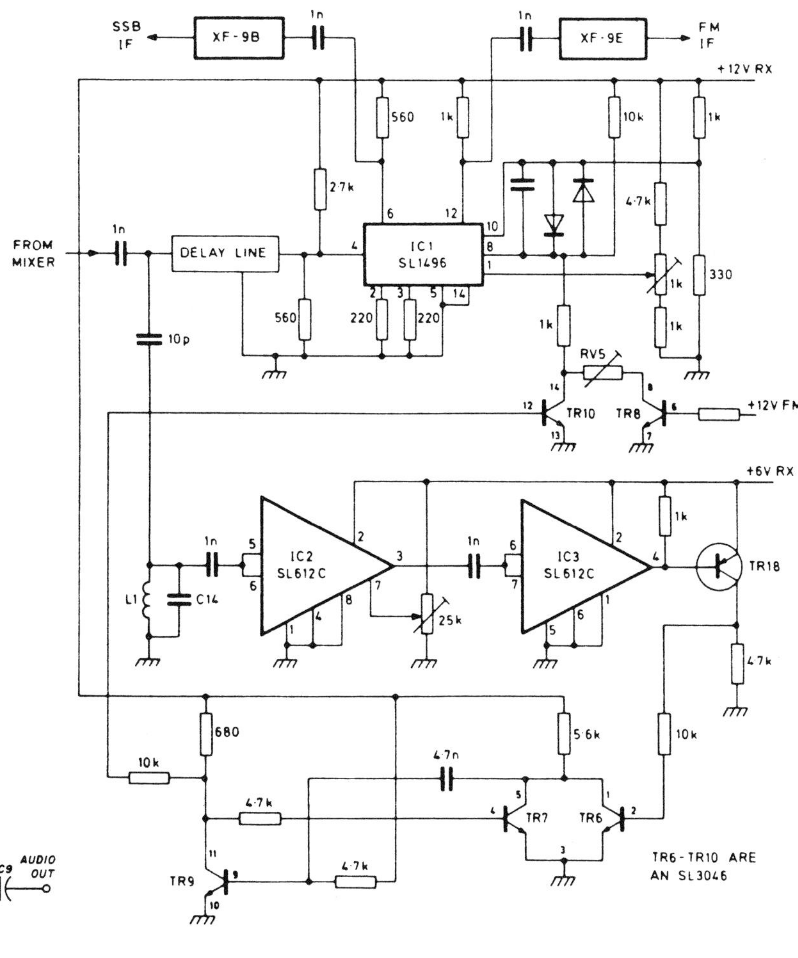

NOISE BLANKER—Noise receiver has input via tuned circuit to prevent local oscillator leak from mixer triggering system. The noise IF amplifier consists of an SL612C (IC2) and an SL613C (IC3) which acts as a detector. Gain control is applied to IC2 to set the blanking level. The pulse outputs from the detector in IC3 are buffered by PNP transistor TR18 to a simple monostable (TR6, TR7, and TR9) with a 10-ms pulse. This pulse operates the noise gate. A 400-ns delay line between the mixer and the noise gate ensures the system is blanked before the pulse that triggers the monostable system arrives at the noise gate.—"Plessey Applications IC Handbook," Plessey Semiconductors, Irvine, CA, 1982, pp. 86–88.

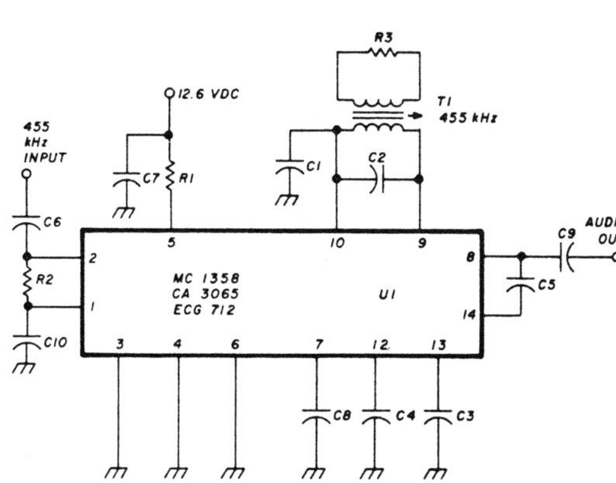

FM DETECTOR—Converts AM receiver to FM operation. The 455-kHz signal from the final IF amplifier of the AM receiver provides input for the detector. The audio output can drive the receiver's AF amplifier stage. T1 should be adjusted for maximum audio response.—J. LaMartina, Add FM to Your Receiver, *Ham Radio,* March 1981, pp. 74–75.

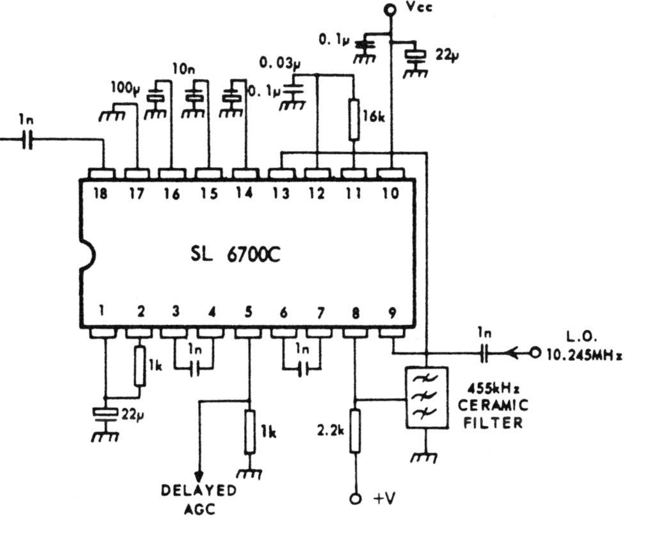

NOISE BLANKER FOR DOUBLE-CONVERSION AM RECEIVER—SL6700C IF amplifier and AM detector IC accept 10.7-MHz input and include mixer, detector, AGC generator, and noise blanker on-chip. The detector is linear, and the power requirements are typically 8 mA at 6 V.—"Plessey Integrated Circuits Databook," Plessey Semiconductor, Irvine, CA, 1983, pp. 183–185.

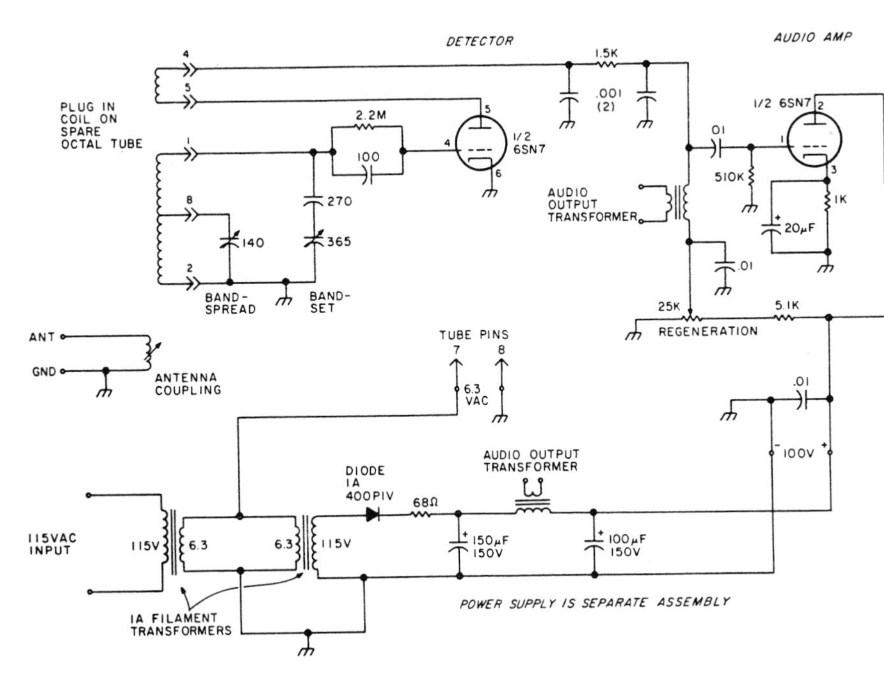

ONE-TUBE RECEIVER—Covers 3.5 to 7.5 MHz in single-tube regenerative design. The plug-in coil is wound on a spare octal tube. The main coil is 17 turns wound to a length of 0.75 in; tap six turns from the grounded end at the top of the tube. The regeneration winding is four turns over the end of the main coil.—P. Clower, The Tube Returns, *73 Magazine,* December 1982, pp. 64–69.

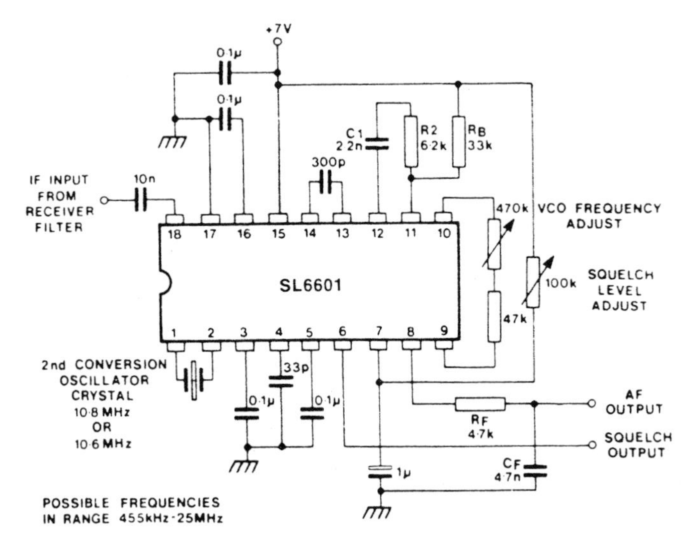

NARROW-BAND FM DETECTOR USING PLL—Operates as double-conversion IF amplifier and detector with first IF frequency of 10.7 MHz and second IF frequency of 100 kHz. The values of R2 and C1 determine the modulation bandwidth; the values used in the diagram give a ±5-kHz maximum deviation and a 3-kHz audio bandwidth.—"Plessey Applications IC Handbook," Plessey Semiconductors, Irvine, CA, 1982, pp. 103–107.

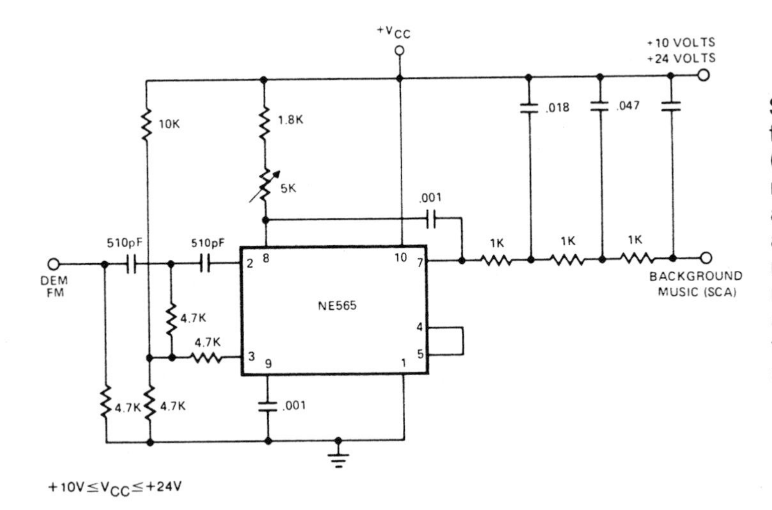

SCA DECODER—Demodulates the frequency-modulated subcarrier (subsidiary carrier authorization) of main FM channel. The SCA signal is a 67-kHz FM subcarrier, which puts it above the frequency spectrum of normal stereo or monaural program material. By connecting the circuit to a point between the FM discriminator and the de-emphasis filter of a standard FM receiver, SCA program material can be received.—"Signetics Analog Applications Manual," Signetics, Sunnyvale, CA, 1979, pp. 319–320.

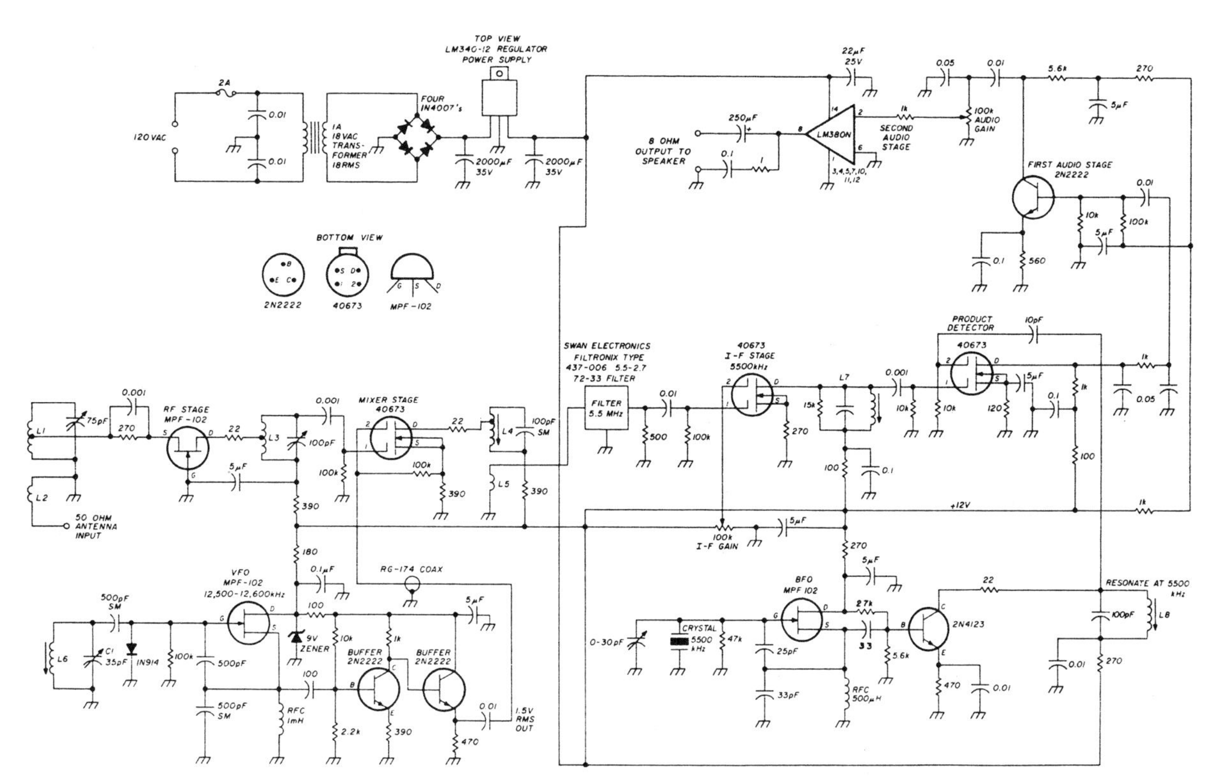

7000- TO 7300-kHz CW/SSB RECEIVER—Incorporates RF amplifier stage for enhanced performance. Output from the RF amplifier goes to the mixer stage tuned to 5.5 MHz, which in turn feeds the IF amplifier stage. The IF amplifier stage features variable gain, and selectivity is increased by a 5.5-MHz filter. The BFO tunes a 1-kHz range and can easily handle SSB. L1 is 35 turns of No. 26 wire on a T-80-2 form tapped 10 turns from the bottom. L2 is three turns tapped from the ground end of L1. L3 is the same as L1. L4 is 25 turns of No. 30 wire on a 9.5-mm ceramic form. L5 is 13 turns of No. 30 wire on the bottom of L4. L6 is seven turns of No. 26 wire on a 9.5-mm ceramic form. L7 is 25 turns of No. 30 wire on a 9.5-mm ceramic form. L8 is the same as L7.—E. Marriner, A Simple 40-Meter Receiver, *Ham Radio,* September 1980, pp. 64–65.

SHORTWAVE CONVERTER—Converts 16-, 19-, 25-, or 31-meter international broadcasting bands to the standard AM broadcast band. For 16-meter operation, L1 should be 1.5 µH, L2 should be 1.8 µH, and the crystal should be 16.3 MHz. For 19-meter operation, L1 should be 1.8 µH, L2 should be 2.2 µH, and the crystal should be 16.3 MHz. For 25-meter operation, L1 should be 2.7 µH, L2 should be 3.9 µH, and the crystal should be 11 MHz. For 31-meter operation, L1 should be 3.9 µH, L2 should be 5.2 µH, and the crystal should be 11 MHz. The converter should be constructed in an "RF-tight" enclosure and shielded connecting cable used in automotive installations.—B. Morrison, Drive-Time SWLing, *73 Magazine,* February 1983, pp. 70–71.

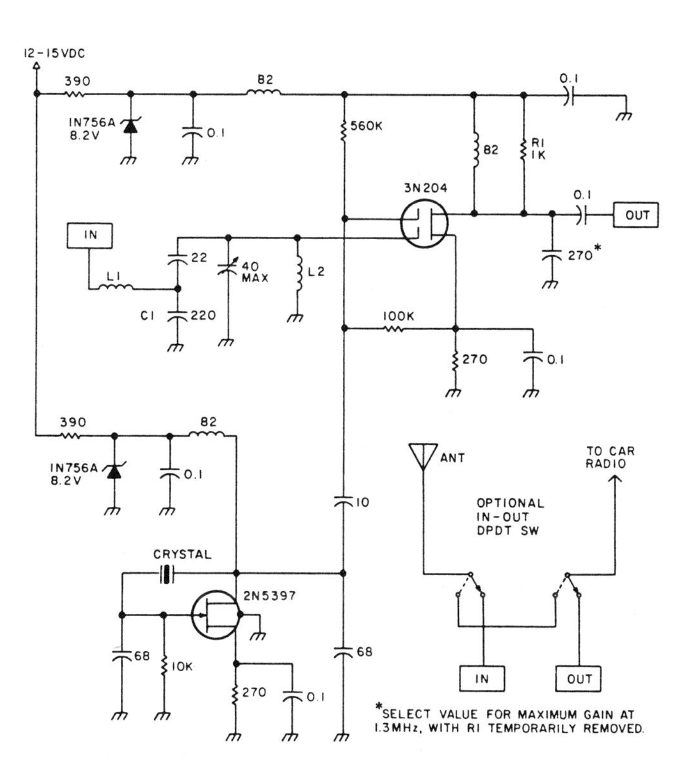

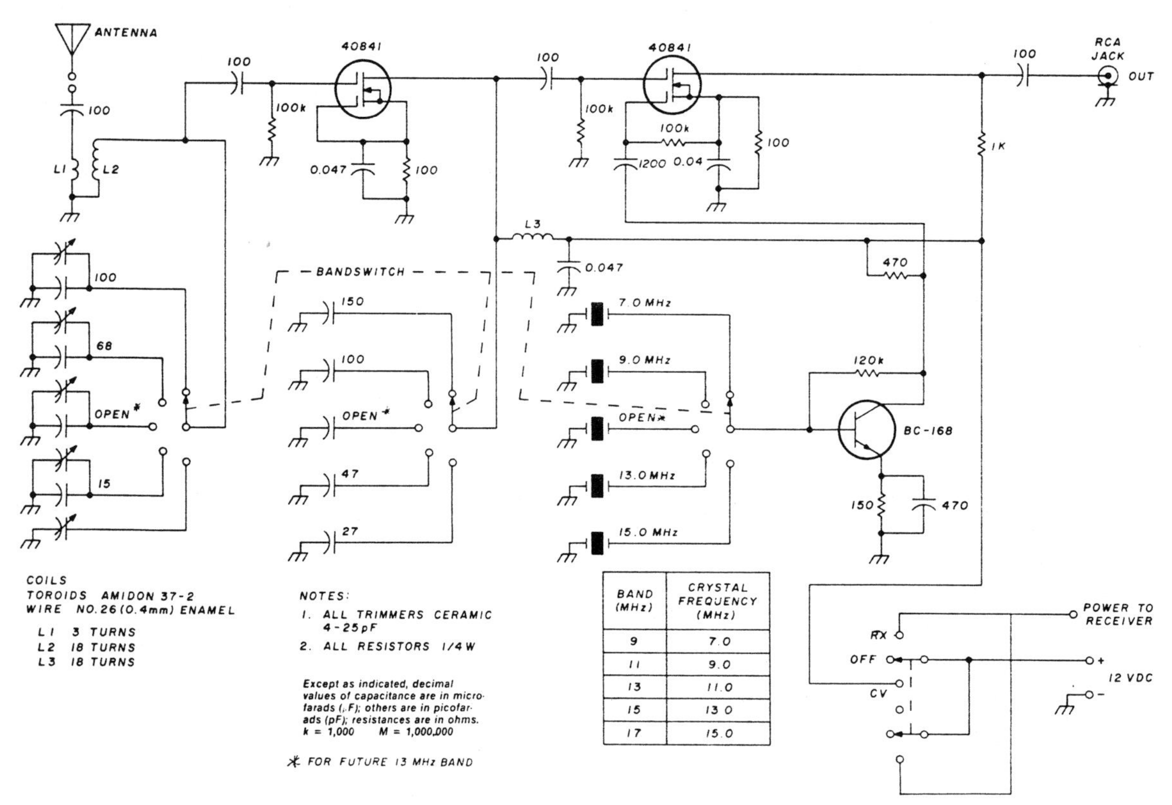

BAND (MHz)	CRYSTAL FREQUENCY (MHz)
9	7.0
11	9.0
13	11.0
15	13.0
17	15.0

SHORTWAVE CONVERTER—Capable of converting the 7- to 22-MHz range to the 2- to 3-MHz range. Crystal replacement and front-end tuning can give a coverage of 3–8 MHz if desired. The current drain is about 10 mA at 12 V, and the circuit is especially suited for mobile or portable use.—J. Perolo, Portable Shortwave Converter, *Ham Radio,* April 1981, pp. 64–66.

BAR GRAPH S-METER—Replaces conventional S-meter with LEDs to indicate strength of received signal. The greater the number of LEDs lit, the greater the received signal's strength. The supply voltage may range from 3 to 15 V.—T. Mealy, TR-7400 behind Bars, *73 Magazine,* March 1981, pp. 46–48.

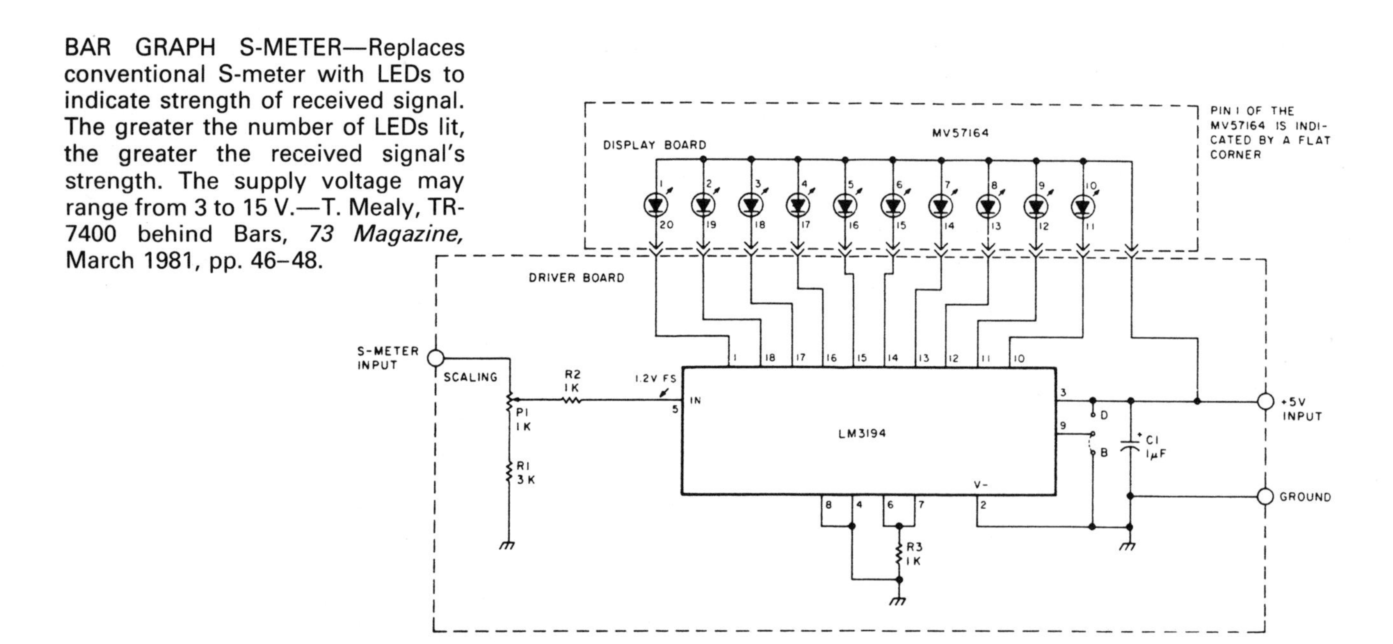

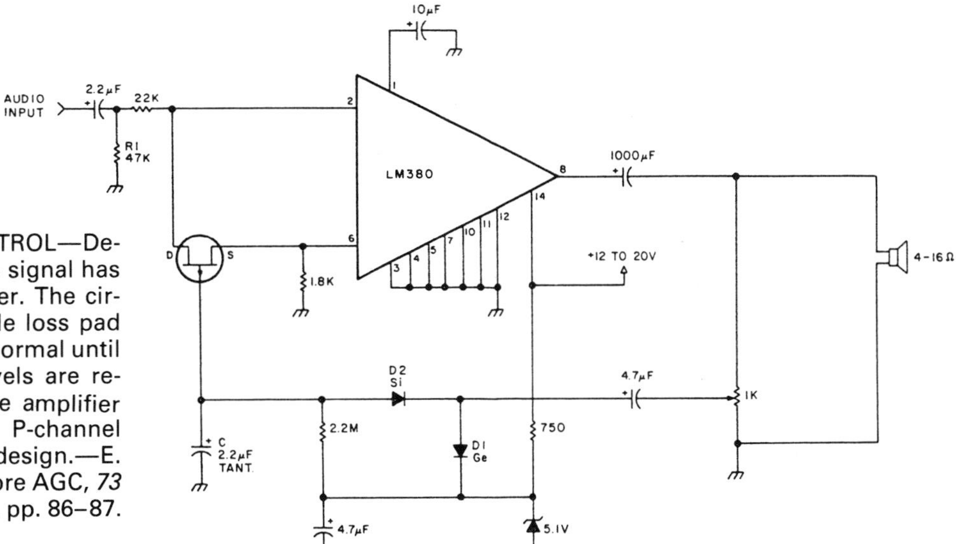

AUTOMATIC GAIN CONTROL—Designed for use after audio signal has been produced by receiver. The circuit operates as a variable loss pad with its amplifier gain at normal until signals above certain levels are received, at which point the amplifier gain is reduced. Most P-channel JFETs will work in the design.—E. Miller, Make Room for More AGC, *73 Magazine,* February 1983, pp. 86–87.

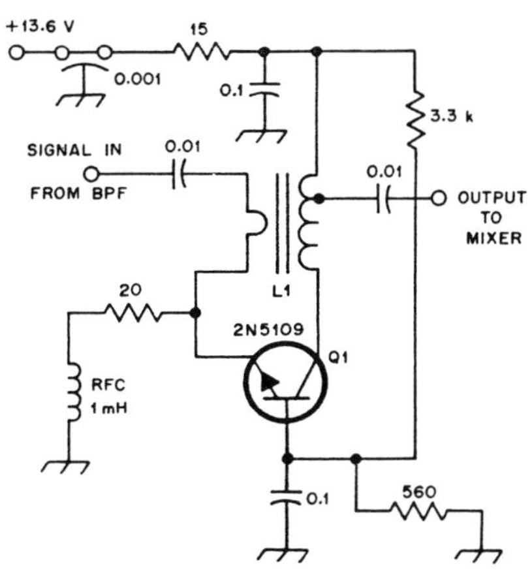

EXCEPT AS INDICATED, DECIMAL VALUES OF CAPACITANCE ARE IN MICROFARADS (μF) ; OTHERS ARE IN PICOFARADS (pF OR μμF); RESISTANCES ARE IN OHMS ; k = 1000

RF AMPLIFIER—Increases sensitivity of HF receivers. The measured specifications at 14 MHz include an IMD dynamic range of −37 dB and a blocking dynamic range of 93 dB. L1 is wound on a FB-43-801 bead with No. 28 wire, with a secondary winding of 15 turns tapped four turns from the supply end. The one-turn primary is polarized, and should be wound on the end of the coil furthest from the transistor.—W. Cooper, A One-Transistor RF Amplifier, *QST,* August 1984, pp. 46–47.

AM DEMODULATOR—Amplifying and limiting of AM carrier is accomplished by ULN2209. The ULN2209 provides a 55-dB gain and symmetrical limiting above 400 μV. The limited carrier is then applied to the detector at the carrier ports to provide the desired switching function. The signal is then demodulated by the MC1496K acting as a synchronous AM demodulator. The MC1496K attenuates the carrier frequency because of the balanced nature of the device.—"Signetics Analog Applications Manual," Signetics, Sunnyvale, CA, 1979, pp. 189–190.

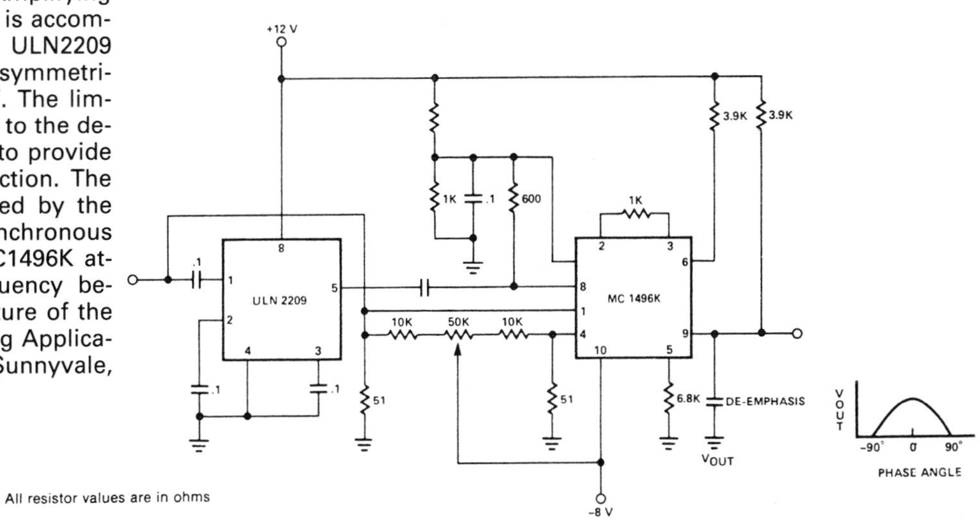

All resistor values are in ohms

CRYSTAL CALIBRATOR—Oscillator produces marker points at selectable 100- or 10-kHz points up to 30 MHz. The CD4001 quad NOR device provides the active part of the 100-kHz oscillator. Its output drives the 74C160 decade counter to produce the 10-kHz calibration points. C2 is used to zero-beat output against a frequency standard such as stations WWV and WWVH. Q1 may be any small NPN power transistor.—G. Moynahan, An Improved Crystal Calibrator, *CQ*, November 1980, pp. 42–43.

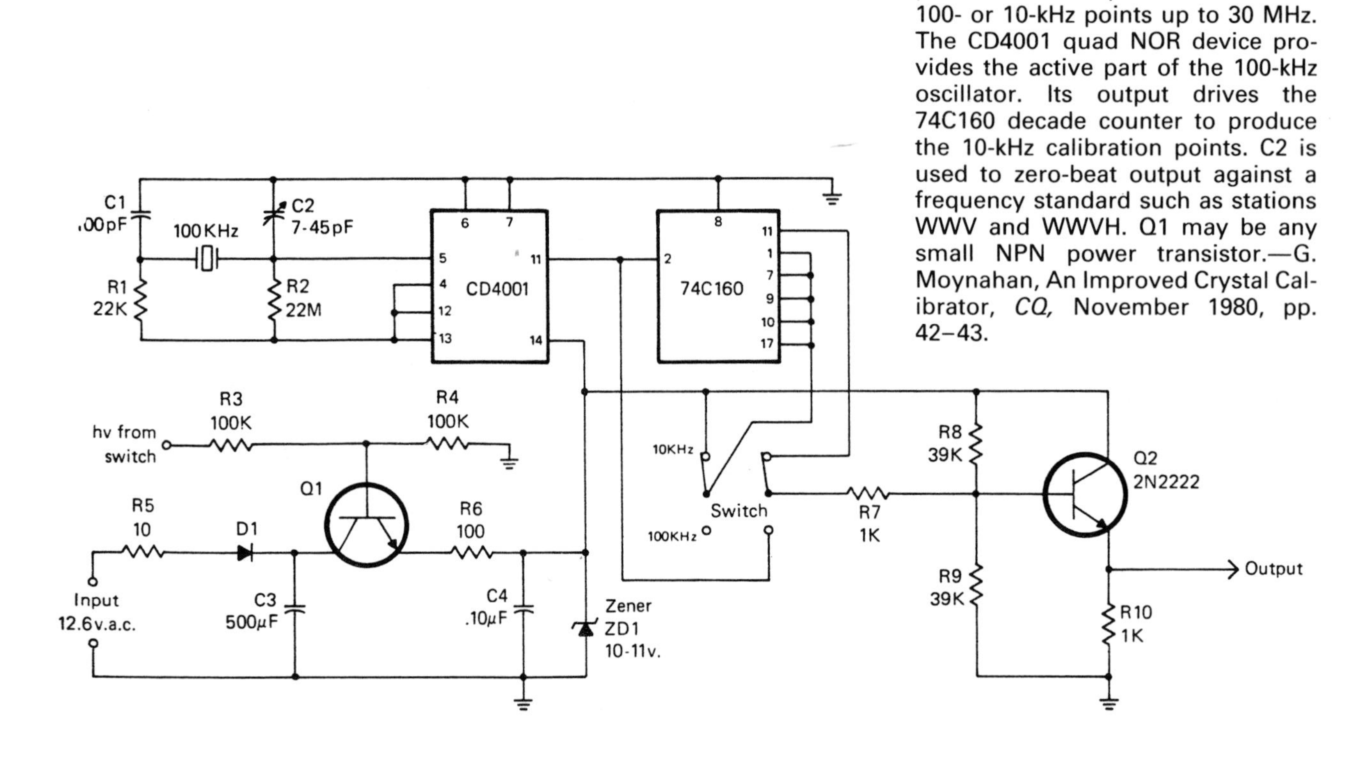

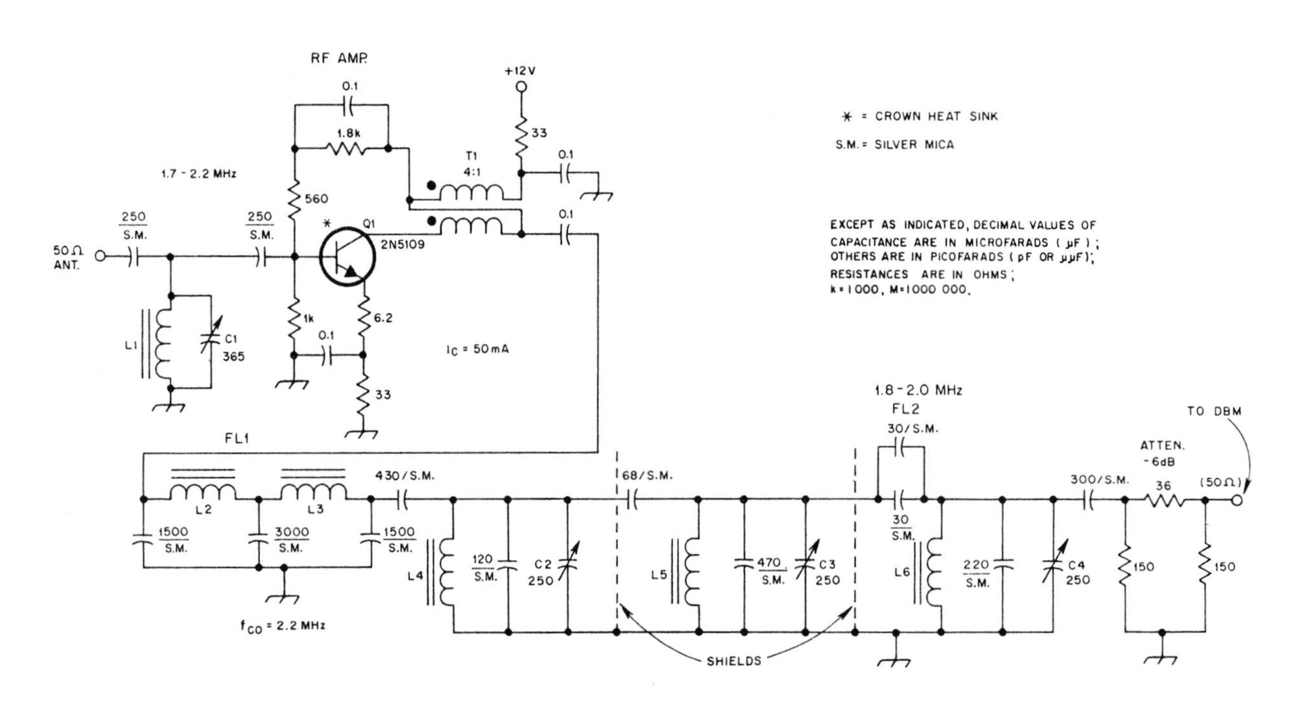

HIGH-LEVEL RF AMPLIFIER STAGE—Used ahead of mixer stage of receiver and operates throughout HF range. The 2N5109 has a high standing collector current and handles large signals easily. The measured noise figure at 30 MHz is 5 dB. L1 consists of 41 turns of No. 24 wire on a T68-2 toroid form. L2 and L3 are both 26 turns of No. 24 wire on T68-2 toroid forms. L4, L5, and L6 consist of 43 turns on No. 24 wire on T68-6 toroid forms. T1 has 16 bifilar turns of No. 26 wire on an Ft50-43 ferrite core.—"The Radio Amateur's Handbook," American Radio Relay League, Newington, CT, 1981, pp. 8-41 to 8-43.

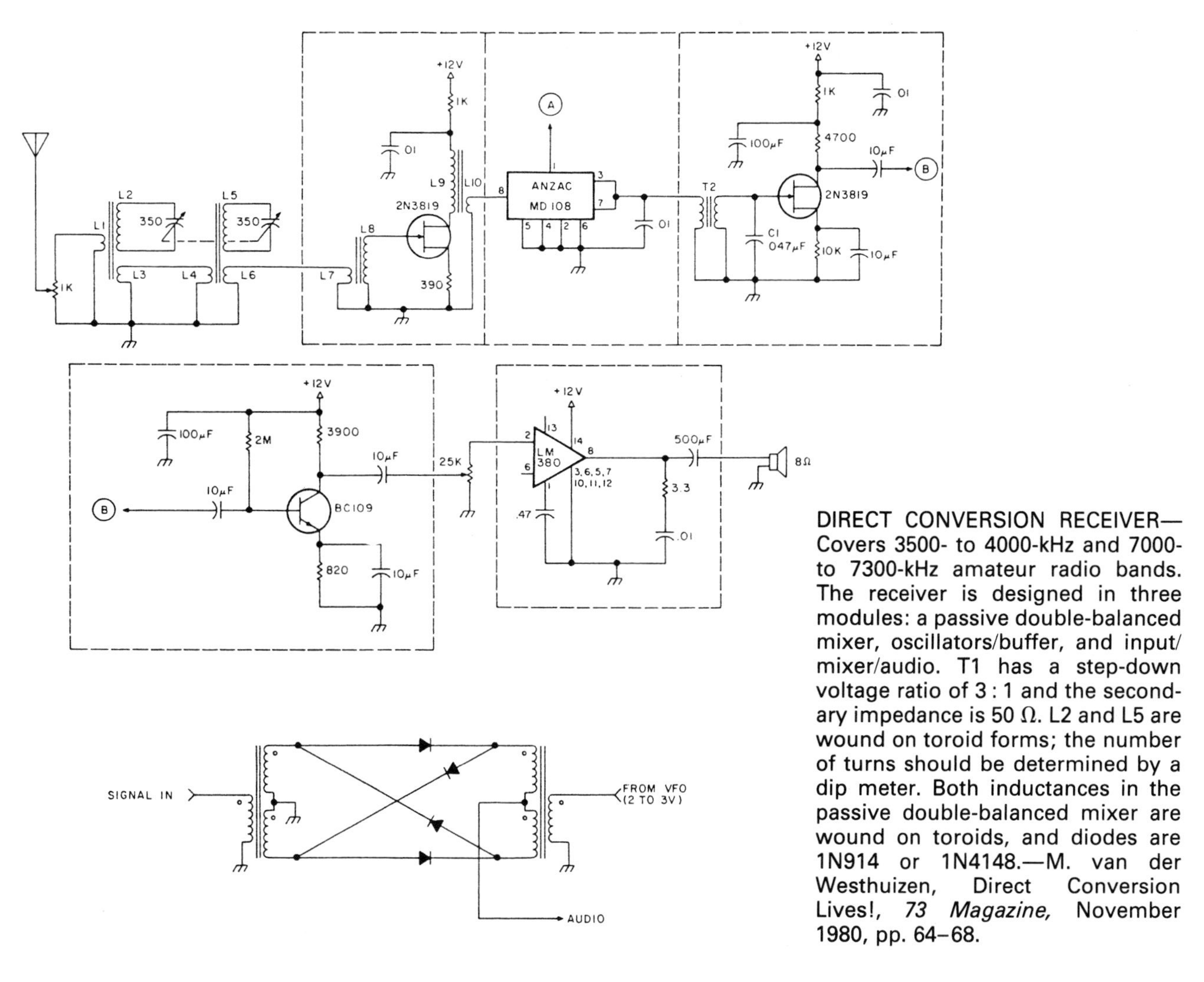

DIRECT CONVERSION RECEIVER—Covers 3500- to 4000-kHz and 7000- to 7300-kHz amateur radio bands. The receiver is designed in three modules: a passive double-balanced mixer, oscillators/buffer, and input/mixer/audio. T1 has a step-down voltage ratio of 3 : 1 and the secondary impedance is 50 Ω. L2 and L5 are wound on toroid forms; the number of turns should be determined by a dip meter. Both inductances in the passive double-balanced mixer are wound on toroids, and diodes are 1N914 or 1N4148.—M. van der Westhuizen, Direct Conversion Lives!, *73 Magazine,* November 1980, pp. 64–68.

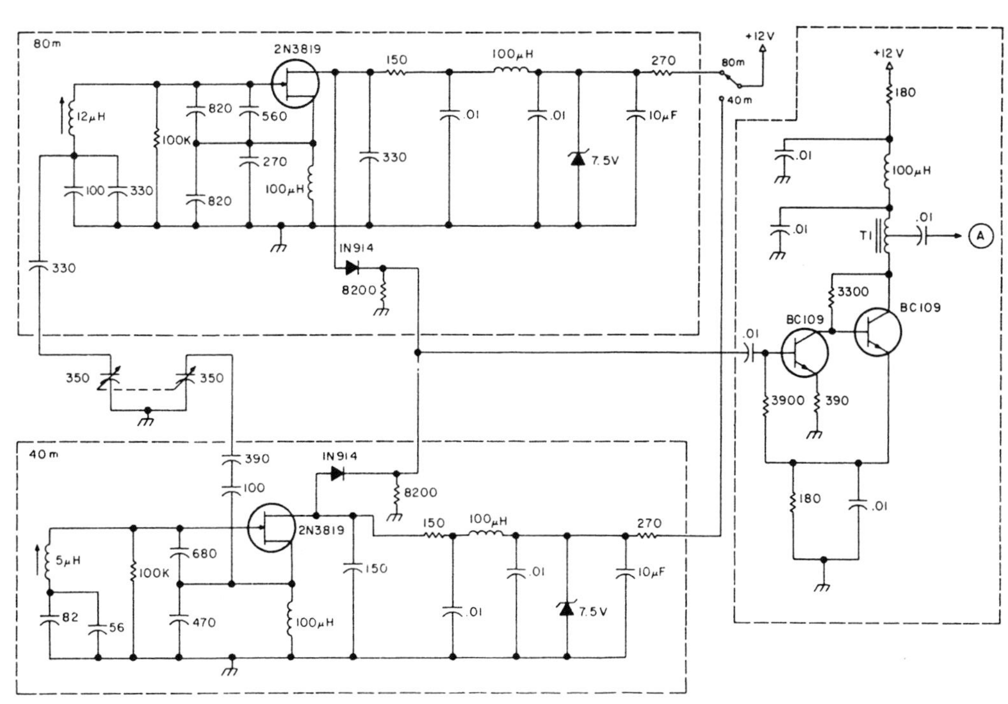

HIGH-PERFORMANCE AUTOMATIC LEVEL CONTROL—Uses NE570 compandor IC; gain is inversely proportional to input level so that 20-dB drop in input level produces 20-dB increase in gain. The output remains fixed at a constant level. C_{RECT} determines the time constant of the circuit. Resistor RX is not needed if a response at very low input levels is not needed.—"Signetics Linear LSI Data and Applications Manual," Signetics, Sunnyvale, CA, 1985, pp. 9-160 to 9-162.

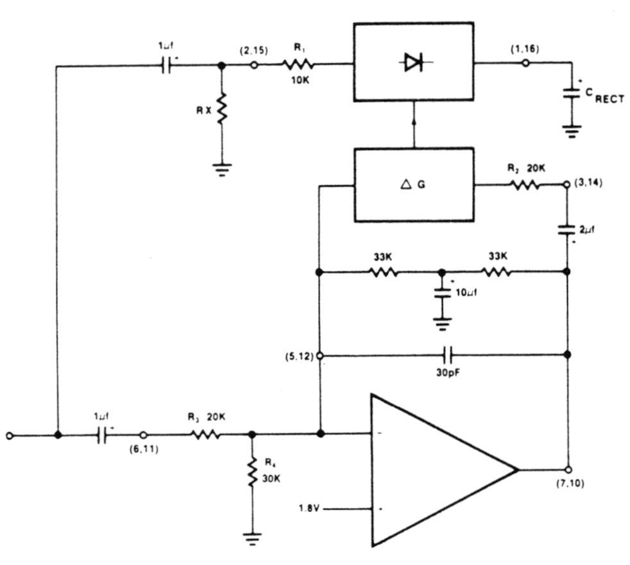

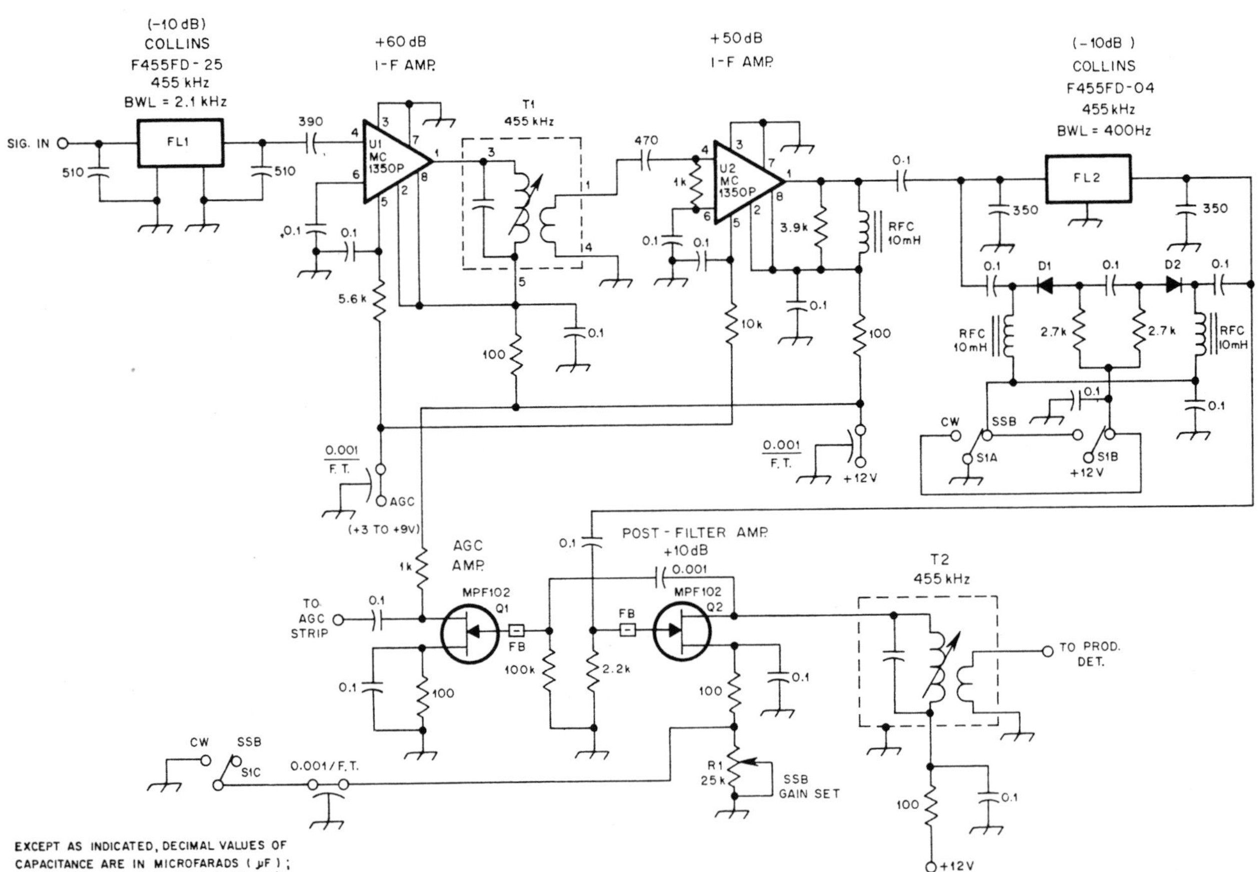

EXCEPT AS INDICATED, DECIMAL VALUES OF CAPACITANCE ARE IN MICROFARADS (µF); OTHERS ARE IN PICOFARADS (pF OR µµF); RESISTANCES ARE IN OHMS; k = 1000, M = 1000 000.

FB = FERRITE BEAD

F.T. = FEEDTHROUGH

DUAL MECHANICAL FILTER IF STRIP—Uses both 2.1-kHz and 400-Hz mechanical filters for optimized reception of both SSB and CW signals. If SSB reception only is desired, a second 2.1-kHz mechanical filter may be used in place of the 400-Hz unit. Q2 (MPF102) compensates for the approximately 10 dB of insertion loss caused by the second mechanical filter (FL2). The exact value of T2 is not critical, but it should have a 20 : 1 impedance step-down ratio for going into a diode-type detector. RF energy is sampled at the drain of Q2 so that the AGC will be relatively constant for both SSB and CW operation.—"The Radio Amateur's Handbook," American Radio Relay League, Newington, CT, 1981, pp. 8-43 to 8-45.

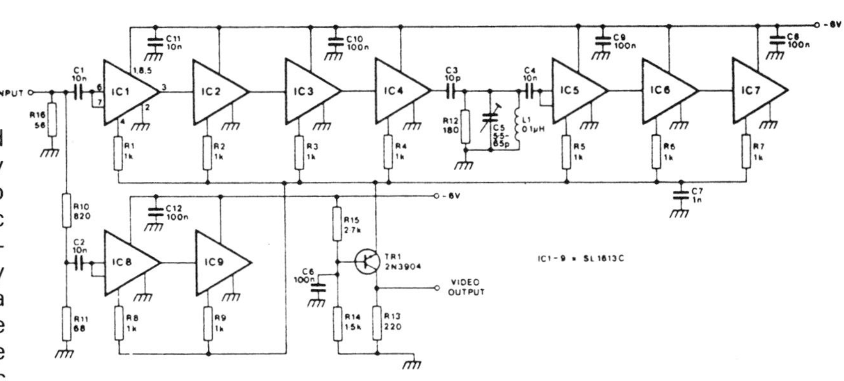

LOGARITHMIC IF STRIP—Designed for use in professional and military pulse radars and features ability to handle signals over wide dynamic range without loss of amplitude information. The center frequency can be either 30 or 60 MHz with a bandwidth of up to 15 MHz. The noise figure is 4 dB. All ICs are SL521 military-grade logarithmic amplifiers.—"Plessey Applications IC Handbook," Plessey Semiconductors, Irvine, CA, 1982, pp. 29–39.

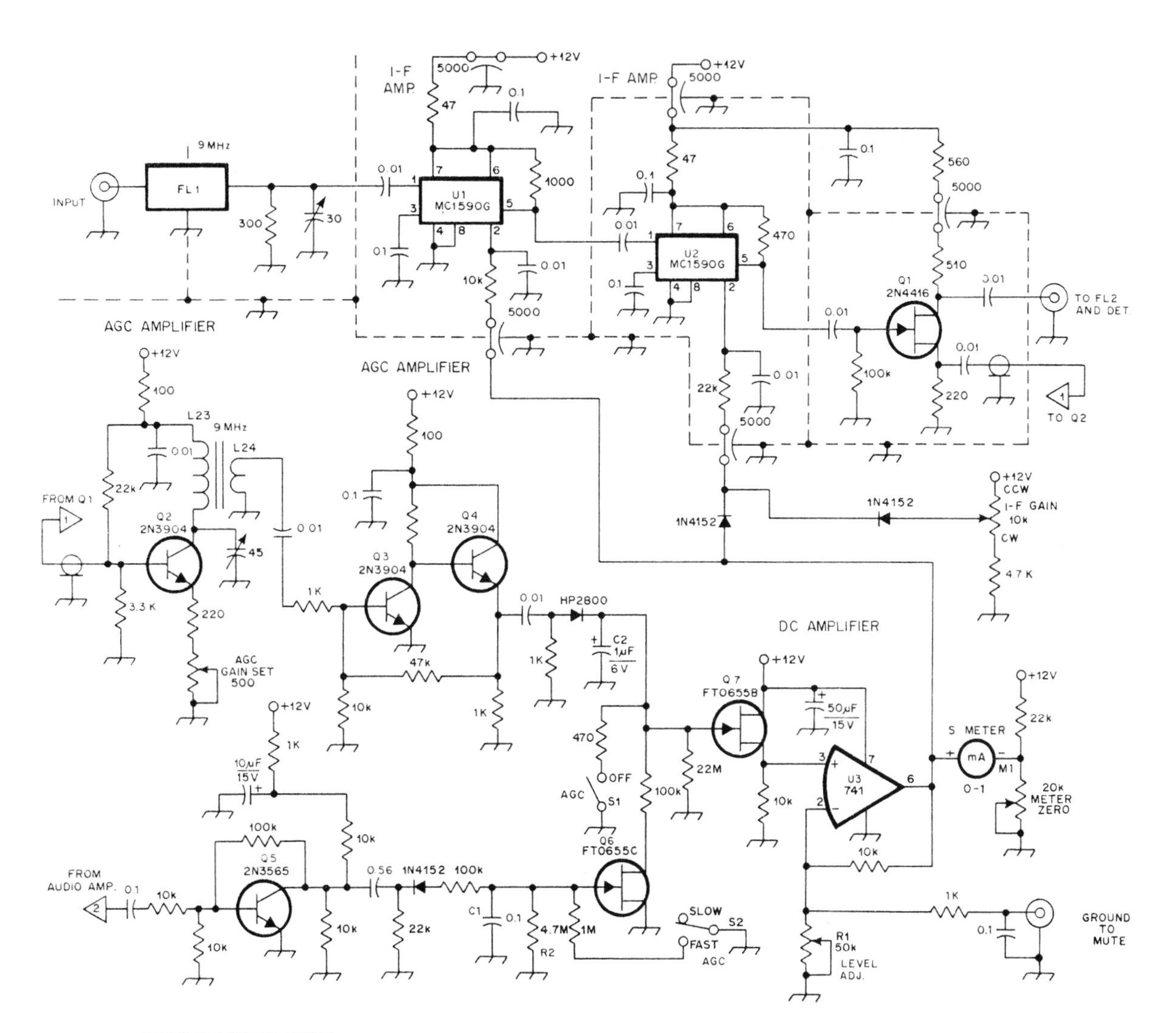

EXCEPT AS INDICATED, DECIMAL VALUES OF CAPACITANCE ARE IN MICROFARADS (μF); OTHERS ARE IN PICOFARADS (pF OR μμF); RESISTANCES ARE IN OHMS; k = 1000, M = 1 000 000

HIGH-PERFORMANCE IF AMPLIFIER AND AGC SYSTEM—Provides "full hang" AGC characteristics. The AGC is defeated by S1. R1 at U3 should be set for +5 V at pin 6 at U3 with the AGC off. The receiver is virtually silent after a strong signal disappears from the passband; it returns to full gain in approximately 50 ms.—"The Radio Amateur's Handbook," American Radio Relay League, Newington, CT, 1981, pp. 8-46 to 8-48.

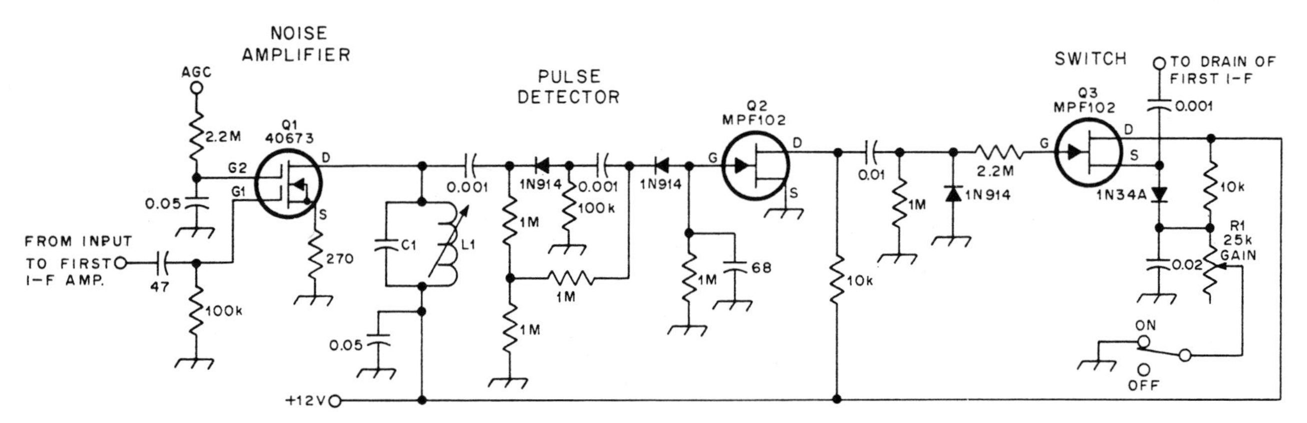

IF NOISE BLANKER—Used ahead of high-selectivity section of receiver. Noise pulses are amplified and rectified, with the resulting negative-going DC pulses used to cut off the amplifier stage during the pulse. C1 and L1 are turned to the receiver's IF frequency.—"The Radio Amateur's Handbook," American Radio Relay League, Newington, CT, 1981, pp. 8-25 to 8-27.

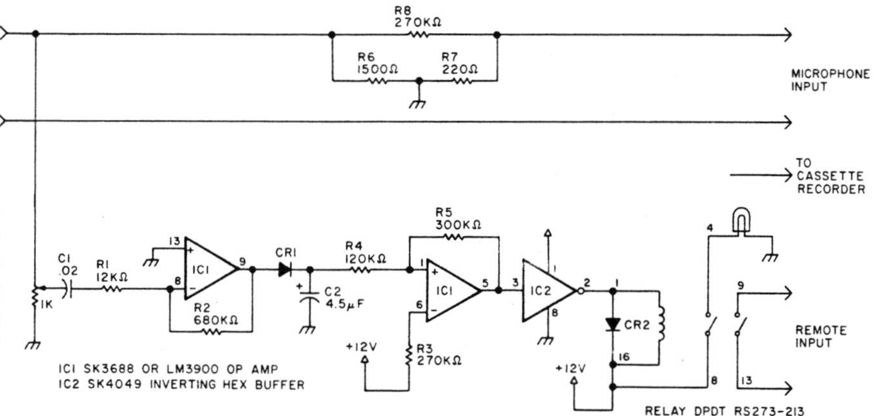

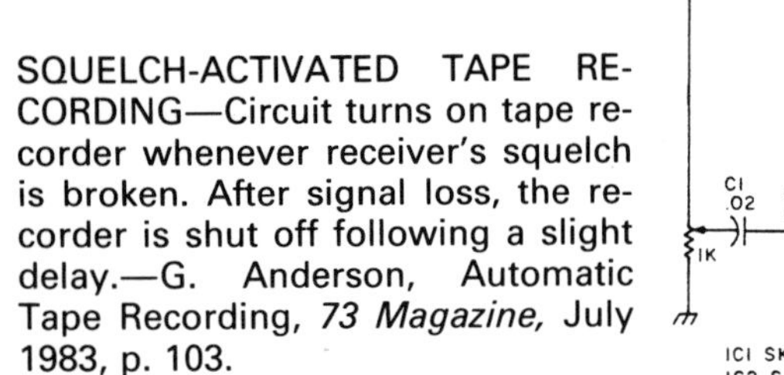
SQUELCH-ACTIVATED TAPE RECORDING—Circuit turns on tape recorder whenever receiver's squelch is broken. After signal loss, the recorder is shut off following a slight delay.—G. Anderson, Automatic Tape Recording, *73 Magazine,* July 1983, p. 103.

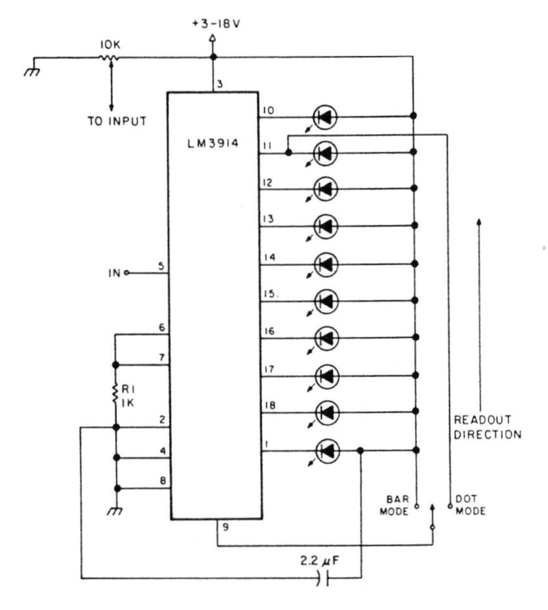

LED S-METER—LEDs indicate strength of received signal. Individual LEDs may be used or a MV57164 LED display may be substituted. A 1.2-V reference voltage is developed in the LM3914 and compared with the voltage applied to the input. Each increase of 0.13 V at the input will result in another LED being turned on.—G. Patterson, Modern-Eyes the S-Meter, *73 Magazine,* July 1984, pp. 24–25.

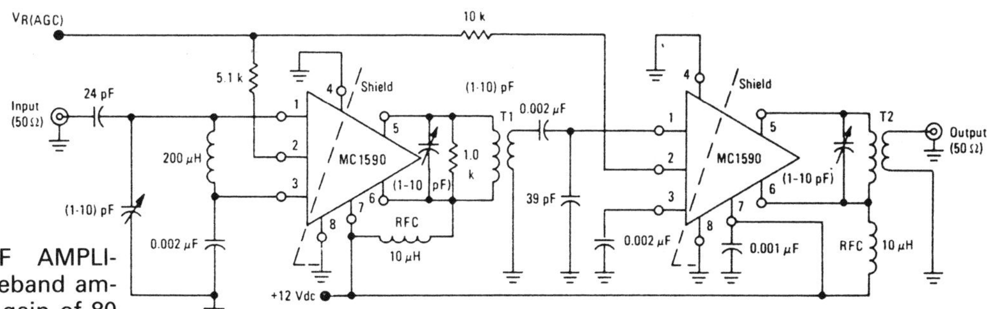

TWO-STAGE 60-MHz IF AMPLIFIER—Dual MC 1590 wideband amplifiers with AGC deliver gain of 80 dB with a bandwidth of 1.5 MHz. T1 consists of a primary of 15 turns of No. 22 wire (0.25-in ID air core) and a secondary of four turns of No. 22 wire. T2 consists of a primary of 10 turns of No. 22 wire (0.25-in ID air core) and a secondary of two turns of No. 22 wire.—"Motorola Linear Integrated Circuits Databook," Motorola, Phoenix, AZ, 1979, p. 6-58.

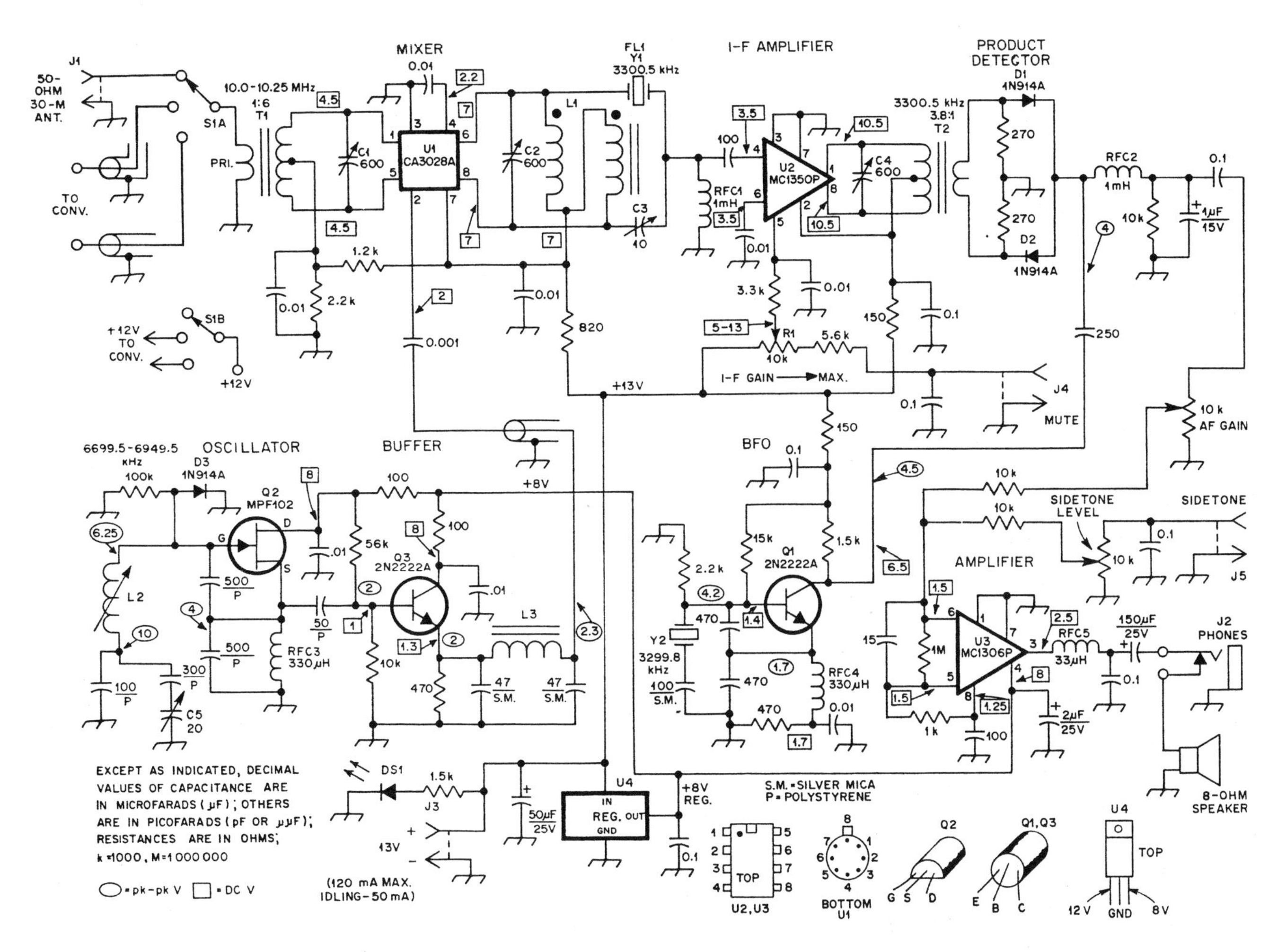

TUNABLE 10-MHz IF STRIP—Covers a tuning range of 10,000–10,300 kHz with selectivity enhanced by crystal filtering. It may be used in VHF converters or as a 10-MHz receiver. The gain is controlled by R1. L1 is eight bifilar turns of No. 28 wire wound on an FT37-61 ferrite core. L2 is a Miller 42A686CB1 or equivalent. L3 is 12 turns of No. 26 wire wound on an FT50-61 ferrite core.—G. Collins, A Tunable IF for VHF Converters, *QST,* May 1982, pp. 32–34.

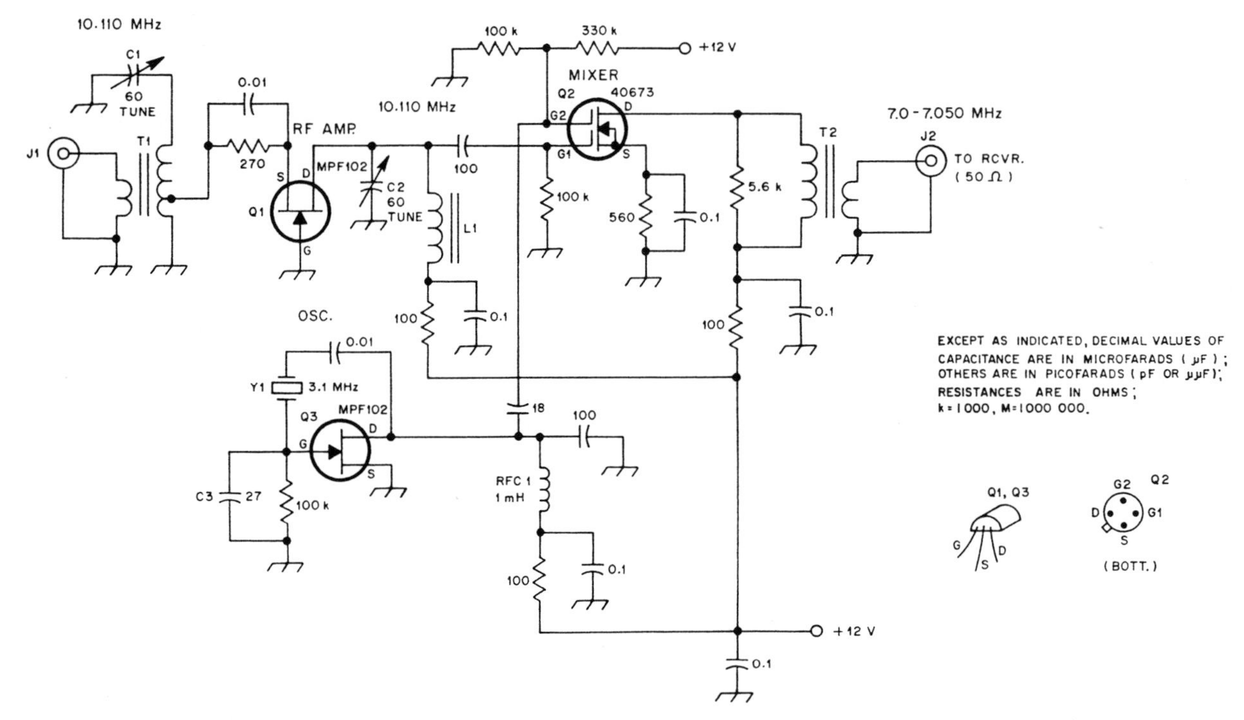

30-METER RECEIVING CONVERTER—Converts signals received in 10,100- to 10,150-kHz amateur radio band to 7000- to 7050-kHz range of 40-meter amateur radio band. L1 is a 6-μH inductor formed from 38 turns of No. 30 wire wound on a T50-6 toroid form. T1 has the same winding on the secondary as L1. Tap the source of Q1 eight turns above the grounded end. The primary has three turns of No. 30 wire. T2 has 15 primary turns of No. 24 wire on an FT50-43 toroid form. The secondary consists of two turns of No. 24 wire.—D. DeMaw, A VXO CW Rig for 30 Meters, *QST,* November 1983, pp. 31–34.

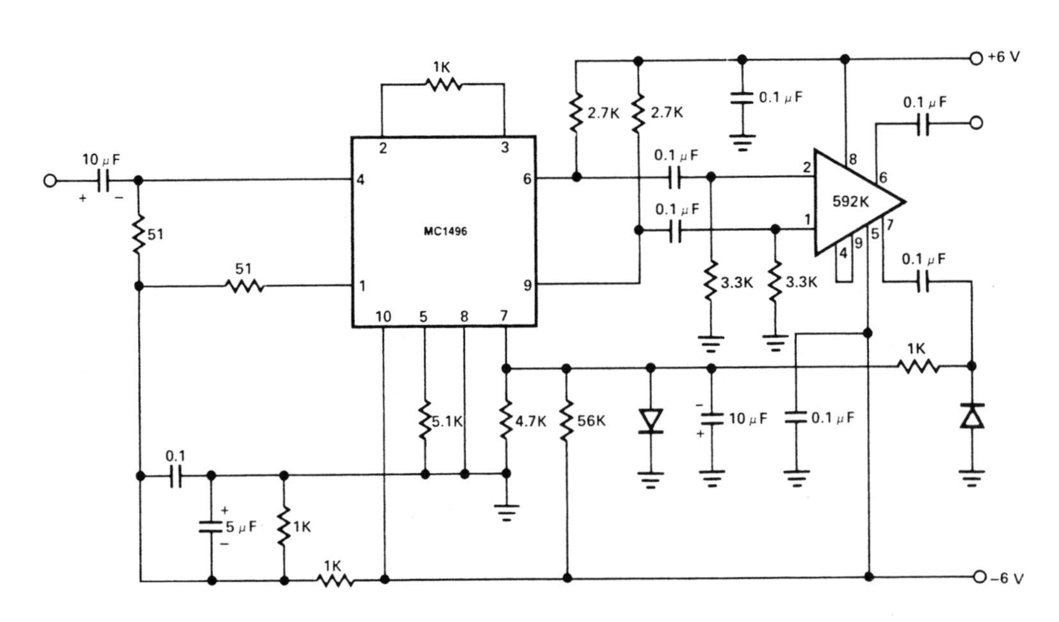

WIDEBAND AGC AMPLIFIER—Signal is fed to signal input of MC1496 and RC-coupled to the NE592. Unbalancing the carrier input of the MC1496 causes the signal to pass through unattenuated. Rectifying and filtering one of the NE592 outputs produces a DC signal which is proportional to the AC signal amplitude. After filtering, this control signal is applied to the MC1496, causing its gain to change.—"Signetics Analog Applications Manual," Signetics, Sunnyvale, CA, 1979, pp. 141–142.

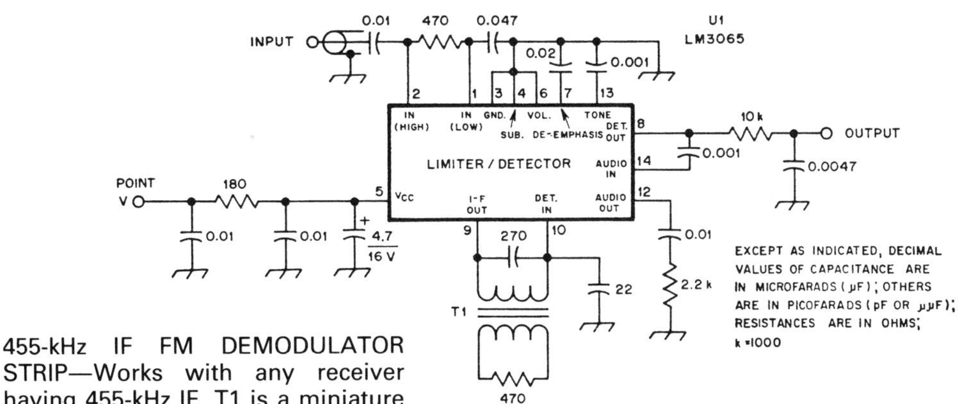

455-kHz IF FM DEMODULATOR STRIP—Works with any receiver having 455-kHz IF. T1 is a miniature 455-kHz IF transformer with a rated inductance of 680 μH. The limiter/detector IC is an LM3065 or equivalent.—B. Heil, Experience 10-Meter FM Operation, *QST,* August 1981, pp. 22–26.

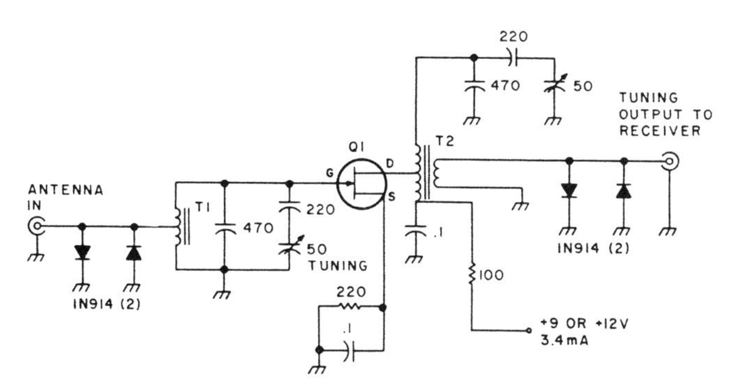

160-METER PREAMPLIFIER—Covers 1800–2000 kHz and is designed especially for use with indoor receiving antennas. Q1 is a general-purpose FET (Radio Shack 276-2036 or equivalent). T1 is an Amidon T50-2 toroid core with approximately 50 turns tapped at six turns. T2 is the same as T1 except that it is tapped at the center and has a six-turn link over the ground end.—J. Geist, Top-Notch for Top End, *73 Magazine,* May 1982, pp. 26–30.

WWV CONVERTER—Converts WWV's 10-MHz signal to 3500- to 4000-kHz amateur radio band. L1, L2, and L3 are wound on a 0.25-in-diameter slug-tuned coil form. Both L1 and L2 are wound counterclockwise. L1 is 12 turns of No. 24 wire and L2 is four turns of No. 24 wire. L3 is 35 turns of No. 32 wire wound clockwise. The crystal frequency is determined by subtracting the desired output frequency from the 10-MHz WWV frequency.—R. Schlegel, WWV-To-80 Meter Converter, *73 Magazine,* January 1981, pp. 48–49.

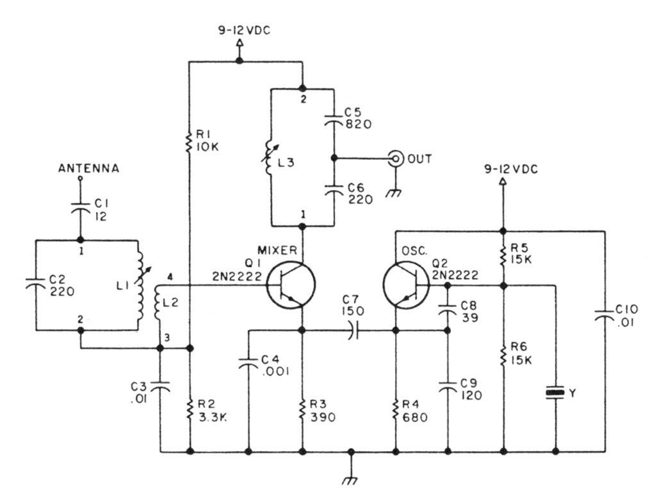

18

Repeater Circuits

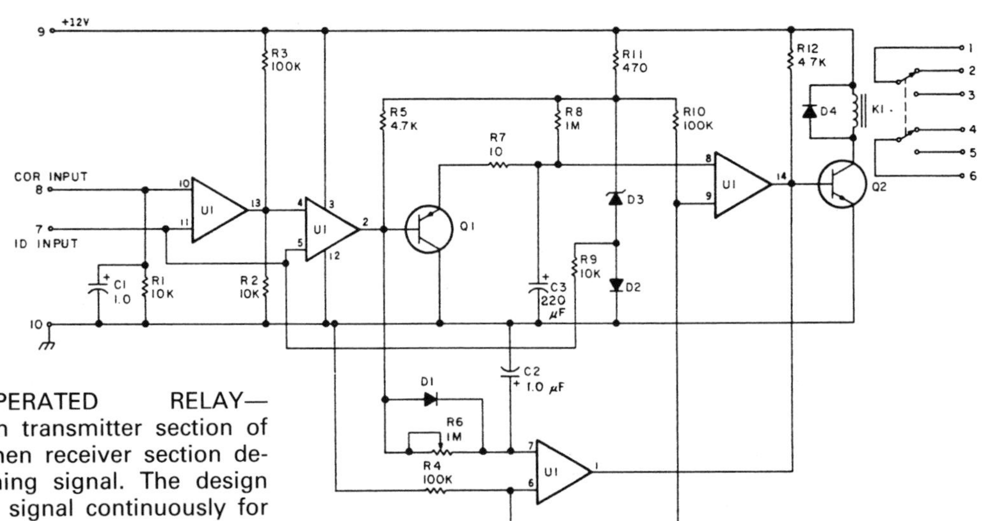

CARRIER-OPERATED RELAY—Switches on transmitter section of repeater when receiver section detects incoming signal. The design will relay a signal continuously for up to 3 minutes; after that the repeater will shut down until the input signal breaks for at least 200 ms. U1 is an LM339N. Q1 is an 2N2905A or 2N4402. Q2 is a 2N2270 or 2N4400. D1 and D2 are each 1N914 or equivalent. D3 is a 5.1-V 400-mW zener and D4 is a 1-A 100-V rectifier (1N4002 or equivalent). K1 is 12-V relay (NC2D-JP-DC12V or equivalent).—F. Kalmus, The COR of a Reliable Repeater, *73 Magazine*, August 1983, pp. 24–25.

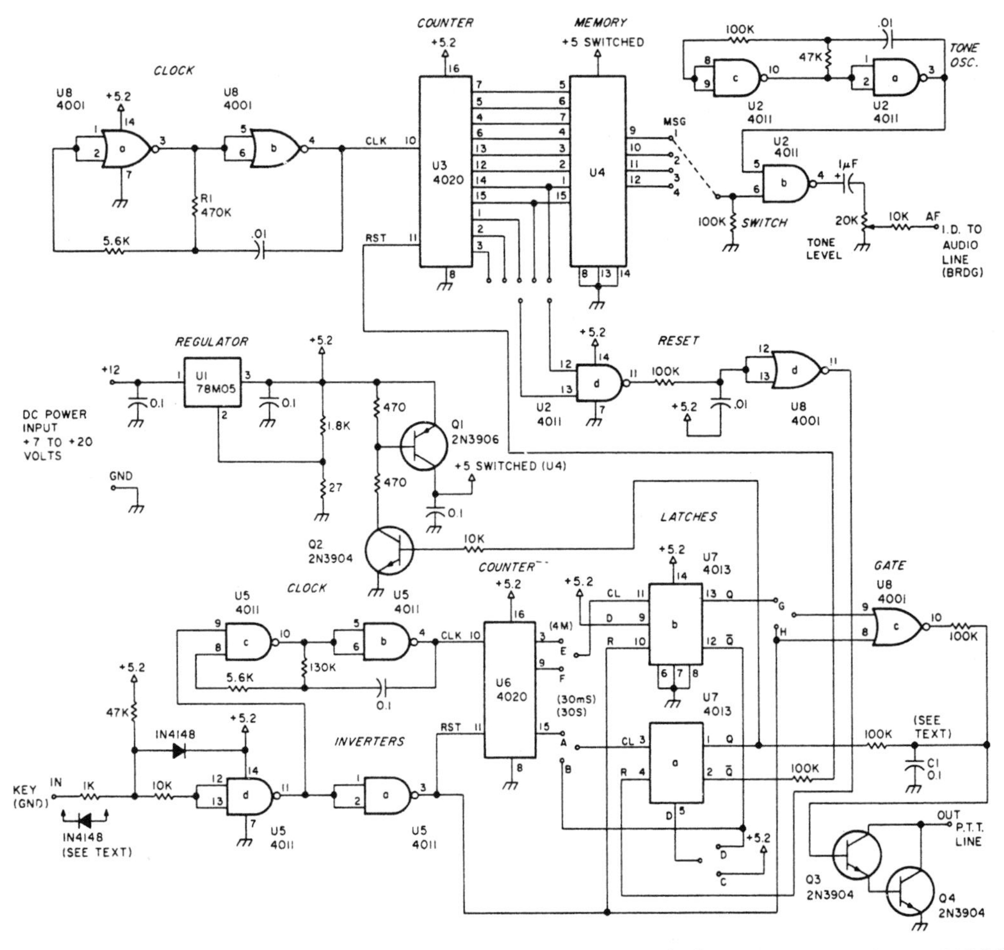

REPEATER IDENTIFIER/TIMER—U4 is 246 × 4 (1-kb) bipolar PROM. The circuit can generate CW messages up to 256 bits in length. Time-out, the identification interval, and the speed of the CW message can be adjusted as desired. The addition of another PROM can increase the message capacity to 512 bits.—D. Jarvis and B. Pepper, Perfect Timing, *73 Magazine*, September 1984, pp. 22–26.

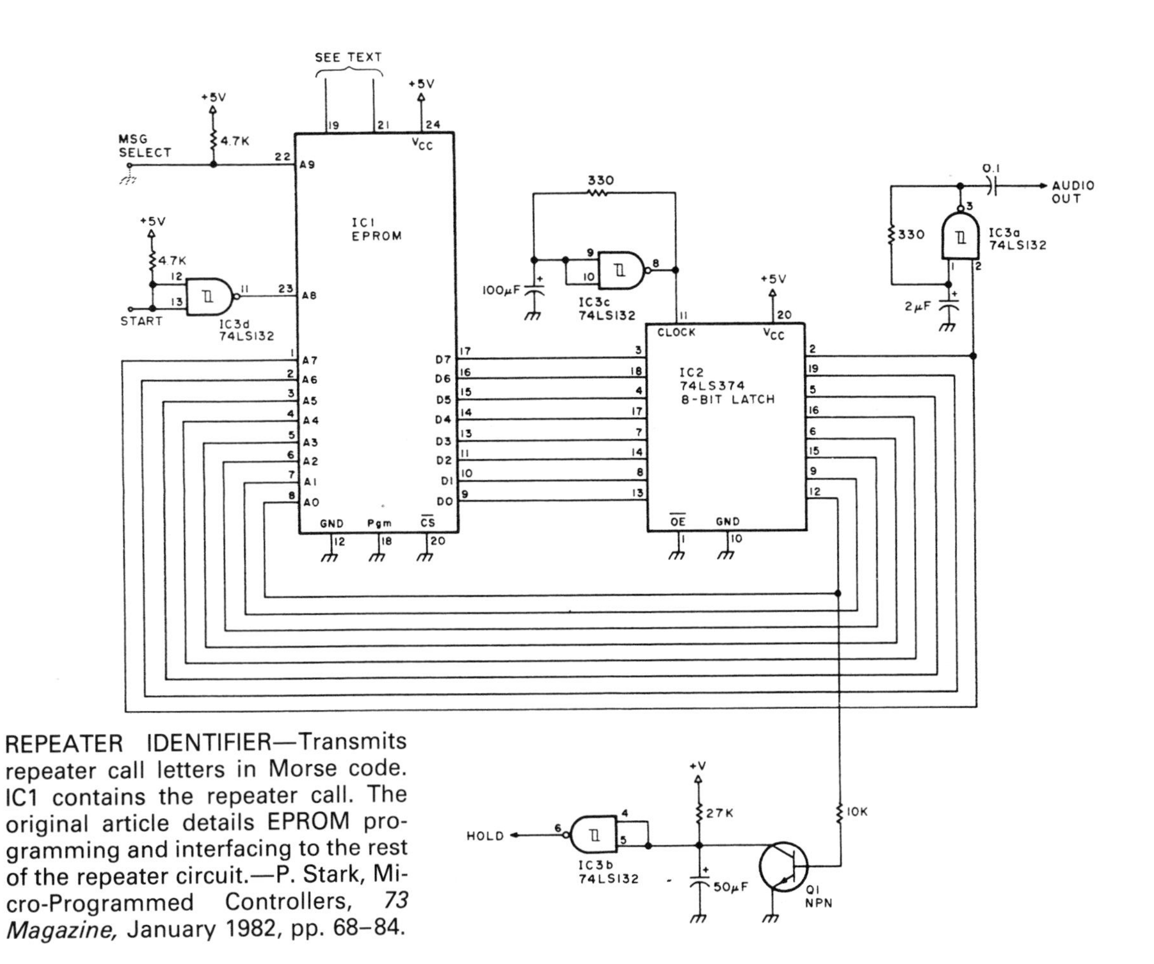

REPEATER IDENTIFIER—Transmits repeater call letters in Morse code. IC1 contains the repeater call. The original article details EPROM programming and interfacing to the rest of the repeater circuit.—P. Stark, Micro-Programmed Controllers, *73 Magazine,* January 1982, pp. 68–84.

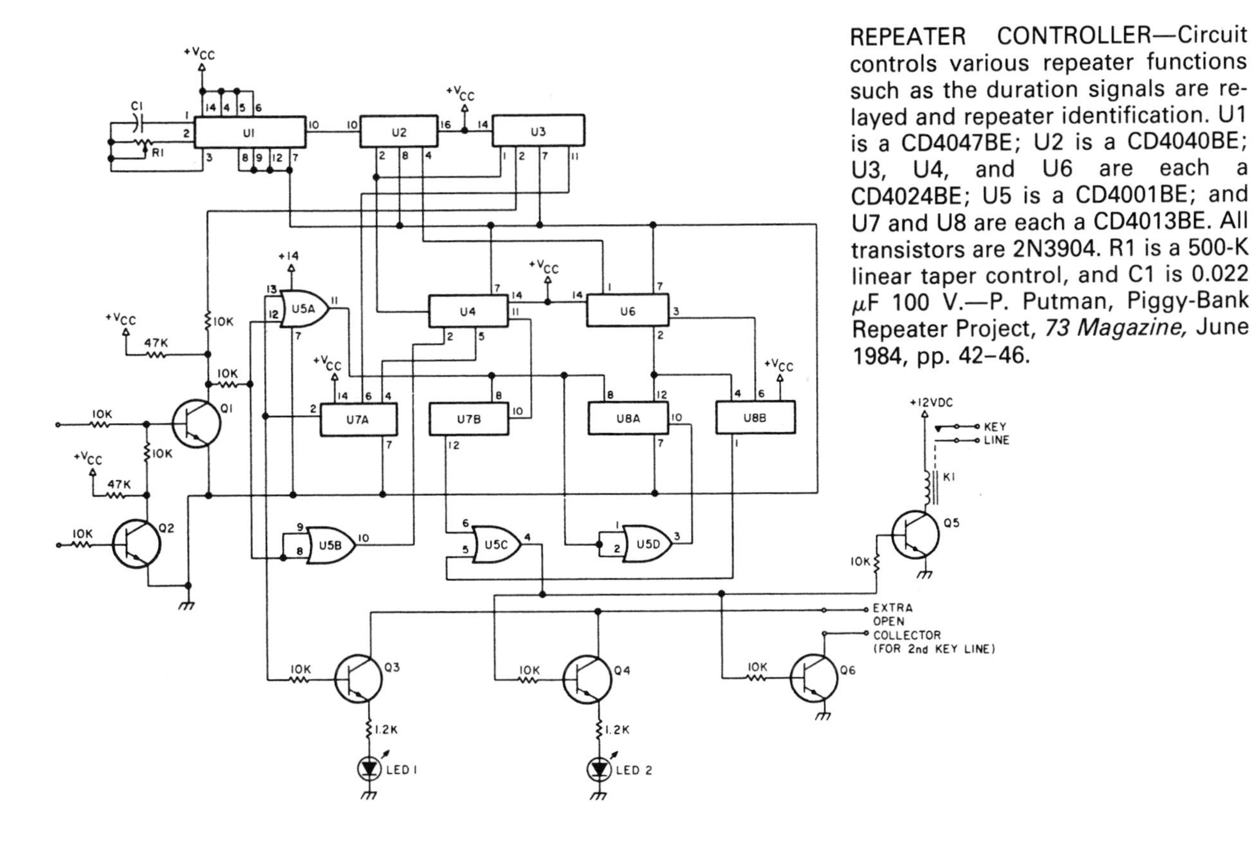

REPEATER CONTROLLER—Circuit controls various repeater functions such as the duration signals are relayed and repeater identification. U1 is a CD4047BE; U2 is a CD4040BE; U3, U4, and U6 are each a CD4024BE; U5 is a CD4001BE; and U7 and U8 are each a CD4013BE. All transistors are 2N3904. R1 is a 500-K linear taper control, and C1 is 0.022 μF 100 V.—P. Putman, Piggy-Bank Repeater Project, *73 Magazine,* June 1984, pp. 42–46.

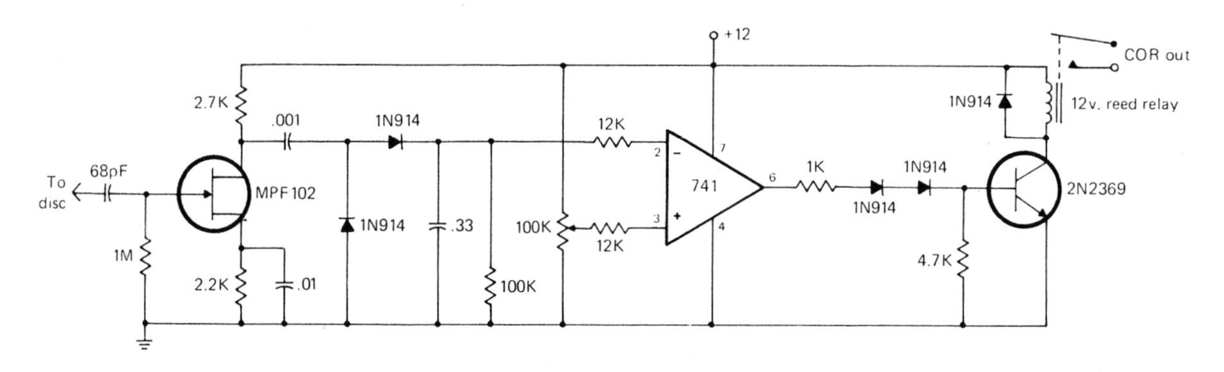

NOISE-OPERATED COR—COR incorporates its own noise amplifier and is activated by FM receiver's squelch, which is noise-activated. The input is provided by the receiver's discriminator, and noise is amplified by the MPF 102 and rectified by the 1N914 diodes. The COR operates independently of the squelch setting.—P. Hughes, A Noise Operated COR (Carrier Operated Relay), *CQ,* February 1981, pp. 18–19.

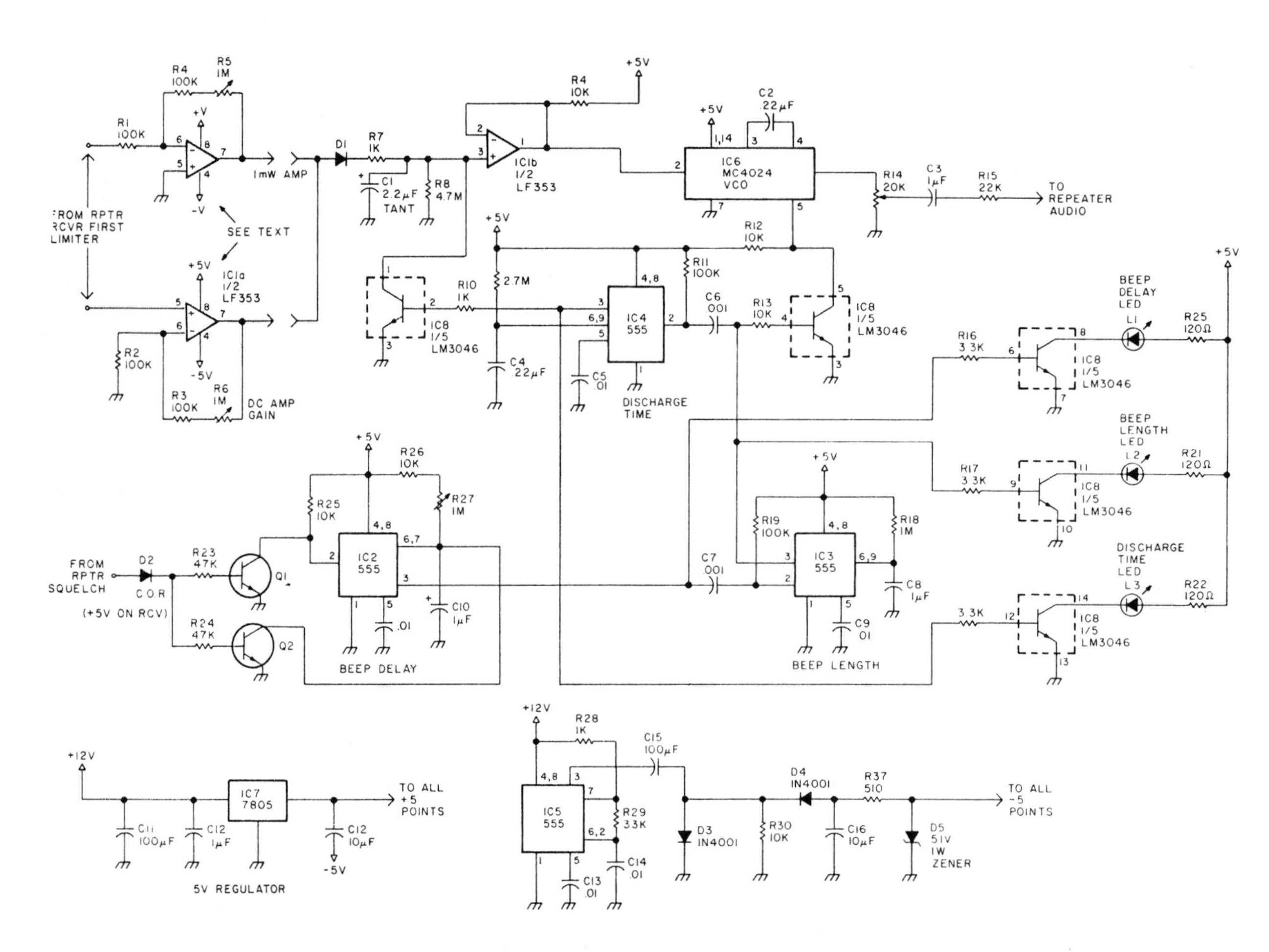

AUDIBLE S-METER—Returns "beep" tone proportional to received signal strength. The tone frequency range is 800 Hz for 0.2 μV and approximately 2800 Hz for about 1.0 μV. The circuit first samples the limiter voltage, then amplifies it and feeds it to a voltage controlled oscillator which generates the beep. Q1 and Q2 are general purpose NPN transistors such as 2N3392.—C. Kollar, Sound Off!, *73 Magazine,* January 1984, pp. 28–30.

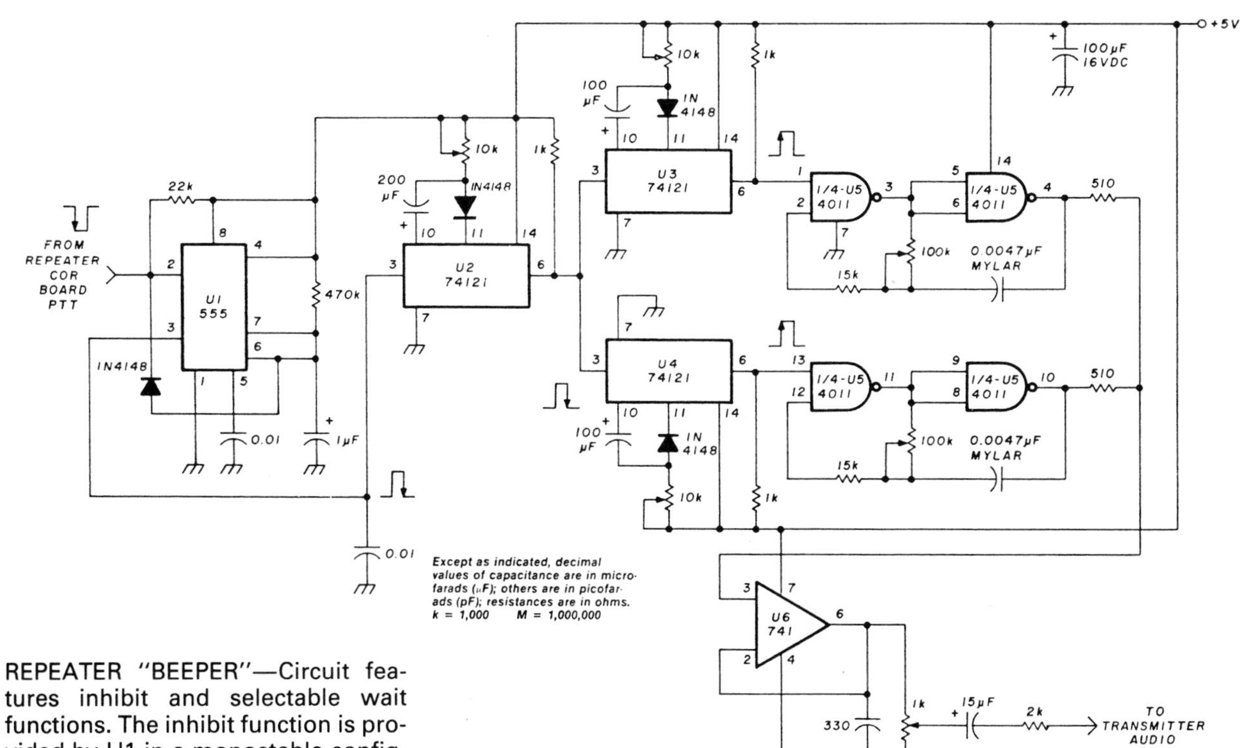

REPEATER "BEEPER"—Circuit features inhibit and selectable wait functions. The inhibit function is provided by U1 in a monostable configuration. It is made retriggerable by CR1 between pins 6 and 2. When the squelch from the repeater is broken, the PTT signal on the COR board (normally high) is forced low. This makes pin 2 on U1 go high. As long as the COR board keeps the PTT signal low, the output of U1 remains high and no timing function is permitted. The wait function is implemented by the 74121 used in a monostable configuration. The audio output is adjustable from 1000 Hz to 10 kHz.—A. Torres, Super Beep Circuit for Repeaters, *Ham Radio,* July 1981, pp. 48–50.

PRIVATE LINE™ TONE GENERATOR—Produces subaudible tones used by Private Line™ tone-coded squelch system to limit access to repeater systems to authorized users. The circuit may be adapted for use as a pulse generator, a programmable reference, or a tone generator for a multiple-tone FSK modem.—M. Di Julio, The Last PL Generator, *73 Magazine,* March 1981, pp. 50–52.

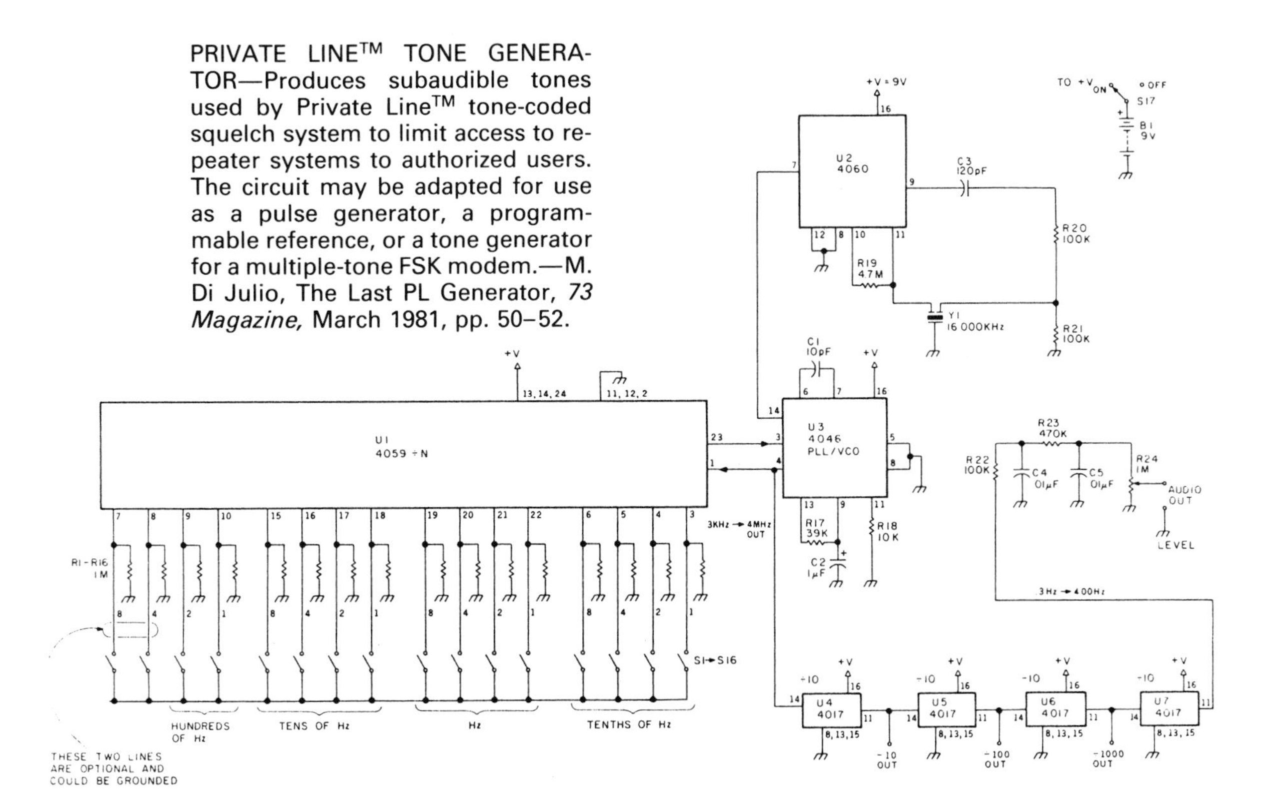

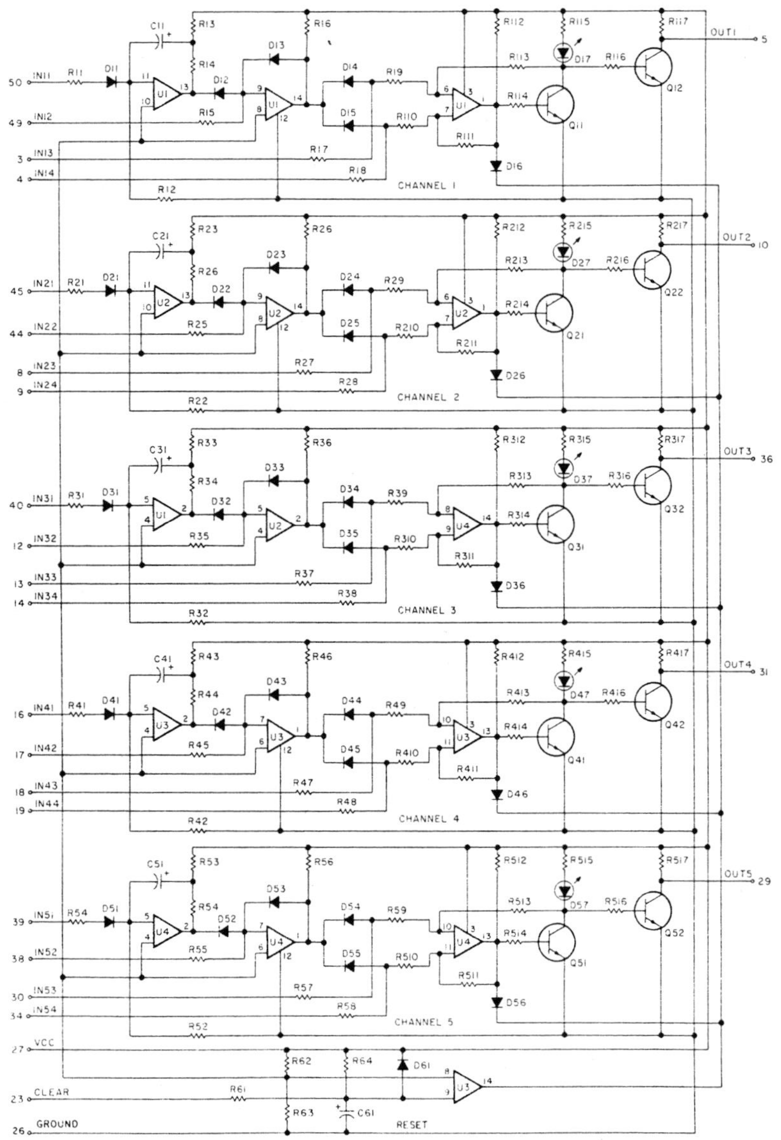

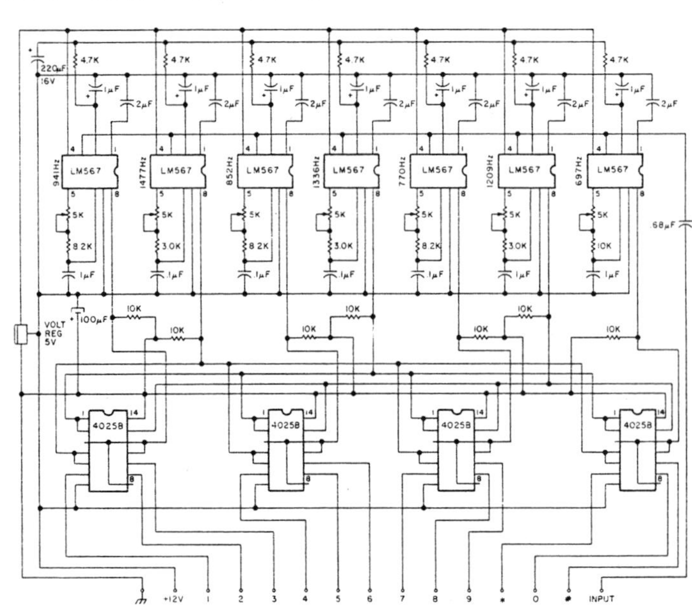

DUAL-TONE TWELVE-OUTPUT REPEATER CONTROL BOARD—Accepts three-digit commands to control such functions as access to repeater, autopatch, station identification, changing antennas, etc. The unit consists of separate tone decoder and access control sections. U1 through U4 on the access control board are all LM339, while all transistors are NPN types rated at 1A and 45 V.—F. Kalmus, Ironclad Repeater Control, *73 Magazine,* December 1983, pp. 100–103.

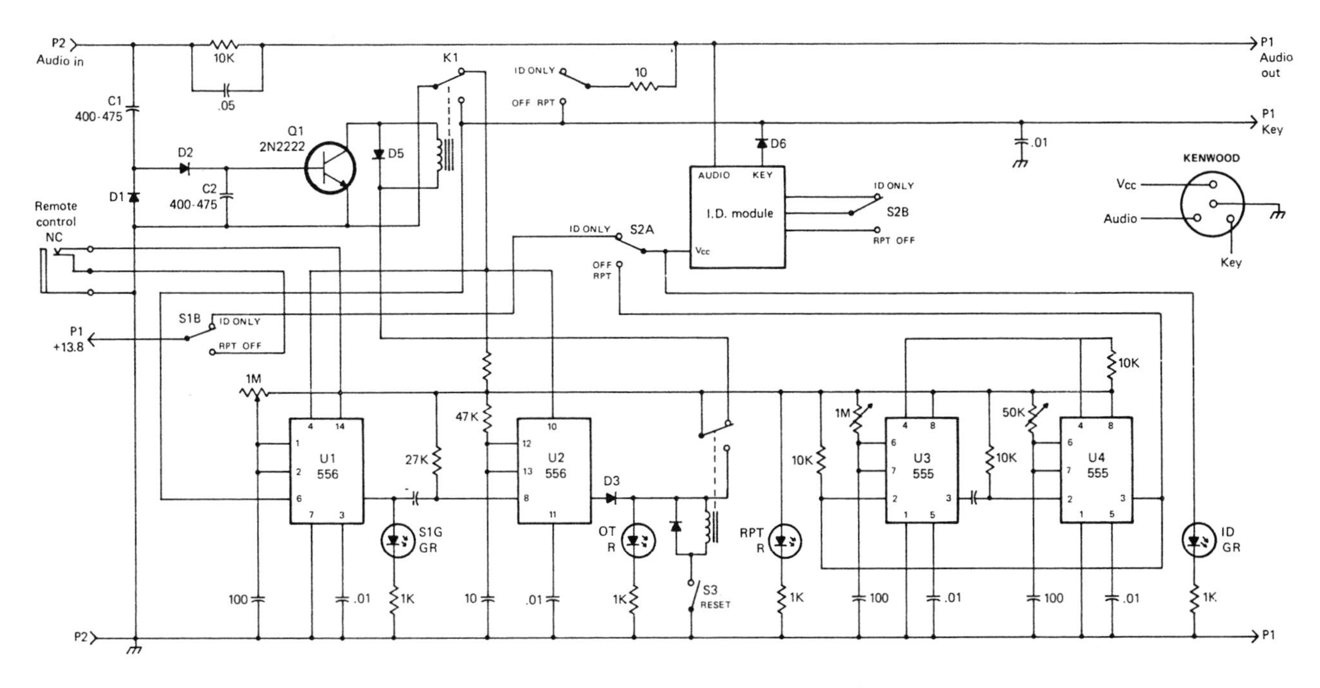

"INSTANT" REPEATER—Converts any modern transceiver/transmitter and receiver/second transceiver into temporary repeater station. The repeater is voice-activated on retransmit and can be set for retransmission times of 20 seconds to 3 minutes. All diodes are 1-A 50-V PIV types. Both relays are 12-V 10-mA SPDT types. The I.D. module is an "Autocode" PROM device. Separate antennas should be used for receive and transmit; they should be separated physically by an odd multiple of one-quarter wavelength at the transmit frequency.—W. Becker, The Instant Repeater, *CD,* March 1983, pp. 20–22.

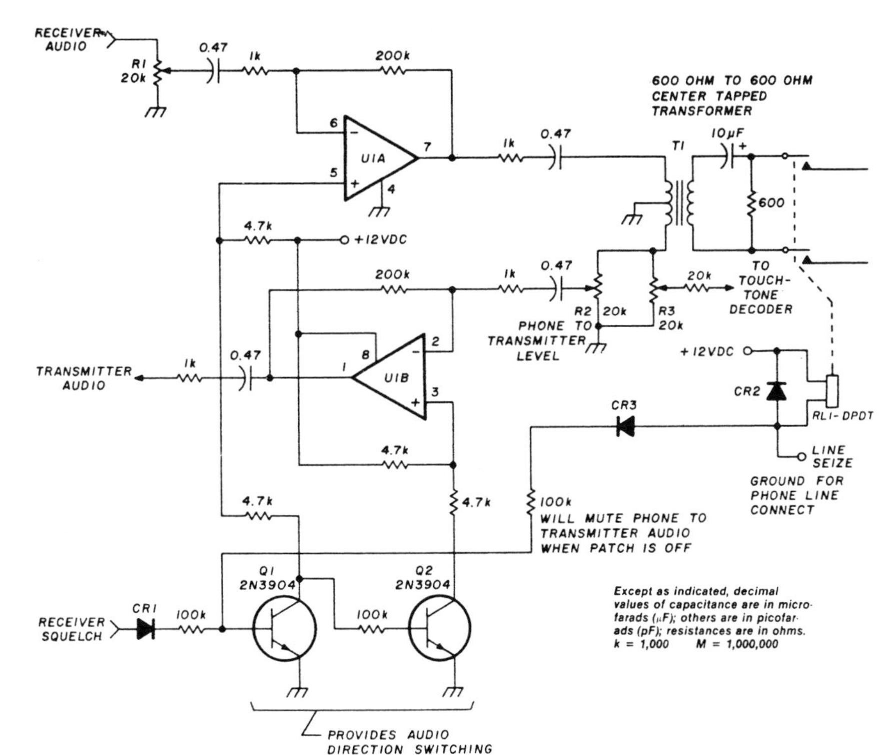

REPEATER AUTOPATCH—Allows connection of repeater station to telephone network so stations accessing repeater can communicate through telephone network. The circuit requires mobile radiotelephone stations to access the telephone system through the TouchTone™ standard telephone dialing tones and for the autopatch to be followed by a TouchTone™ decoder circuit. R1 adjusts the receiver-to-telephone line audio, and R2 adjusts the telephone-to-transmitter audio level. R3 allows the adjustment of audio to a TouchTone™ decoder. T1 isolates the circuit from the telephone network. RL1 acts as the ON/OFF hook interface. U1A and U1B are each LF353 or equivalent.—R. Wright, Autopatch Phone Line Interface, *Ham Radio,* January 1984, pp. 91–93.

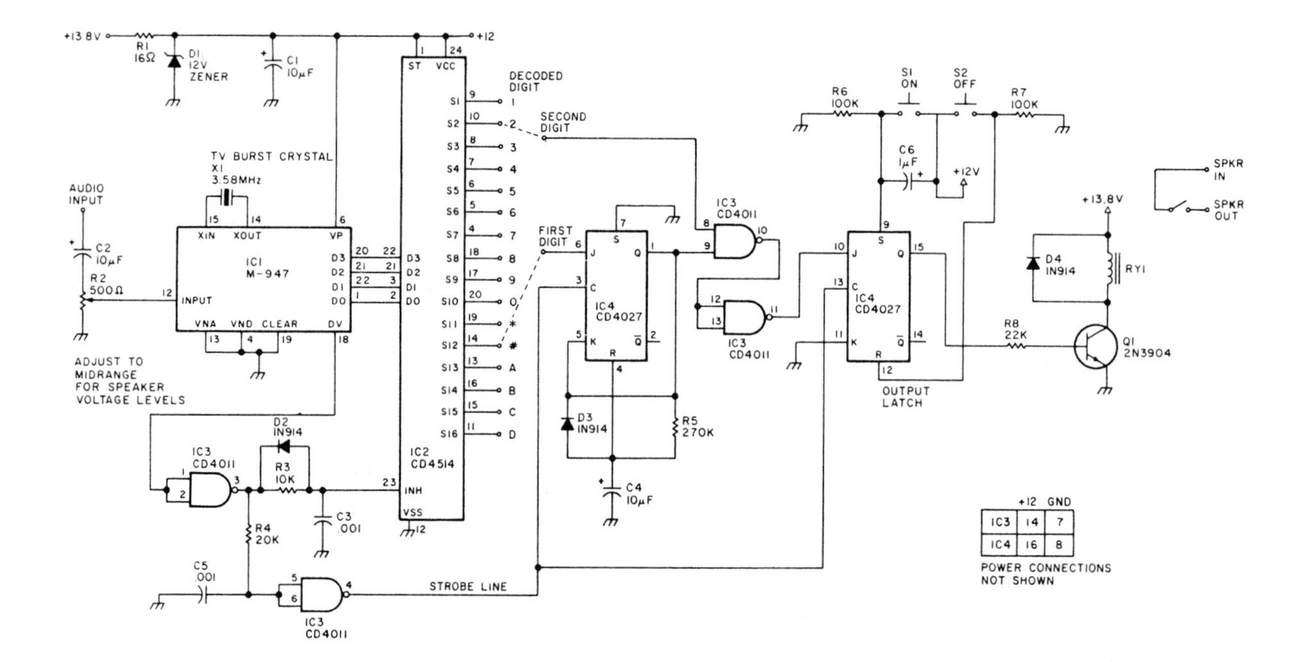

TWO-DIGIT TONE-OPERATED SQUELCH—Uses two standard telephone dialing tones to open squelch of receiver. The unit is built from separate tone squelch and four-digit sequence detector sections. Once the receiver is enabled, it will remain so until S2 (the OFF switch) is pressed.—R. Rumbolt, World's Fair Super Squelch, *73 Magazine,* October 1983, pp. 86–87.

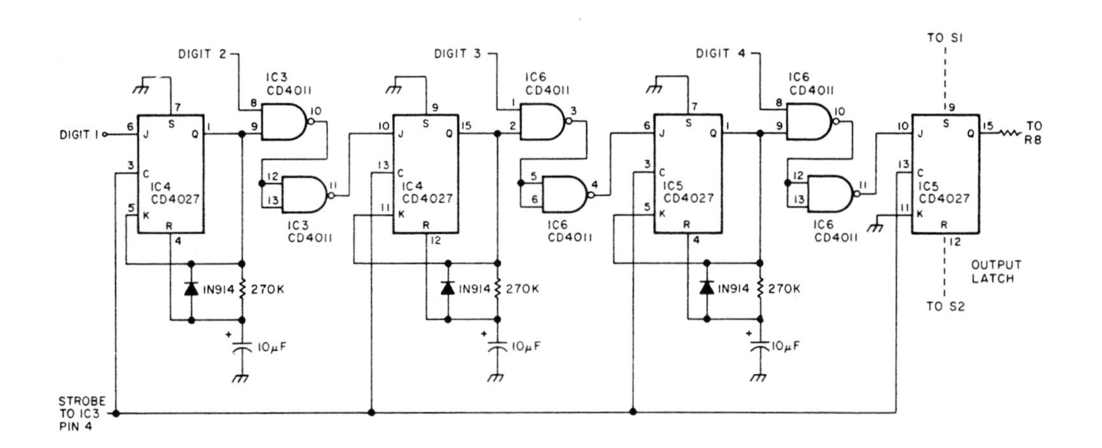

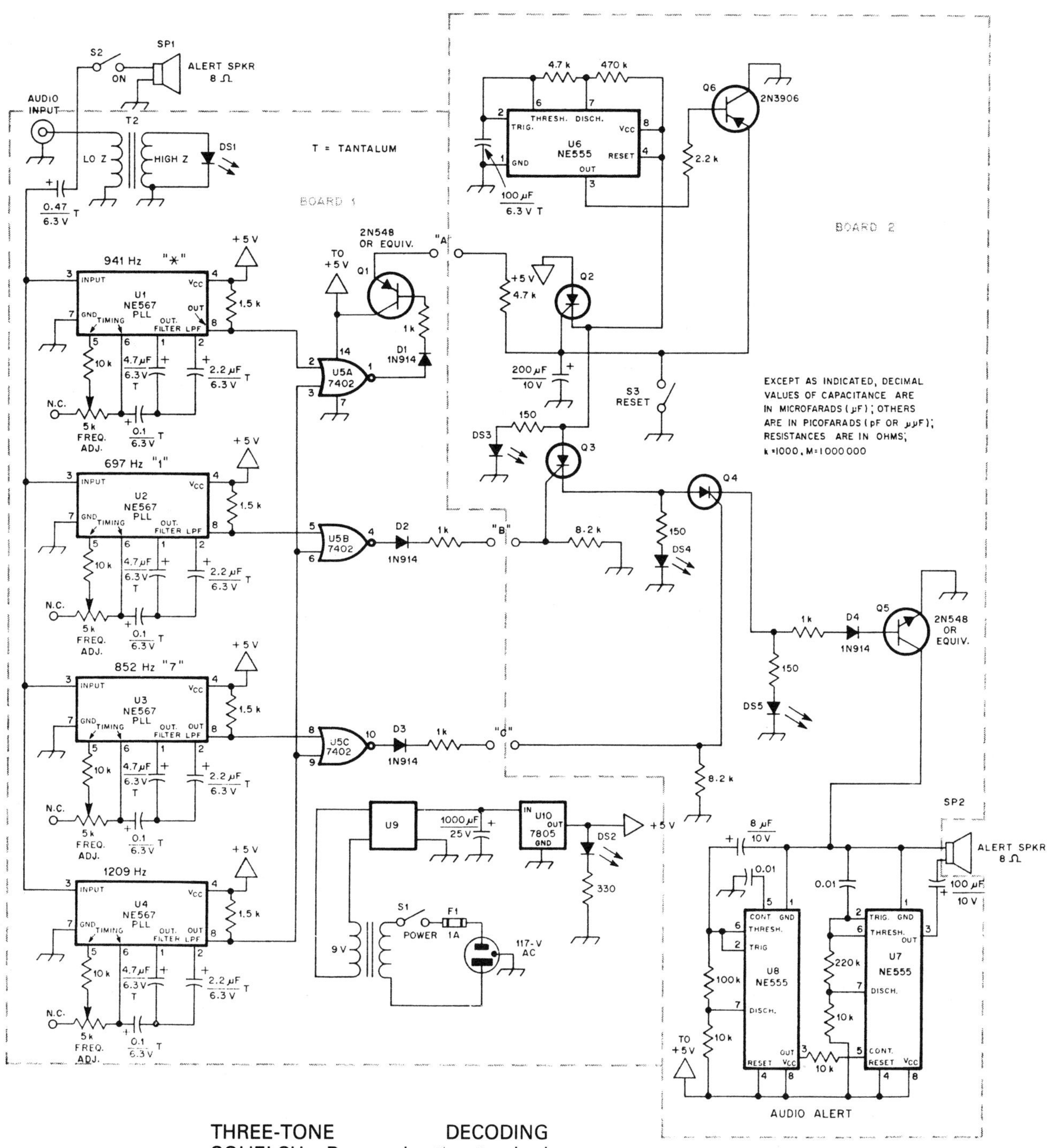

THREE-TONE DECODING SQUELCH—Responds to standard telephone dialing tones and sounds siren for 45 seconds with automatic shutoff. It requires three tones for access and incorporates false alert suppression. LEDs give a visual indication if the unit has been activated in the operator's absence.—T. Tanner, The London Tone Alert, *QST,* November 1983, pp. 35–37.

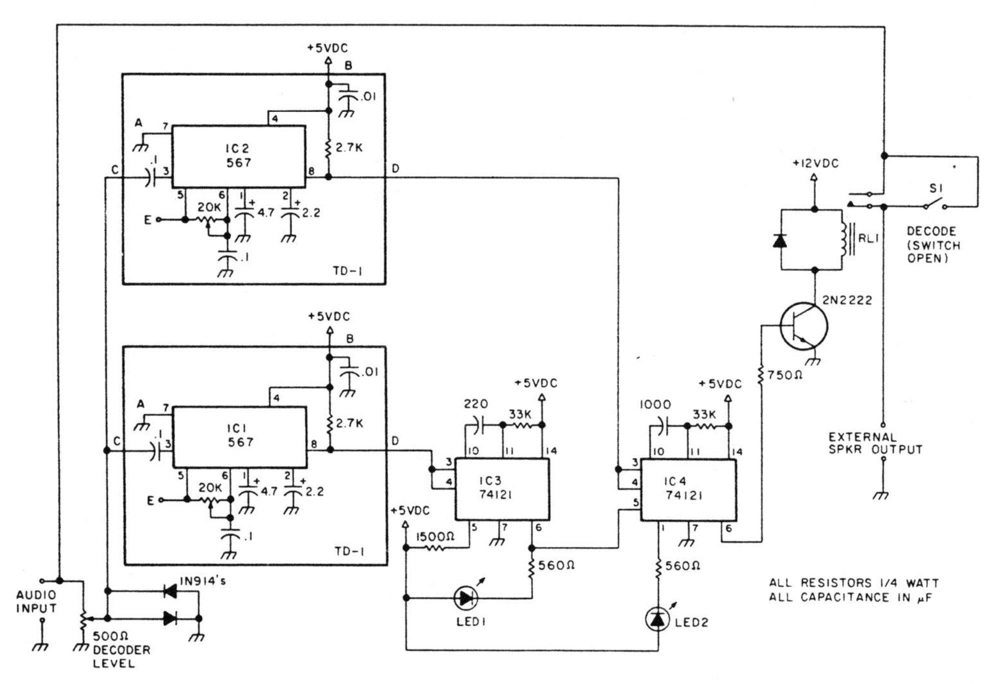

TWO-TONE SEQUENTIAL SQUELCH—Silences receiver until proper pair of tones is received. The design responds to a 585-Hz/652-Hz tone pair. TD-1 is a Ramsey Electronics tone decoder module.—T. Bowman, A Two-Tone Squelch to Solve Your Scanner Woes, *73 Magazine,* November 1983, pp. 84–85.

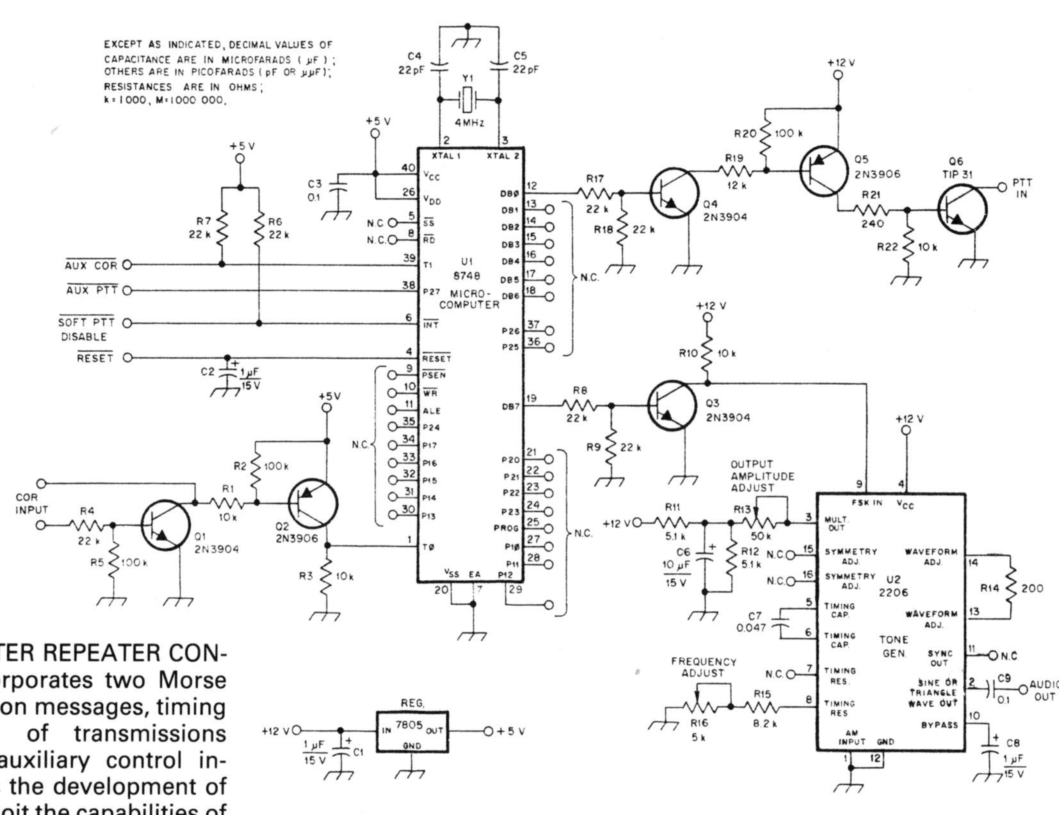

MICROCOMPUTER REPEATER CONTROLLER—Incorporates two Morse code identification messages, timing functions, end of transmissions "beeps," and auxiliary control inputs. It requires the development of software to exploit the capabilities of the hardware.—S. Freeberg, The Microcomputer Repeater Controller, *QST,* December 1983, pp. 26–31.

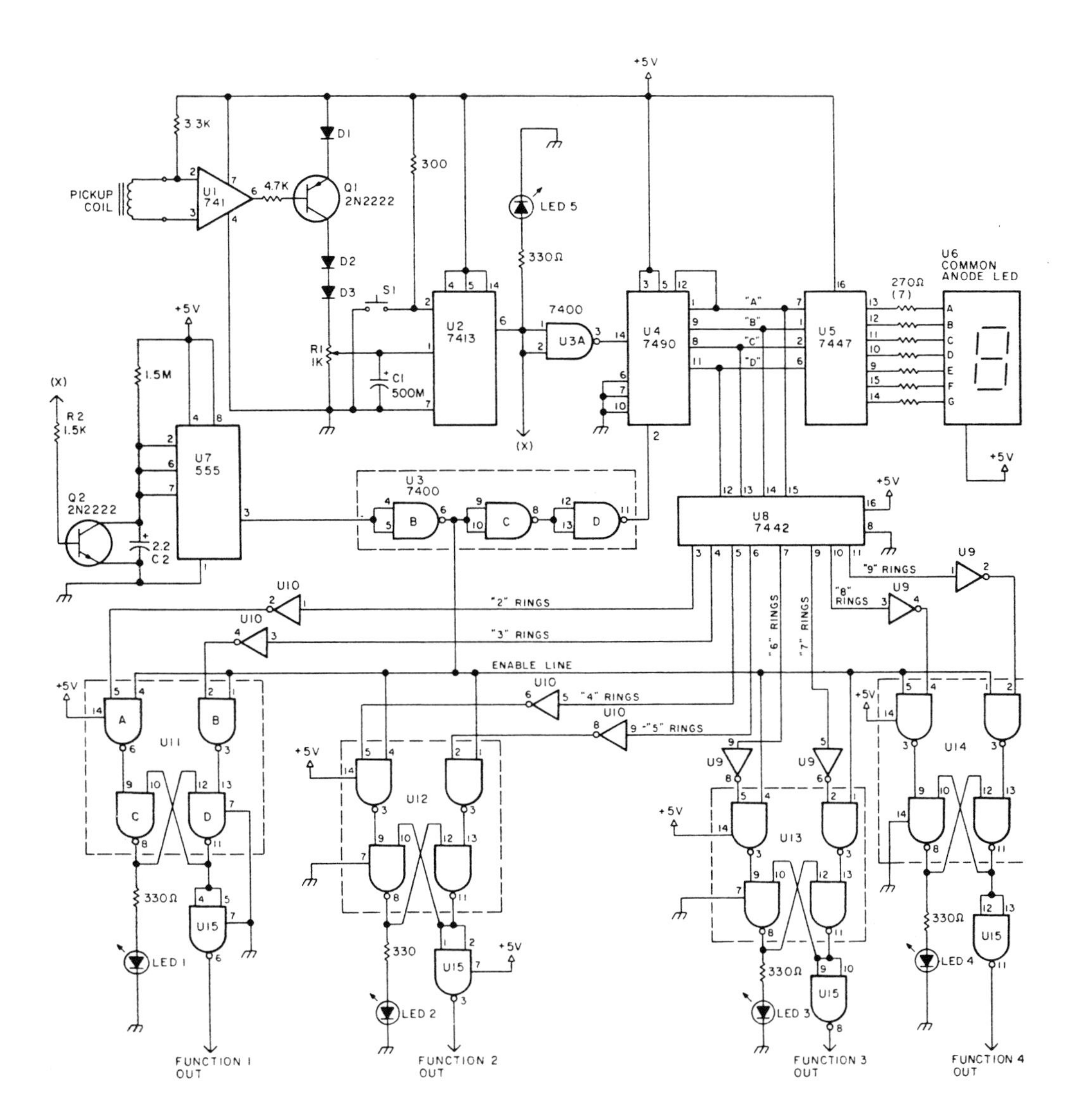

TELEPHONE-BASED REPEATER CONTROL SYSTEM—Provides four independent ON/OFF outputs and no direct connection to telephone system. The circuit monitors the telephone for an incoming ring. If two rings are received, function 1 is turned on. If three rings are received, function 1 is turned off. If four rings are received, function 2 is turned on. If five rings are received, function 2 is turned off. If six rings are received, function 3 is turned on. If seven rings are received, function 3 is turned off. If eight rings are received, function 4 is turned on. If nine rings are received, fucntion 4 is turned off. U9 and U10 are each 7404, and U11 through U14 are all 7400.—T. Krohn, Superman's Repeater Control System, *73 Magazine,* February 1983, pp. 32–35.

19

Single Sideband Circuits

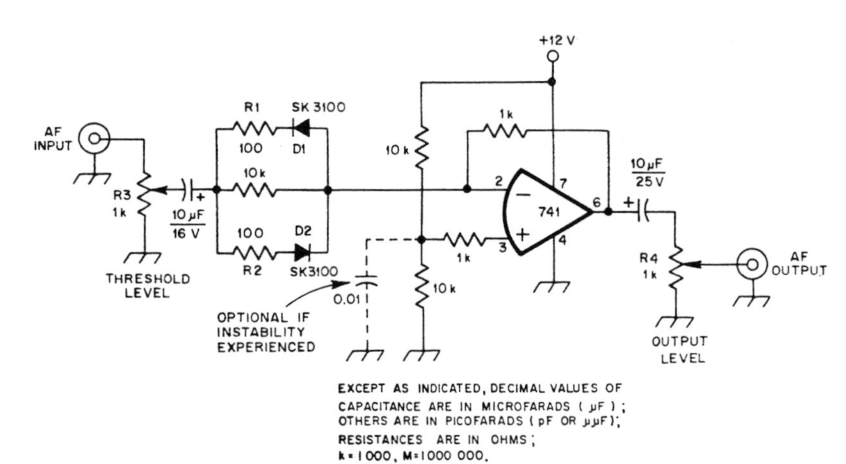

SPEECH DECOMPRESSOR—Restores natural-sounding quality to received SSB signals that have been heavily processed during transmission. Background noise and receiver "hiss" are also reduced.—E. Nichols, Try this Speech "Decompressor," *QST,* December 1983, pp. 24–25.

PRODUCT DETECTOR—Circuit is broadband for entire high-frequency range and has sensitivity of 3.0 μV and dynamic range of 90 dB when operated at intermediate frequency of 9 MHz. If it is operated at intermediate frequencies down to 50 kHz, the 0.1-μF capacitors at pins 7 and 8 should be increased to 1.0 μF. The circuit may also be used as an AM detector by introducing a carrier signal at the carrier input and an AM signal at the SSB input.—"Motorola Linear Integrated Circuits Databook," Motorola, Phoenix, AZ, 1979, p. 6-98.

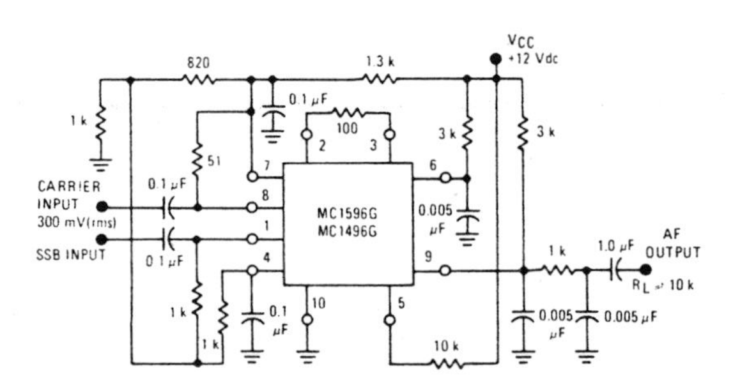

MICROPHONE EQUALIZER—Adjusts response of microphone to operator's voice characteristics. The active elements in the circuit allow the boosting or cutting of bass and treble frequencies in the range of ±15–30 dB. DS1 is an LED; its size and color are not critical. Q1 is a 2N2222 or equivalent. T1 is a 2- to 10-K audio interstage transformer. U1 and U2 are each 1458 or equivalent.—B. Heil, Equalize Your Microphone and Be Heard, *QST,* July 1982, pp. 11–13.

EXCEPT AS INDICATED, DECIMAL VALUES OF CAPACITANCE ARE IN MICROFARADS (μF) ; OTHERS ARE IN PICOFARADS (pF OR μμF); RESISTANCES ARE IN OHMS ; k = 1 000, M = 1 000 000.

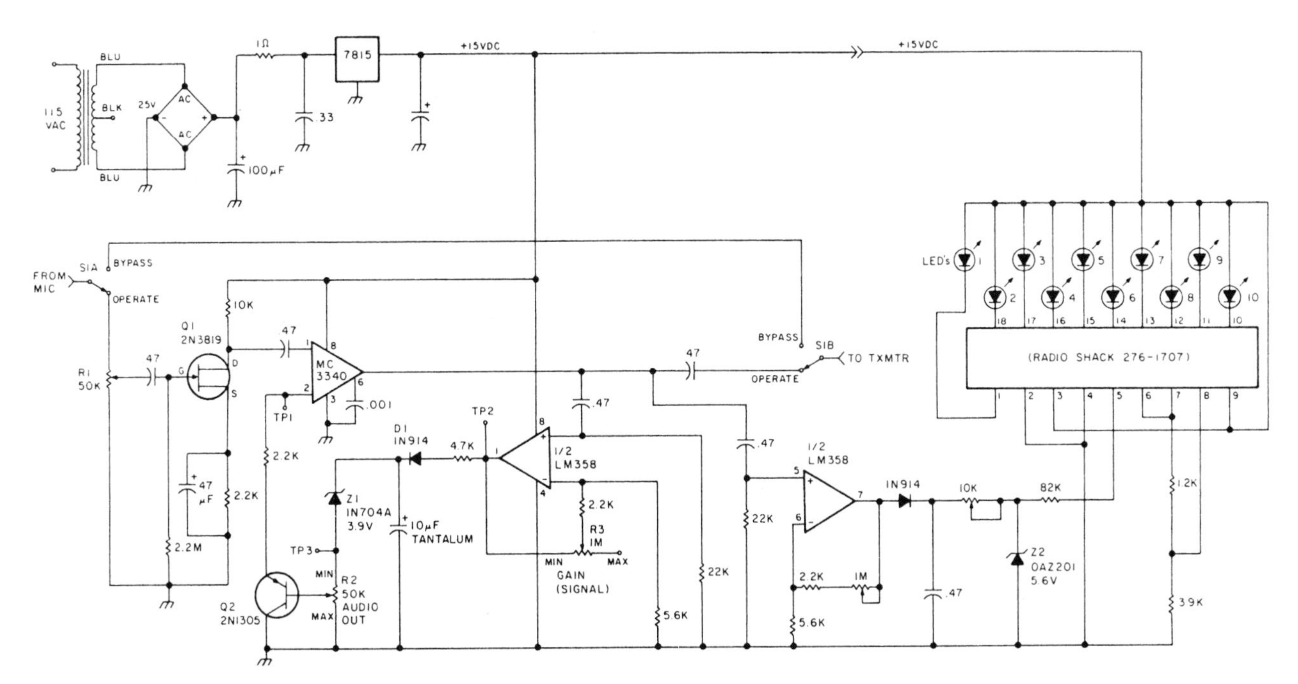

SPEECH EXPANDER/CLIPPER—Processes speech input for SSB transmitter to give higher average output. The OPERATE/BYPASS switch allows the switching of the circuit in and out of the transmitting path.—D. Sladen, Thank You for Listening, *73 Magazine,* January 1984, pp. 86–87.

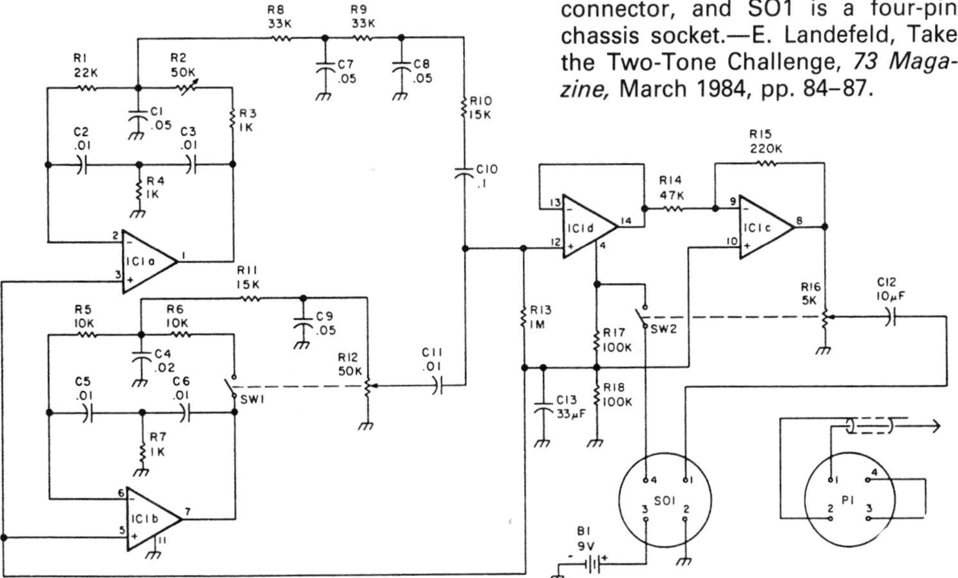

TWO-TONE AUDIO GENERATOR—Generates two-tone audio signal to test SSB transmitters. The audio signal is fed into the microphone input of the SSB transmitter, and the output is displayed on an oscilloscope or signal monitor unit. It can also be used as a sine-wave audio source by turning the balance control (R12) fully counterclockwise. IC1 is a LM324 quad op amp, P1 is a four-pin connector, and SO1 is a four-pin chassis socket.—E. Landefeld, Take the Two-Tone Challenge, *73 Magazine,* March 1984, pp. 84–87.

SPEECH PROCESSOR/COMPRESSOR—Designed for use with SSB transmitters, circuit delivers higher average modulation to transmitter. The low frequency roll-off is at about 400 Hz. If the transmitter microphone has a particularly low output level, R6 should be changed to 150 K.—A. Massa, The Masher, *73 Magazine,* March 1982, pp. 76–77.

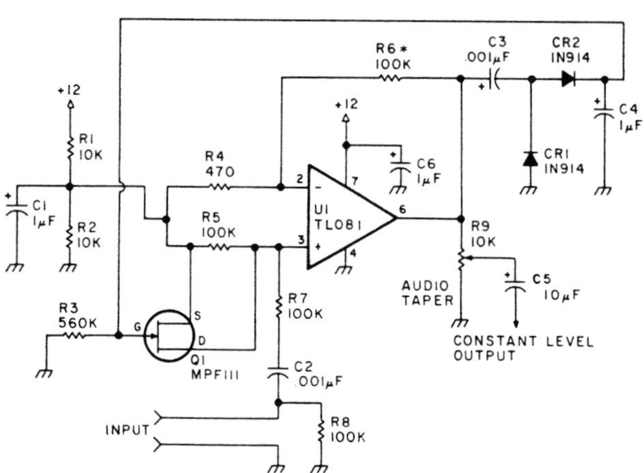

SPEECH COMPRESSOR—MC1590 wideband amplifier with AGC drives base of MPS6517. It operates a common-emitter with a voltage gain of approximately 20. R1 varies the Q point of the transistor. R_x controls the charging time constant, and C_x is involved in both the charge and the discharge. Typical values for an operating range of 1–10 kHz are 1.5 K and 0.68 μF.—"Motorola Linear Integrated Circuits Databook," Motorola, Phoenix, AZ, 1979, p. 6-58.

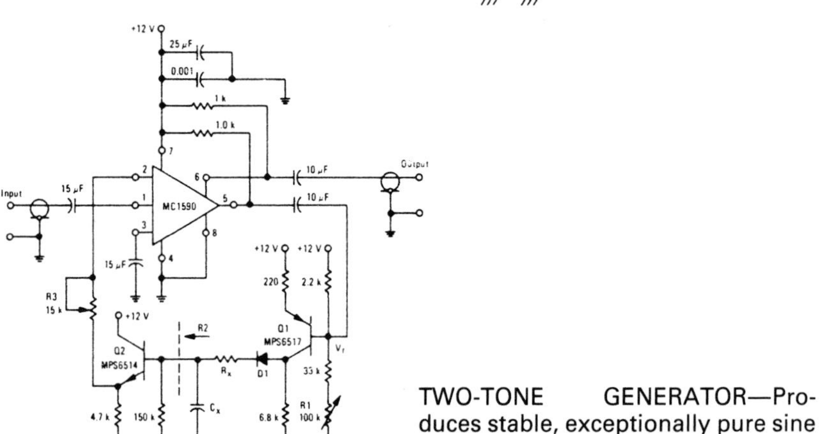

TWO-TONE GENERATOR—Produces stable, exceptionally pure sine waves for testing SSB transmitters. All distortion products are at least 60 dB below the primary output. The unit is designed to plug directly into the microphone jack of the transmitter under test.—P. Clower, Penn's Two-Tone Gadget, *73 Magazine,* August 1984, pp. 21–30.

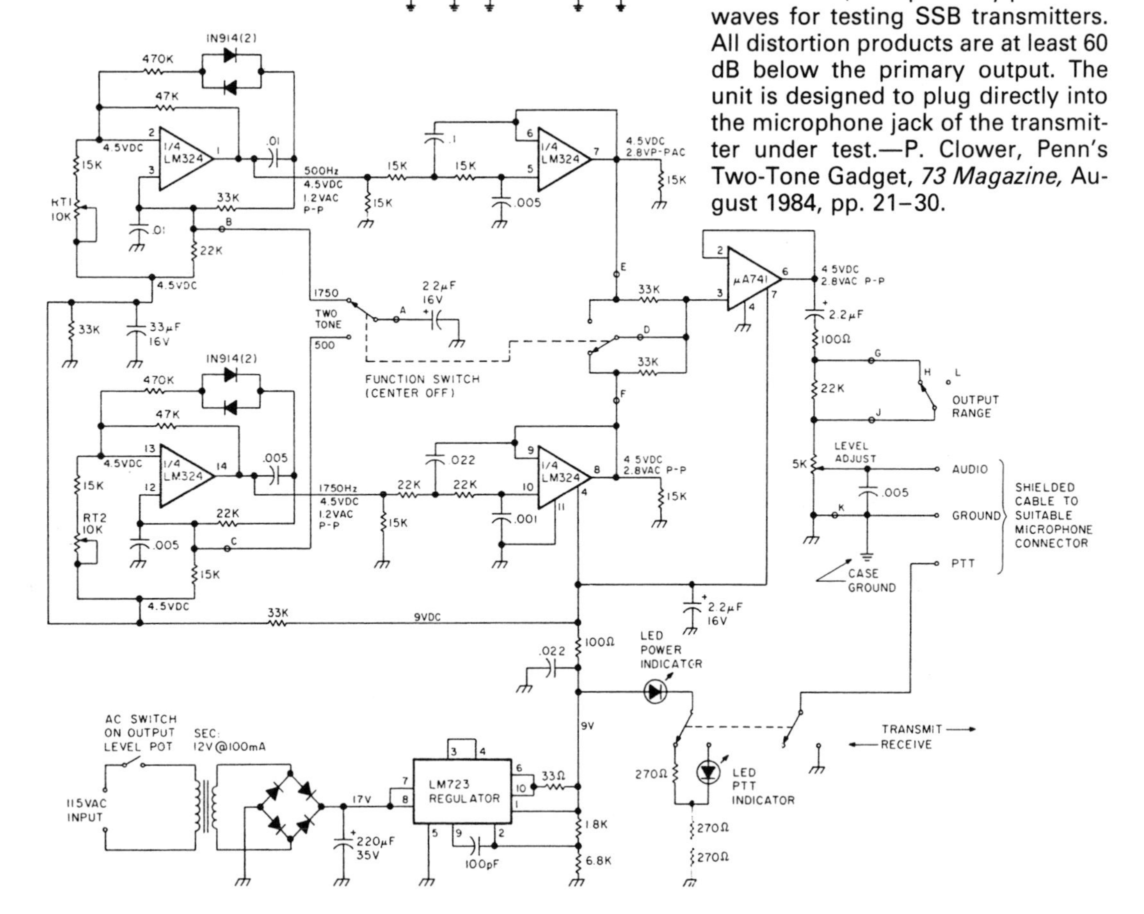

SSB OVERDRIVE INDICATOR—DS1 remains lit as long as SSB signal from transmitter/transceiver is "clean." DS2 starts to flash if the SSB signal is distorting and splattering. DS3 is a "pilot" indicating the unit is on. Q1 is a 2N2646 or equivalent. RFC is a 1-mH inductor. D1 and D2 are each 1N914. DS1 and DS2 are a T-1-3 and T-1-4, respectively (different colors). DS3 can be a pilot LED. The RF sample is taken from the transmission line through an SO-239 connector with an M-358 T fitting attached.—J. Kennicott, SSB Overdrive Indicator, *QST,* November 1981, p. 45.

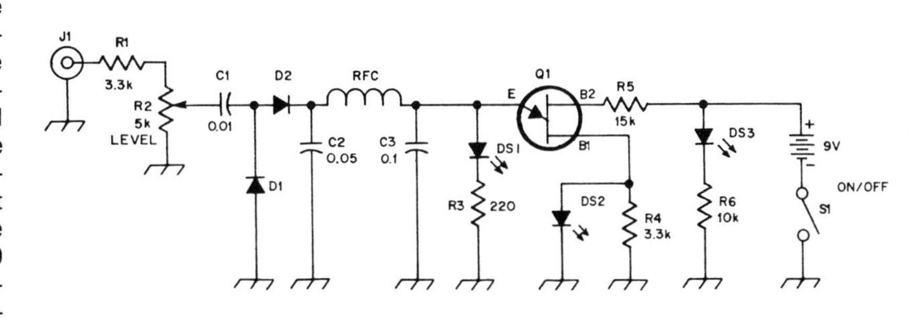

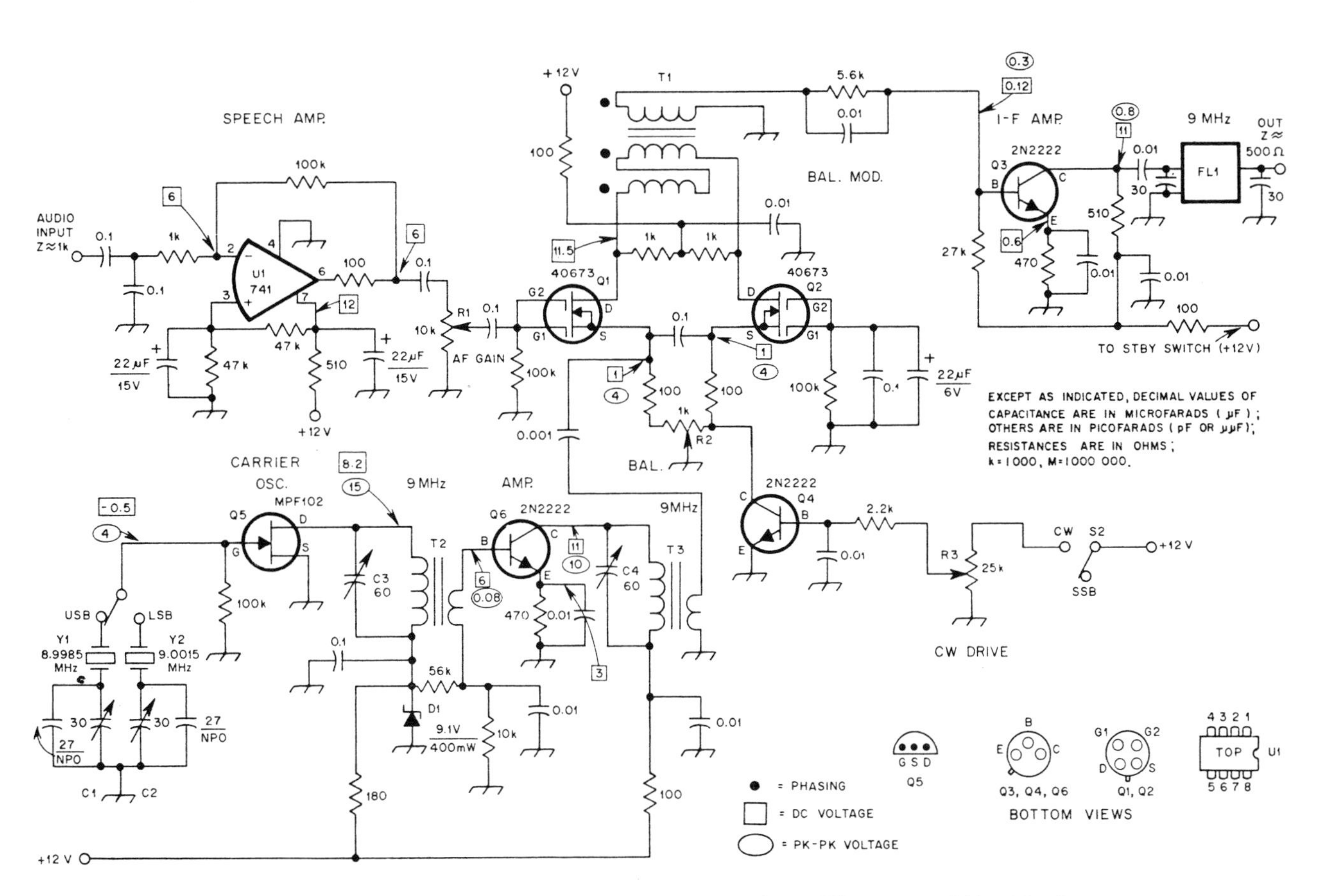

9-MHz SSB GENERATOR—9-MHz output can be heterodyned to desired frequency. The 741 op amp serves as a speech amplifier, with Q1 and Q2 together acting as a balanced modulator. T1 provides the necessary 180° phase difference for the drains of Q1 and Q2. R2 compensates for the difference in operating levels between Q1 and Q2. FL1 is a Spectrum International 9-MHz-type XF-9A crystal lattice filter. T1 consists of 15 trifilar turns of No. 26 wire on a FT-50-61 toroid form. T2 consists of 44 turns of No. 26 wire on a T50-2 iron core; the link has 10 turns of No. 30 wire over the D1 end of the primary. T3 is 44 turns of No. 26 wire on a T50-2 iron core; the link has 22 turns of No. 30 wire over the "cold" end of the primary.—"The Radio Amateur's Handbook," American Radio Relay League, Newington, CT, 1981, pp. 12-27 to 12-28.

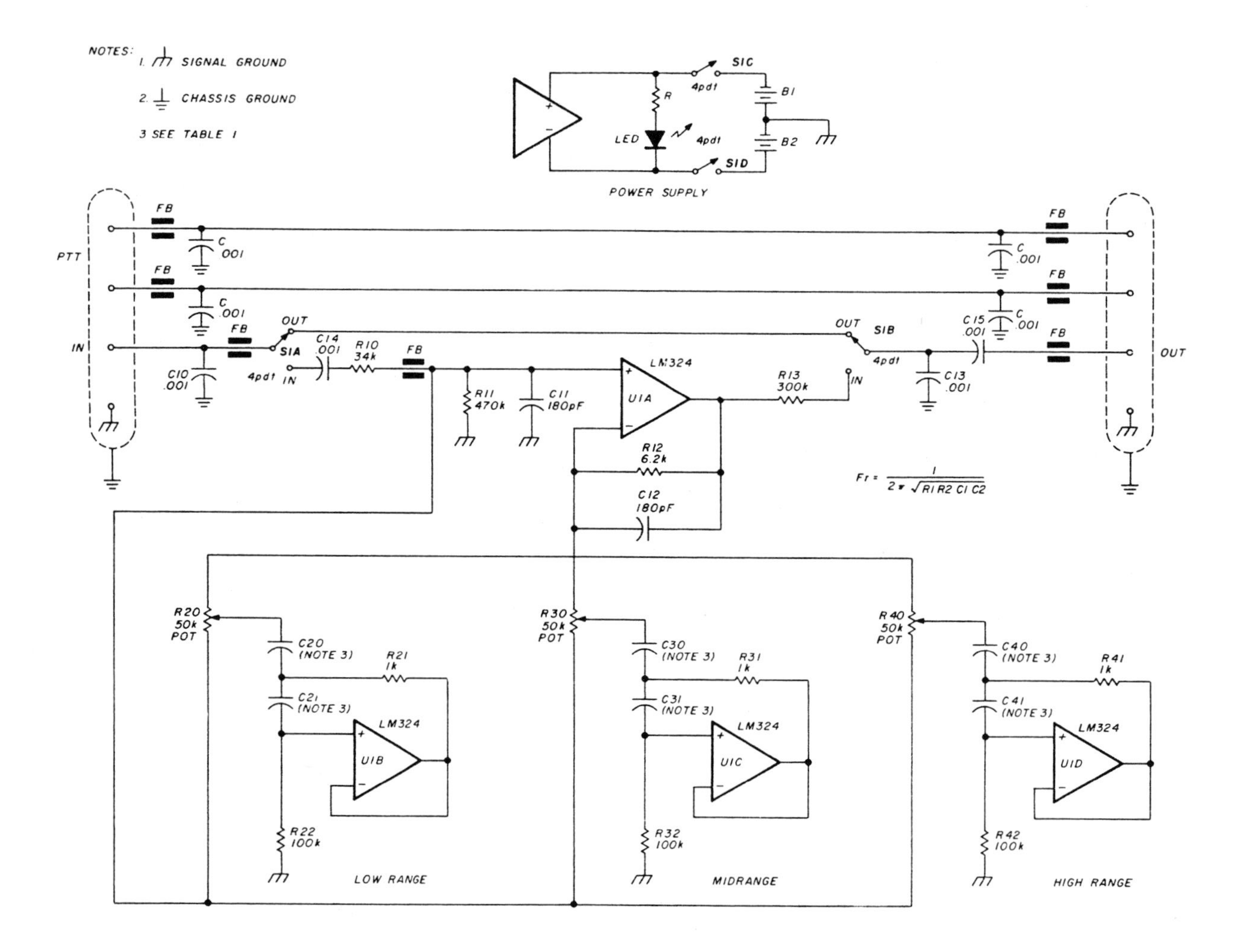

VOICE-BAND EQUALIZER—Allows adjustment of microphone response in low, midrange, and high audio frequencies. The frequencies may be boosted or attenuated over a range of 24 dB. The circuit may be powered from two 9-V batteries. The table accompanying the drawing gives suggested values for various components.—R. Bradley, The Voice-Band Equalizer, *Ham Radio,* October 1980, pp. 50–55.

filter	resonant frequency (Hz)	capacitor (fig. 2)	value (μF)
low frequencies	335	C21	0.2
		C21	0.01
	410	C20	0.15
		C21	0.01
	521	C20	0.15
		C21	0.0062
	1299	C30	0.05
		C31	0.003
mid-range frequencies	1412	C30	0.047
		C31	0.0022
	1959	C40	0.033
		C41	0.002
high frequencies	2905	C40	0.033
		C41	0.001
	3558	C40	0.02
		C41	0.001

20

Telephone Circuits

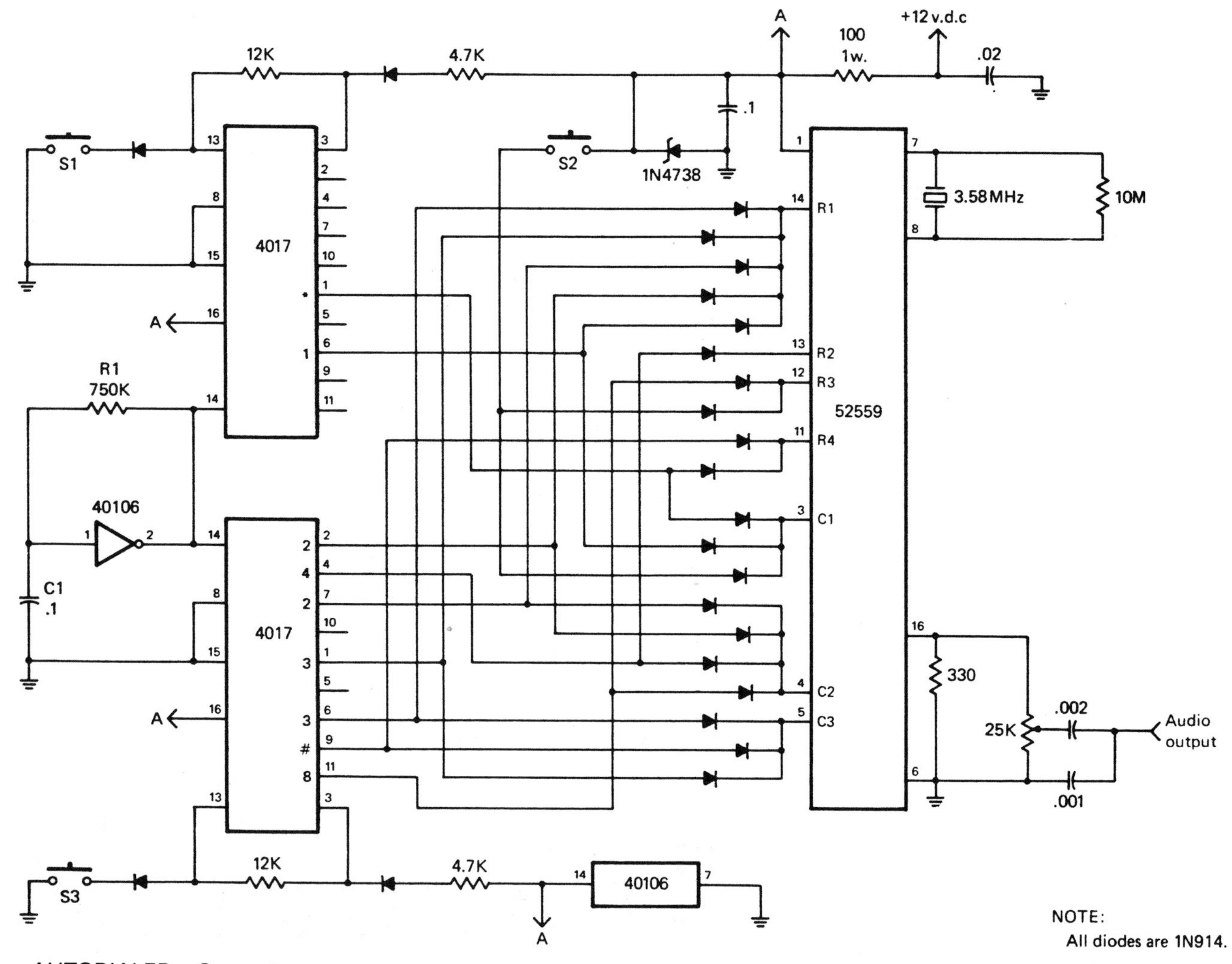

NOTE:
All diodes are 1N914.

AUTODIALER—Generates seven-digit telephone number and four-digit access codes using standard touch dialing tones automatically by pressing proper switch. S1 and S2 generate access code tones, and S3 generates the tones for the telephone number itself. The circuit makes use of an S2559 digital tone generator device; the programming of the desired codes and numbers is done through diodes. The original article includes a programming example and details.—E. Rate, Build Your Own Autodialer for V.H.F. Mobiling, *CQ*, July 1983, pp. 18–21.

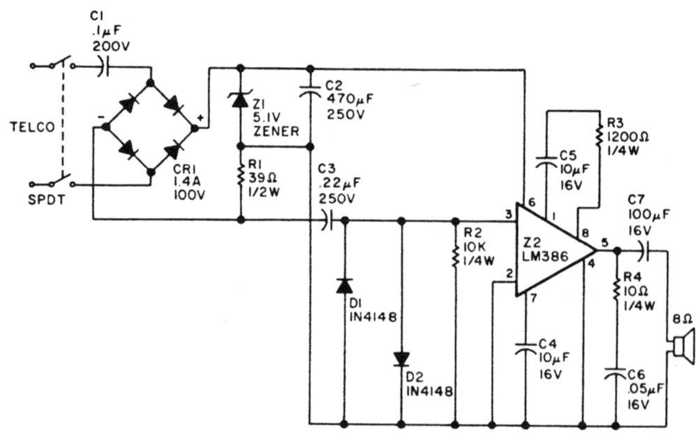

TELEPHONE LINE MONITOR—Amplifies telephone line audio to drive speaker.—E. Sherrill, Telephone Line Monitor, *73 Magazine*, December 1983, p. 115.

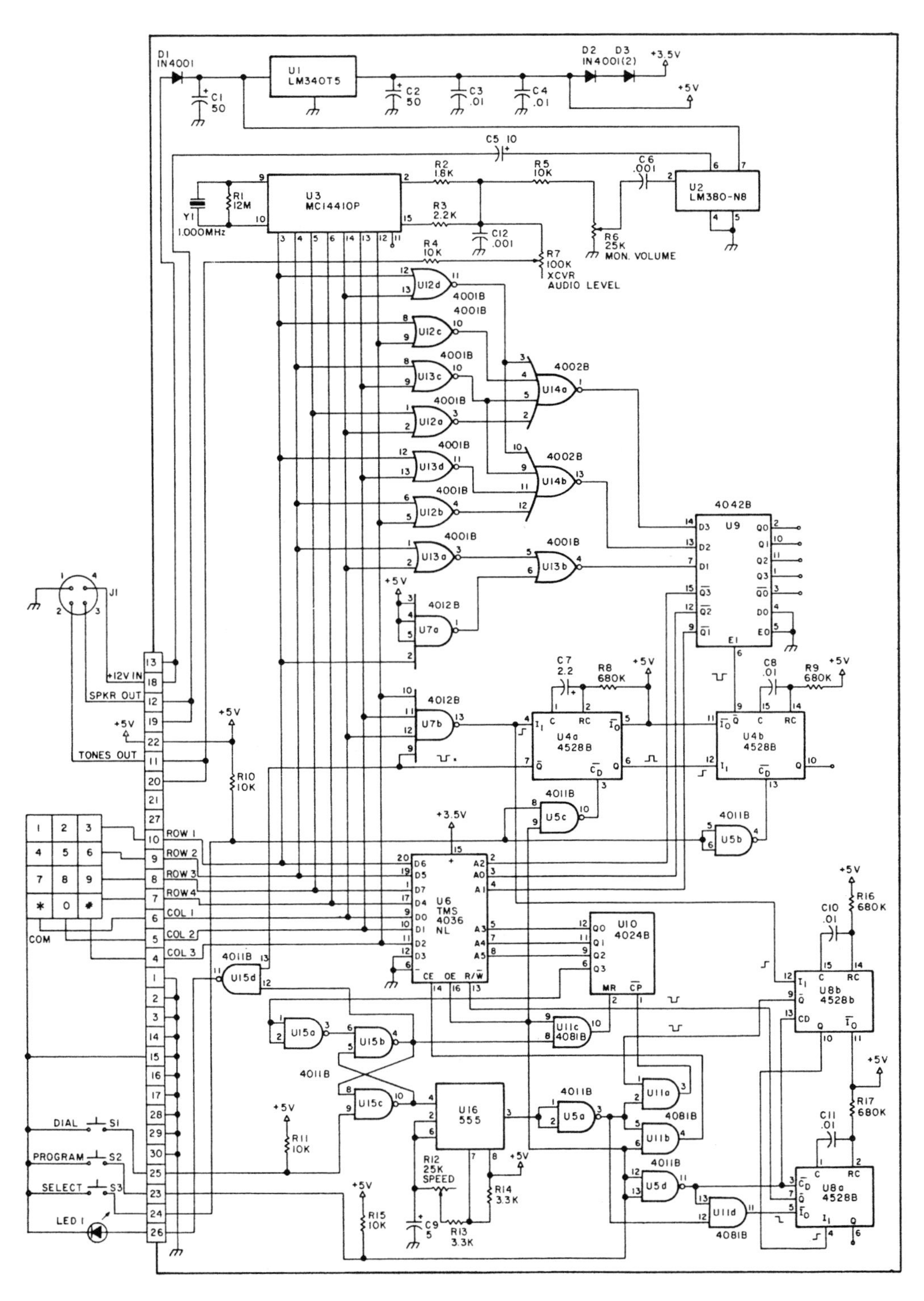

AUTODIALER—Designed for use with mobile amateur radio equipment; up to eight telephone numbers can be stored and selected by single button. Standard telephone dial tones are used, and the unit may be reprogrammed without hardware changes.—H. Batie, The Forgetful Autodialer Puzzle, *73 Magazine,* January 1983, pp. 40–44.

DTMF DECODER—Dual-tone multi-frequency (DTMF) decoder allows controlling of repeater functions by audio tones. The incoming audio is fed in parallel to high tone and low tone filter groups. Each filter has the correct bandpass response for the associated tone group. Each filter's output is fed to an op amp limiter. Limiter output is an approximate square wave and goes to the MK5102/3. Decoder output is a 4-bit binary code, so a 4- to 16-line decoder provides the necessary 16 outputs. The decoder strobe output is buffered. Q1 through Q5 are all 2N2222A or equivalent. DS1 through DS4 are red LEDs, while DS5 is a green LED. R14 through R17 are all 470 or 1000 Ω.—J. Jarrett, The DTMF "Easy-Ceiver," *QST,* January 1982, pp. 25–29.

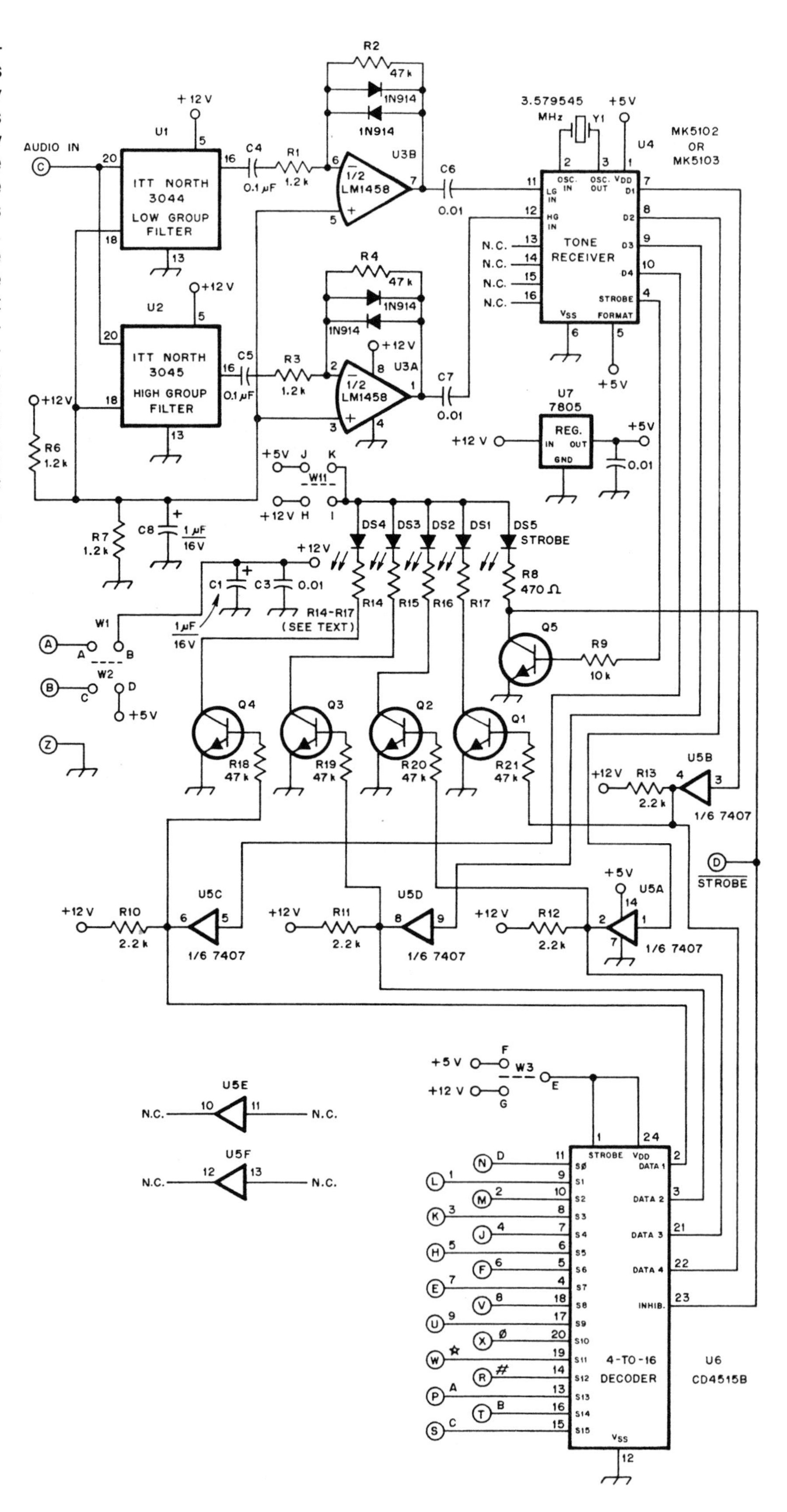

TEN-NUMBER REPERTORY TONE DIALER—Stores ten 16-digit telephone numbers and generates standard dialing tones for them. Memory retention is provided by the 3-V battery. The diode bridge (D1 through D4) protects the circuit from telephone line polarity reversals.—"Mostek 1984/85 Microelectronics Data Book," Mostek Corporation, Carrollton, TX, 1984, pp. XIV-1 to XIV-10.

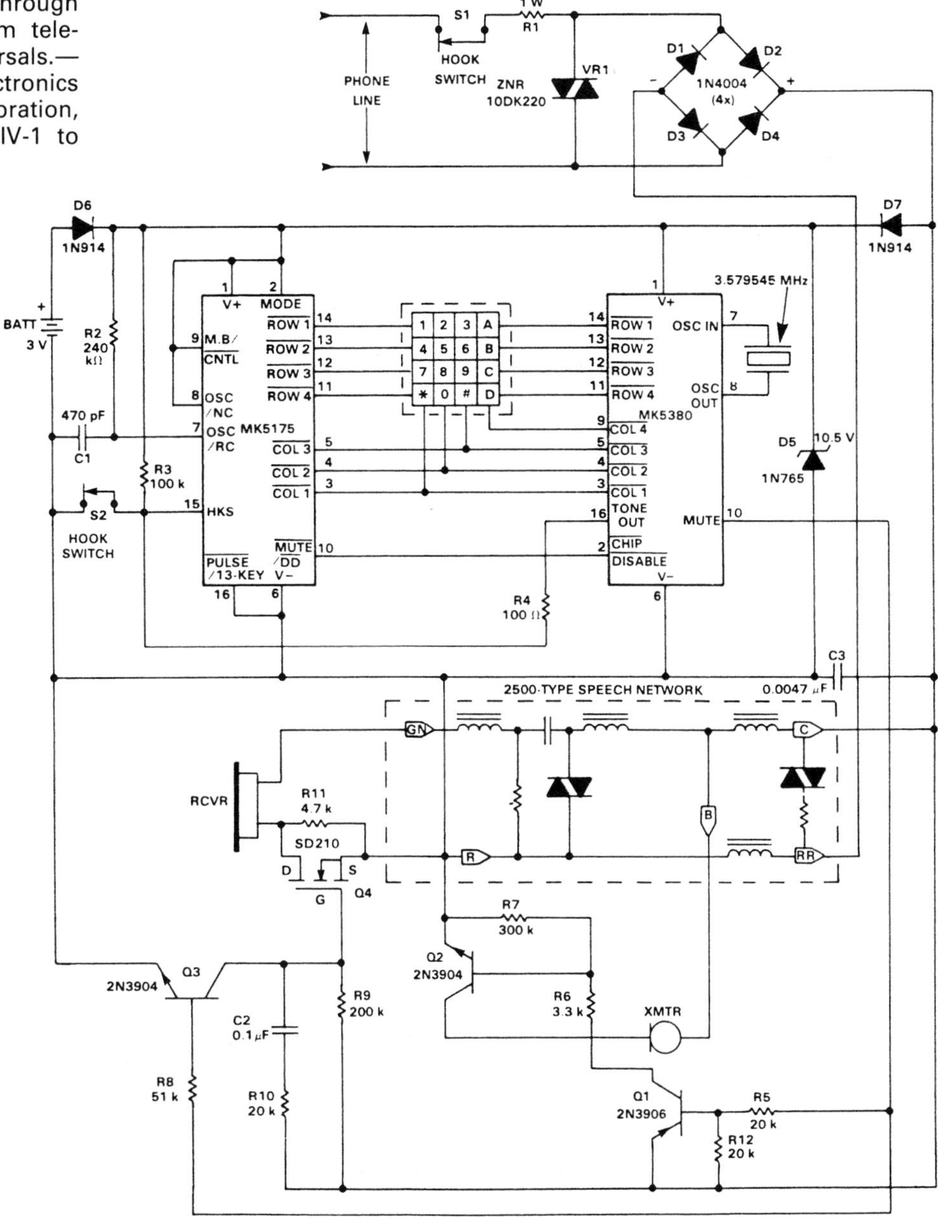

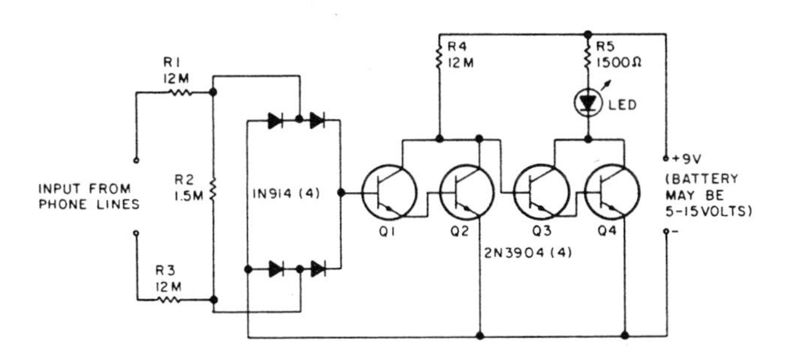

TELEPHONE OFF-HOOK INDICATOR—LED flickers when another person is dialing or talking on extension and also provides visual ring indicator. The LED glows steadily when the phone is off its hook.—E. Fruitman, Telephone Off-Hook Indicator, *73 Magazine,* January 1984, p. 115.

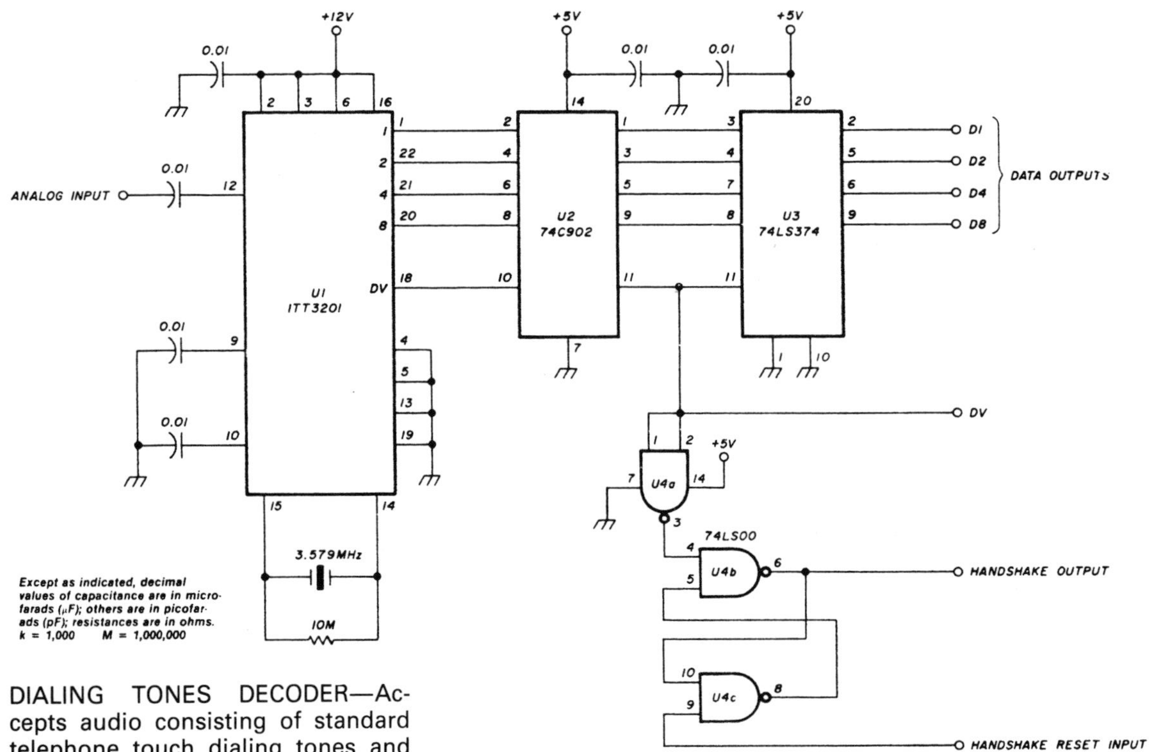

DIALING TONES DECODER—Accepts audio consisting of standard telephone touch dialing tones and produces a 4-bit TTL-level output. The output can be used in conjunction with a microprocessor for remote control applications. The analog input of the 3201 has high impedance. If the input signal is a valid DTMF tone pair, the 3201 produces an output on the four data lines. During the time a tone is being received and decoded, the DV (data valid) output is high. The handshake circuit allows a microprocessor to tell the difference between a newly received DTMF input and a previously stored word.—J. Hinshaw, An Improved Touch-Tone® Decoder, *Ham Radio,* December 1982, pp. 24–28.

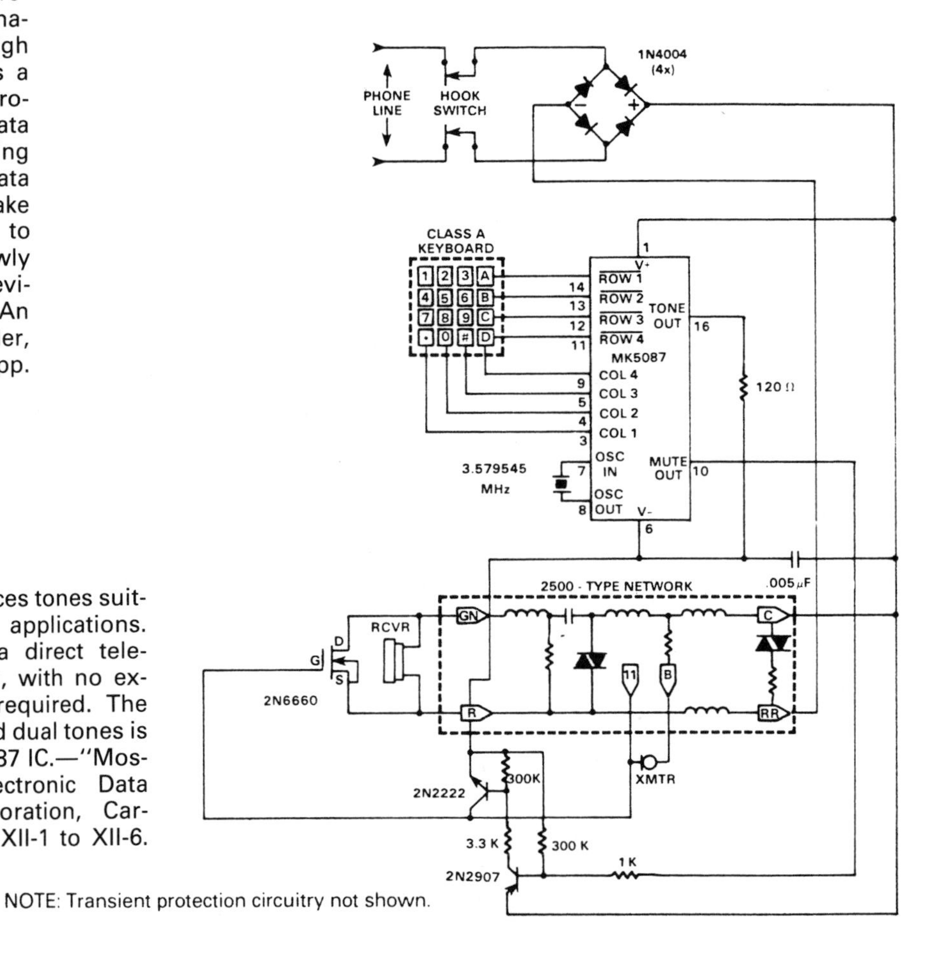

DTMF DIALER—Produces tones suitable for DTMF dialing applications. The circuit provides a direct telephone line application, with no external power supply required. The regulation of single and dual tones is provided on the MK5087 IC.—"Mostek 1984/85 Microelectronic Data Book," Mostek Corporation, Carrollton, TX, 1984, pp. XII-1 to XII-6.

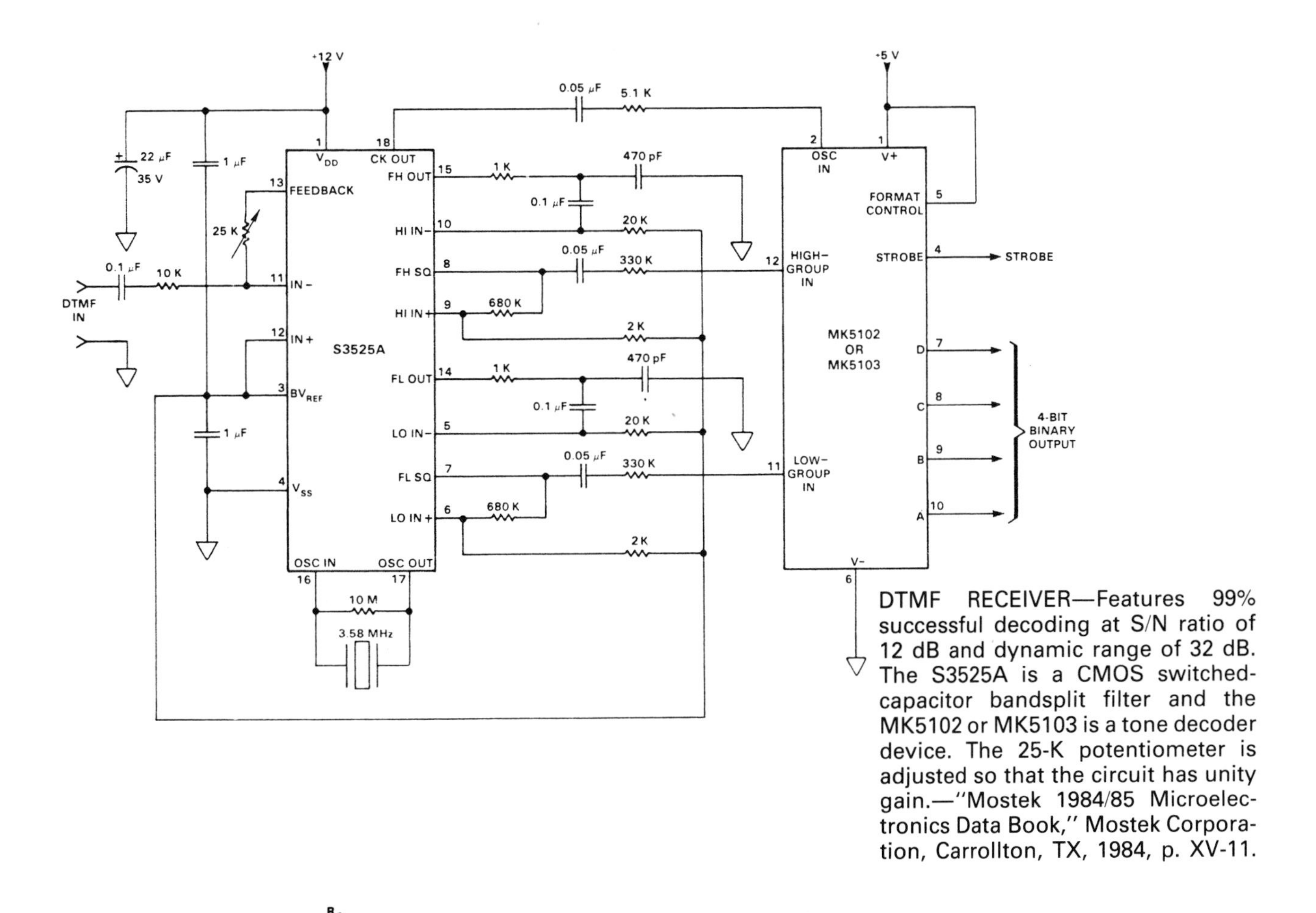

DTMF RECEIVER—Features 99% successful decoding at S/N ratio of 12 dB and dynamic range of 32 dB. The S3525A is a CMOS switched-capacitor bandsplit filter and the MK5102 or MK5103 is a tone decoder device. The 25-K potentiometer is adjusted so that the circuit has unity gain.—"Mostek 1984/85 Microelectronics Data Book," Mostek Corporation, Carrollton, TX, 1984, p. XV-11.

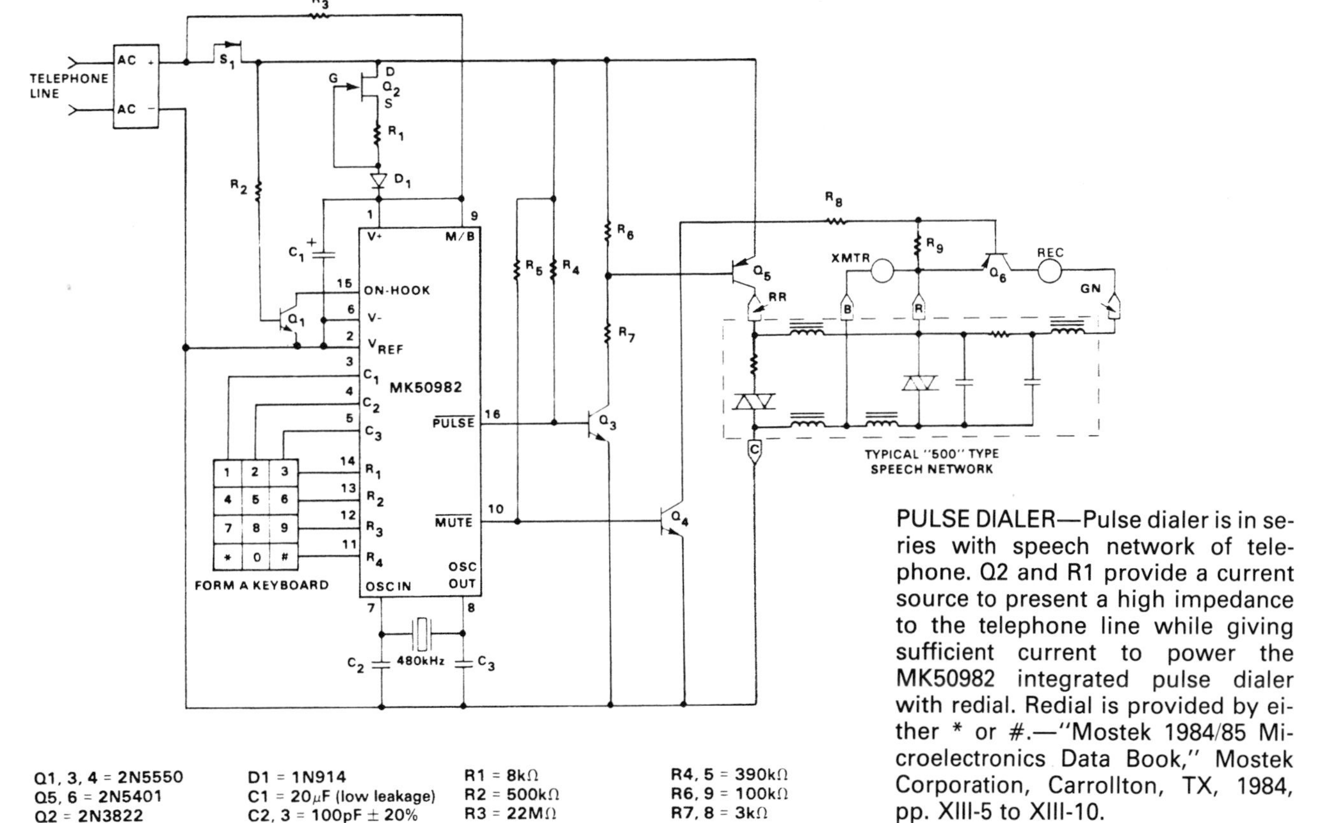

PULSE DIALER—Pulse dialer is in series with speech network of telephone. Q2 and R1 provide a current source to present a high impedance to the telephone line while giving sufficient current to power the MK50982 integrated pulse dialer with redial. Redial is provided by either * or #.—"Mostek 1984/85 Microelectronics Data Book," Mostek Corporation, Carrollton, TX, 1984, pp. XIII-5 to XIII-10.

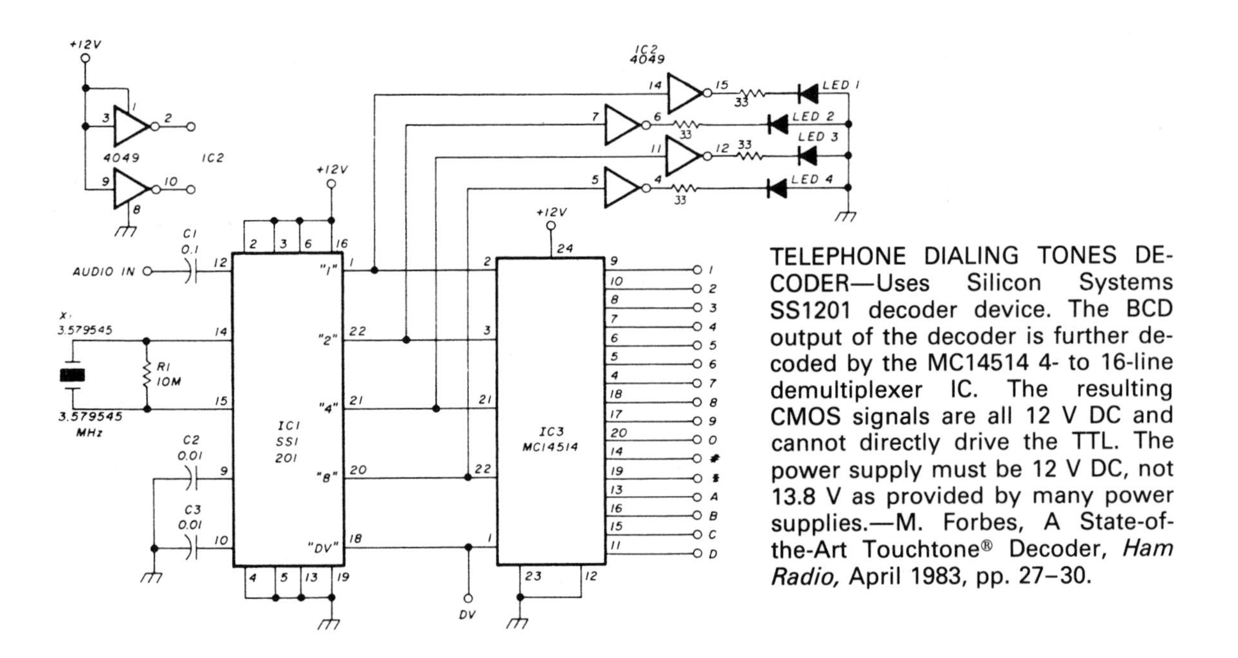

TELEPHONE DIALING TONES DECODER—Uses Silicon Systems SS1201 decoder device. The BCD output of the decoder is further decoded by the MC14514 4- to 16-line demultiplexer IC. The resulting CMOS signals are all 12 V DC and cannot directly drive the TTL. The power supply must be 12 V DC, not 13.8 V as provided by many power supplies.—M. Forbes, A State-of-the-Art Touchtone® Decoder, *Ham Radio,* April 1983, pp. 27–30.

PHONE PATCH—Allows connection of transceiver or transmitter/receiver combination to telephone system. Telephones in this system operate in simplex rather than duplex. It requires a transmitter or transceiver with phone patch input.—P. Danzer, A $10 Phone Patch, *73 Magazine,* February 1981, pp. 68–69.

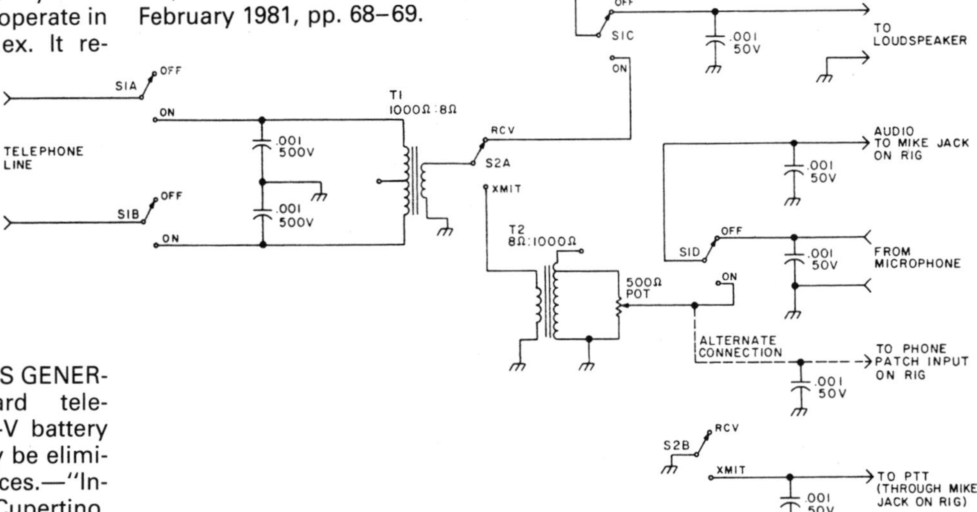

PORTABLE DIALING TONES GENERATOR—Produces standard telephone dialing tones. A 6-V battery may also be used; D4 may be eliminated in such circumstances.—"Intersil Data Book," Intersil, Cupertino, CA, 1981, p. 7-35.

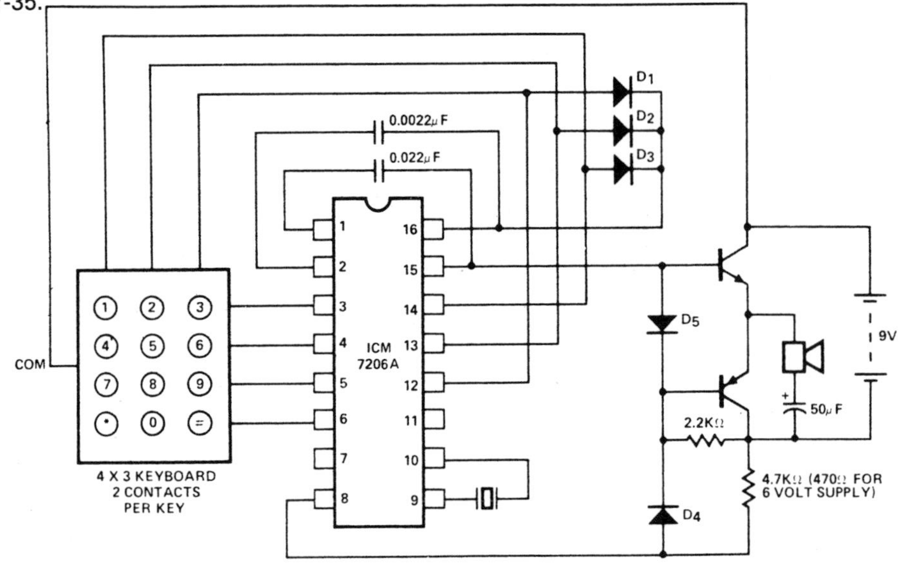

ELECTRONIC GENERATION OF DIALING TONES—U1 is hex inverter (7404, 4049, or equivalent) while U2 is hex buffer (7407, 7417, 4050, or equivalent). Logic inputs drive the MK5087 tone dialer IC to produce tones.—"Mostek 1984/85 Microelectronic Data Book," Mostek Corporation, Carrollton, TX, 1984, p. XII-13.

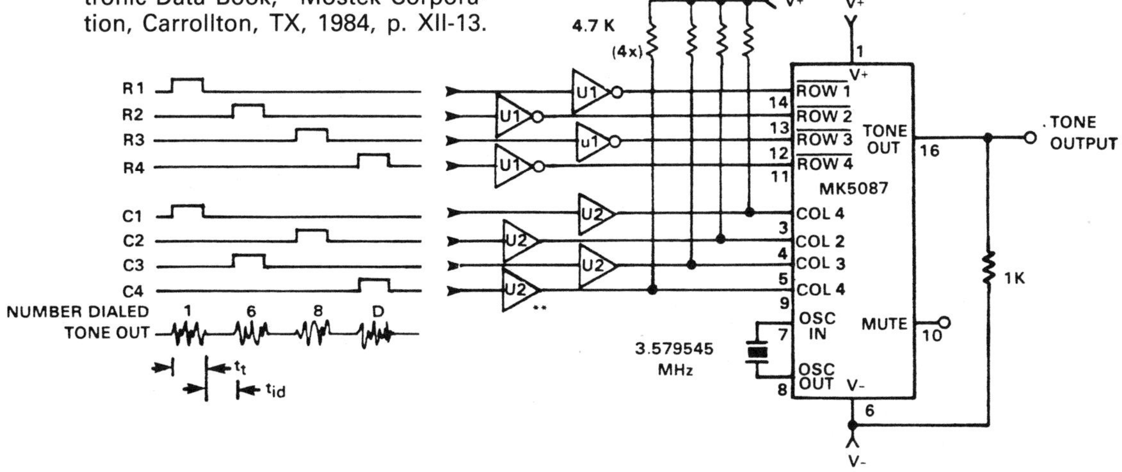

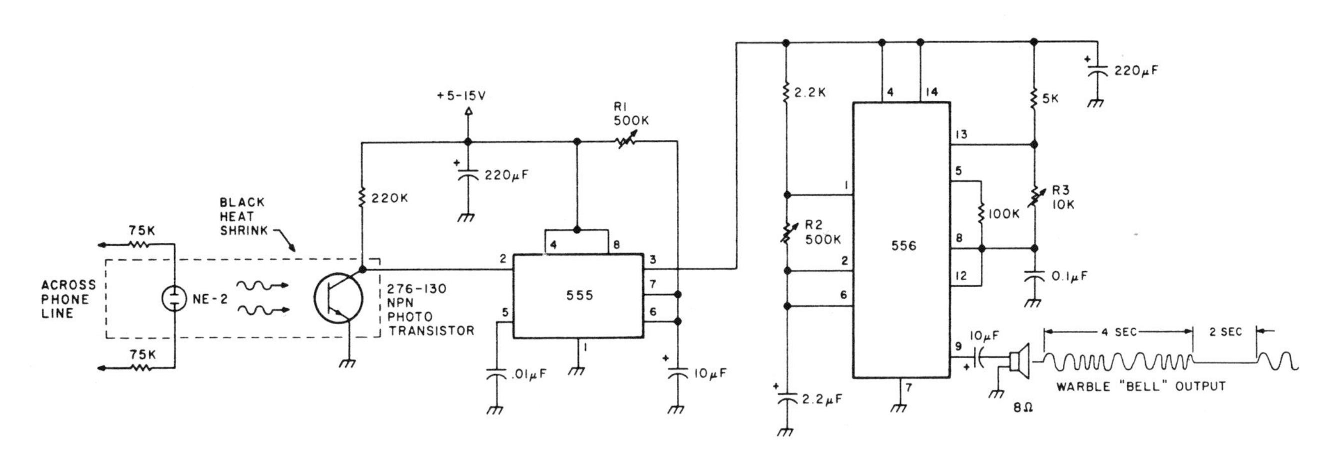

ELECTRONIC TELEPHONE BELL—Speaker emits distinctive warble tone when ring pulses are applied to telephone line. R1 adjusts the duration of the output; R2 and R3 control the tone's duty cycle and frequency. The neon bulb and transistor are coupled with heat-sink tubing to form an opto-isolator.—J. Mairs, Electronic Phone Bell, *73 Magazine,* September 1982, p. 92.

21

Teleprinter Circuits

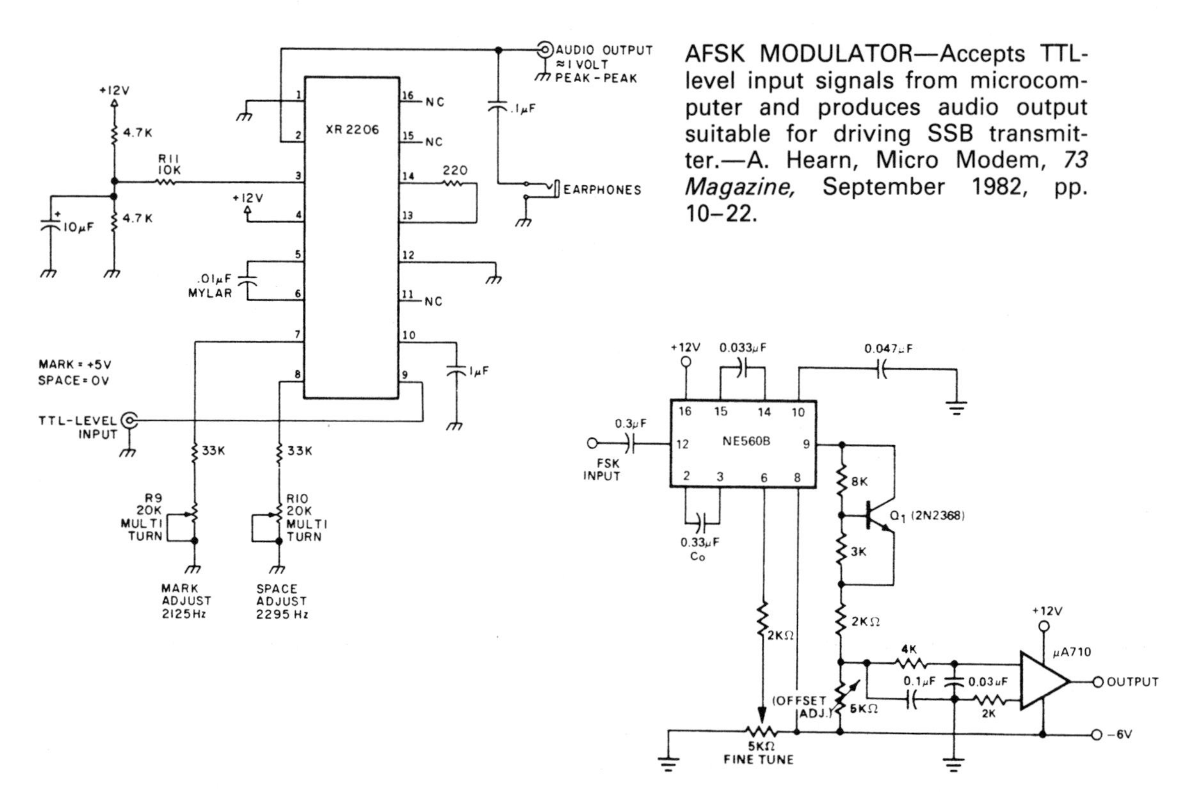

AFSK MODULATOR—Accepts TTL-level input signals from microcomputer and produces audio output suitable for driving SSB transmitter.—A. Hearn, Micro Modem, *73 Magazine*, September 1982, pp. 10–22.

FSK DEMODULATOR—Demodulates FSK audio tones and provides shifting DC voltage to initiate mark and space frequencies. The circuit was designed to match Bell® 103C and 103D Data Phones. System functions by locking on and tracking the output frequency of the receiver. The input frequency shift is translated within the 560B to a DC voltage of about 60-mV amplitude that appears at pin 9. This output is amplified, conditioned, and applied to the voltage comparator (710) that provides compatible outputs for interfacing with digital logic systems.—"Signetics Analog Applications Manual," Signetics, Sunnyvale, CA, 1979, pp. 313–315.

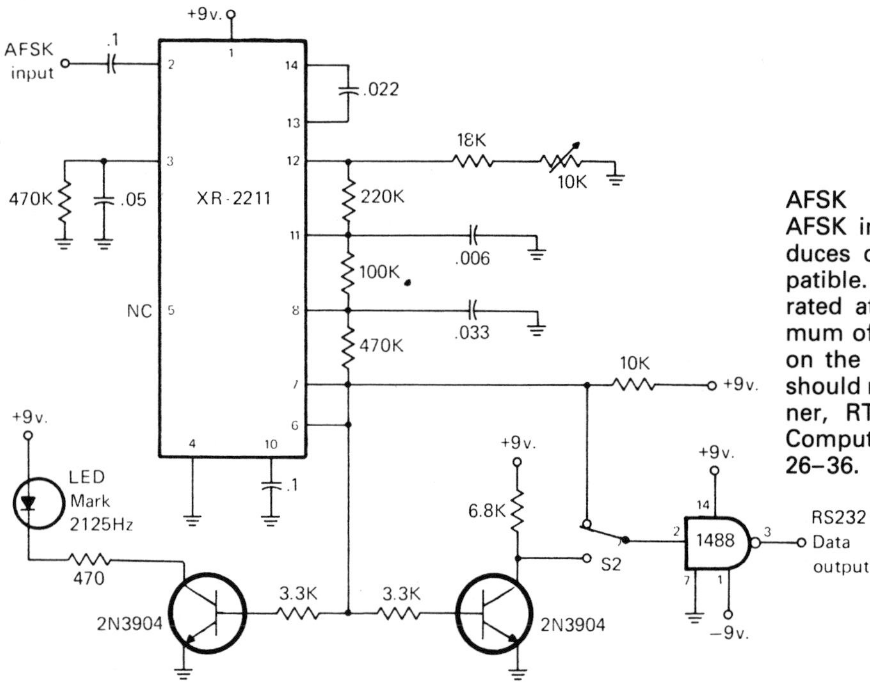

AFSK DEMODULATOR—Takes AFSK input from receiver and produces output which is RS232-compatible. The input of the XR2211 is rated at 20 K and requires a minimum of 10 mV RMS for full limiting on the weakest signals. Input level should not exceed 3 V RMS.—J. Botner, RTTY and the IBM Personal Computer, *CQ*, November 1983, pp. 26–36.

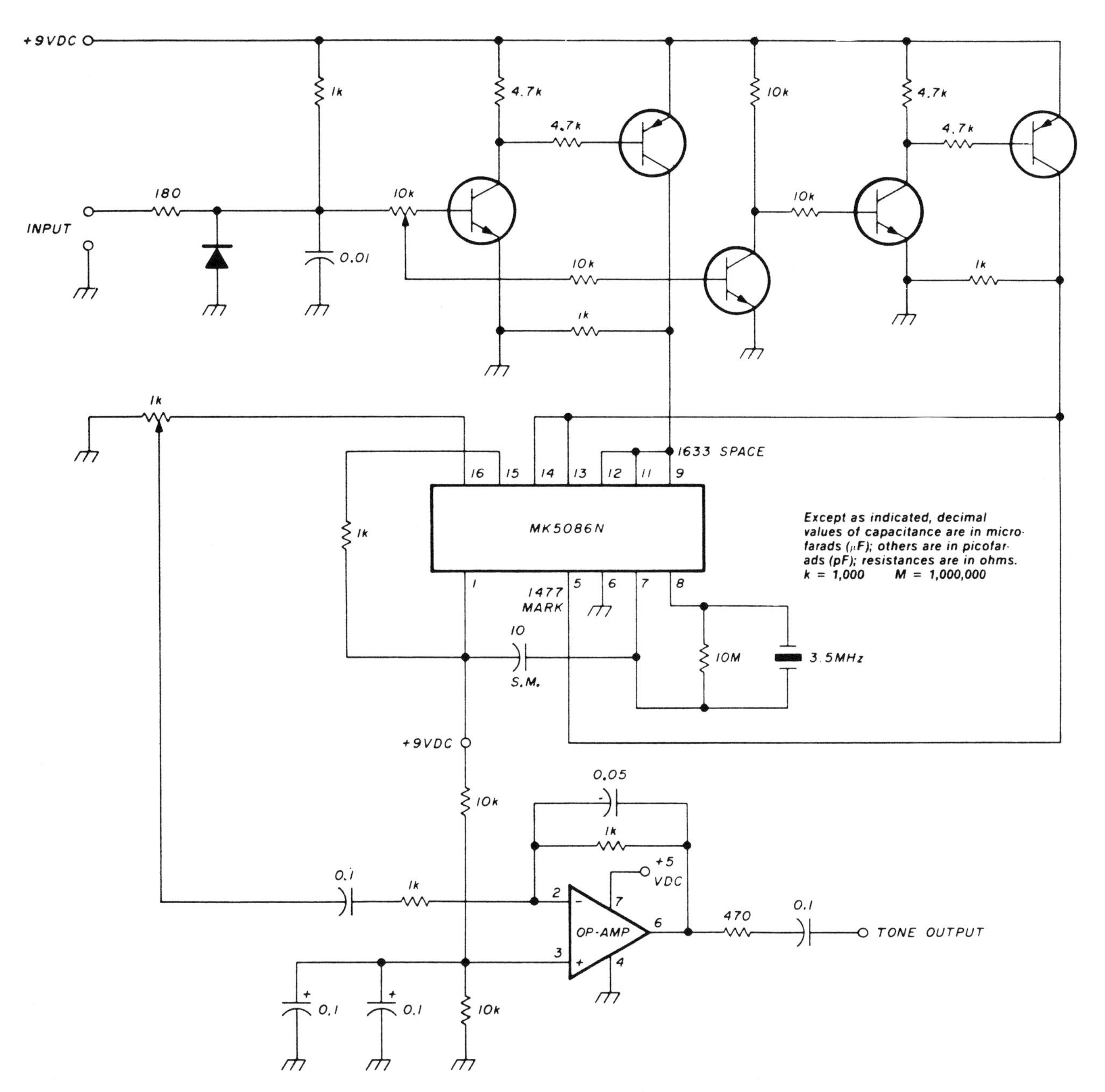

AFSK TONE GENERATOR—MK5086N tone generator device generates tones for AFSK transmission. Pin 9 produces a 1633-Hz space tone, and pin 5 produces a 1470-Hz mark tone. Keyboard operation can be simulated by connecting pins 13 and 14 to pin 5 and pins 11 and 12 to pin 9. The op amp section provides a small amount of necessary gain. The capacitor across pins 2 and 6 of the op amp will serve as a low-pass filter.—D. Nagel, FSK Tone Generator Using an Integrated Tone Dialer, *Ham Radio,* April 1983, pp. 88–89.

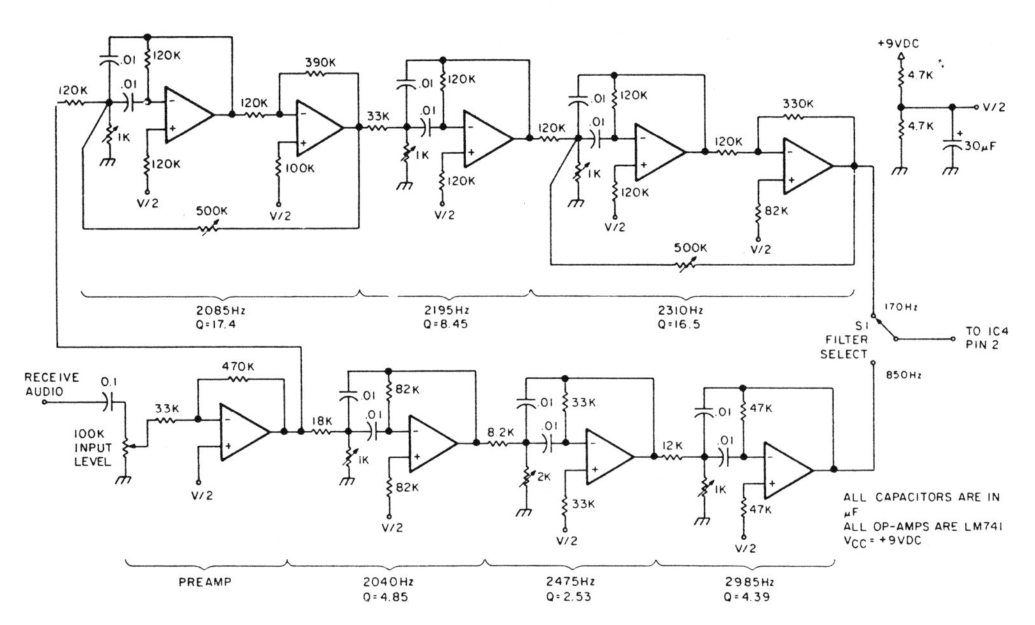

MICROCOMPUTER INTERFACE FOR RTTY—Designed for use with Timex/Sinclair 1000 microcomputer. With the appropriate assembly language program, the circuit allows the transmission and reception of Baudot code at 60, 66, 75, and 100 WPM at a 170- or 850-Hz shift as well as the reception of 425-Hz shift transmissions. LED indicators H and L simplify tuning through the quick identification of high and low tones. The matching active bandpass filter operates at 170 and 850 Hz for best reception.—R. Morrow, Ntty Grtty RTTY, *73 Magazine,* September 1984, pp. 38–46.

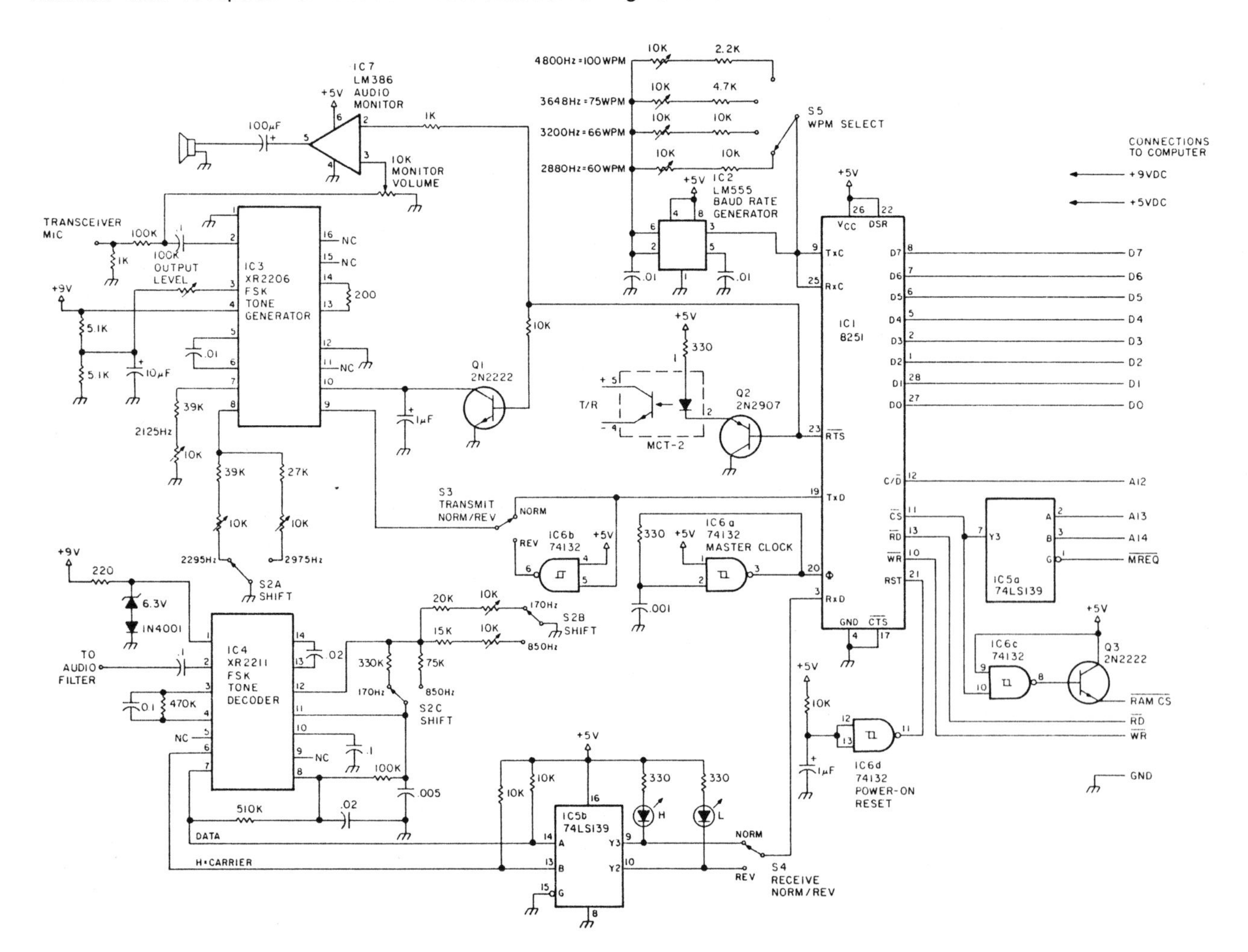

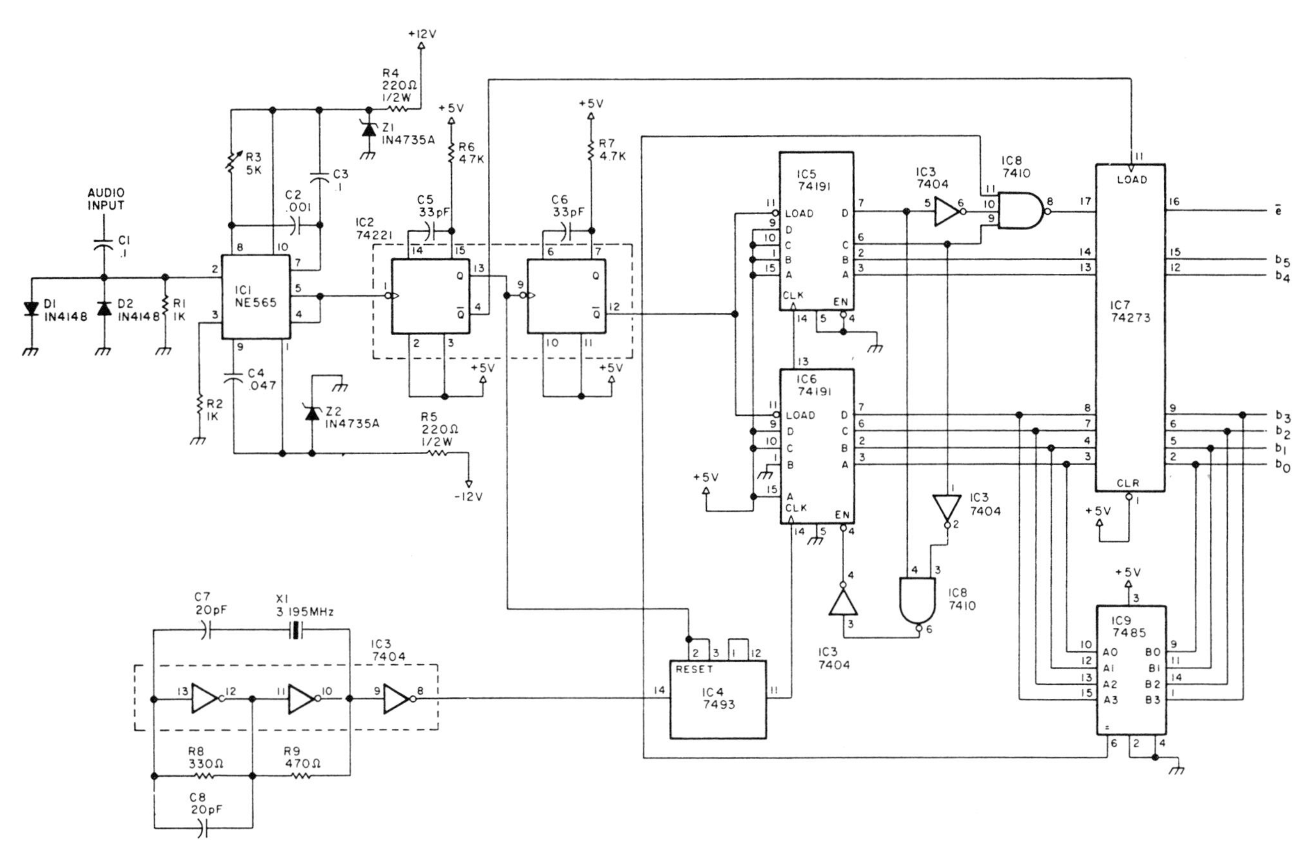

TUNING INDICATOR—LEDs assist in tuning of radioteleprinter signals. The circuit requires at least 10 mV peak-to-peak for good tracking of the input signal. The display is a row of 20 LEDs; each one will light up when the strongest audio frequency present is within its particular range.—J. Langner, RTTY Tuning: The New Solution, *73 Magazine,* March 1983, pp. 10–14.

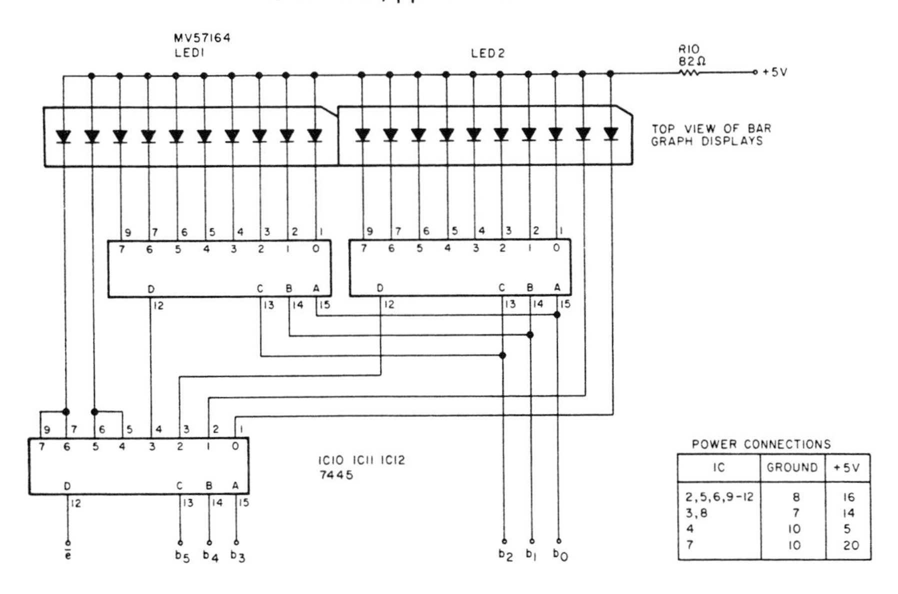

POWER CONNECTIONS

IC	GROUND	+5V
2,5,6,9–12	8	16
3,8	7	14
4	10	5
7	10	20

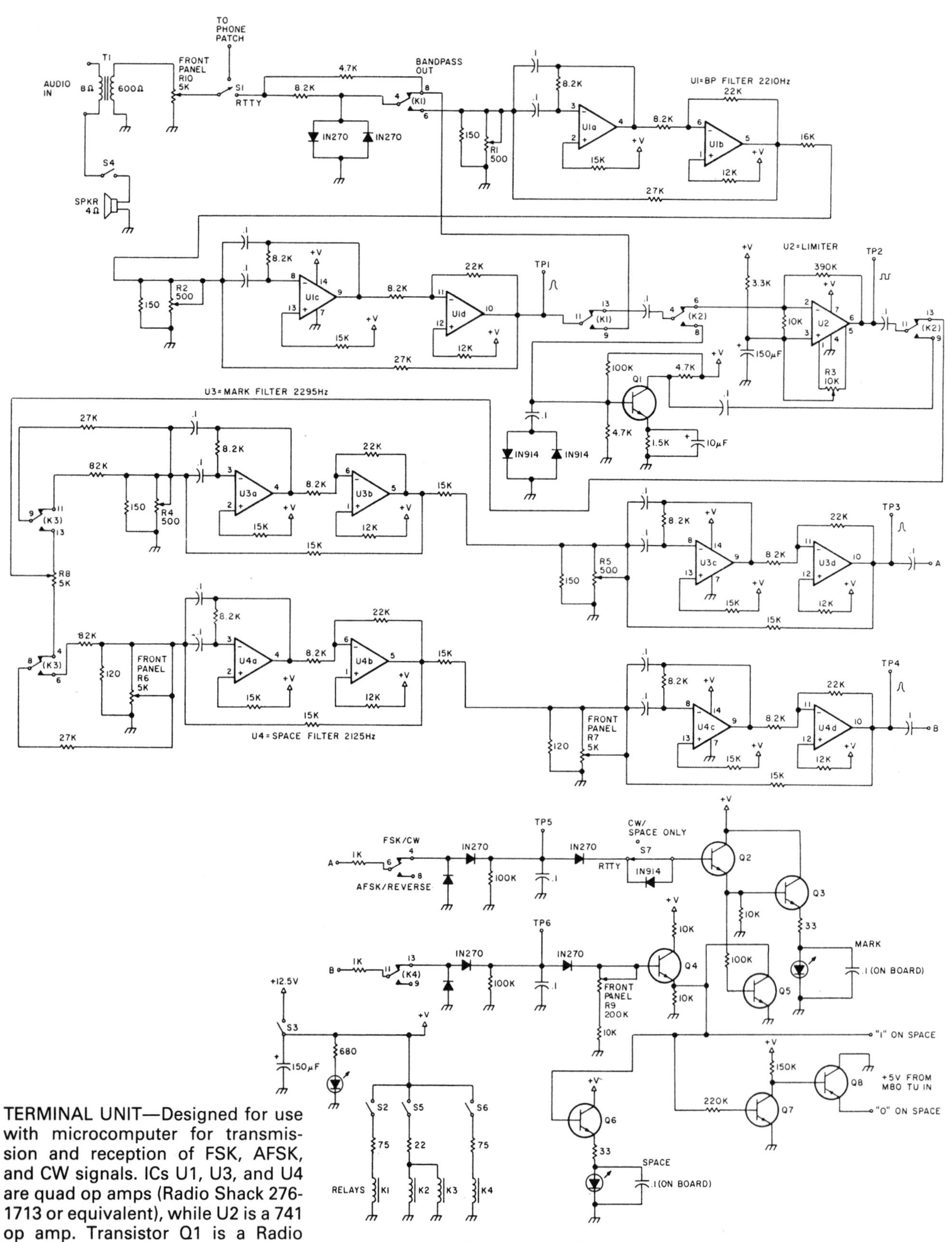

TERMINAL UNIT—Designed for use with microcomputer for transmission and reception of FSK, AFSK, and CW signals. ICs U1, U3, and U4 are quad op amps (Radio Shack 276-1713 or equivalent), while U2 is a 741 op amp. Transistor Q1 is a Radio Shack 276-2030 or equivalent and transistors Q2 through Q7 are all 2N2222. Transistor Q8 is a 2N1305. Relays K1 through K4 are 12-V DIP types, and all LEDs are Radio Shack 276-021 or equivalent. The microcomputer used with this circuit will need appropriate software for the transmission and reception of signals.—F. Williams, The Terminal Unit, *73 Magazine,* April 1984, pp. 70–82.

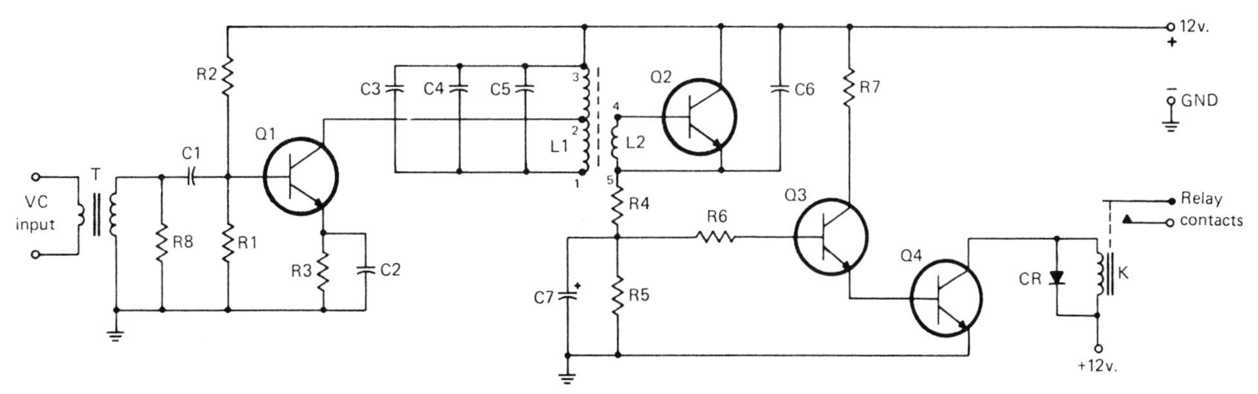

TELEPRINTER AUTOSTART DETECTOR—Automatically starts and stops teleprinter machines and allows unattended reception of messages. AFSK keying must be used, with standard frequencies of 2125 Hz for mark and 2295 Hz for space. C1 is a 0.05-μF 100-V ceramic, C2 is a 0.1-μF 100-V ceramic, and C3 is a 0.0022-μF 100-V ceramic. C4 is a 0.01-μF 100-V, C5 a 0.056-μF 100-V, and C6 a 0.022-μF 50-V ceramic. C7 is a 220-μF electrolytic rated at 35 V. CR is a 1N4001 1-A 50 PIV silicon diode. K is an SPDT relay with a 12-V coil. L1 is a 88-μH toroid, while L2 is 33 turns of No. 26E wire wound on L1. Q1 through Q4 are all 2N2222. R1 and R6 are both 10 K. R2 and R4 are both 100K. R5 is 22 K. R7 is 3.3 K, and R8 is 1 K. T is 8 to 1000 Ω, center-tapped.—B. Kretzman, "Autostart on RTTY," *CQ,* November 1984, pp. 50–52.

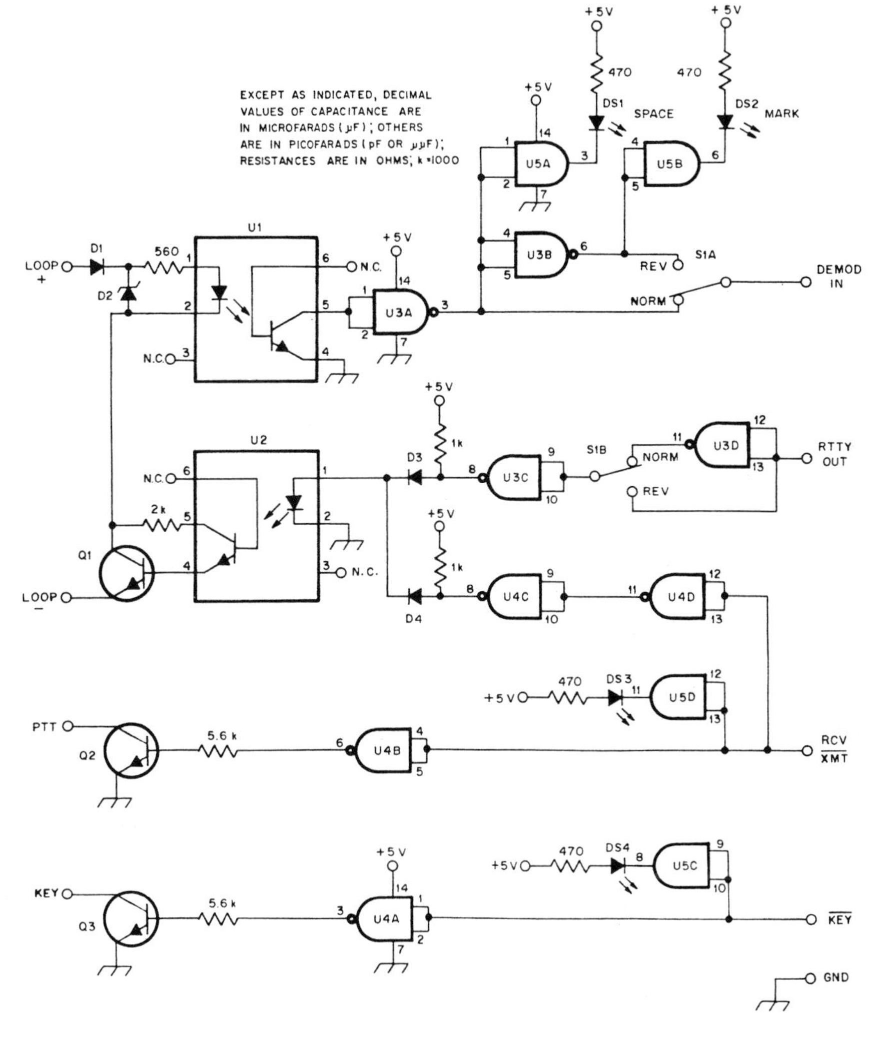

60-mA LOOP TELEPRINTER INTERFACE—Interfaces TTL-level teleprinter system to 60-mA teleprinter. The isolation between the 150-V DC loop and the TTL circuits is provided by optoisolators U1 and U2 (MCT-2). U3 and U4 are each 7400 and U5 is a 7408. Q1 is an MPS-A42, while Q2 and Q3 are each 2N2222. D1 is a 1N4820, D2 is a 5-V, 500-mW zener, and D3 and D4 are each 1N4148.—S. Nafziger, A Universal RTTY Current-Loop Interface, *QST,* January 1984, pp. 17–18.

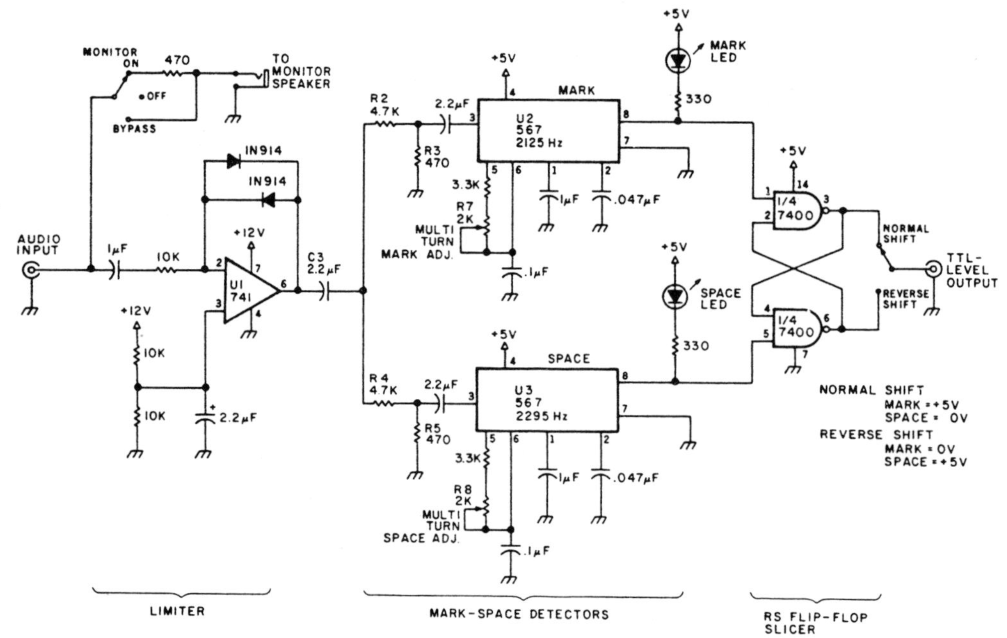

TTL-LEVEL OUTPUT DEMODULATOR—Takes audio input of radioteleprinter signals from receiver and produces TTL-level output for microcomputers. The LEDs indicate mark and space signals.—A. Hearn, Micro Modem, *73 Magazine,* September 1982, pp. 10–22.

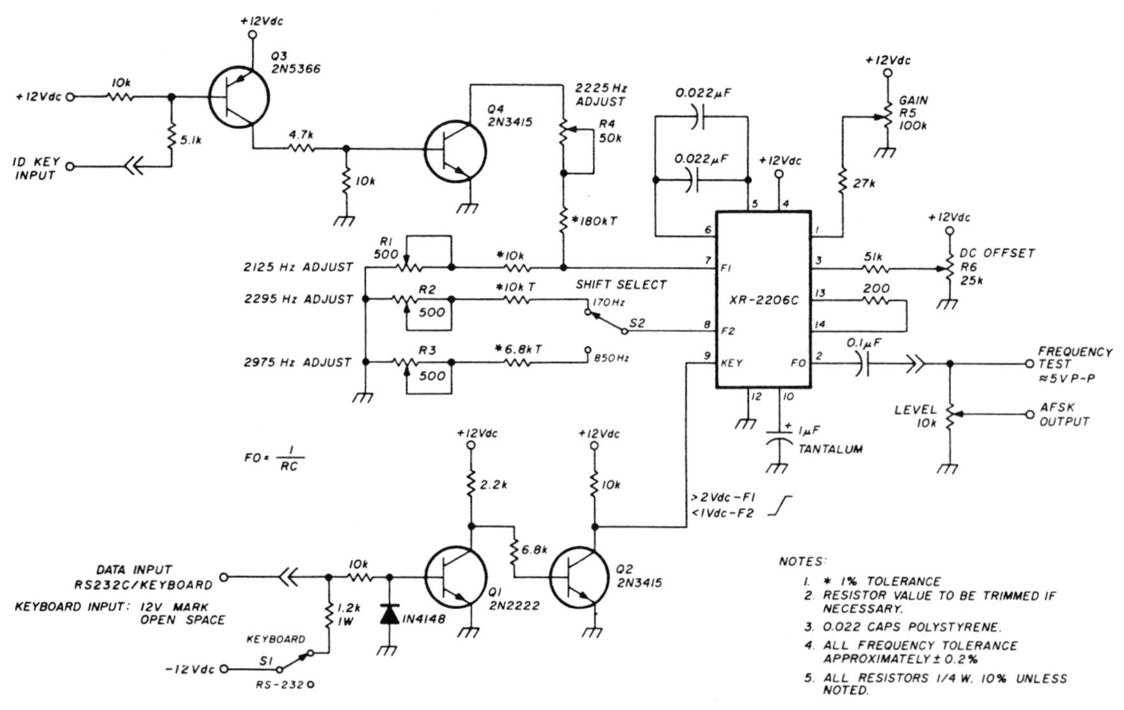

AFSK GENERATOR—Built around XR-2206C function generator IC, circuit offers excellent frequency stability and low-distortion output. Q3 and Q4 form an optional CW identifier section; if the feature is not needed, it may be deleted. Q1 and Q2 act as an input level translator and switch for either RS232C or mechanical keyboard input. The values of the timing capacitors are critical; they must be polystyrene for maximum stability. The tuning procedure is described in detail in the original article and requires a frequency counter and oscilloscope.—G. Boldenow, An Accurate and Practical AFSK Generator, *Ham Radio,* August 1980, pp. 56–58.

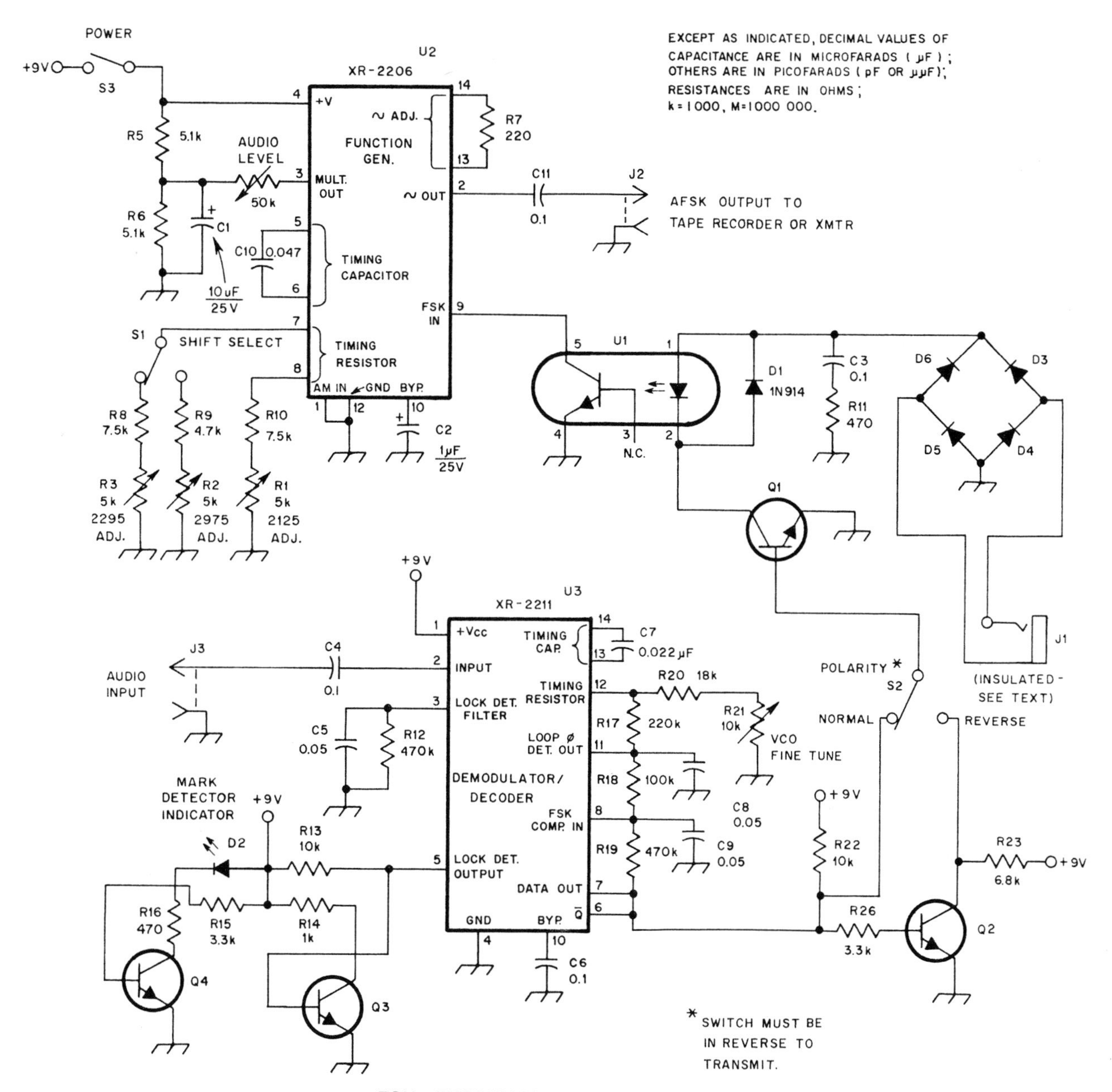

FSK TERMINAL UNIT—Interfaces teleprinter unit to transmitter and provides 850- and 170-Hz frequency shifts. The circuit interfaces to a standard 60-mA simplex loop. The XR-2206 generates stable, low-distortion sine waves. Switching in different values of the resistance determines the mark and space frequencies. Keying is accomplished by the presence or absence of current in the loop via optoisolator U1 (OPI-2150, HEP-P5000, Motorola 4N28, or equivalent). Q1 is a 2N5655, MJE340, or TIP-48. Q2, Q3, and Q4 are all 2N2222. D2 is any general-purpose LED. D3-D6 are all 1N4003.—M. Di Julio, A State-of-the-Art Terminal Unit for RTTY, *QST*, December 1980, pp. 20–22.

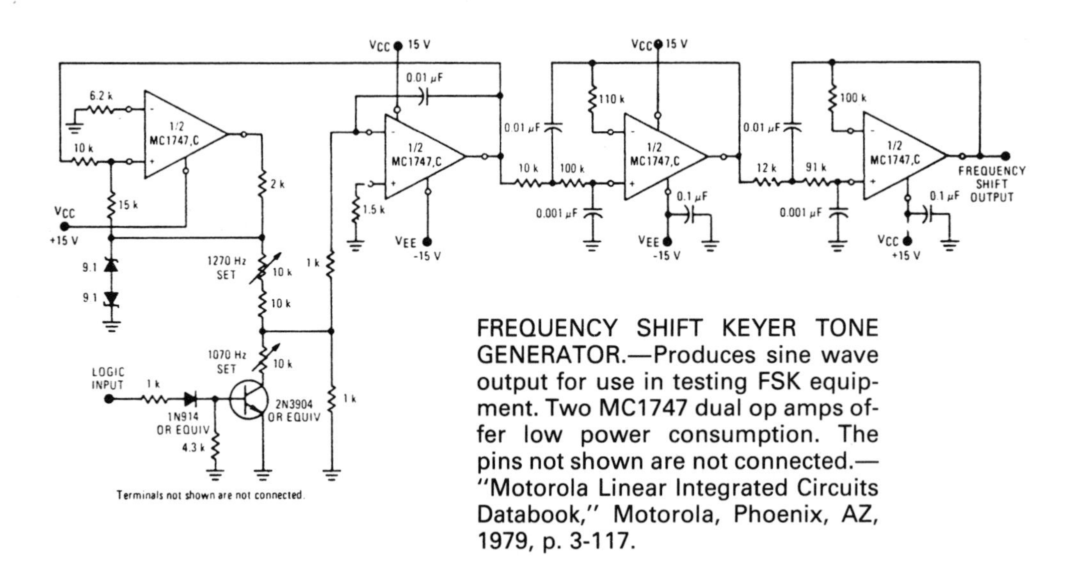

FREQUENCY SHIFT KEYER TONE GENERATOR.—Produces sine wave output for use in testing FSK equipment. Two MC1747 dual op amps offer low power consumption. The pins not shown are not connected.—"Motorola Linear Integrated Circuits Databook," Motorola, Phoenix, AZ, 1979, p. 3-117.

22

Television and Video Circuits

10-GHz VIDEO MODULATOR—Designed for transmission of television signals in 10-GHz amateur radio band. The circuit can be used with a Gunn diode transmitter or a modified microwave motion detector. The audio subcarrier is 4.5 MHz.—E. Sullivant, 10 GHz ATV the Easy Way, *CQ*, July 1984, pp. 62–64.

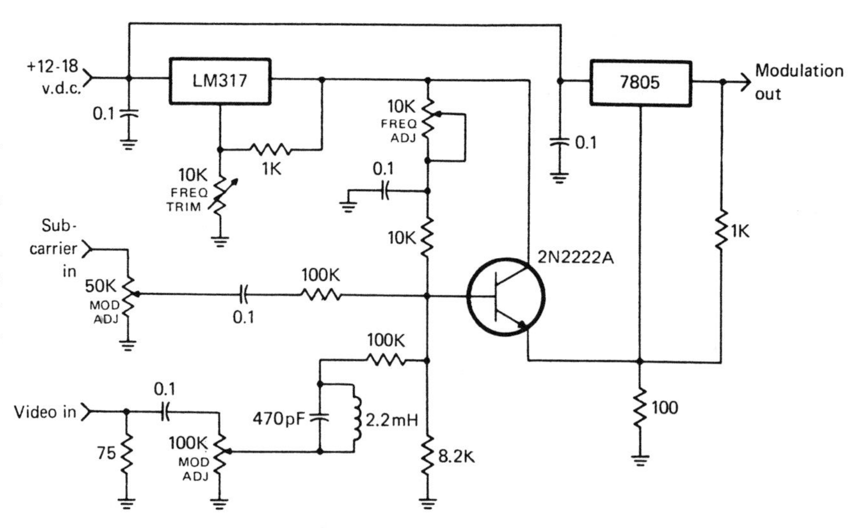

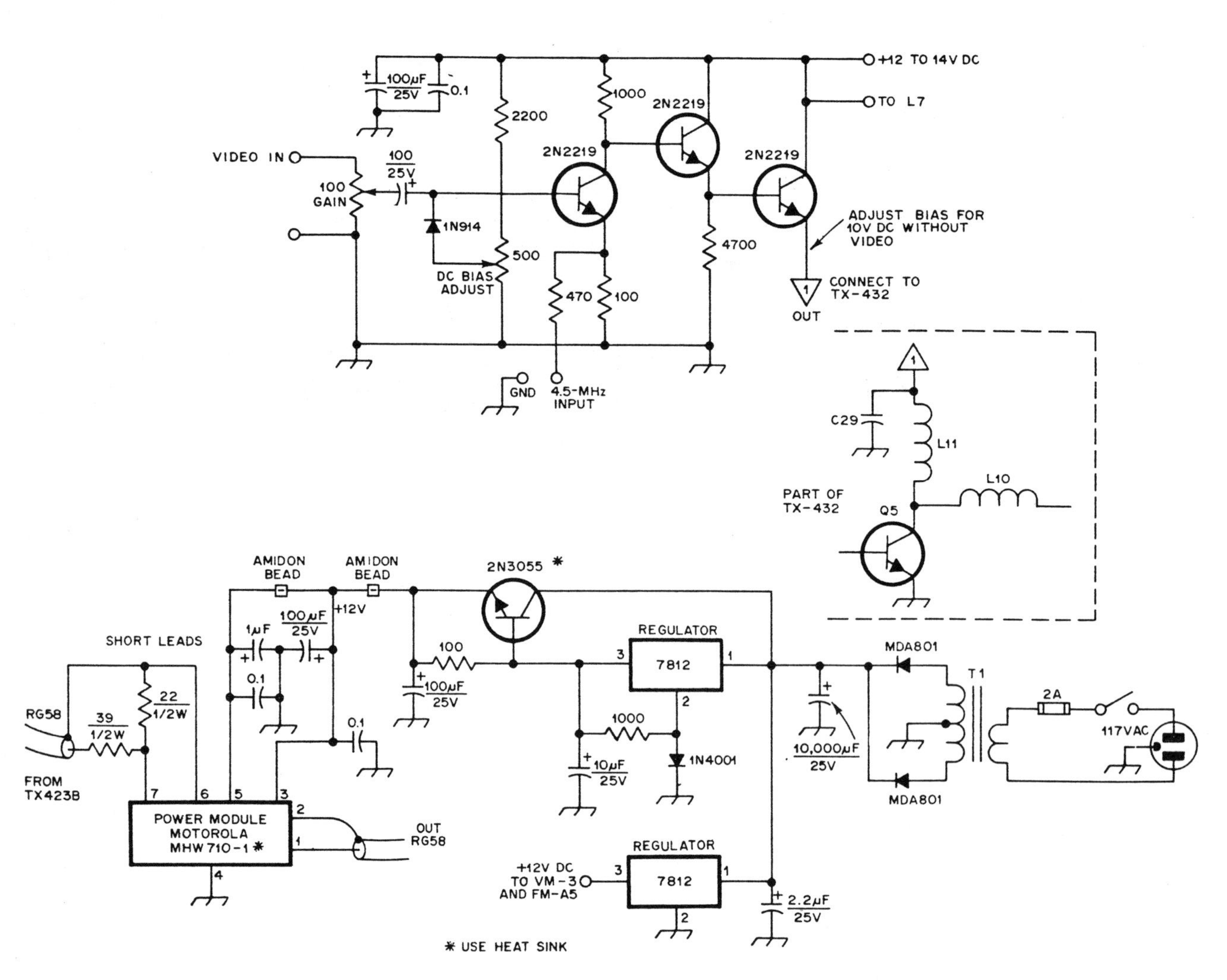

SOLID-STATE VIDEO MODULATOR—Based upon VHF Engineering TX-432 and Motorola MHW-710 RF modules, this circuit is designed for use on 432-MHz amateur radio band but may be adapted for closed-circuit and similar purposes. T1 consists of two 25-V ct filament transformers with their primaries in series and secondaries paralleled (Triad F-41X or equivalent).—"The Radio Amateur's Handbook," American Radio Relay League, Newington, CT, 1981, pp. 14-30 to 14-31.

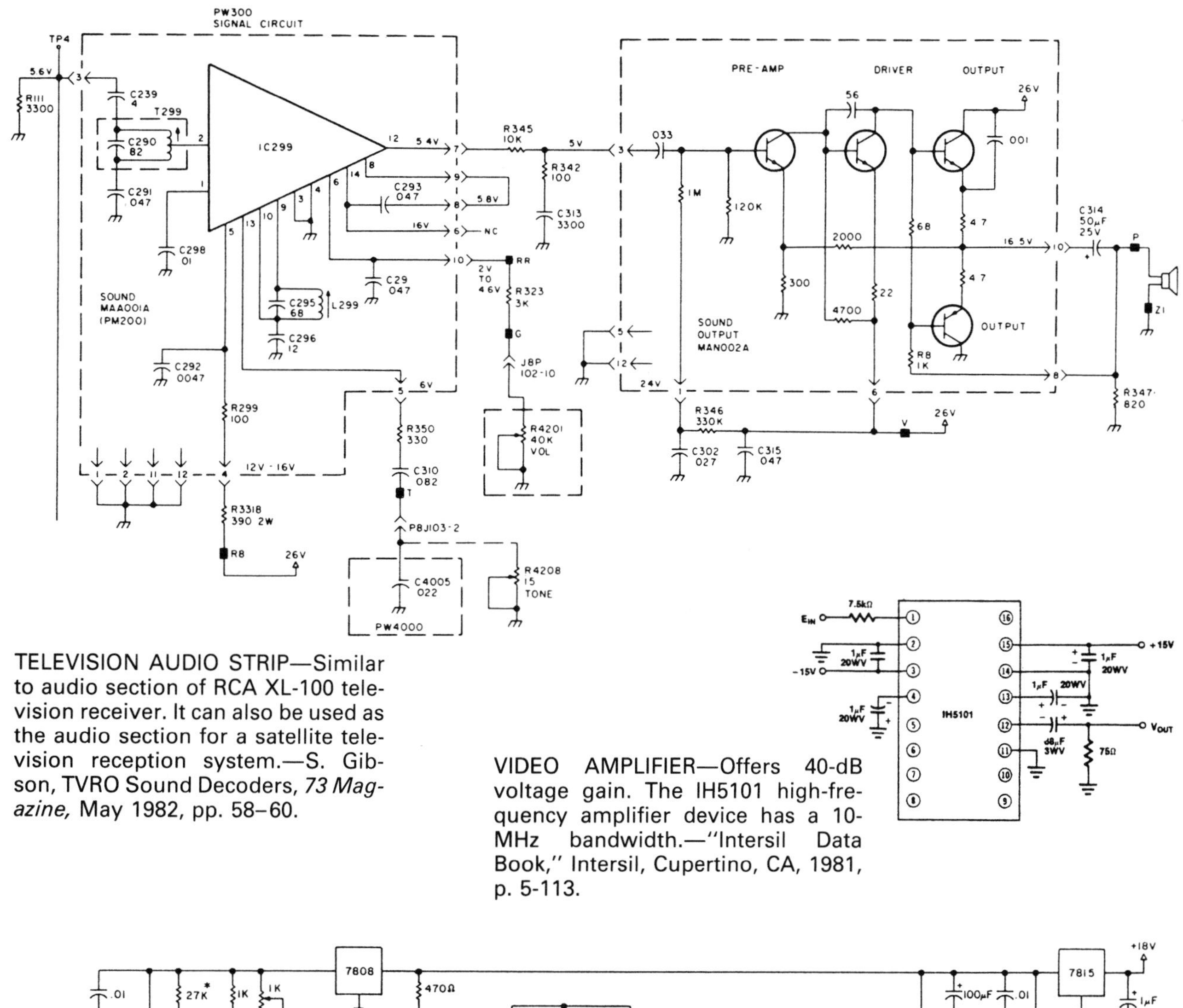

TELEVISION AUDIO STRIP—Similar to audio section of RCA XL-100 television receiver. It can also be used as the audio section for a satellite television reception system.—S. Gibson, TVRO Sound Decoders, *73 Magazine,* May 1982, pp. 58–60.

VIDEO AMPLIFIER—Offers 40-dB voltage gain. The IH5101 high-frequency amplifier device has a 10-MHz bandwidth.—"Intersil Data Book," Intersil, Cupertino, CA, 1981, p. 5-113.

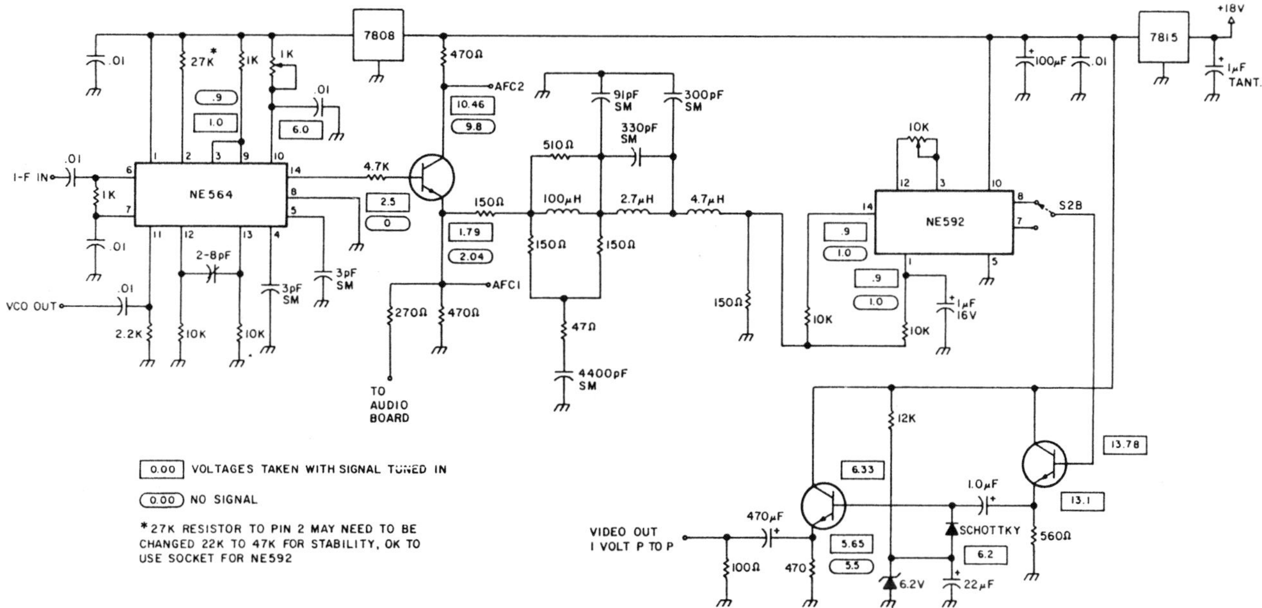

VIDEO DEMODULATOR STRIP—Designed as part of receiving system for satellite television receiver, this demodulator accepts 70-MHz input. The NE564 should be carefully selected to perform at 70 MHz. All transistors are 2N2222. The Schottky diode should be an MDB101.—S. Mitchell and R. Christian, 'Lite Receiver IV, *73 Magazine,* May 1982, pp. 48–56.

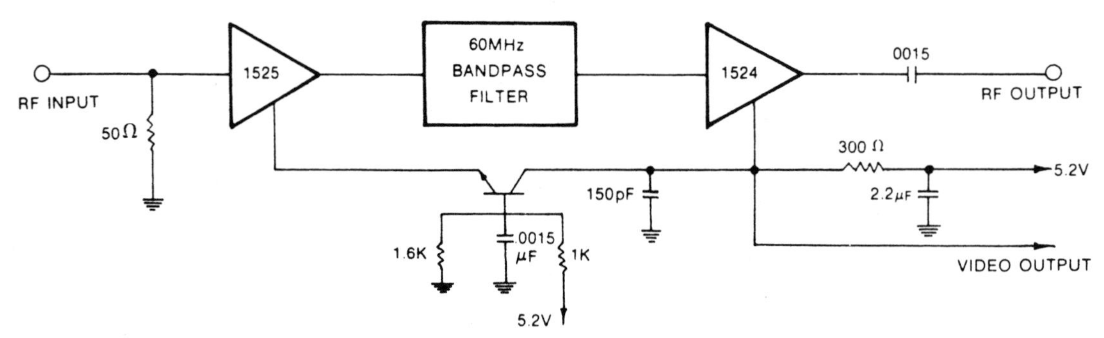

60-MHz BANDPASS FILTER SECTION—Increases dynamic range of log strip. The technique for increasing the dynamic range is done by the attenuator at the input inside the SL1525 quad limiting RF amplifier device. Normally the log response limits when the input signal exceeds that necessary to produce a full video output from the first stage. However, the parallel stage is being fed with an attenuated signal, so it will continue to give a video output change.—"Plessey Integrated Circuits Databook," Plessey Semiconductor, Irvine, CA, 1983, pp. 63–67.

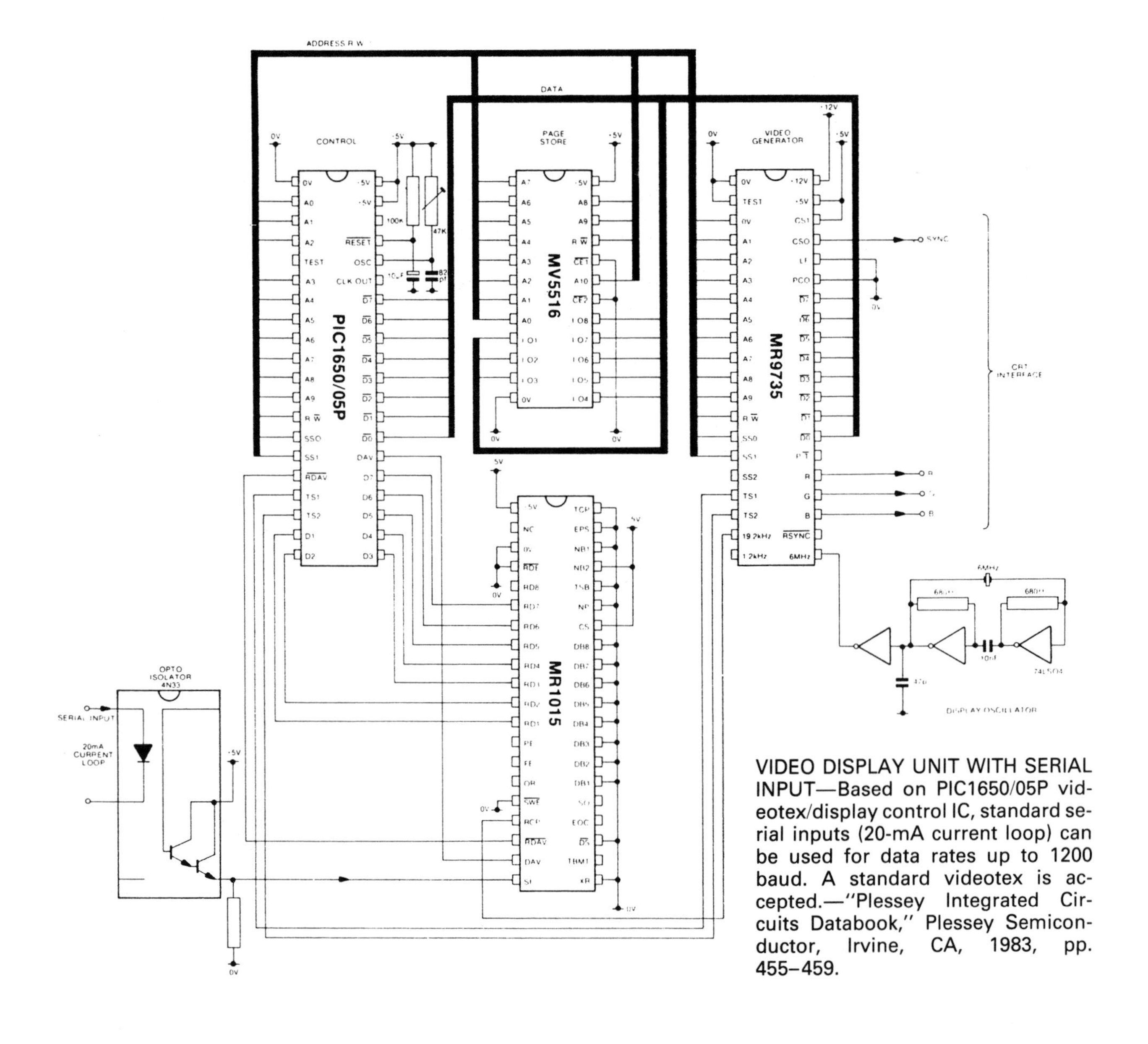

VIDEO DISPLAY UNIT WITH SERIAL INPUT—Based on PIC1650/05P videotex/display control IC, standard serial inputs (20-mA current loop) can be used for data rates up to 1200 baud. A standard videotex is accepted.—"Plessey Integrated Circuits Databook," Plessey Semiconductor, Irvine, CA, 1983, pp. 455–459.

23

Test Equipment Circuits

AUDIO SIGNAL GENERATOR—Design features adjustable output level. Parts values are generally noncritical—within 10% of those specified will generally be acceptable. L1 must be a toroid inductor. The circuit should not be used with very-high-voltage circuits although it is safe for almost all solid-state circuits and most tube circuits.—A. MacLean, Construct this All-American Audio Signal Generator, *73 Magazine,* December 1982, pp. 116–120.

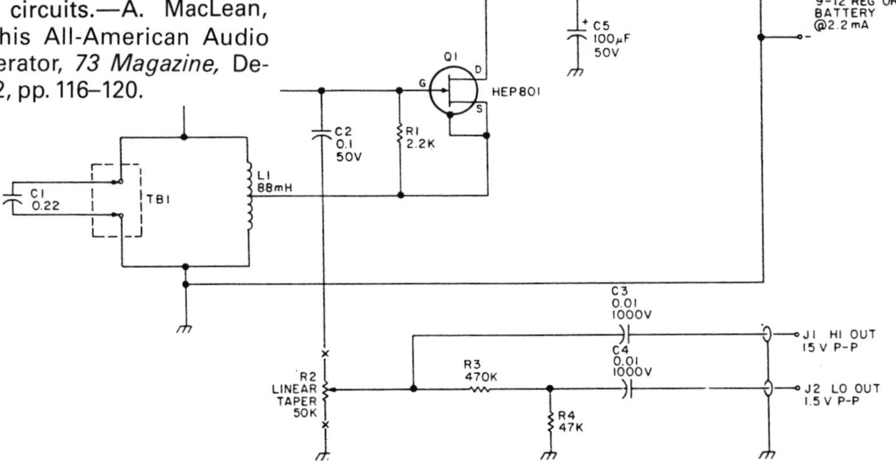

CONTINUITY CHECKER—Gives audible indication of resistance up to approximately 2000 Ω. It is useful for testing semiconductor junctions since with practice one can determine audibly differences in resistance of only a few ohms.—C. Schoeffler, The Meterless Ohmmeter, *73 Magazine,* August 1981, pp. 36–37.

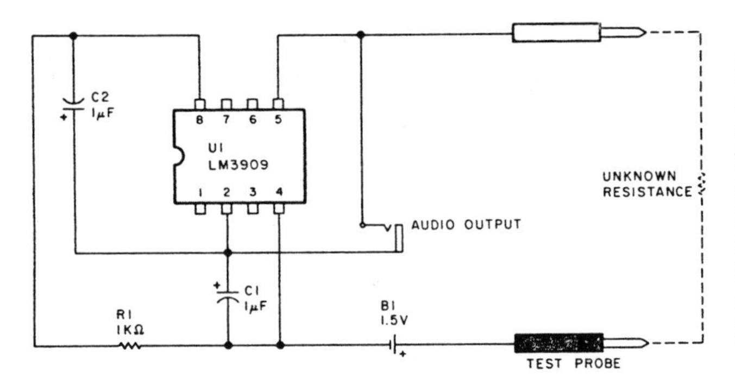

CAPACITANCE METER—Measures 200 to 10 μF in six ranges with accuracy to within 5%. Potentiometers R1 and R2 act as trimmers for the adjustment of the lower two ranges. Meter calibration can be done with capacitors of known values. The 555 is used in the monostable mode. One side of the capacitor under test is switched between the positive and negative supply terminals by the 555 at a rate determined by the resistors (R_A and R_B) and C1. When connected to the negative terminal, the capacitor charges to a value near the supply voltage. When switched to the positive terminal, the capacitor discharges and M1 indicates the average discharge current value, which is a function of the capacitance being tested.—H. Neben, A Simple Capacitance Meter You Can Build, *QST,* January 1983, pp. 34–37.

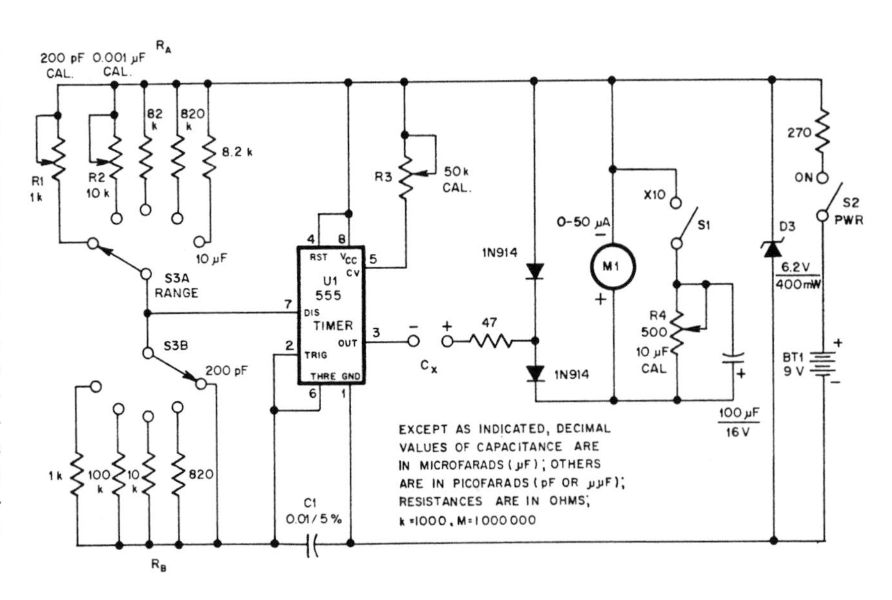

DIP METER—Unique design uses only single coil to cover 3.9–21 MHz. Three additional I/O jacks are provided: RF (to drive a frequency counter or mixer), AF (to monitor AM signals or detect carrier ripple), and PWR (for an external power source). Meter tuning is very sharp at higher frequencies. L1 is 41 turns of No. 16 wire on a 1⁄16-in form. D1 through D3 are each 1N103, 1N56, or equivalent. RFC1 is 500-μH RF choke, while RFC2 is a 56-μH RF choke.—G. Brizendine, Construct the Cyclops Dip Meter, *73 Magazine,* June 1983, pp. 26–28.

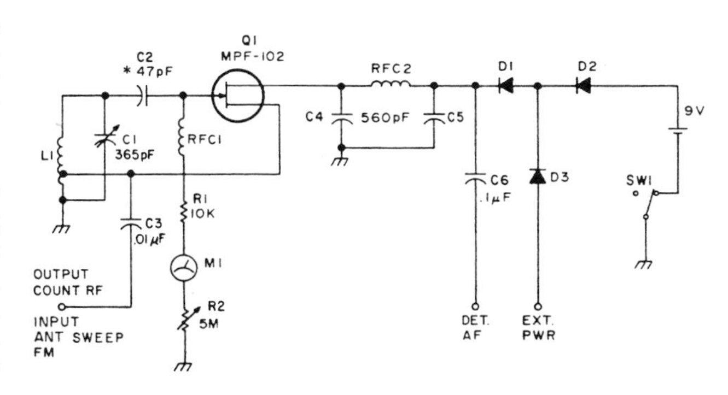

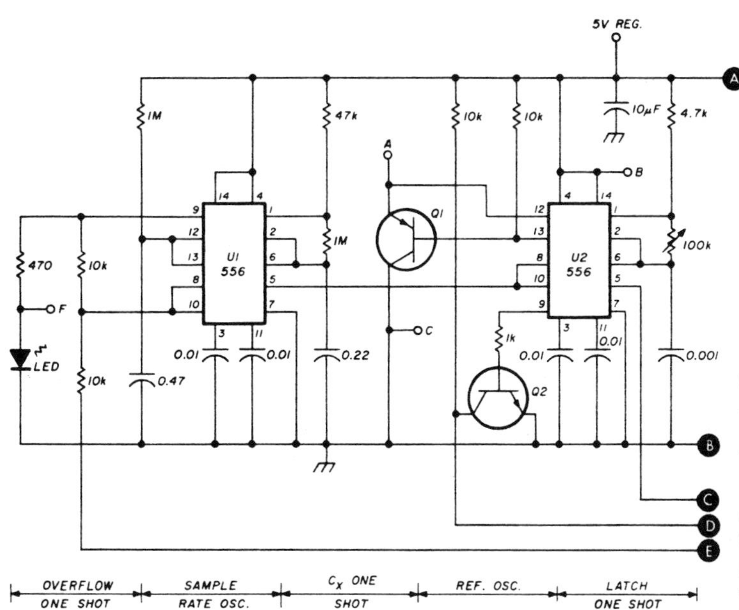

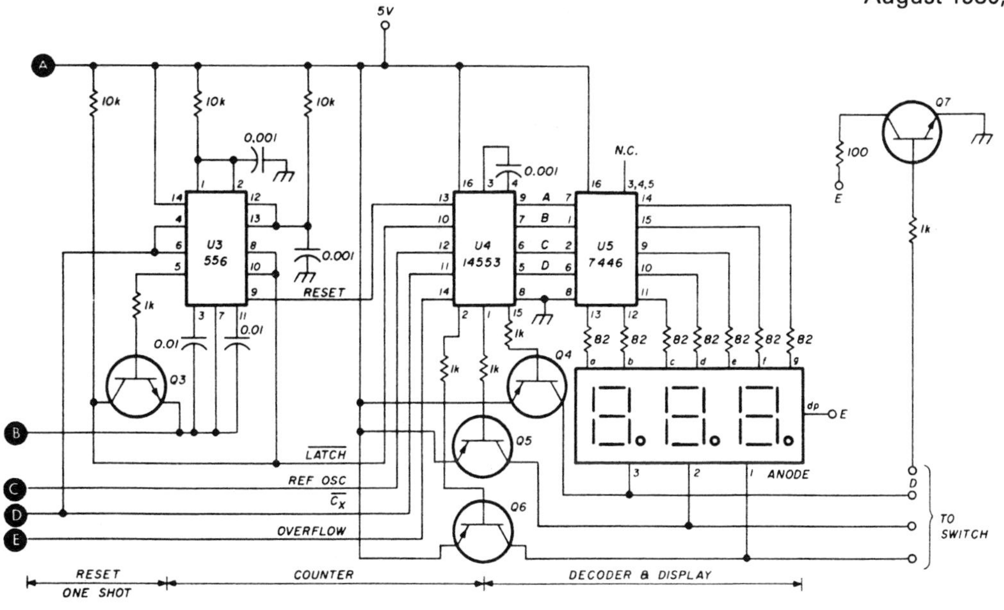

DIGITAL CAPACITANCE METER—Measures capacitors from 0.001 to 999 μF in six ranges with accuracy of approximately 1%. An LED flashes to indicate an overrange measurement. The circuit requires 100 mA from a regulated 5-V source. Q1, Q4, Q5, and Q6 are each 2N3906. Q2, Q3, and Q7 are each 2N3904. Calibration is made with capacitors of known value and the adjustment of the 100-K potentiometer.—M. Kitchens, Digital Capacitance Meter, *Ham Radio,* August 1980, pp. 66–71.

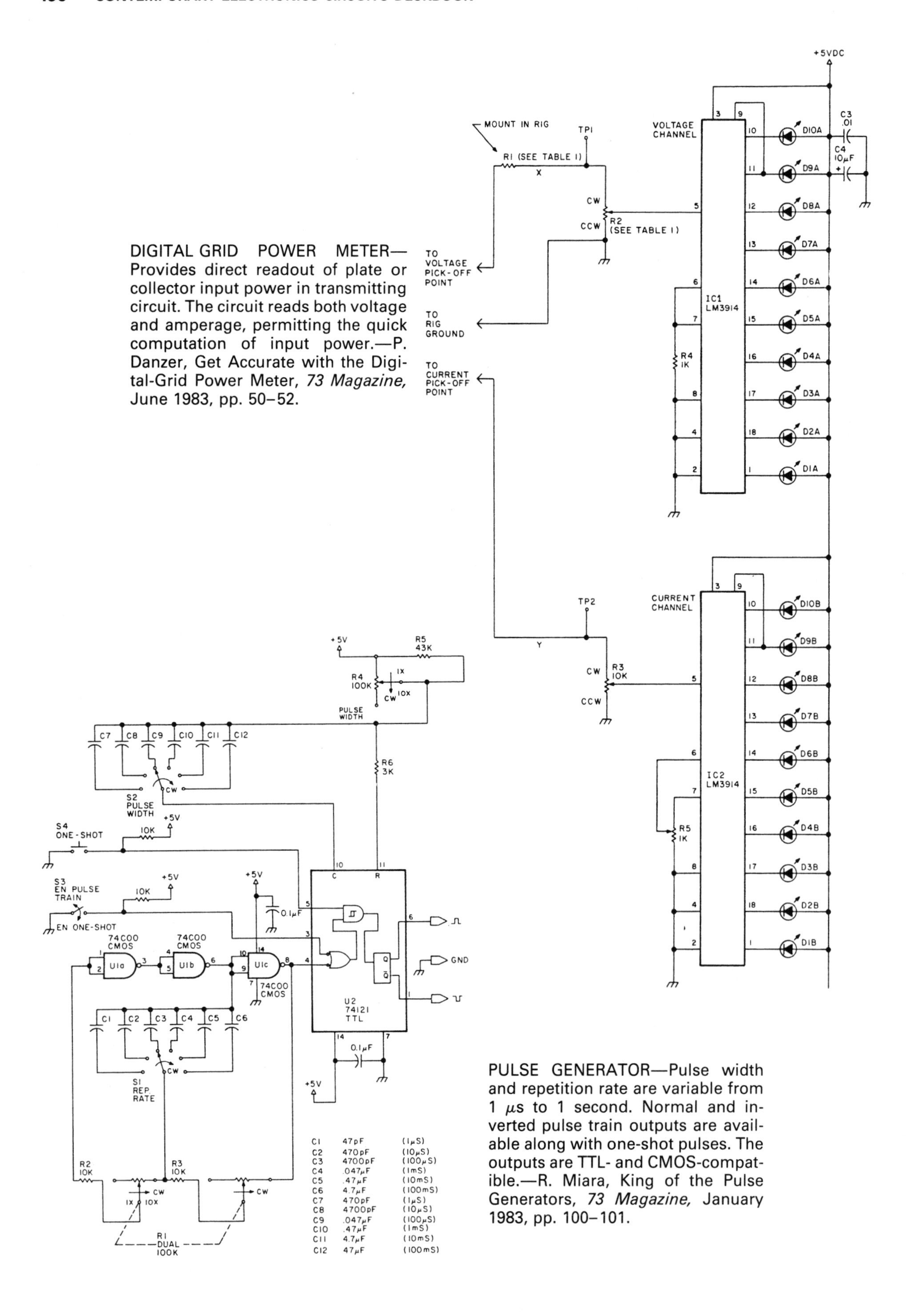

DIGITAL GRID POWER METER—Provides direct readout of plate or collector input power in transmitting circuit. The circuit reads both voltage and amperage, permitting the quick computation of input power.—P. Danzer, Get Accurate with the Digital-Grid Power Meter, *73 Magazine,* June 1983, pp. 50–52.

PULSE GENERATOR—Pulse width and repetition rate are variable from 1 μs to 1 second. Normal and inverted pulse train outputs are available along with one-shot pulses. The outputs are TTL- and CMOS-compatible.—R. Miara, King of the Pulse Generators, *73 Magazine,* January 1983, pp. 100–101.

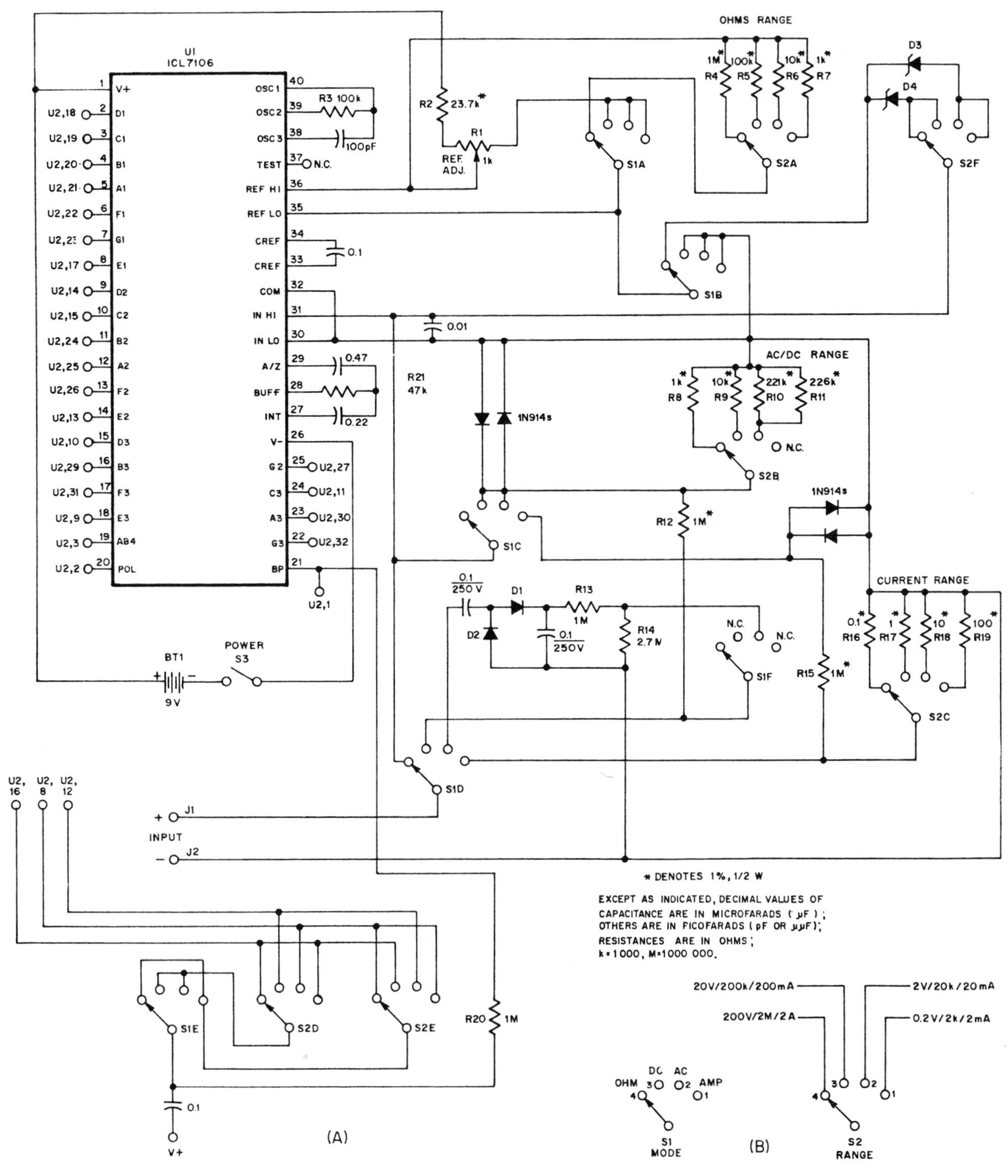

DIGITAL MULTIMETER—Allows measurement of AC/DC voltages and currents as well as resistances as low as 1 Ω. The circuit also features automatic zero adjustment and automatic polarity indication. The circuit is designed around an ICL7106 ADC device. An internal regulator maintains the voltage between V+ (pin 1) and COMMON (pin 32) at 2.8 V. This is used to supply a reference voltage by connecting a divider (R1 and R2) between those pins. By adjusting the divider, the reference voltage can be set to the required 100-mV value. D1 and D2 are each 1N4007, D3 is a 1N5228B, and D4 is a 1N5231B. U2 is a Hamlin 390-23-155 or equivalent 3.5-digit LCD.—B. Shriner and G. Collins, Learning to Work with Integrated Circuits, 1982 Style, *QST,* August 1982, pp. 29–33.

AUDIO LOGIC PROBE—Lets users "hear" a logic level 1 and audibly determine the voltage of that logic level within a few tenths of a volt in the range of approximately 2.2 to 12 V.—H. Batie, The Sweet Sounding Probe, *73 Magazine,* July 1980, p. 84.

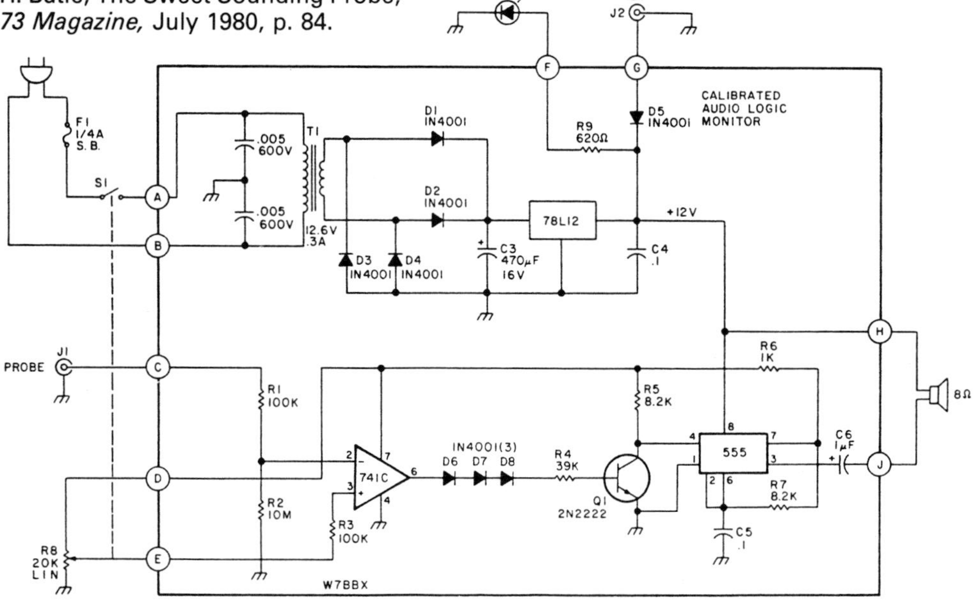

DUAL-GATE MOSFET DIP METER—Covers 2.3–200 MHz and allows checking of resonant circuits. The meter operates by supplying RF power to the circuit under test; when the circuit under test and the dip meter are tuned to the same frequency, they are coupled together and meter M1 "dips." The values of L1, C1, and C2 vary according to the frequency range the meter operates over (see the table that accompanies the drawing). L1 is wound on a Millen 45005 form.—"The Radio Amateur's Handbook," American Radio Relay League, Newington, CT, 1981, pp. 16-20 to 16-23.

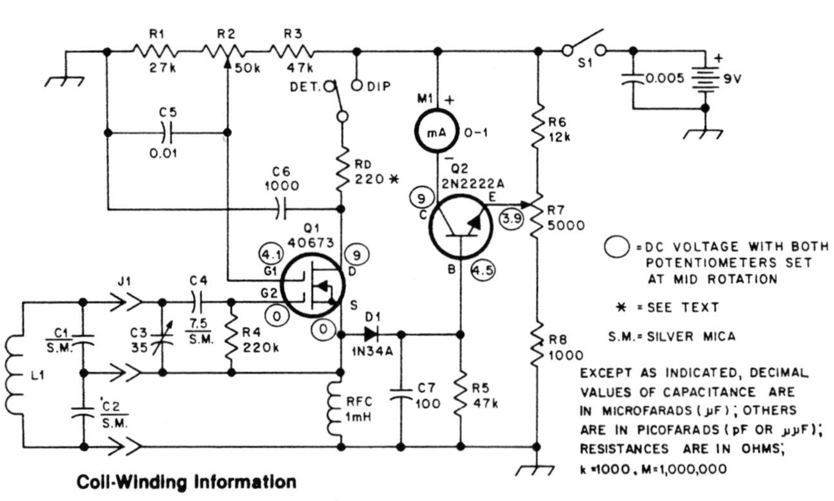

Coil-Winding Information

Freq. Range MHz	C1 pF	C2 pF	L1 Turns
2.3-4	15	15	71-1/2
3.4-5.1	33	10	39-1/2
4.8-8	10	33	25-1/2
7.9-13	10	33	14-1/2
12.8-21.2	10	33	6-1/2
21-34	10	33	4-1/2
34-60	10	33	2-1/2
60-110	10	33	*
90-200	not used	not used	**

*denotes a 1-1/2-turn coil of no. 18 enam. wire wound on a 1/2-inch (13-mm) form spaced 1/8 inch (3 mm) between turns. It should be placed so that the coil is near the top of the coil form.

**denotes a hairpin loop made from flashing copper, 3/8-inch (9.5-mm) wide × 1-7/8-inch (89 mm) total length.

All other coils are wound with no. 24 enam. wire.

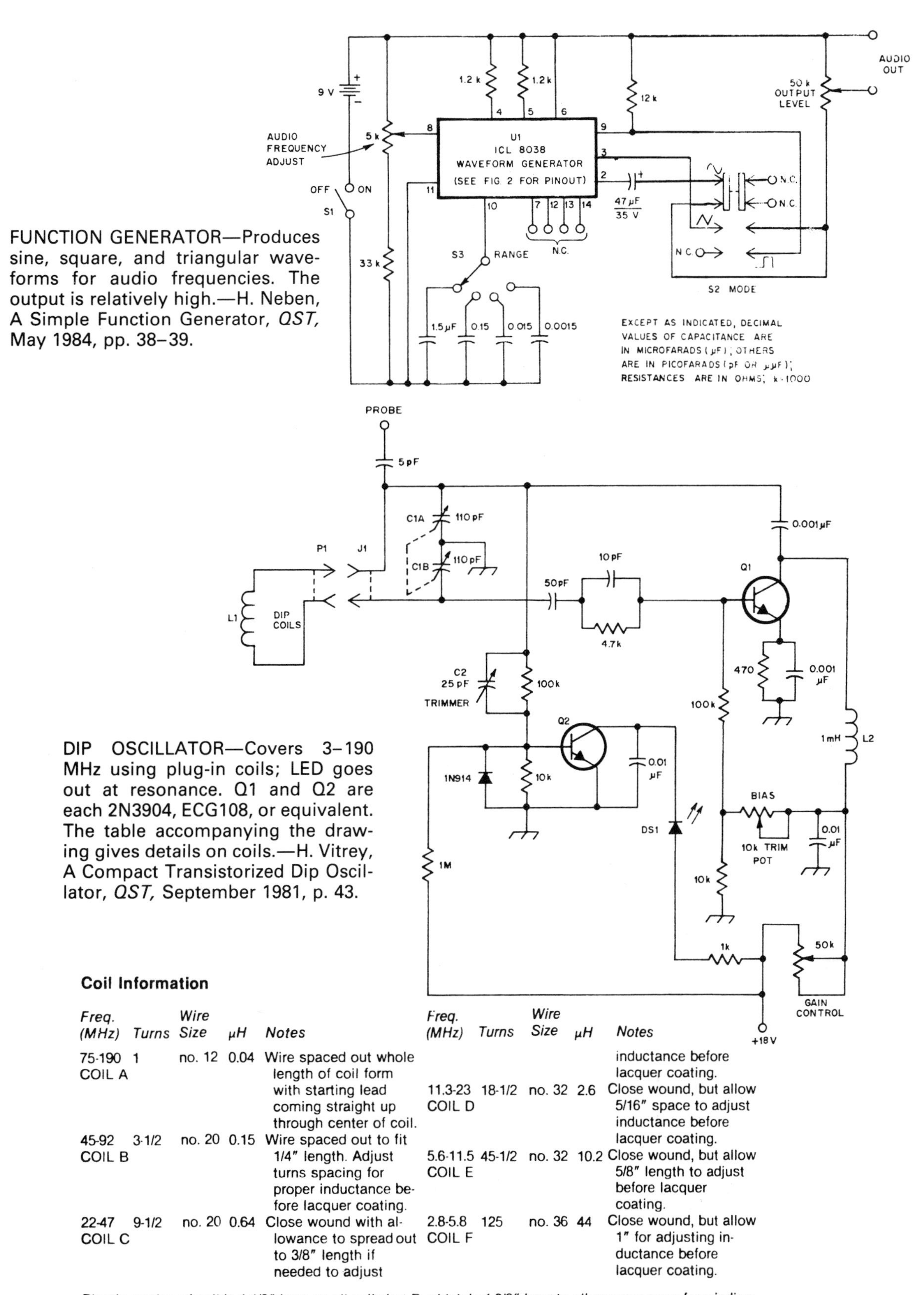

FUNCTION GENERATOR—Produces sine, square, and triangular waveforms for audio frequencies. The output is relatively high.—H. Neben, A Simple Function Generator, *QST,* May 1984, pp. 38–39.

DIP OSCILLATOR—Covers 3–190 MHz using plug-in coils; LED goes out at resonance. Q1 and Q2 are each 2N3904, ECG108, or equivalent. The table accompanying the drawing gives details on coils.—H. Vitrey, A Compact Transistorized Dip Oscillator, *QST,* September 1981, p. 43.

Coil Information

Freq. (MHz)	*Turns*	*Wire Size*	*μH*	*Notes*
75-190 COIL A	1	no. 12	0.04	Wire spaced out whole length of coil form with starting lead coming straight up through center of coil.
45-92 COIL B	3-1/2	no. 20	0.15	Wire spaced out to fit 1/4″ length. Adjust turns spacing for proper inductance before lacquer coating.
22-47 COIL C	9-1/2	no. 20	0.64	Close wound with allowance to spread out to 3/8″ length if needed to adjust inductance before lacquer coating.
11.3-23 COIL D	18-1/2	no. 32	2.6	Close wound, but allow 5/16″ space to adjust inductance before lacquer coating.
5.6-11.5 COIL E	45-1/2	no. 32	10.2	Close wound, but allow 5/8″ length to adjust before lacquer coating.
2.8-5.8 COIL F	125	no. 36	44	Close wound, but allow 1″ for adjusting inductance before lacquer coating.

Plastic portion of coil is 1-1/8″ long on all coils but F, which is 1-3/8″ long to allow more room for winding. All coils start with lead coming straight up through center of coil. Top of plastic tube is slotted 1/8″. Hook wire through this slot, and start winding. End winding at proper length by drilling two holes 1/4″ apart. Feed wire in one hole and out the other, and end coil by running this wire straight down and by soldering to outside of RCA plug. Coil adjustment would be less critical by using seven coils, thus having more overlap.

HIGH-IMPEDANCE FET VOLTMETER—Measures DC voltages in seven ranges from 0.5 to 500 V. The input impedance is 11 M, and accuracy is within 10%. The zero setting of the meter is adjusted through R10; R9 controls the sensitivity of the meter. U1 is an LF353N dual JFET op amp.—G. Collins, Some Basics for Equipment Servicing, *QST,* January 1982, pp. 38–41.

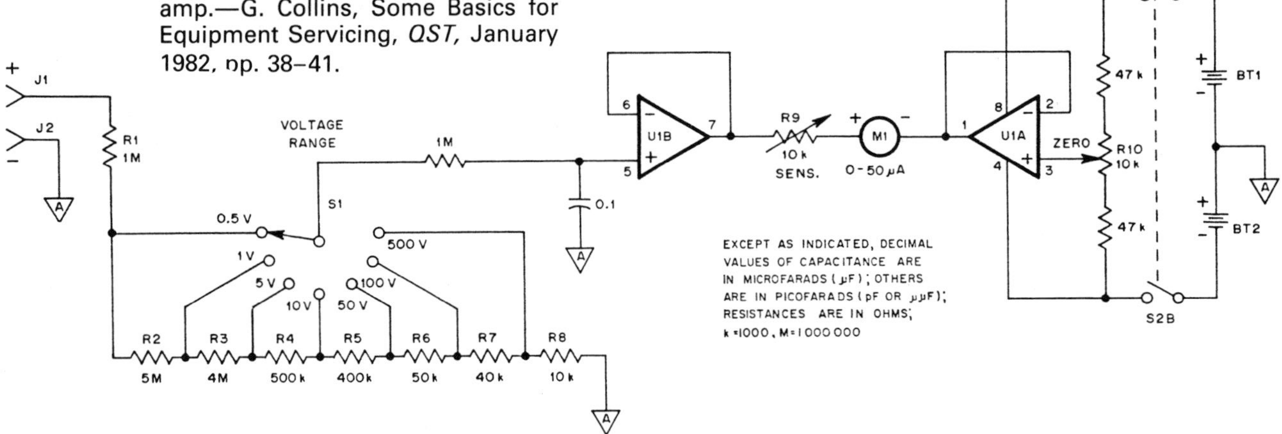

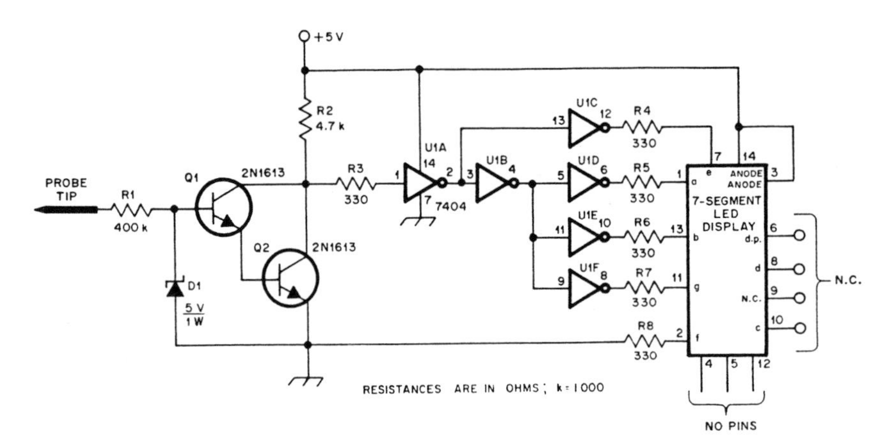

LOGIC PROBE—Logic level 0 will produce 0 on the LED display, logic level 1 will produce 1 on the display, and pulsing level will produce P.—R. Crawford, A Simple Logic Probe, *QST,* August 1983, p. 40.

LOGIC PROBE—Has input impedance of 1 M. A 2N2222 or similar NPN transistor may be substituted for the MPSA13.—T. Roth, Put Together the Shawnee Logic Probe, *73 Magazine,* October 1983, p. 66.

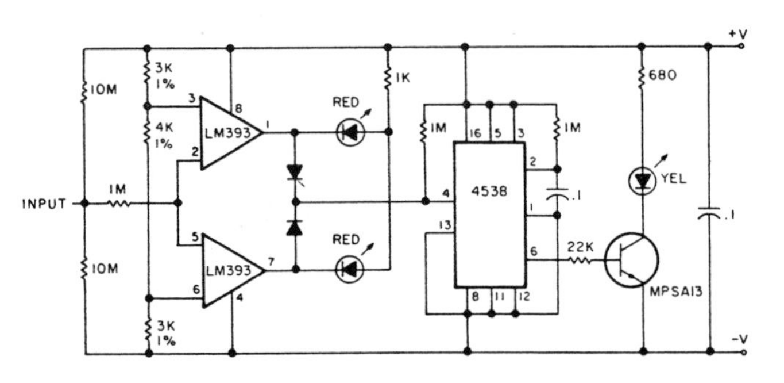

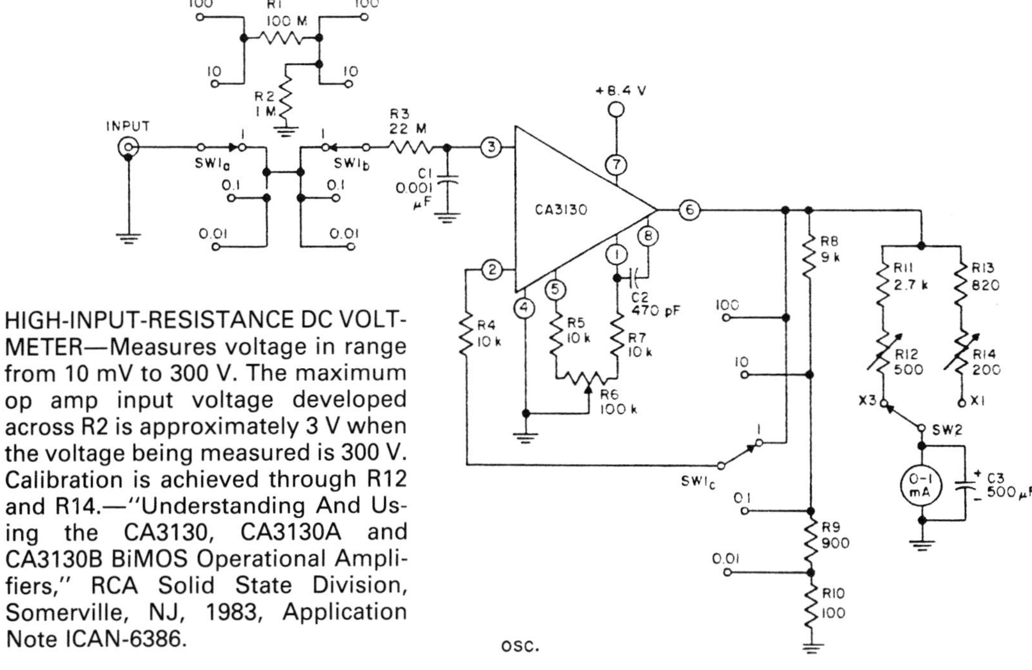

HIGH-INPUT-RESISTANCE DC VOLTMETER—Measures voltage in range from 10 mV to 300 V. The maximum op amp input voltage developed across R2 is approximately 3 V when the voltage being measured is 300 V. Calibration is achieved through R12 and R14.—"Understanding And Using the CA3130, CA3130A and CA3130B BiMOS Operational Amplifiers," RCA Solid State Division, Somerville, NJ, 1983, Application Note ICAN-6386.

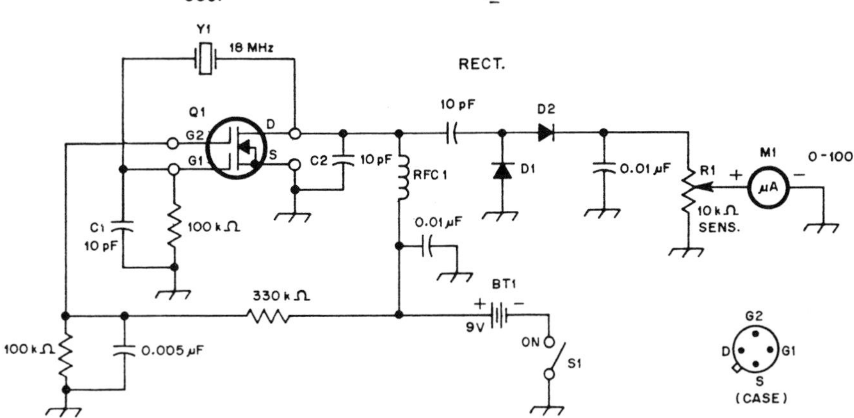

MOSFET TESTER—Simple crystal oscillator uses MOSFET under test in circuit covering 15–21 MHz (Y1 must be fundamental cut crystal in that range). RF energy from the oscillator drain is rectified by a diode doubler (D1 and D2). The resulting DC voltage is monitored at M1. If the MOSFET is defective, there will be no meter deflection. D1 and D2 are each 1N34A or equivalent.—D. DeMaw, Some Basics for Equipment Servicing, *QST,* December 1981, pp. 11–14.

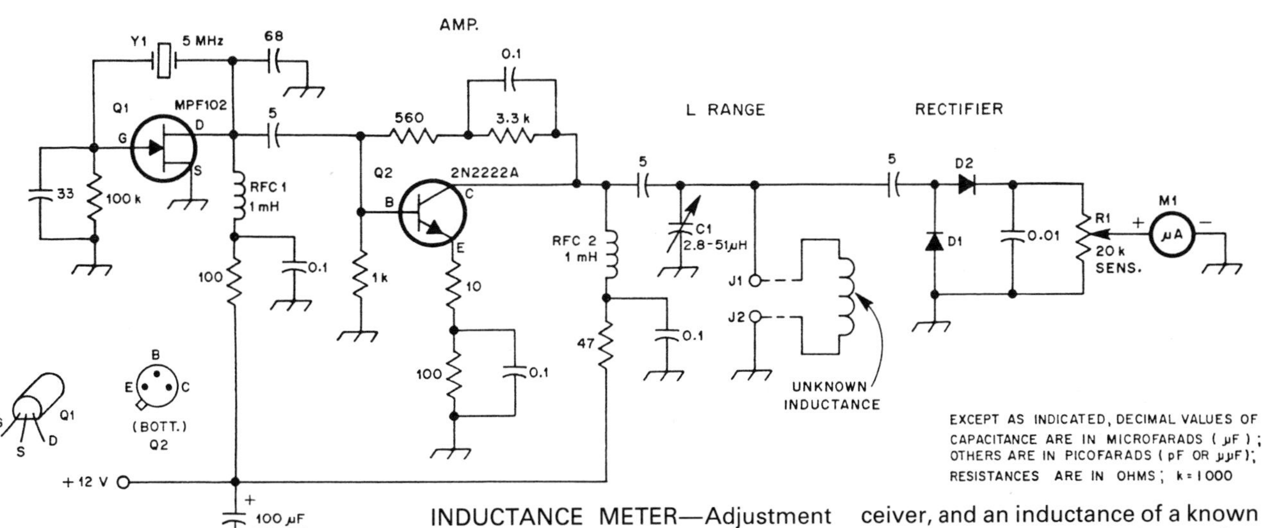

INDUCTANCE METER—Adjustment of C1 allows computation of value of inductor across terminals J1 and J2. The meter is calibrated using a dip meter, an accurately calibrated receiver, and an inductance of a known value.—D. DeMaw, Understanding Coils and Measuring Inductance, *QST,* October 1983, pp. 23–26.

SILICON DIODE CHECKER—Allows charting peak inverse voltage (PIV) of silicon diode connected across points labeled with "x" when used in conjunction with voltmeter on 1000-V scale having sensitivity of 5000 Ω/V or more. The input voltage is gradually increased from zero, and the resulting graph will indicate the diode breakdown voltage.—N. Johnson, Simple Diode Tester, *Ham Radio,* April 1983, p. 90.

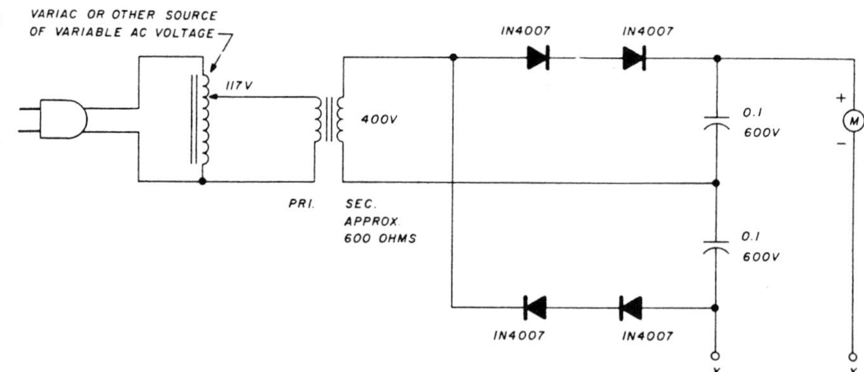

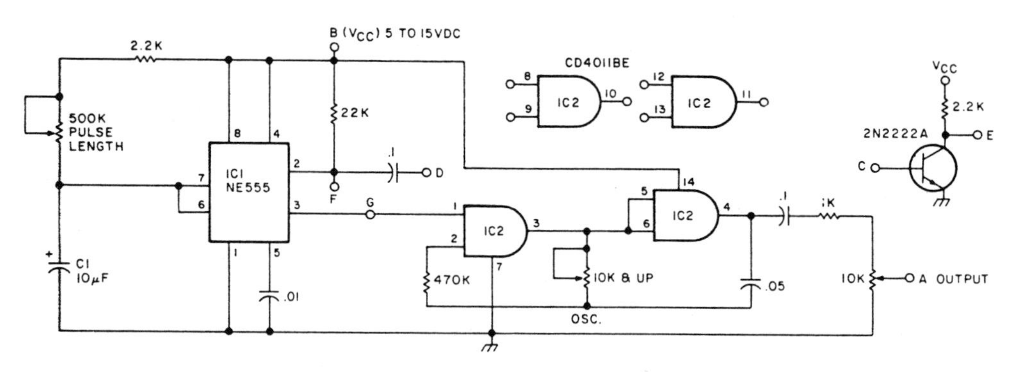

MULTIFUNCTION TEST INSTRUMENT—Combines TTL/CMOS logic probe, pulse stretcher, variable signal generator, timer, and other functions in single test instrument. The original article tells how to use the circuit for 10 different test functions.—M. Strange, A Perfect "10," *73 Magazine,* November 1982, pp. 10–12.

HIGH-IMPEDANCE METER AMPLIFIER—Used for measuring small voltages with VTVM or other high-impedance voltmeter. The four ferrite beads indicated by the letters "fb" are Micrometals No. 73 or No. 75. The unit can also be used as an amplifier for RF probes.—B. Parmley, An Amplifier for High-Impedance Voltmeters, *QST,* December 1983, pp. 41–42.

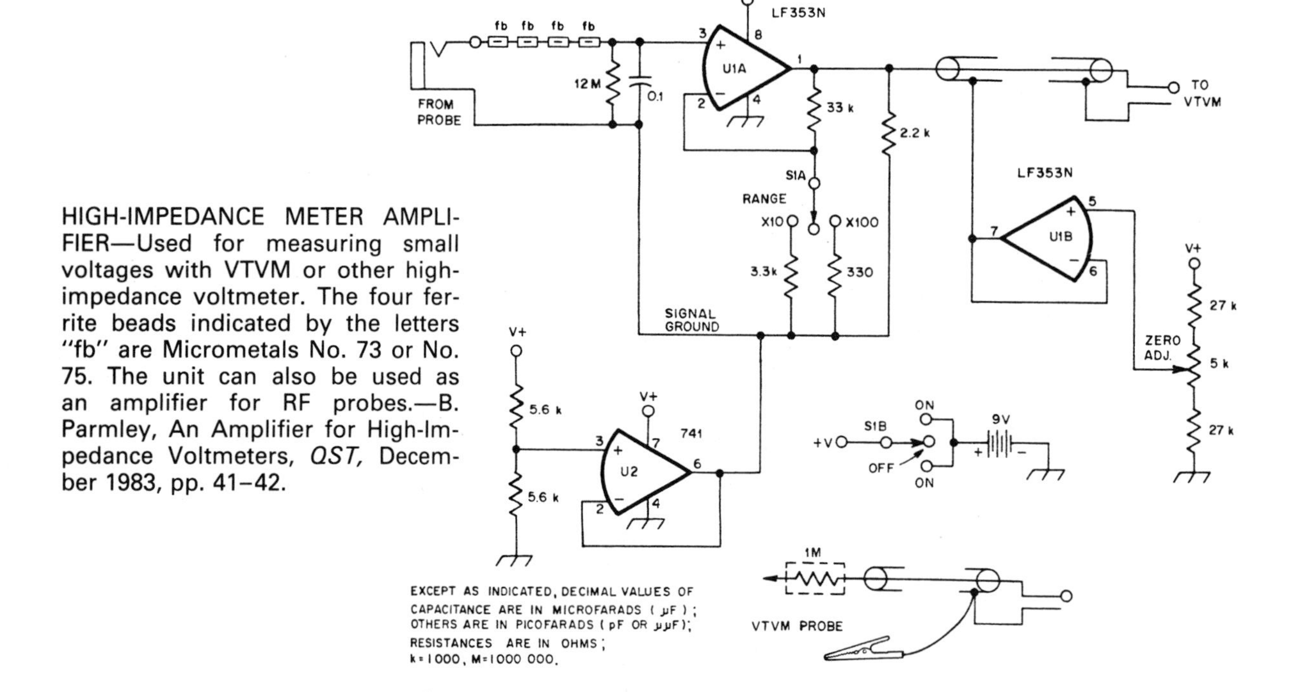

RESISTOR SUBSTITUTION BOX—Provides resistance values from 1 to 9,999,999 Ω in 1-Ω steps. R1 is 1 Ω, R2 is 2 Ω, R3 and R4 are 3 Ω, R5 is 10 Ω, R6 is 20 Ω, and R7 and R8 are both 30 Ω. R9 is 100 Ω, R10 is 200 Ω R11 and R12 are 300 Ω, R13 is 1 K, R14 is 2K, and R15 and R16 are both 3 K. R17 is 10 K, R18 is 20 K, R19 and R20 are 30 K, and R21 is 100 K. R22 is 200 K, R23 and R24 are 300 K, R25 is 1 M, R26 is 2 M, and R27 and R28 are each 3 M.—A. Guzman, R_x and C_x, *73 Magazine,* January 1981, pp. 84–85.

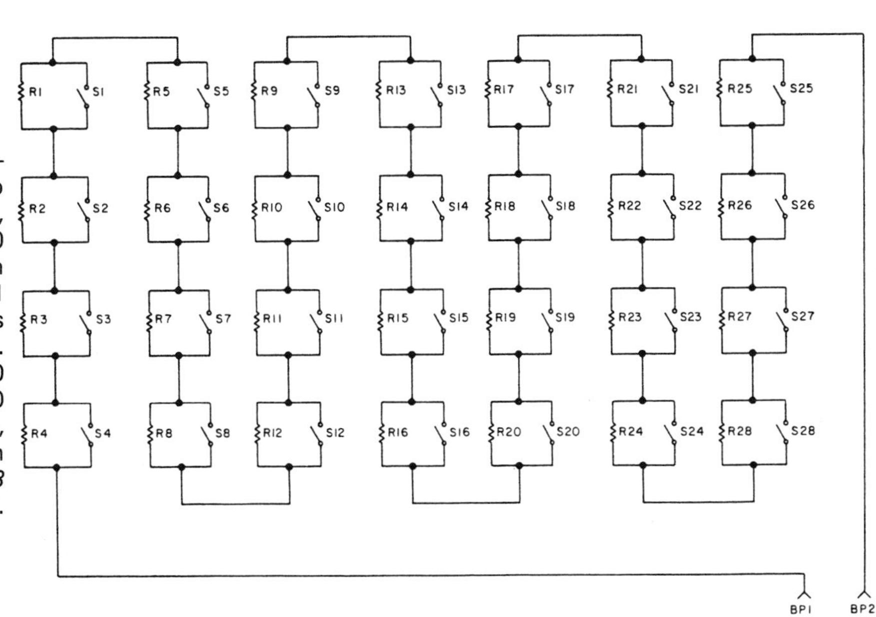

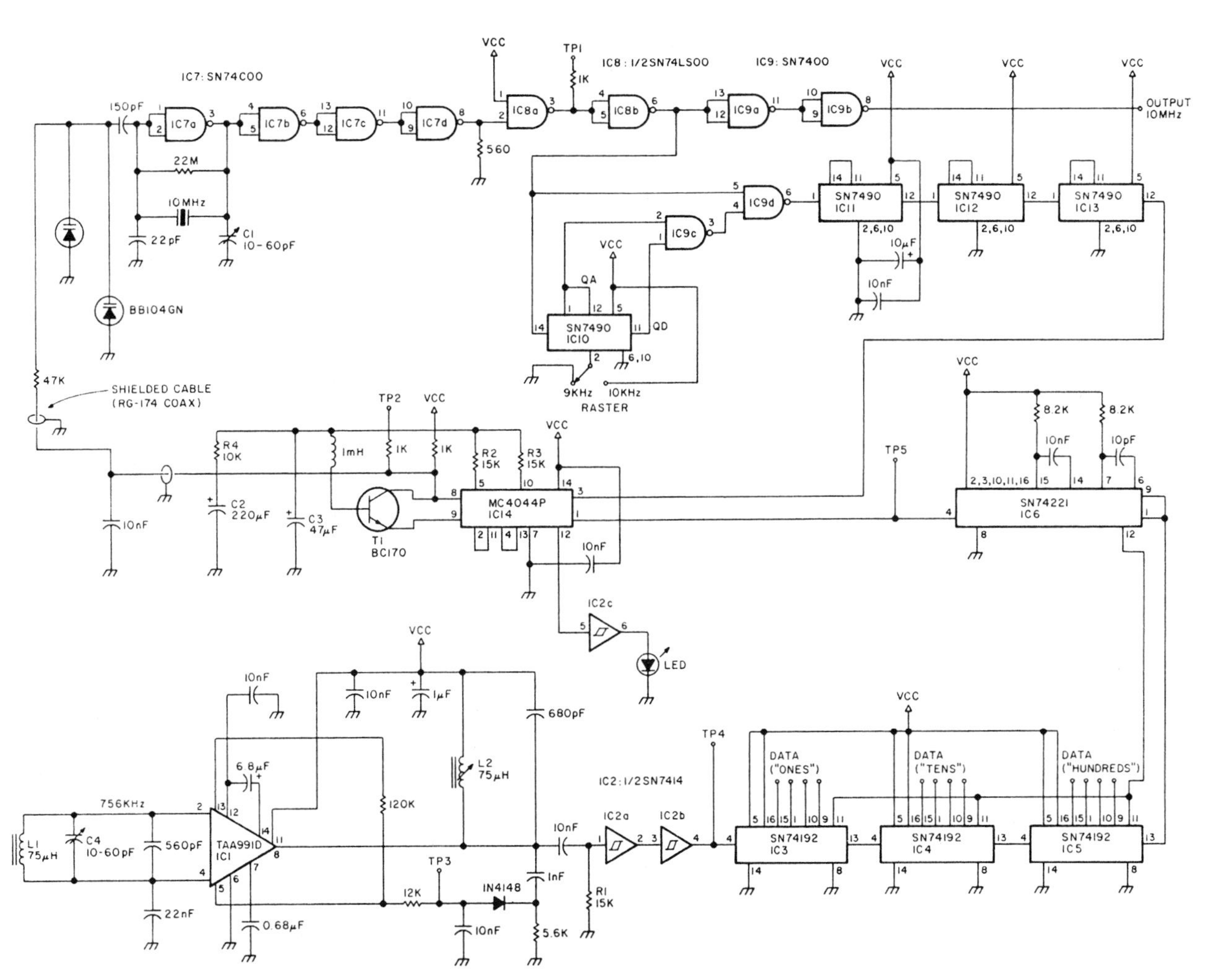

PRECISION FREQUENCY REFERENCE—10-MHz crystal oscillator phase-locks onto AM broadcast band station, producing virtually drift-free frequency/time signal generator for frequency counters, calibration, etc. Short term stability is at least 0.01 PPM and is higher in the long term. The circuit is programmable for either the North American 10-kHz AM broadcast channel standard or the 9-kHz channel spacing used elsewhere in the world.—R. Beyer, More Stable than a Rock, *73 Magazine,* July 1983, pp. 32–40.

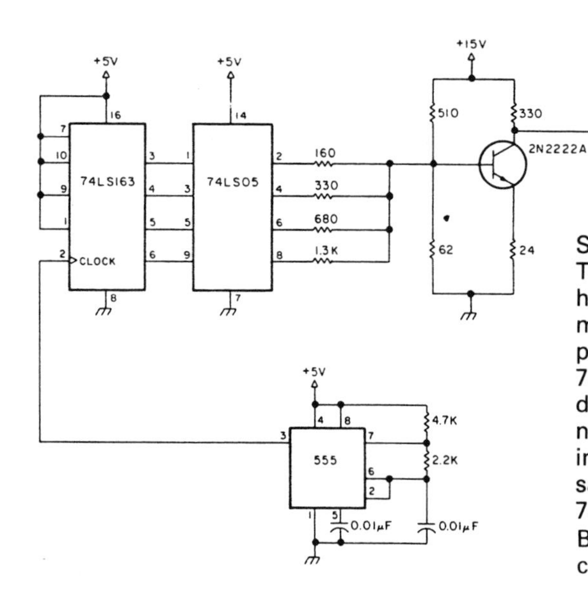

SAWTOOTH WAVEFORM GENERATOR—Provides sawtooth waveform having peak voltage of 10–11 V and minimum of 2–3 V. The 555 timer provides a 16-kHz clock for the 74LS163 counter, which in turn drives the 74LS05 DAC with the binary digits from 0 to 15 in a repeating sequence. This produces a sawtooth waveform from the 74LS05.—B. Myers, Transistors: A Biased Approach, *73 Magazine,* December 1984, pp. 34–36.

LINEAR, SELF-CALIBRATING OHMMETER—Linear-scale ohmmeter self-calibrates against internal standards for accuracy. The meter has six decade ranges from 100 Ω to 10 M full scale. All fixed resistors should be 5% tolerance or better.—F. Noble, A Linear, Self-Calibrating Ohmmeter, *QST,* September 1983, pp. 28–30.

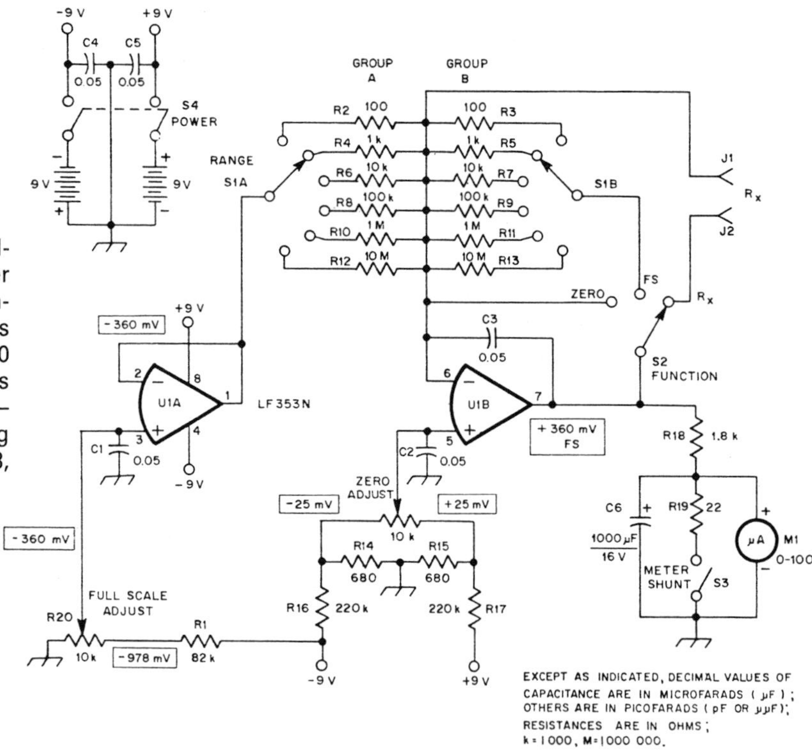

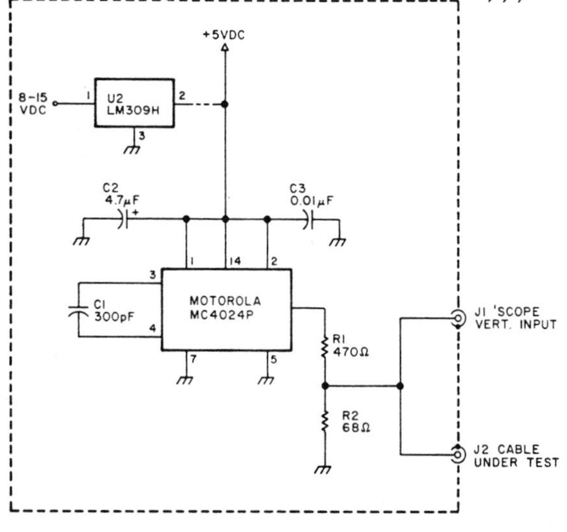

SQUARE WAVE SIGNAL GENERATOR—Designed for use with oscilloscope to test coaxial cable. The circuit uses the MC4024P VCO IC, not the CMOS 4024 device. The frequency can be altered by inserting a 1-K potentiometer at the dashed lines.—J. Carr, Find Fault with Your Coax, *73 Magazine,* October 1984, pp. 10–14.

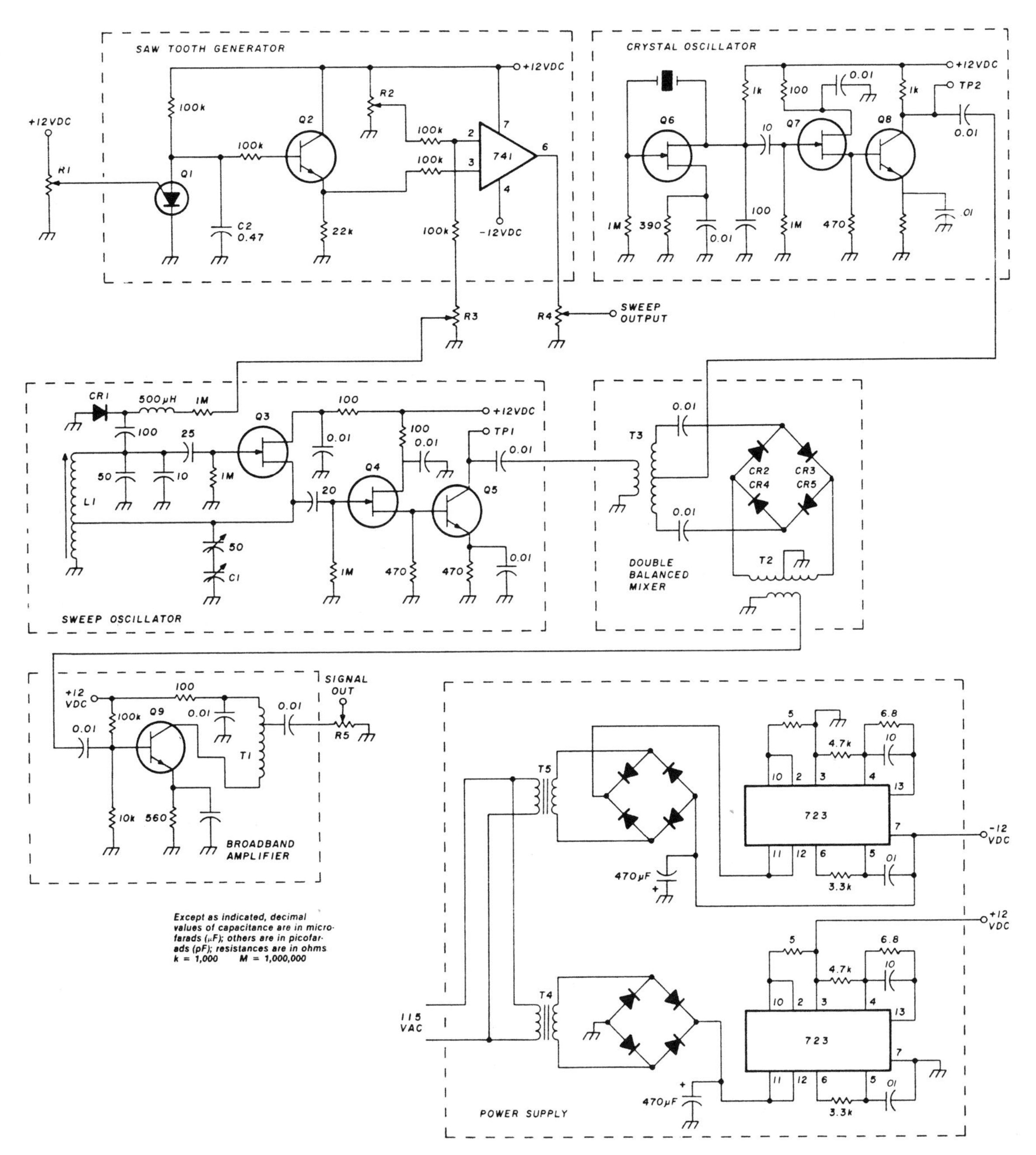

SWEEP GENERATOR—Useful for testing multipole filters and similar tuned circuits in conjunction with oscilloscope. The instrument is constructed on separate circuit boards (indicated by the dashed lines). C1 is a 25-pF variable capacitor. C2 is a 0.47-pF low-leakage capacitor. CR1 is an MV2203 tuning diode. CR2 through CR5 are all 1N914 or equivalent. L1 is 36 turns of No. 26 wire, tapped at 12 turns wound on a 0.5-in slug-tuned form. Q1 is a unijunction transistor (276-2029 or equivalent). Q2, Q5, Q8, and Q9 are all 2N2222. Q3, Q4, Q6, and Q7 are all MPF 102. R1 through R4 are all 25-K potentiometers. R5 is a 10-K potentiometer. T1, T2, and T3 are all 12 turns of No. 30 wire wound on a T50–72 form core. T4 and T5 are 12 V 300 mA (Radio Shack No. 273-1385 or equivalent).—H. Sievers, Stable Wideband Sweep Generator, *Ham Radio,* June 1981, pp. 18–21.

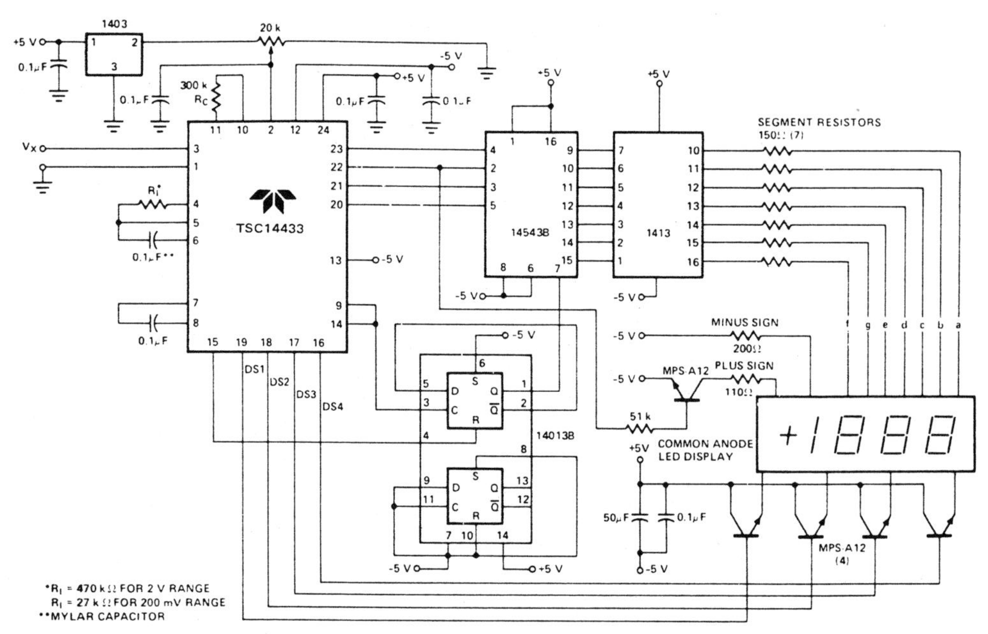

3.5-DIGIT VOLTMETER—Based on TSC14433 analog-to-digital converter device used in conjunction with common-anode displays, system requires 2.5-V reference. The full scale may be adjusted to 1.999 V or 199.9 mV. The display flashes when the input is above the specified range. R1 should be 470 K for a 2-V range and 27 K for a 200-mV range.—"Data Acquisition Design Handbook," Teledyne Semiconductor, Mountain View, CA, 1984, p. 7-119.

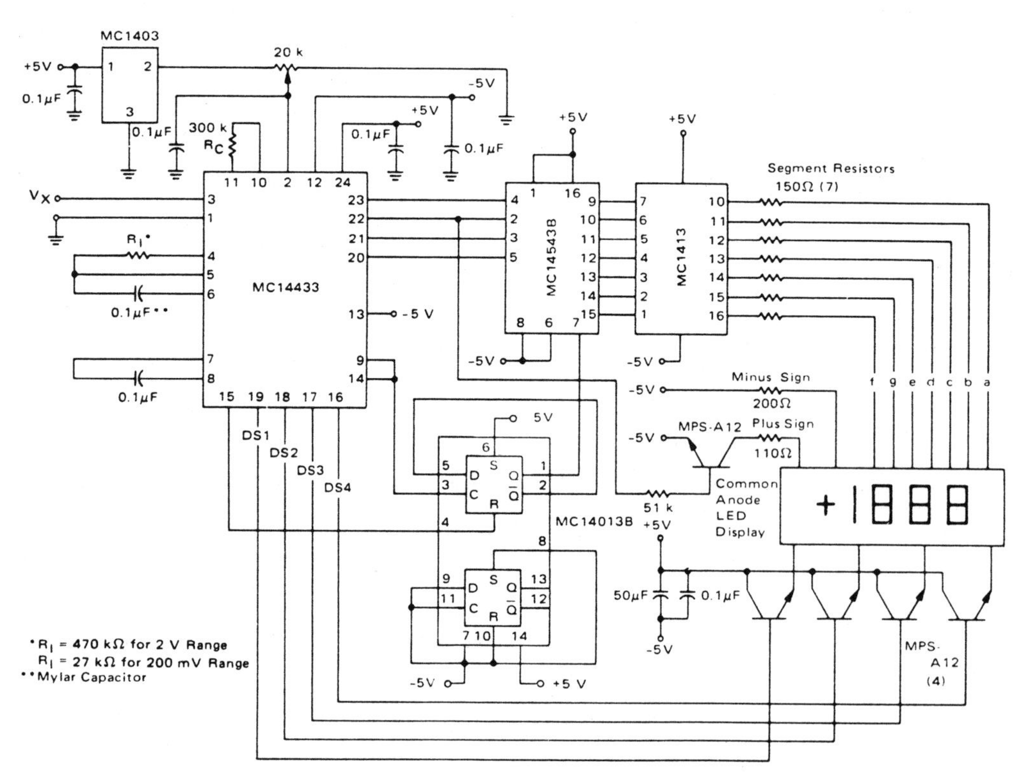

3.5-DIGIT VOLTMETER-Can be calibrated to read 199.9 mV or 1.999 V by value of R1; 470 K for 1.999 V and 27 K for 199.9 mV. If R_C is equal to 300 K, the clock frequency is 66 kHz and the conversion time is approximately 250 ms. If the input is above the circuit's range, the display flashes. The capacitor marked with** should be Mylar type.—"Motorola Linear Integrated Circuits Databook," Motorola, Phoenix, AZ, 1979, p. 4–45.

0- TO 200-V DIGITAL VOLTMETER—Measures DC and RF voltages. The taps on the 10-M resistive ladder between +IN and −IN select maximum voltage limits of 1.999, 19.99, and 199.9 V. The matching RF probe allows the measuring of RF voltages.—S. Creason, Creason's Do-It DVM, *73 Magazine,* June 1984, pp. 26–28.

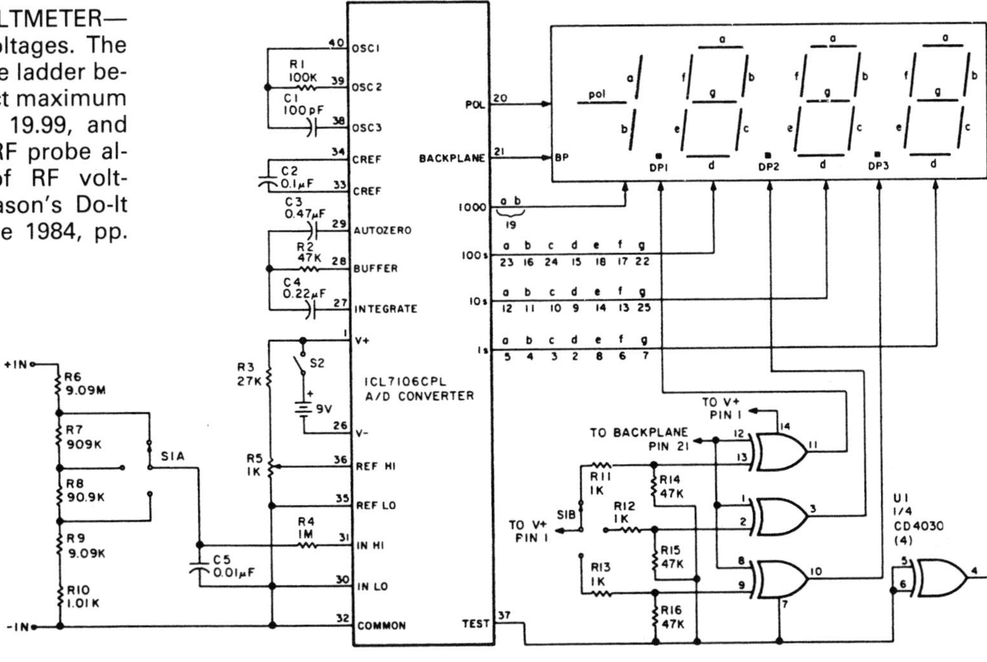

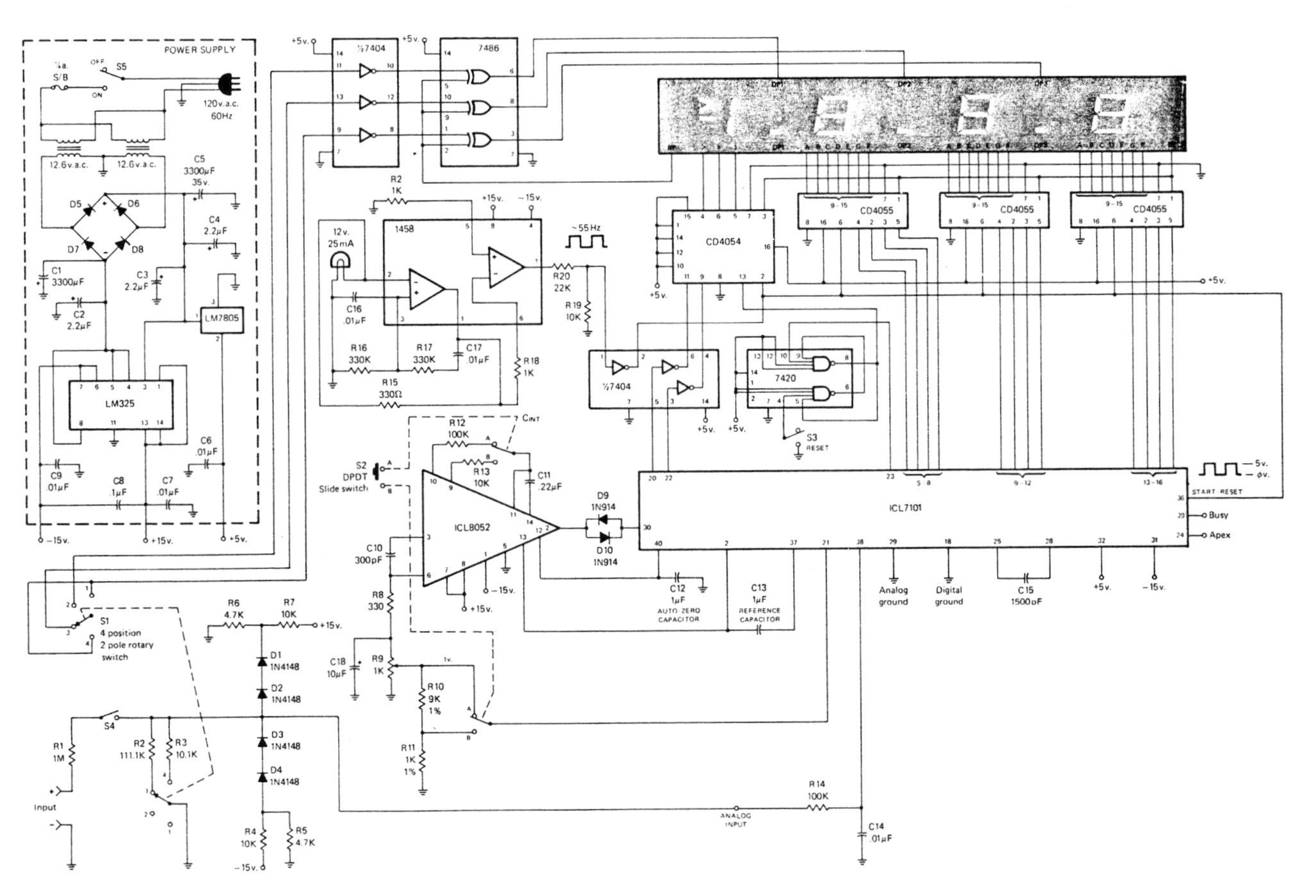

3.5-DIGIT DVM—Measures positive and negative input voltages from 200 mV to 200 V. The circuit incorporates overload protection and allows the resetting of the counter once overload has happened. S1 and S2 select voltage ranges of 0.200, 2, 20, and 200 V.—V. Martin and J. McCullin, How to Build a 3.5 Digit Digital Voltmeter (DVM), *CQ,* March 1982, pp. 44–50.

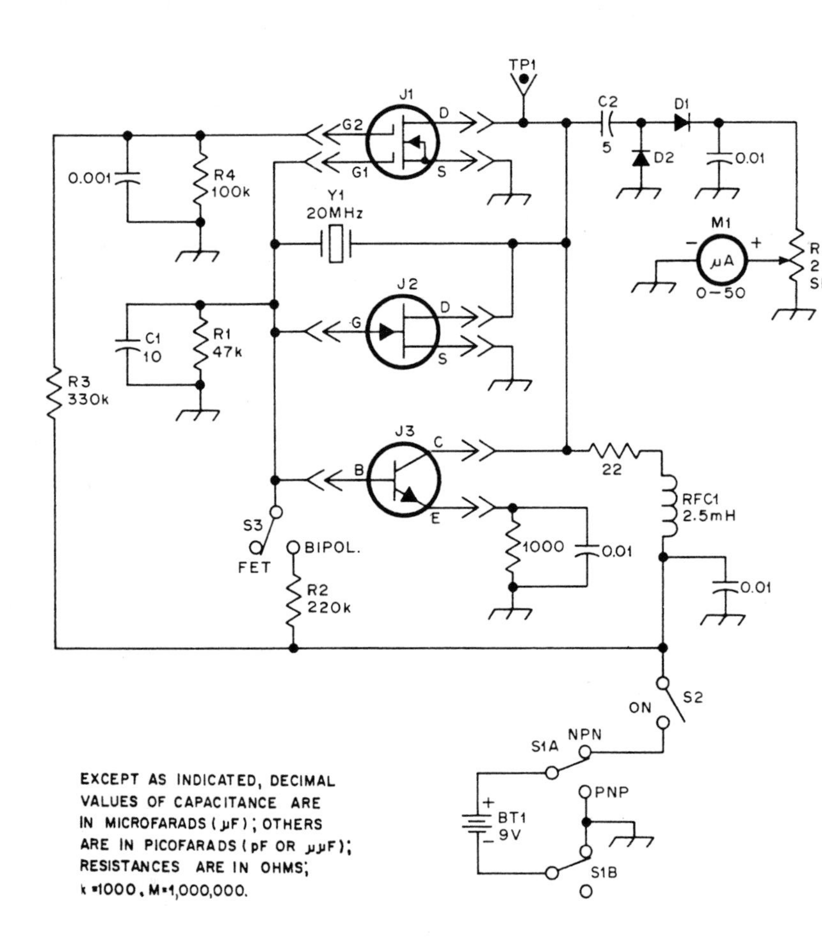

TRANSISTOR TESTER—Suitable for testing NPN, PNP, JFET, and dual-gate MOSFET transistors but not audio or high-power RF transistors. Meter deflection indicates the condition of the transistor. No deflection will take place if the transistor is open, shorted, or extremely leaky. The greater the meter reading, the more active the transistor is at the operating frequency.—"The Radio Amateur's Handbook," American Radio Relay League, Newington, CT, 1981, pp. 16-25 to 16-27.

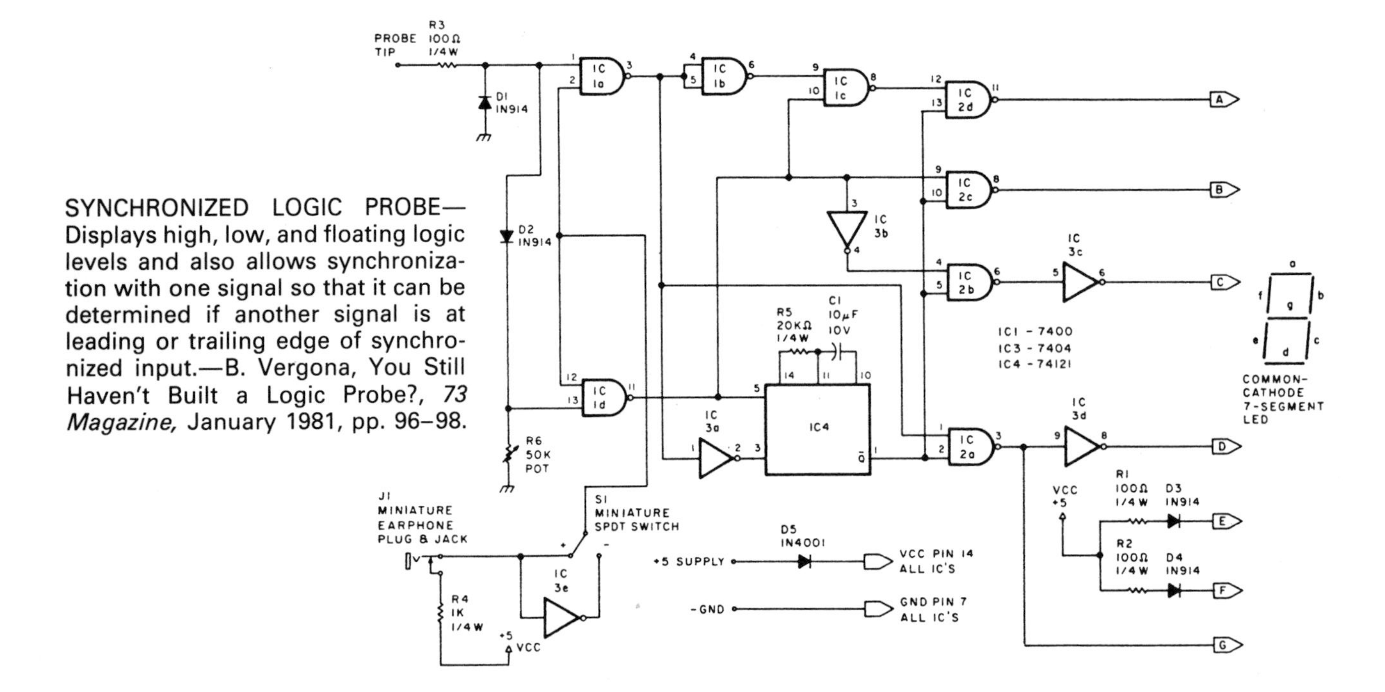

SYNCHRONIZED LOGIC PROBE—Displays high, low, and floating logic levels and also allows synchronization with one signal so that it can be determined if another signal is at leading or trailing edge of synchronized input.—B. Vergona, You Still Haven't Built a Logic Probe?, *73 Magazine,* January 1981, pp. 96–98.

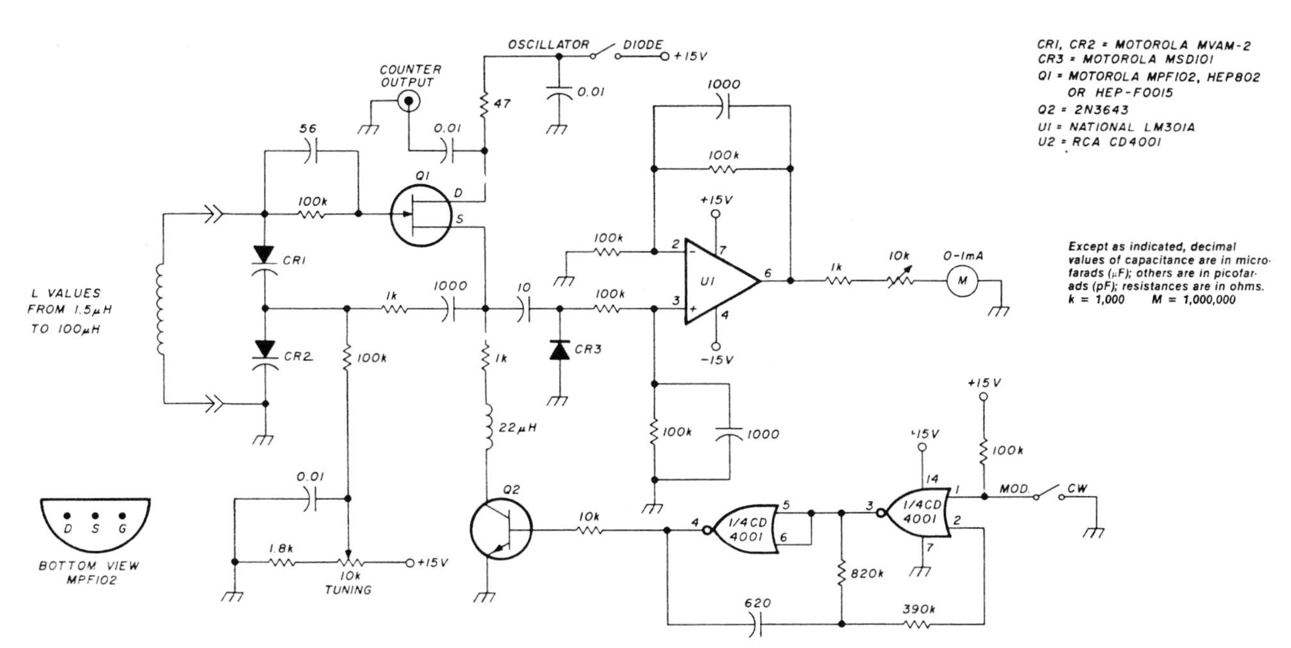

VARICAP-TUNED FET DIP METER WITH 1-KHz MODULATION—Operates over 2–32 MHz. A 1000-Hz square wave oscillator is designed around the 4001 CMOS device, and the LM301A (U1) is the meter's DC amplifier. The circuit was designed to accept plug-in coils with 0.75-in-spaced banana pins; the coils were borrowed from an old tube grid-dip meter.—H. Olson, A New Look at Dip Meters, *Ham Radio,* August 1981, pp. 25–28.

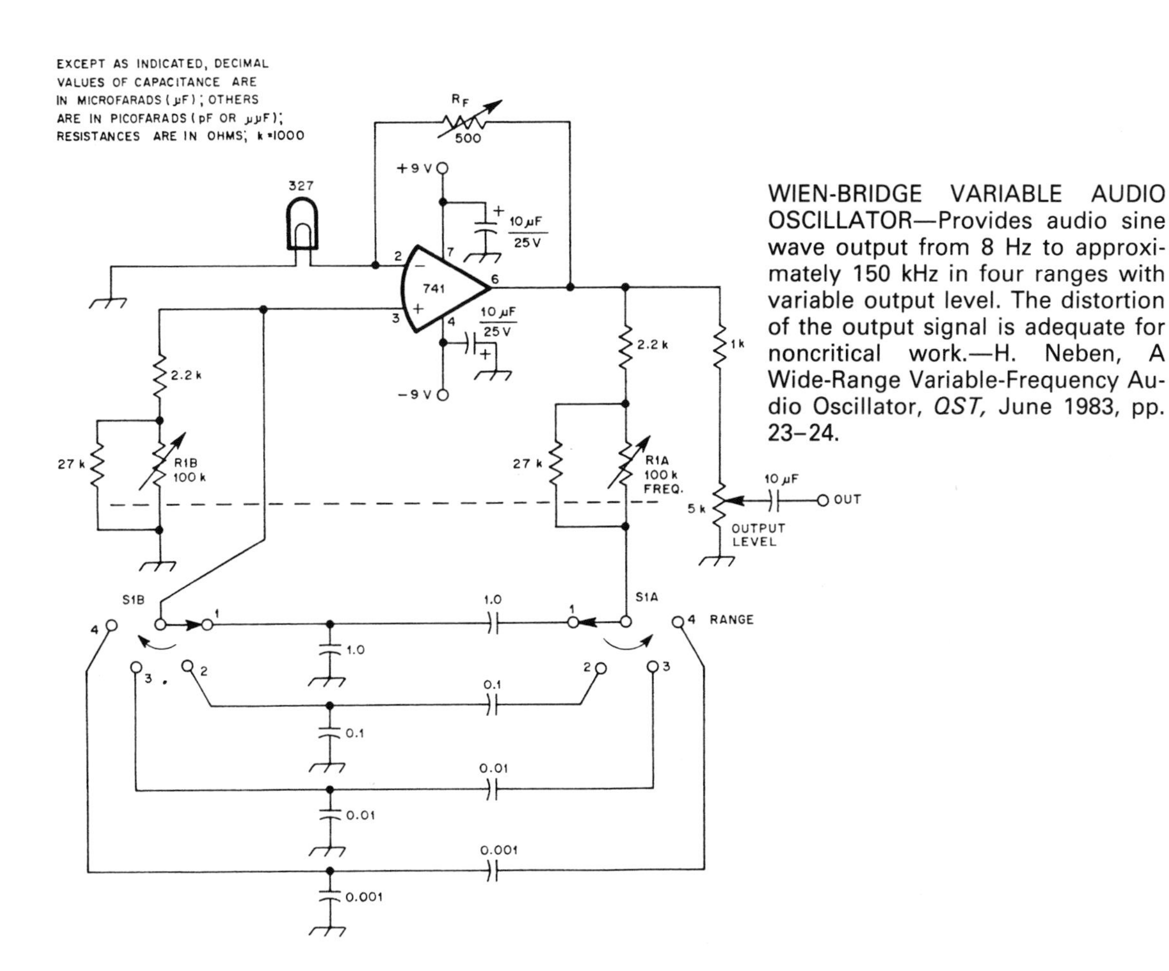

WIEN-BRIDGE VARIABLE AUDIO OSCILLATOR—Provides audio sine wave output from 8 Hz to approximately 150 kHz in four ranges with variable output level. The distortion of the output signal is adequate for noncritical work.—H. Neben, A Wide-Range Variable-Frequency Audio Oscillator, *QST,* June 1983, pp. 23–24.

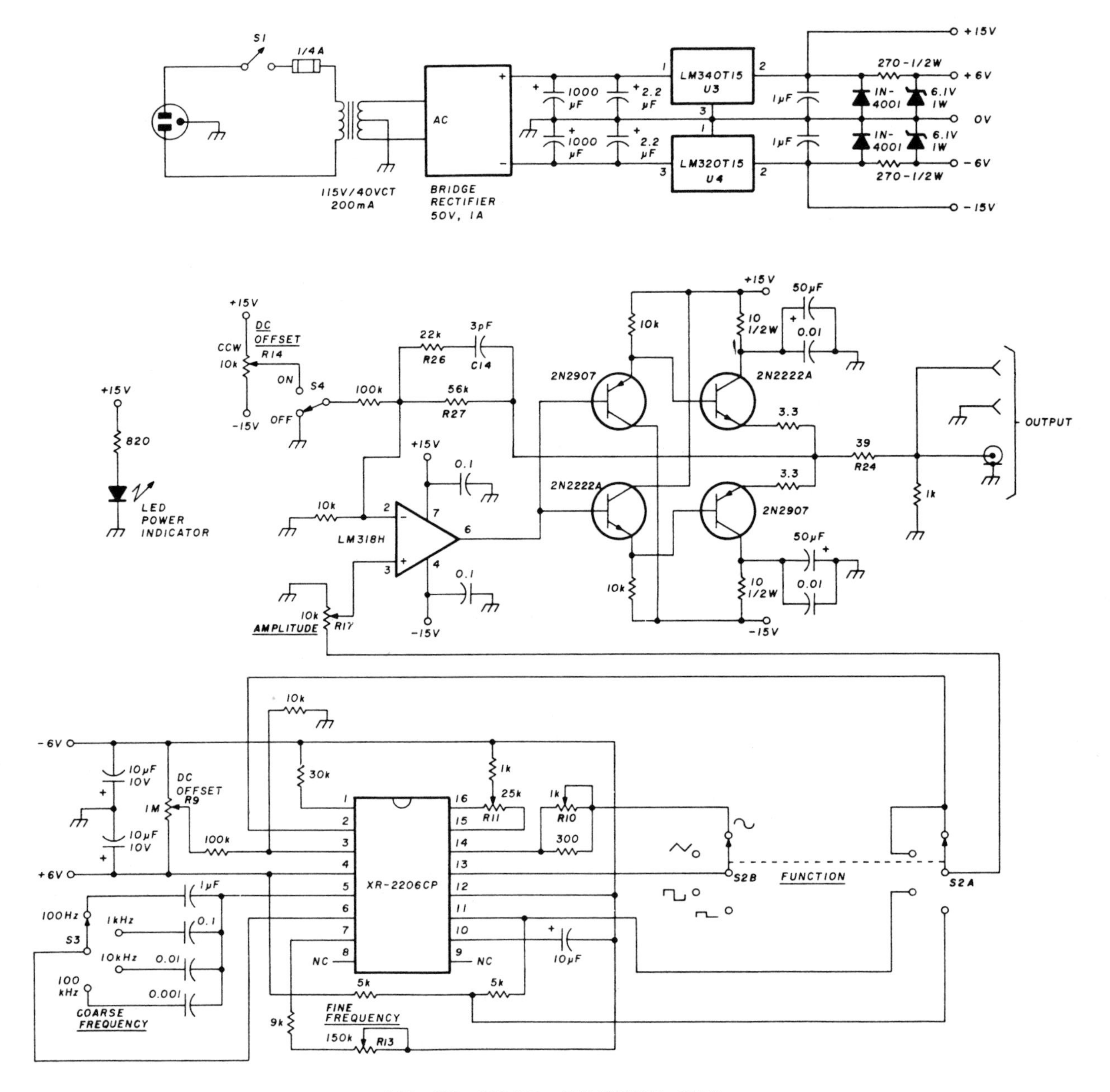

1-Hz TO 100-kHz FUNCTION GENERATOR—Produces sine waves, square waves, triangular waves, and 50% duty cycle positive pulse. The sinusoidal distortion is less than 3% over the entire frequency range. The circuit is built around the XR-2206 device and is capable of AM and FSK modulation. Output waveforms are not of laboratory quality but are suitable for noncritical applications.—F. Getz, Integrated Circuit Function Generator, *Ham Radio,* August 1980, pp. 30–32.

24

Timer Circuits

TIMER WITH LED DISPLAY—Can be adjusted to indicate time from seconds to over 10 minutes by adjusting R1. IC1 is a 555 timer, IC2 is a 7490, and IC3 is a 7447. C1 is 100 μF, R2 is 1 K, R3 through R9 are each 330 Ω, and R1 is a 1-M variable resistor.—G. Fait, Construct the Minuteman Timer, *73 Magazine,* February 1984, pp. 14–15.

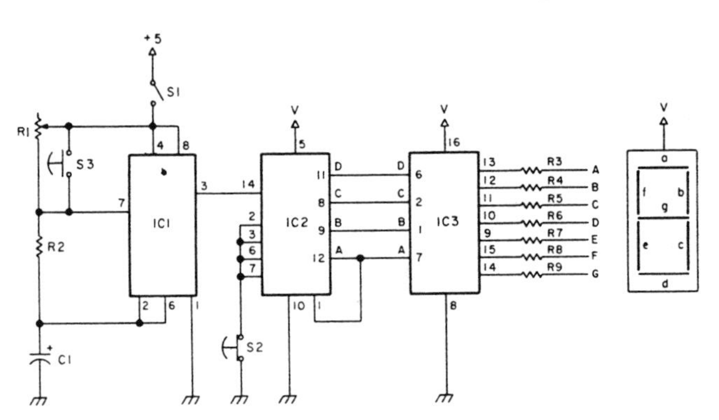

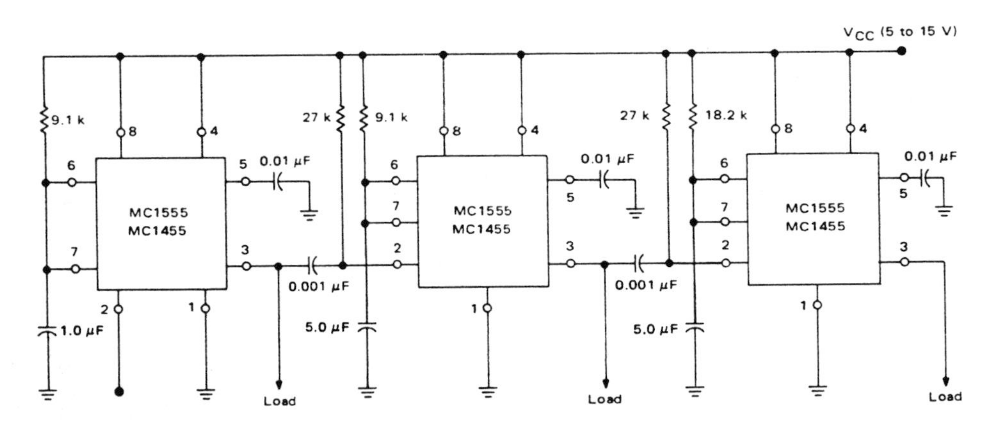

SEQUENTIAL TIMER—Multiple MC1555 timers drive each other for sequential timing. The first timer runs for 10 ms and then switches low momentarily, starting the second timer. The second timer then runs for 50 ms, and so forth.—"Motorola Linear Integrated Circuits Databook," Motorola, Phoenix, AZ, 1979, p. 6-50.

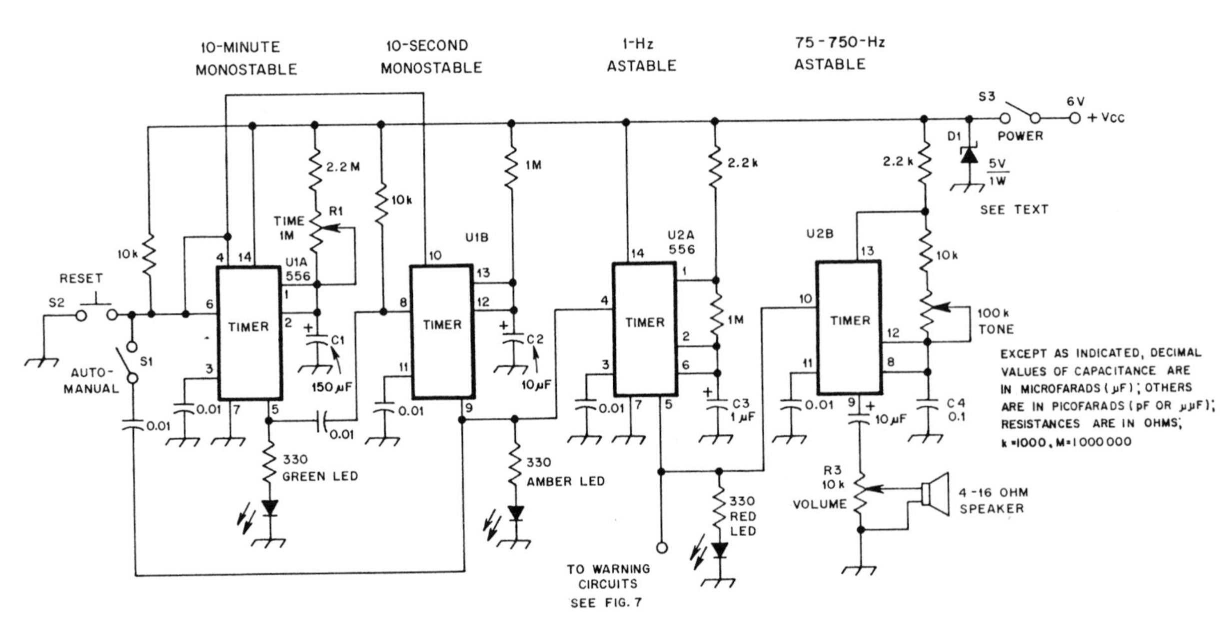

10-MINUTE TIMER—Designed to give audible indication every 10 minutes. U1A and U1B and U2A and U2B are both 556 dual timer devices.—L. Cebok, A 10-Minute Timer that Won't Quit, *QST,* September 1980, pp. 35–38.

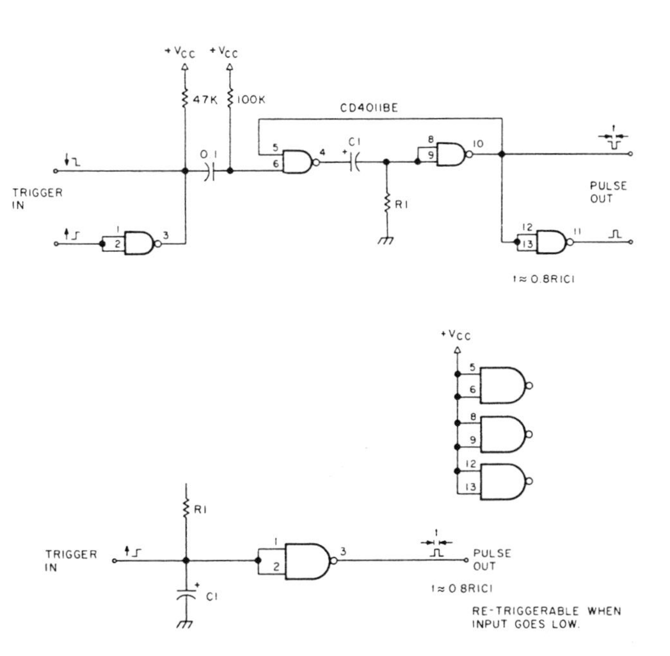

LOW-POWER/RETRIGGERABLE TIMERS—Part *a* shows one-shot timer which is not retriggerable or affected by steady signal on trigger. Time intervals can last up to approximately 10 minutes. The timer shown in part *b* is retriggerable. The output is normally low, going high upon a low trigger. The timing of both circuits depends on the values of R1 and C1; the formula 0.8R1C1 gives the approximate time in seconds.—C. Crichton, Low-Power And Retriggerable Timers from NAND Gates, *73 Magazine,* March 1983, p. 110.

Except as indicated, decimal values of capacitance are in microfarads (μF); others are in picofarads (pF); resistances are in ohms. k = 1,000 M = 1,000,000

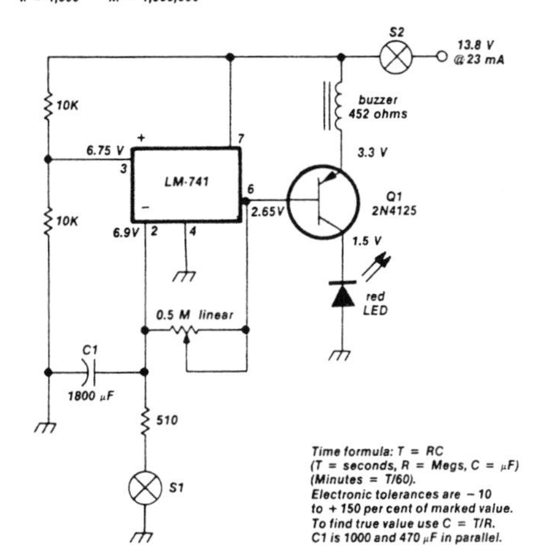

1- TO 15-MINUTE TIMER—LM741 op amp is connected as inverting differential comparator. The reference voltage is taken from the junction of two 10-K resistors. The resulting 6.75 V is connected to the inverting input (pin 3). The control voltage is picked up from output pin 6 of the 741 and passes through the 0.5-M linear potentiometer (which serves as the time setter for the circuit).—D. Tolle, Electronic Timer, *Ham Radio,* March 1982, pp. 65–66.

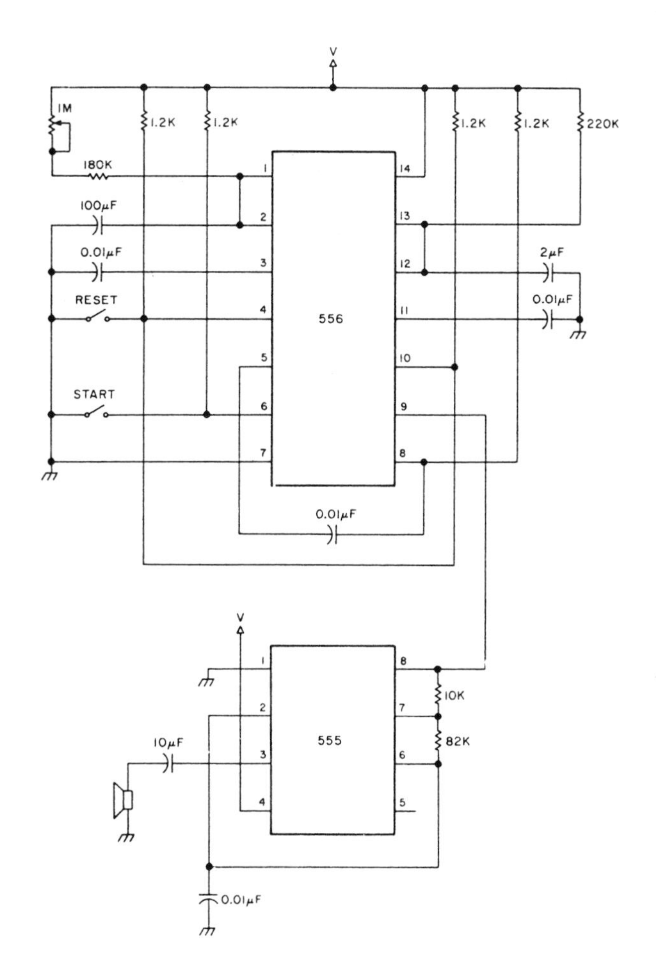

AUDIBLE TIMER—Originally designed as repeater "time-out" warning, circuit can be used whenever audible warning of elapsed time is needed. The circuit as shown can be adjusted from 30 seconds to 3 minutes; by the proper selection of timing capacitors and resistors, timing intervals in excess of 10 minutes may be obtained.—J. Coleman, Repeater Time-Out Warning, *73 Magazine,* July 1980, pp. 120–121.

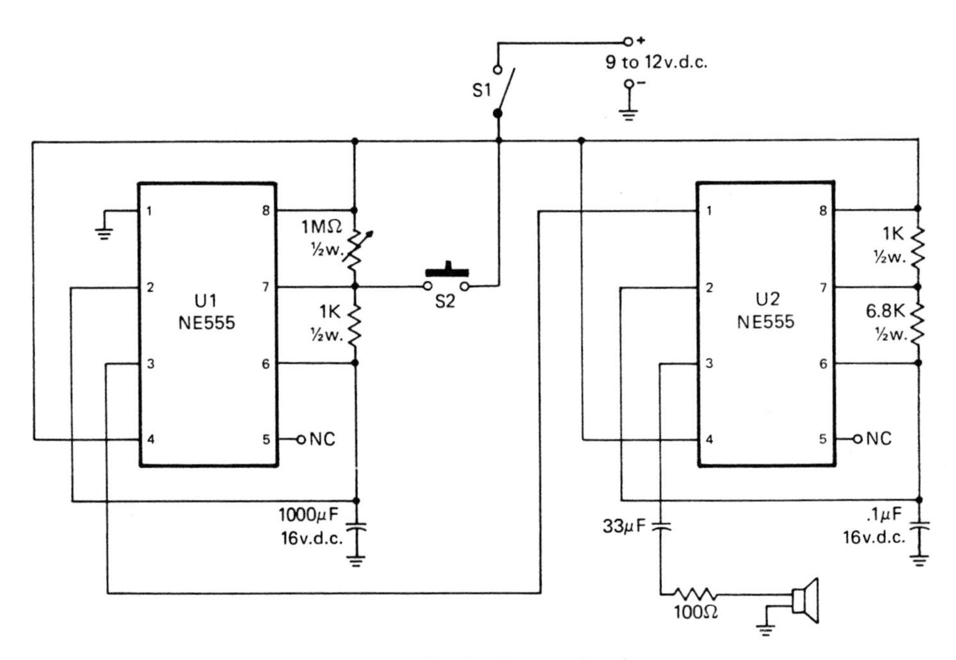

TIMER—Useful for intervals of over 9 minutes through adjustment of 1-M potentiometer. S2 resets the circuit, and a tone sounds when the desired time is reached.—E. Solov, A Simple ID Timer, *CQ,* May 1982, p. 24.

25

Transceiver Circuits

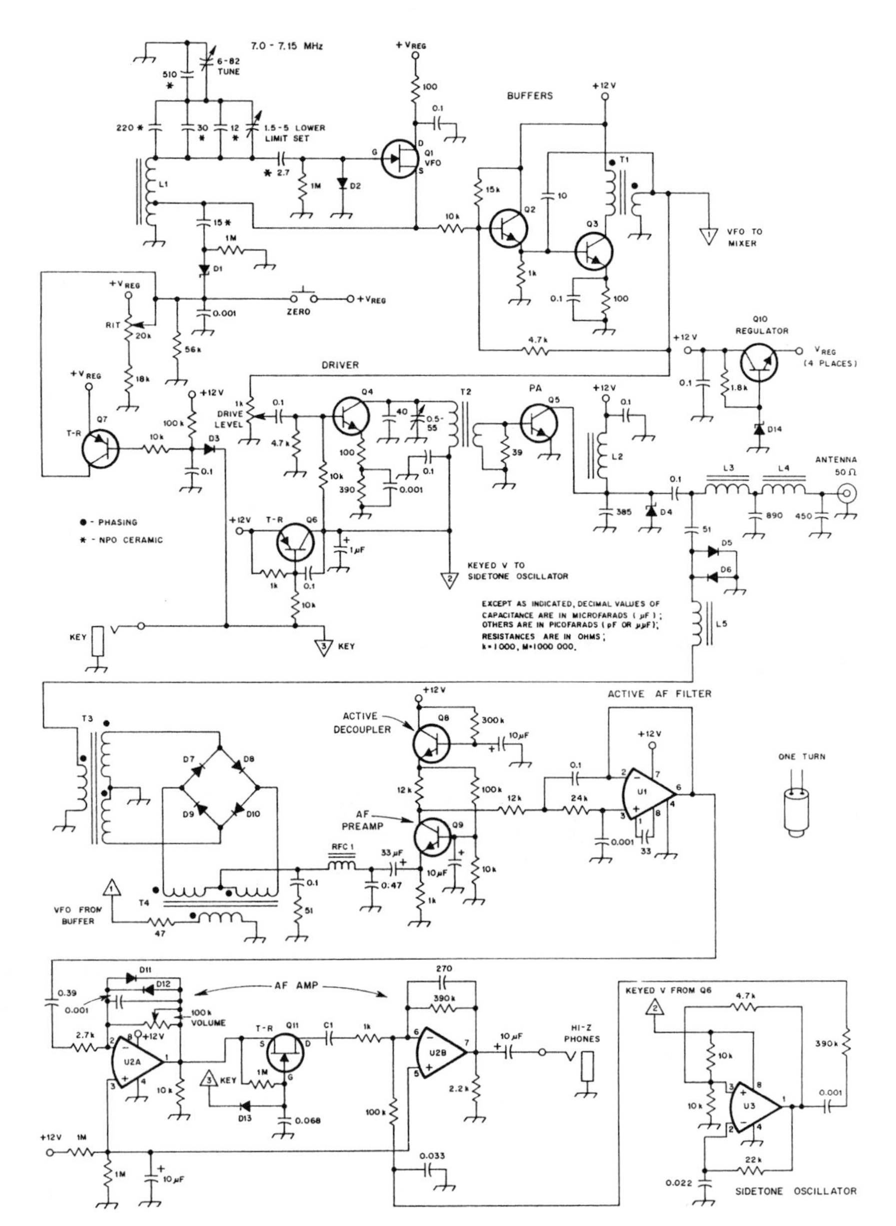

1.5-W 40-METER CW TRANSCEIVER—Covers 7000–7150 kHz with full break-in keying capability. The circuit requires a single 12-V DC power source, and current drain during receive is less than 20 mA. The unit also features incremental tuning of the receive frequency independent of the transmitter frequency. The VFO is a Hartley oscillator with a drift of less than 200 Hz after warm-up. The transmitter is a class C oscillator with an efficiency of roughly 75%. The transmitter output requires a 50-Ω load. The receiver section is a direct conversion type. The receiver mixer is a doubly balanced design. The mixer stage is followed by a diplexer to prevent RF energy from getting into the AF amplifier. The receiver section also includes an active filter with a center frequency of 650 Hz. The keying sidetone is provided through U3. U1 is an LM301, while U2/U3 are LM358N (one section of U3 is not used). Q1 and Q11 are 2N4416. Q2, Q3, and Q10 are all 2N3904 or equivalent. Q4 is a 2N2222. Q5 is a 2N3553 or 2N5859. Q6 and Q7 are both 2N3906. Q8 is a 2N4124 or 2N3565. Q9 is a 2N3565. RFC1 is a 100-μH choke wound on a subminiature ferrite form approximately the size of a 0.25-W resistor. T1 has 15 primary turns and three secondary turns on a BLN-43-2402 core. T2 has 39 primary turns and five secondary turns on a T-44-6 form. T3 and T4 are five trifilar turns on a BLN-43-2402 form. L1 is 26 turns on a T-44-6 form tapped seven turns from its ground end. L2 is 43 turns on a T-50-2 form. L3 and L4 are 19 turns on a T-37-6 core. L5 is 58 turns on a T-37-6 core. C1 is a 1-μF 3-V nonpolarized ceramic. D4 is a 1N973 or equivalent. D14 is a 1N961 or equivalent. Other diodes are 1N914, 1N4152, or equivalent.—R. Lewallen, An Optimized QRP Transceiver, *QST,* August 1980, pp. 14–19.

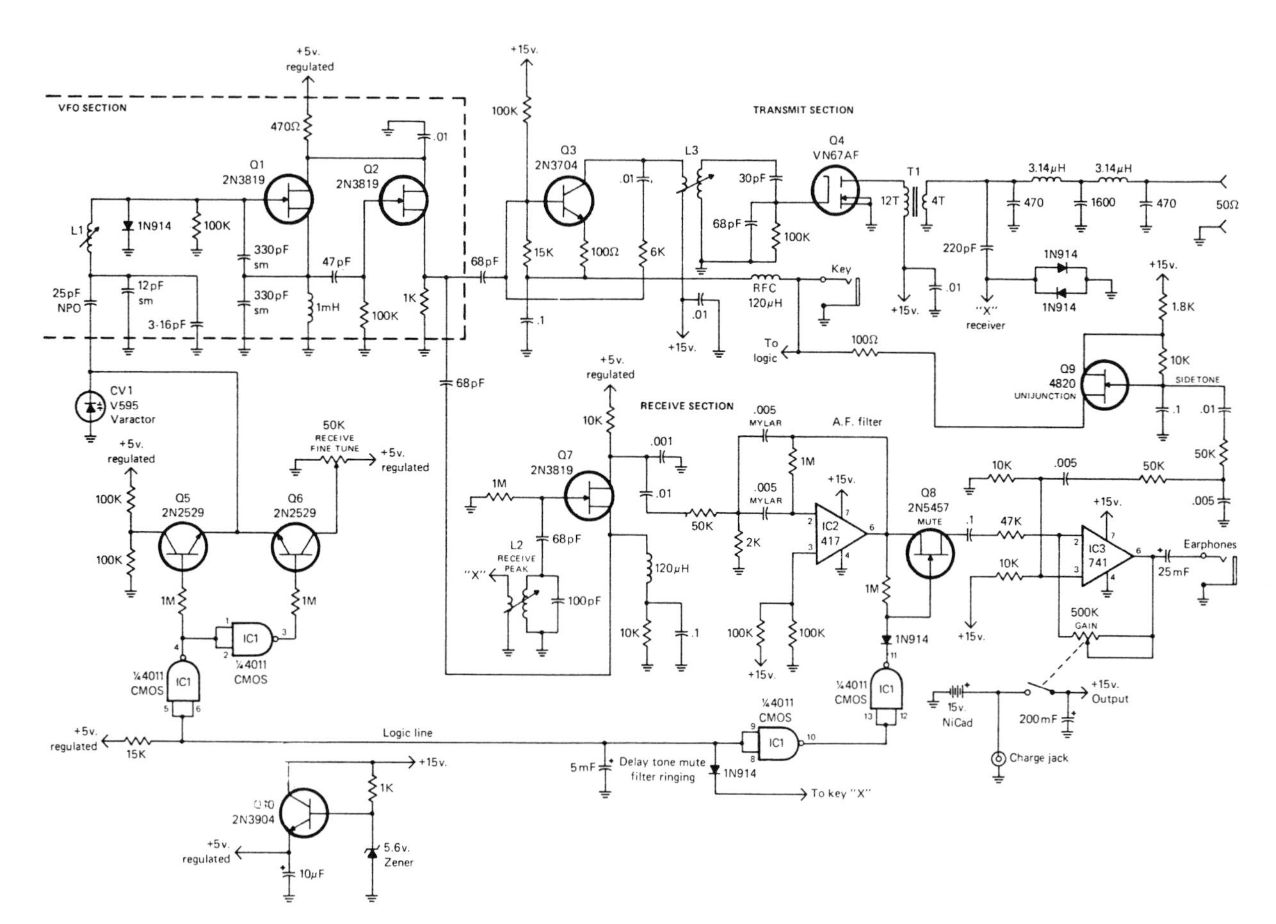

80-METER QRP CW TRANSCEIVER—Covers 3495–3950 kHz at up to 3 W of CW output. IC2 (a 741 op amp) is used in a 700-Hz audio filter (gain and Q of 10) for improved receiver selectivity. The transmitter stage uses VMOS devices because of their low drive requirements. T1 is wound on a T106-1 core using No. 26 wire with 12 turns for the primary and nine turns on the secondary. L1 and L2 are both variable inductors covering 15–100 μH. L3 is a 0.375-in-diameter coil with a value of 10–90 μH; its primary winding consists of 12 turns of No. 30 wire.—S. Defrancesco, Construct Your Own 80 Meter QRP/QSK C.W. Transceiver, *CQ,* June 1983, pp. 13–18.

RF CLIPPER—Designed for addition to SSB transceivers, circuit provides clipping and filtering of RF envelope for higher transmitter efficiency. The effects are similar to AF clipping but the results are superior.—*73 Magazine* Staff, Practical AF and RF Speech Processing, *73 Magazine,* March 1981, pp. 72–74.

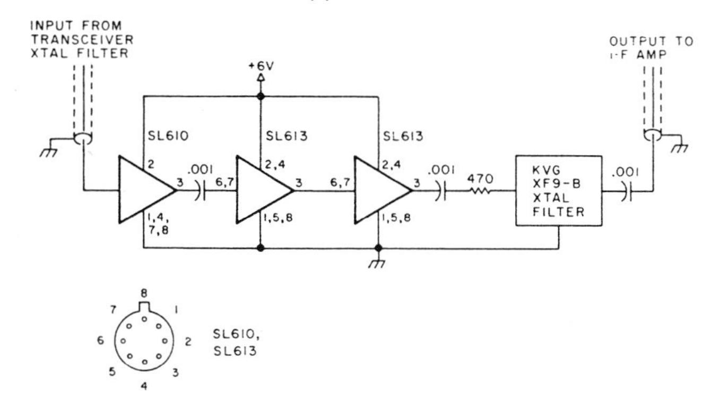

14,250- TO 14,300-kHz SSB TRANSCEIVER—Operates in upper sideband mode in 20-meter amateur radio band with approximately 1.5 W of RF output. A 9-MHz crystal filter serves on both transmit and receive. The complete transceiver is built on several circuit boards and fits in a 12 × 7 × 4 = in^3 enclosure. Individual sections may also be adapted for other projects. The original article gives extensive details on constructing a complete transceiver.—F. Perkins, Action Machine for 20, *73 Magazine,* January 1983, pp. 12–23.

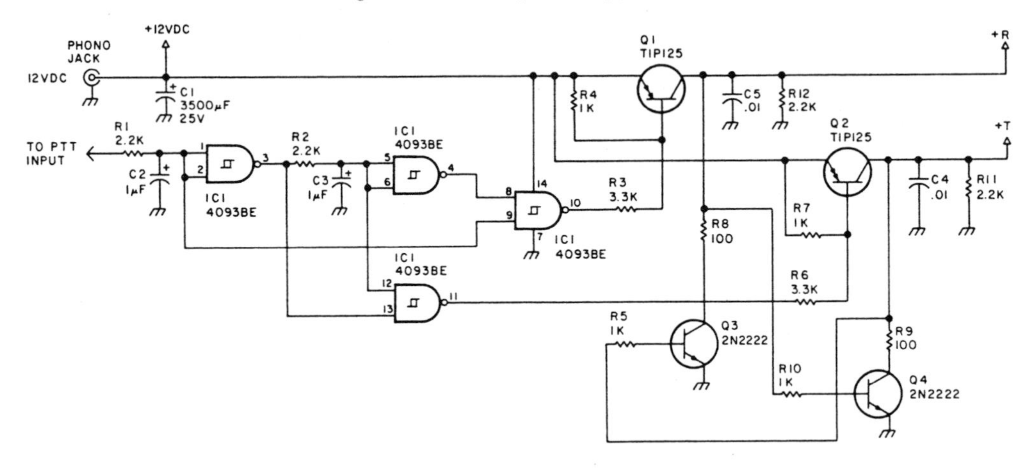

VFO

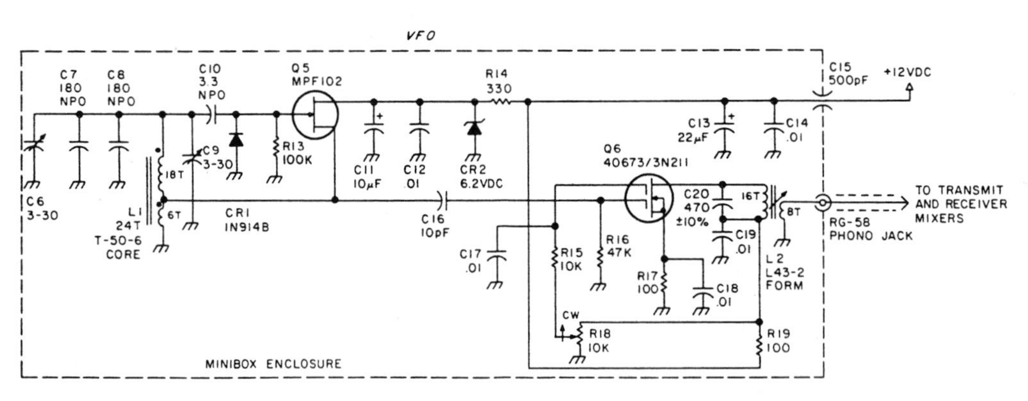

BFO

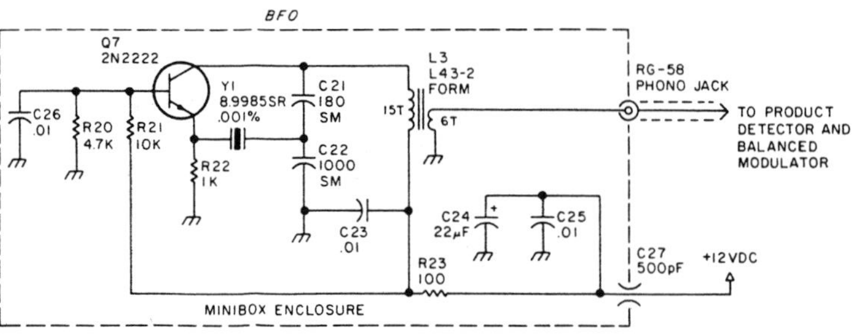

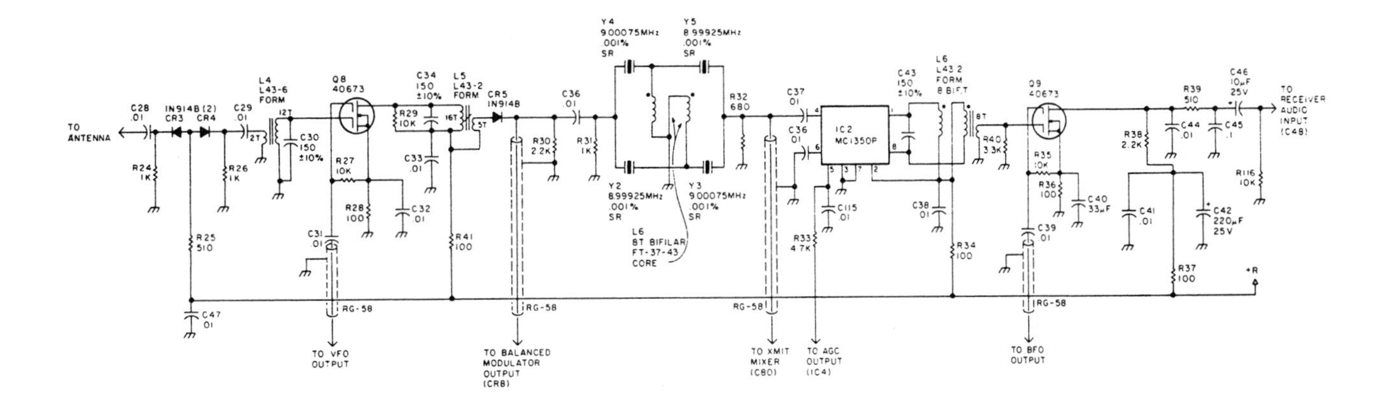
TO ANTENNA
C28 .01
IN914B (2)
CR3
CR4
R24 1K
R25 510
R26 1K
C47 .01
C29 .01
L4 L43-6 FORM
12T
2T
C30 150 ±10%
Q8 40673
R27 10K
R28 100
C31 .01
RG-58
TO VFO OUTPUT
C34 150 ±10%
R29 10K
C33 .01
C32 .01
16T
5T
L5 L43-2 FORM
R41 100
CR5 IN914B
R30 2.2K
RG-58
TO BALANCED MODULATOR OUTPUT (CR8)
C36 .01
R31 1K
Y4 9.00075MHz .001% SR
Y5 8.99925MHz .001% SR
Y2 8.99925MHz .001% SR
Y3 9.00075MHz .001% SR
L6 8T BIFILAR FT-37-43 CORE
R32 680
C37 .01
C36 .01
RG-58
TO XMIT MIXER (C80)
IC2 MC1350P
C115 .01
R33 4.7K
TO AGC OUTPUT (IC4)
C43 150 ±10%
L6 L43-2 FORM 8 BIFT
8T
C38 .01
R34 100
R40 3.3K
Q9 40673
R35 10K
R36 100
C40 33μF
C39 .01
RG-58
TO BFO OUTPUT
R38 2.2K
C44 .01
C41 .01
R37 100
R39 510
C45 .1
C42 220μF 25V
C46 10μF 25V
R116 10K
TO RECEIVER AUDIO INPUT (C48)
+R

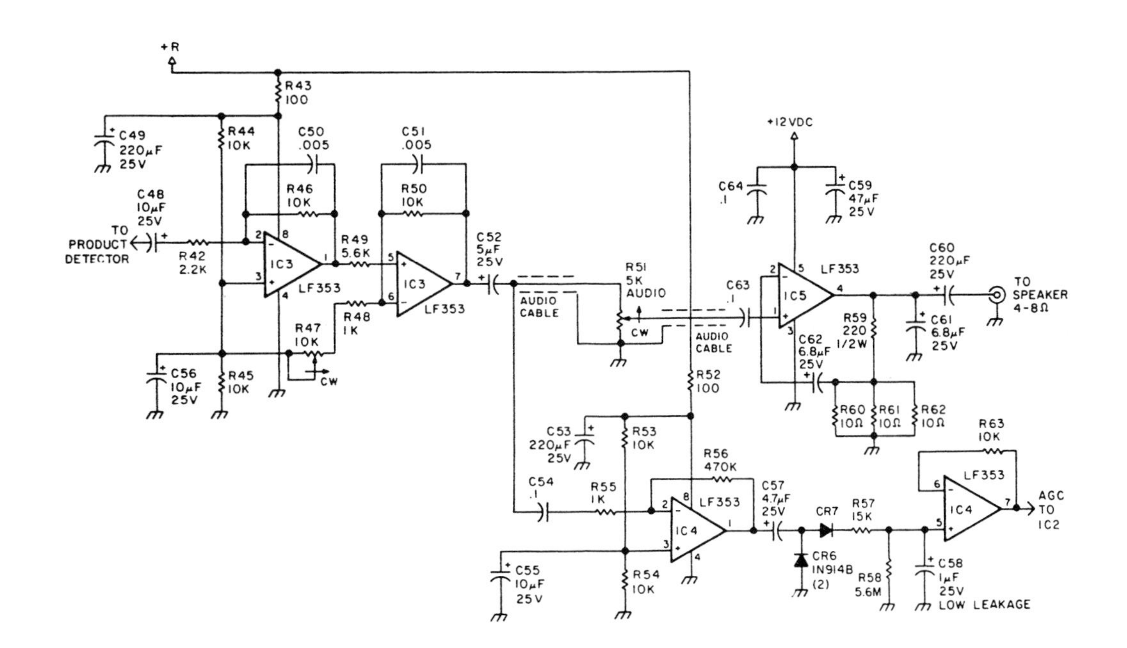
+R
R43 100
C49 220μF 25V
R44 10K
C50 .005
C51 .005
R46 10K
R50 10K
C48 10μF 25V
TO PRODUCT DETECTOR
R42 2.2K
IC3
LF353
R49 5.6K
R48 1.5K
R47 10K
CW
C56 10μF 25V
R45 10K
C52 5μF 25V
AUDIO CABLE
R51 5K AUDIO
CW
AUDIO CABLE
R52 100
+12VDC
C64 .1
C59 47μF 25V
C63 .1
IC5
LF353
C60 220μF 25V
TO SPEAKER 4-8Ω
C61 6.8μF 25V
R59 220 1/2W
C62 6.8μF 25V
R60 10Ω
R61 10Ω
R62 10Ω
C53 220μF 25V
R53 10K
R56 470K
C54 .1
R55 1K
IC4
LF353
R54 10K
C55 10μF 25V
C57 4.7μF 25V
CR7
R57 15K
CR6 IN914B (2)
R58 5.6M
C58 1μF 25V LOW LEAKAGE
R63 10K
IC4
LF353
AGC TO IC2

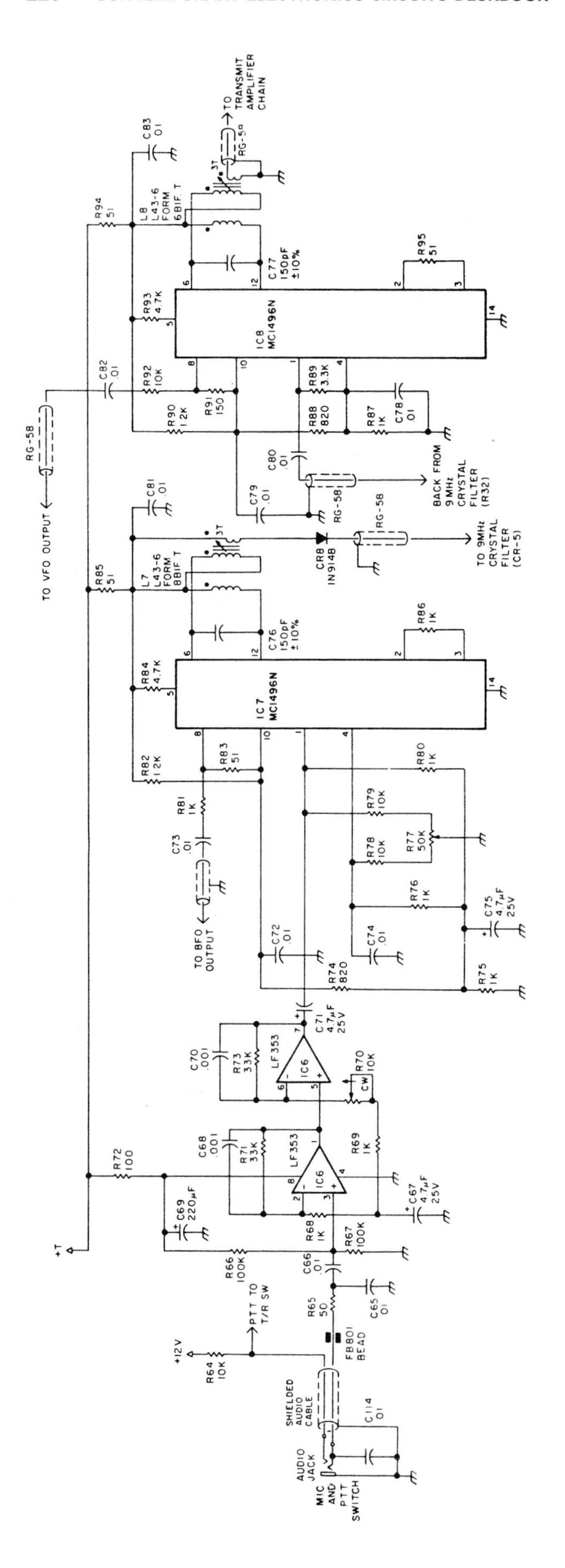
TO TRANSMIT AMPLIFIER CHAIN
TO VFO OUTPUT
RG-58
IC8 MC1496N
IC7 MC1496N
BACK FROM 9MHz CRYSTAL FILTER (R32)
TO 9MHz CRYSTAL FILTER (CR-5)
TO BFO OUTPUT
LF353
MIC AND PTT SWITCH
AUDIO JACK
SHIELDED AUDIO CABLE
PTT TO T/R SW
FB801 BEAD

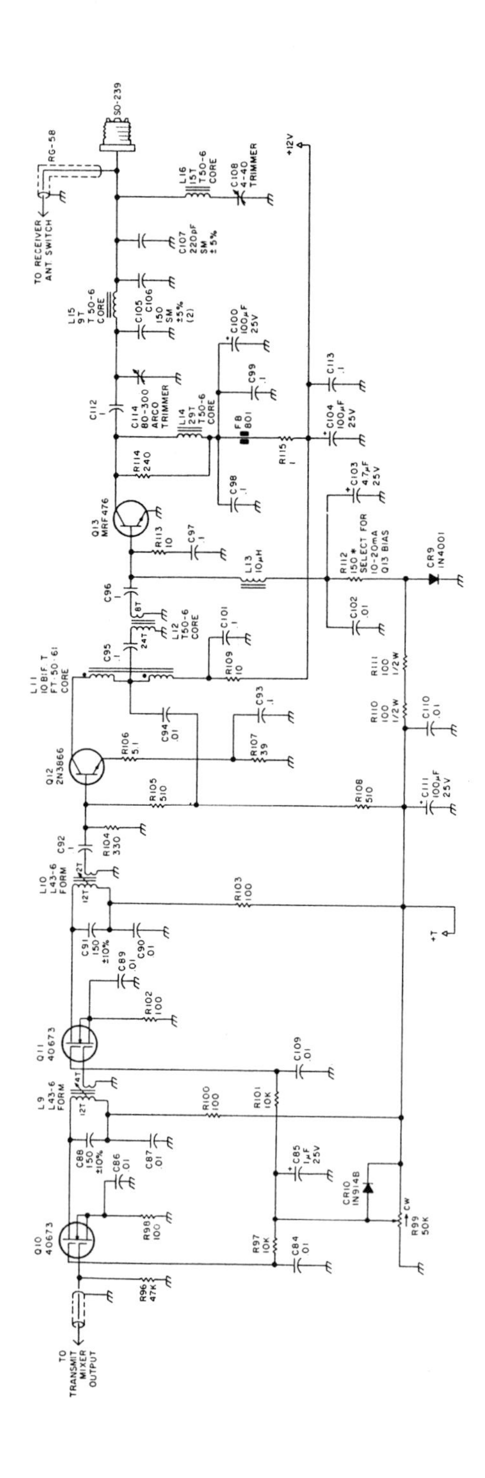
SO-239
TO RECEIVER ANT SWITCH
RG-58
MRF476
2N3866
40673
TO TRANSMIT MIXER OUTPUT
SELECT FOR 10-20mA Q13 BIAS

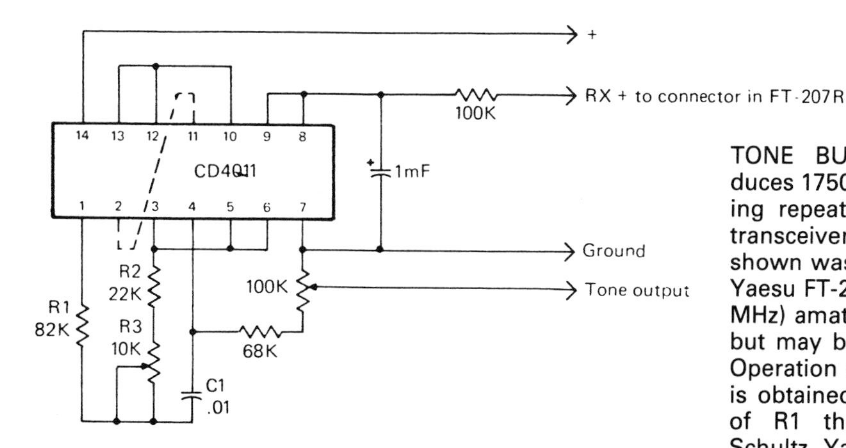

TONE BURST GENERATOR—Produces 1750-Hz tone burst for activating repeaters when PTT switch of transceiver is pressed. The circuit shown was designed for use with a Yaesu FT-207R 2-meter (144- to 148-MHz) amateur radio FM transceiver, but may be adapted to other units. Operation at other tone frequencies is obtained by changing the values of R1 through R3 and C1.—J. Schultz, Yaesu FT-207R Product Review, *CQ,* July 1980, pp. 64–68.

TRANSMIT/RECEIVE SWITCH—Includes sidetone generator and allows section of full break-in or semi-break-in keying by means of S2. U1 is a CD40106BE. The circuit is suggested for RF powers of 100 watts RF or less. It may also be adapted as a carrier-operated relay for repeater service.—D. DeMaw, TR Circuits for Homemade Rigs, *QST,* October 1984, pp. 17–20.

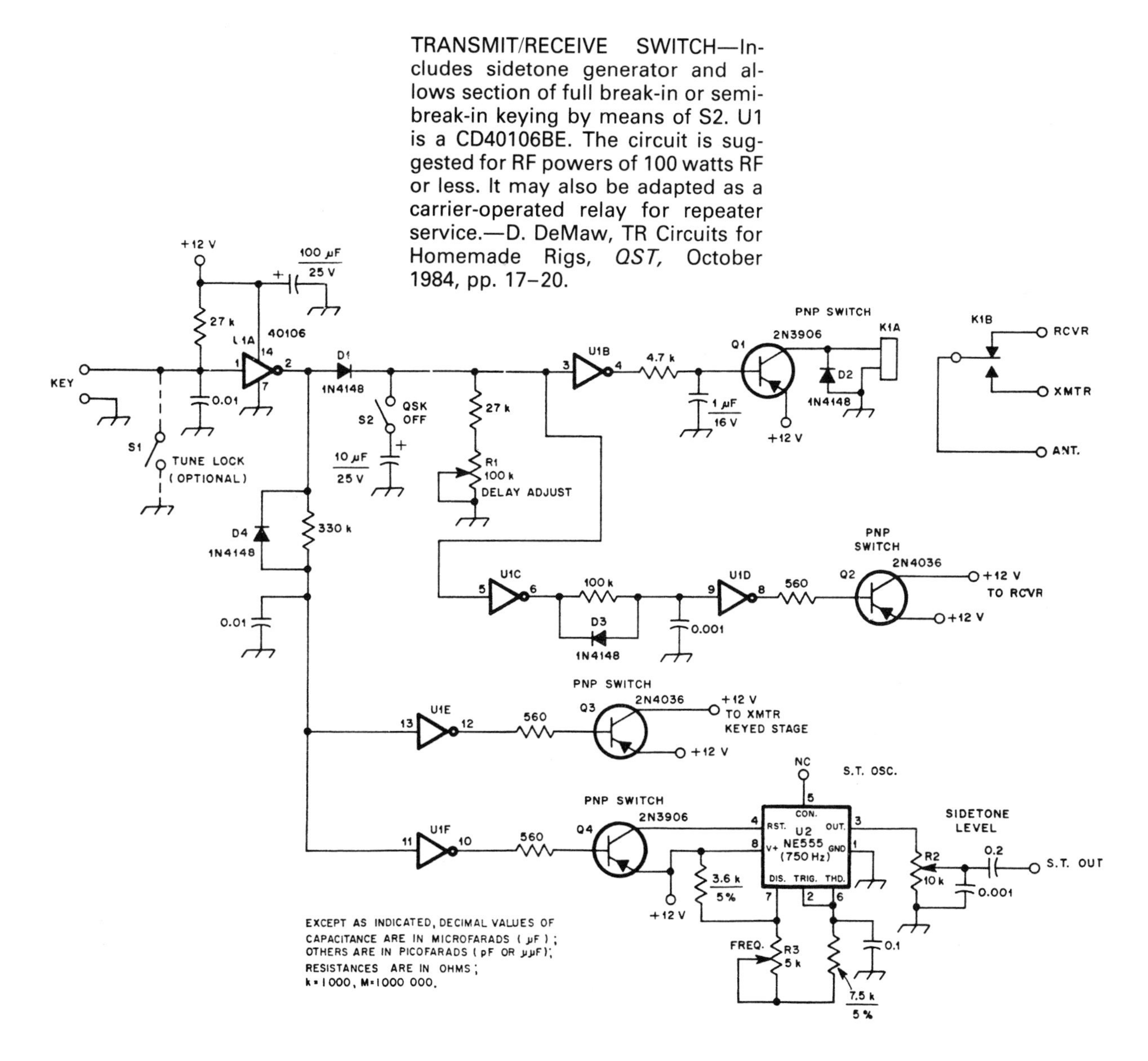

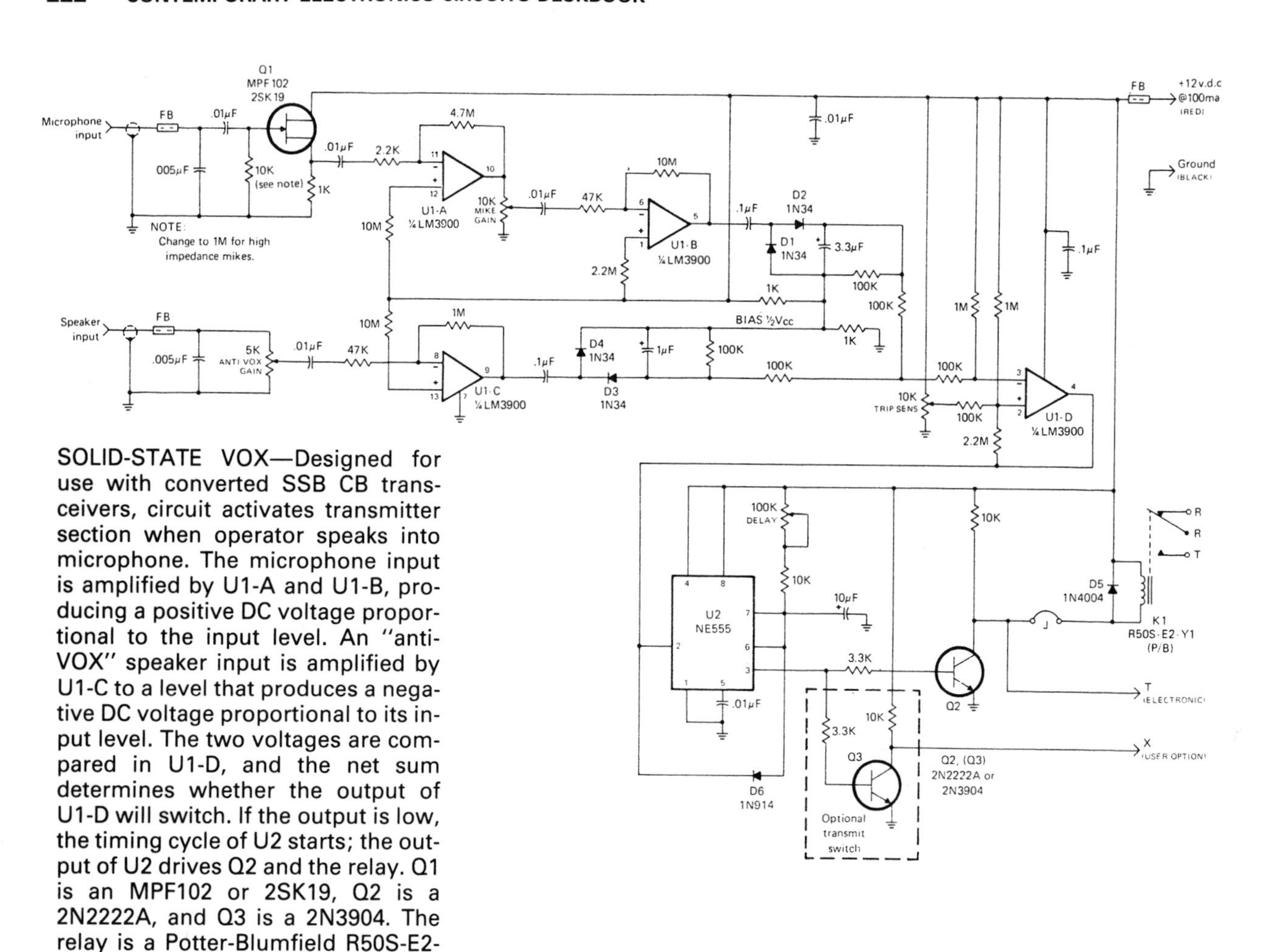

SOLID-STATE VOX—Designed for use with converted SSB CB transceivers, circuit activates transmitter section when operator speaks into microphone. The microphone input is amplified by U1-A and U1-B, producing a positive DC voltage proportional to the input level. An "anti-VOX" speaker input is amplified by U1-C to a level that produces a negative DC voltage proportional to its input level. The two voltages are compared in U1-D, and the net sum determines whether the output of U1-D will switch. If the output is low, the timing cycle of U2 starts; the output of U2 drives Q2 and the relay. Q1 is an MPF102 or 2SK19, Q2 is a 2N2222A, and Q3 is a 2N3904. The relay is a Potter-Blumfield R50S-E2-YA 12 V DC or equivalent.—L. Franklin, How to Build a Deluxe Solid-State Voice Operated Relay, *CQ,* April 1981, pp. 36–39.

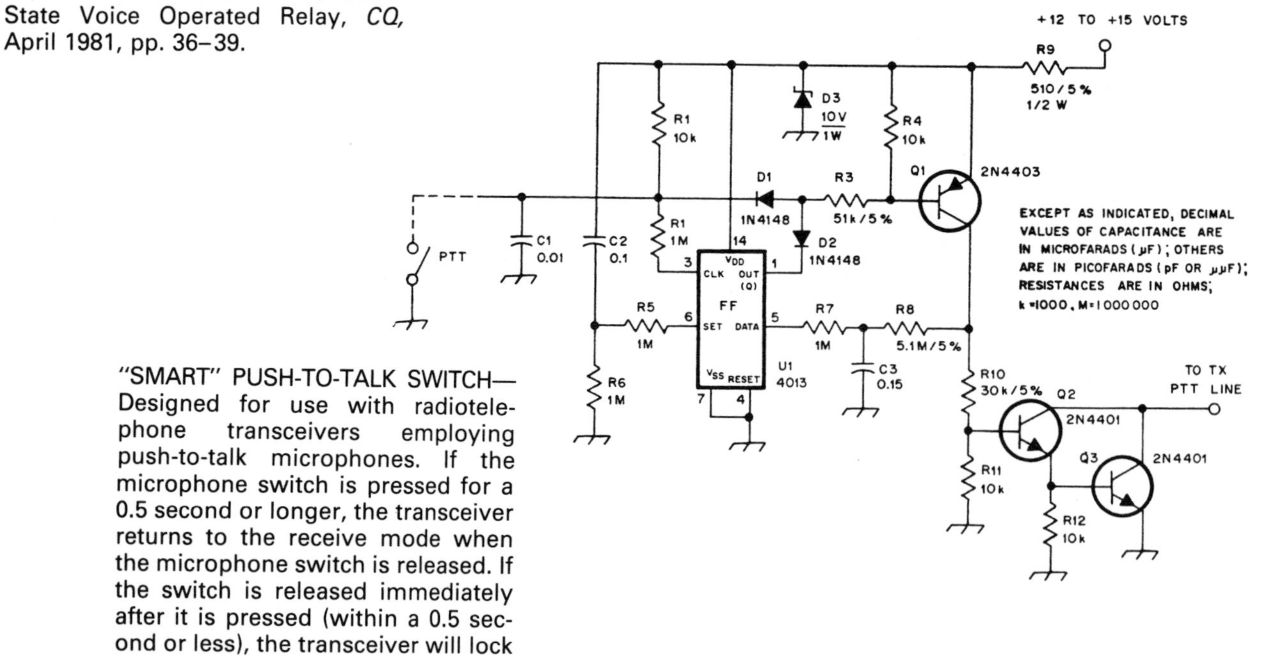

"SMART" PUSH-TO-TALK SWITCH—Designed for use with radiotelephone transceivers employing push-to-talk microphones. If the microphone switch is pressed for a 0.5 second or longer, the transceiver returns to the receive mode when the microphone switch is released. If the switch is released immediately after it is pressed (within a 0.5 second or less), the transceiver will lock in the transmit mode until the microphone switch is again pressed.—K. Stuart, A Smart Push-to-Talk Circuit, *QST,* December 1980, pp. 38–39.

26

Transmitting Circuits

CW TRANSMITTER FOR 14,000–14,350 AND 21,000–21,450 kHz—Delivers up to 20 W when used with amplifier section. In the transmitter, Q1 is a Radio Shack No. 276-2009 and Q2 and Q3 are each Radio Shack No. 276-2038. The crystal frequency may be varied by a 100- to 300-pF variable capacitor indicated by C VXO. The original article provides details on modifying coils for operation on desired bands. In the amplifier section, all transistors are VN67AF VMOS FETs.—M. Oman, Fun-Equipment Revisited, *73 Magazine,* January 1983, pp. 48–51.

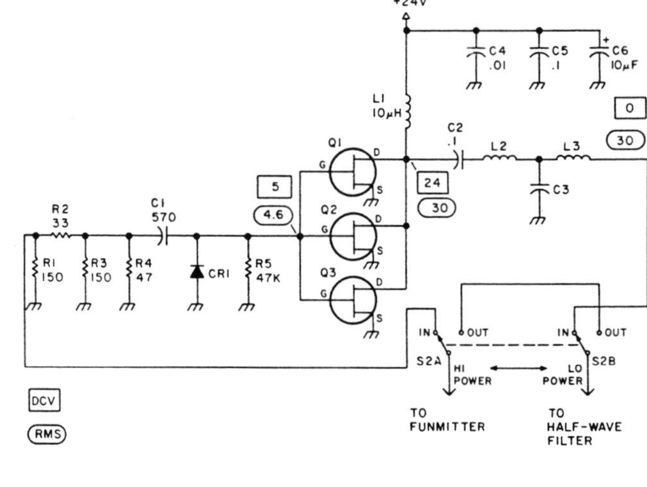

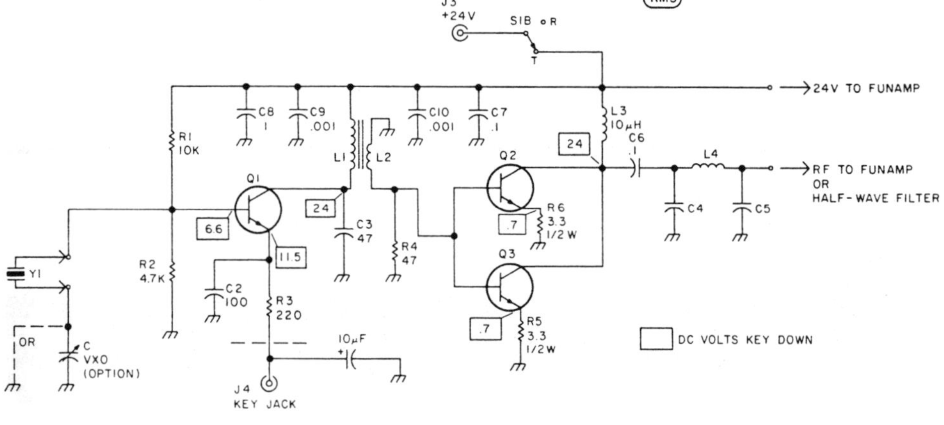

LOW-POWER FM TRANSMITTER—Can be tuned to operate in FM broadcast band (88–108 MHz), 2-meter amateur radio band (144–148 MHz), or higher frequencies above 150 MHz. The frequency is adjusted by values of C5 and L1; the values given in the circuit will place the output near the middle of the FM broadcast band. L1 is four turns of No. 20 wire airwound 0.25 in in diameter to a length of 5 mm. The range of the transmitter is claimed to be several hundred yards.—J. Kretzchmar, Simple FM Transmitter, *73 Magazine,* August 1983, p. 100.

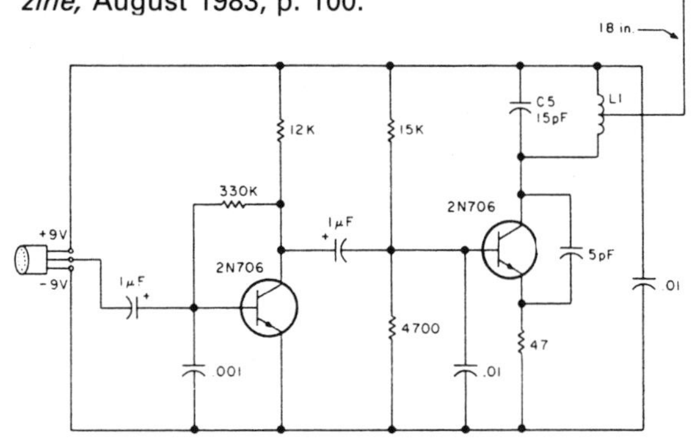

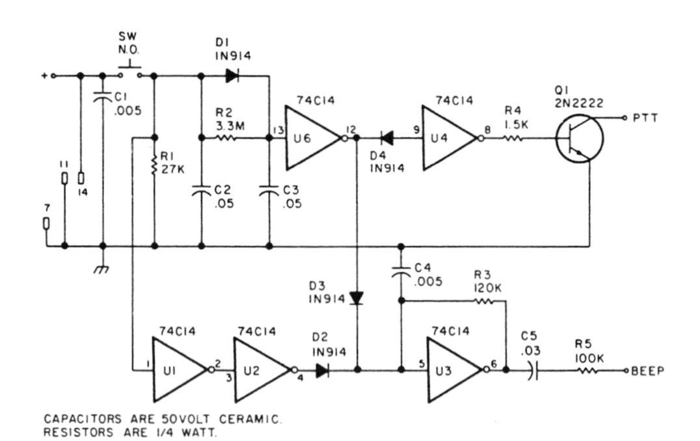

END-OF-TRANSMISSION TONE SIGNAL—Generates "beep" tone when PTT switch of radiotelephone transmitter/transceiver is released to indicate end of transmission. The output may be into a normal microphone input in parallel with the microphone used.—N. Van de Sande, Built the NASA Beeper, *73 Magazine,* March 1984, pp. 88–89.

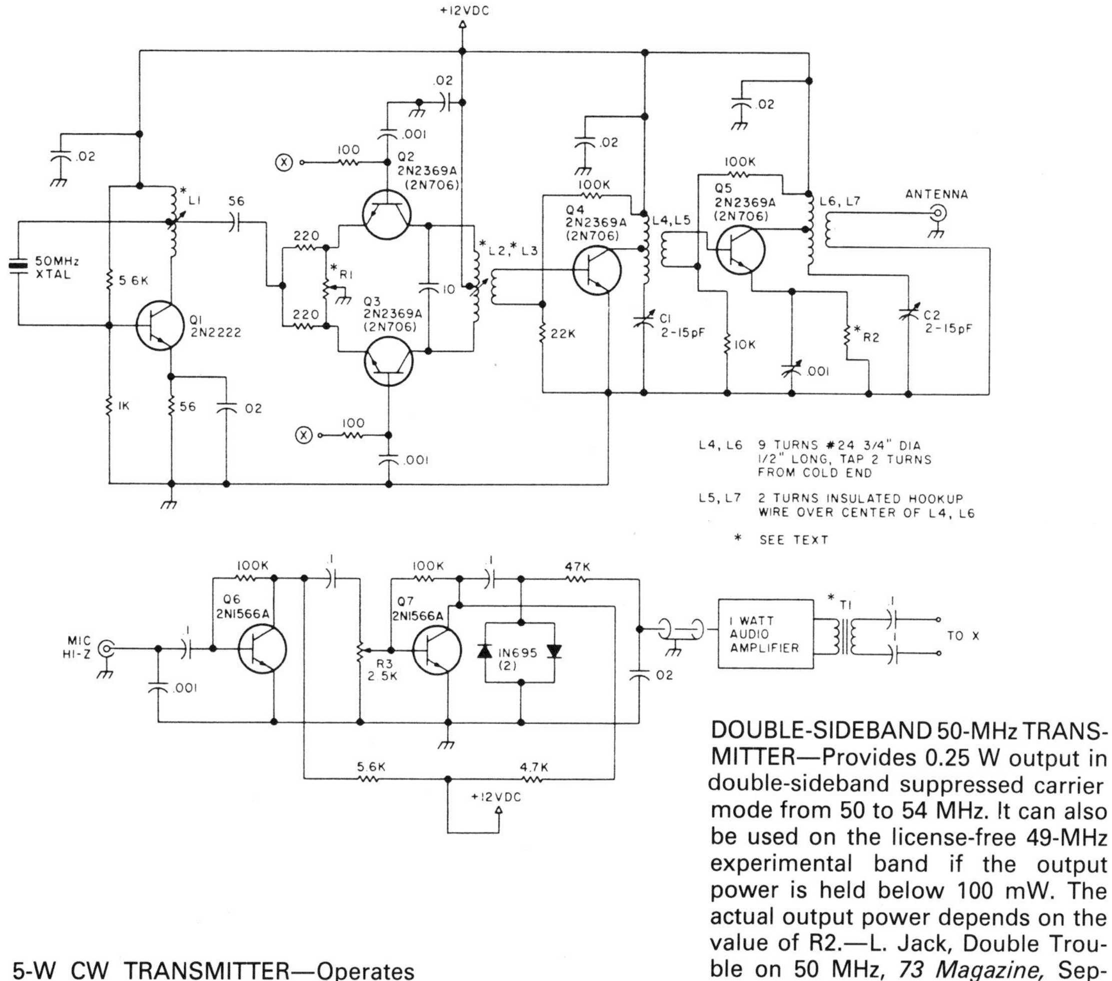

DOUBLE-SIDEBAND 50-MHz TRANSMITTER—Provides 0.25 W output in double-sideband suppressed carrier mode from 50 to 54 MHz. It can also be used on the license-free 49-MHz experimental band if the output power is held below 100 mW. The actual output power depends on the value of R2.—L. Jack, Double Trouble on 50 MHz, *73 Magazine,* September 1982, pp. 58–59.

5-W CW TRANSMITTER—Operates in 3500- to 4000- and 7000- to 7300-kHz amateur radio bands. Q1 operates as a Pierce oscillator, and Q2 and Q3 comprise a class C amplifier with an output impedance of 60 Ω. Y1 is an FT-243 crystal for the desired operating frequency; the transmitter frequency may be varied about 1.5 kHz if C_{OPT} (a 30- to 200-pF variable capacitor) is added. Q1 is an RS-2033, and Q2 and Q3 are each RS-2038. L1 is 8.4 μH for 3500–4000 kHz and 10 μH for 7000–7300 kHz. L2 is five turns wound over the side of L1. L3 is approximately 30 μH. L4 is 2.4 μH for 3500–4000 kHz and 1.2 μH for 7000–7300 kHz.—M. Oman, The Fun-Mitter—A Goof-Proof RF Project, *73 Magazine,* February 1981, pp. 100–103.

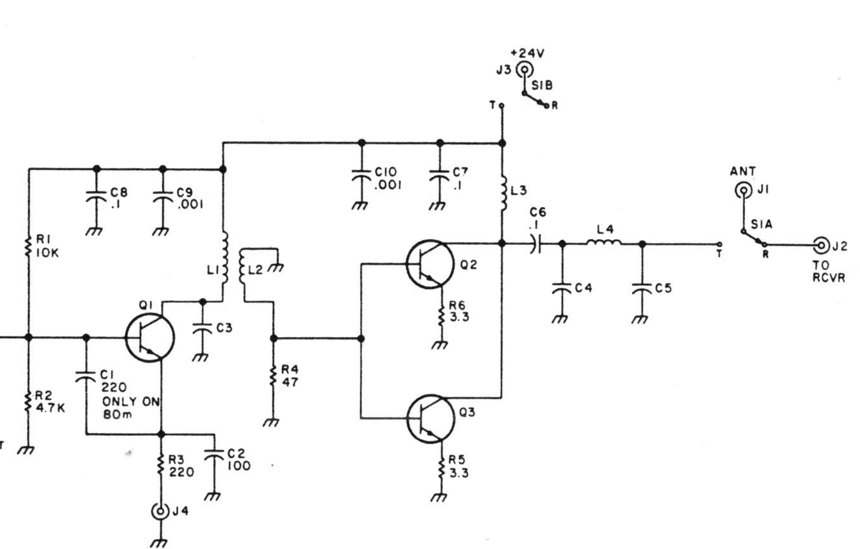

"CUBIC INCHER" 40-METER CW TRANSMITTER—Circuit is capable of delivering over 2 W output on 7000- to 7300-kHz amateur radio band, yet can be built in little over 1 in^3 of space. The design is the basic Pierce crystal oscillator. Q1 is an MRF472 or equivalent. C1 is a 390-, 430-, or 470-pF silver mica or disc ceramic capacitor. C2 is a 51-pF silver mica or ceramic capacitor. C3 is a 4- to 40-pF mica compression trimmer. C4 is formed from 0.01- and 0.001-μF capacitors in parallel. D1 is a 1N914 or equivalent. R1 is 10 K. RFC1 is an FB-43-101 ferrite bead. T1 is wound with No. 26 wire on a T50-2 form, 38 turns on the primary and four turns for the secondary. Y1 is a fundamental crystal for the desired operating frequency.—D. Monticelli, Build a 40-M Cubic Incher, *QST,* July 1982, pp. 34–36.

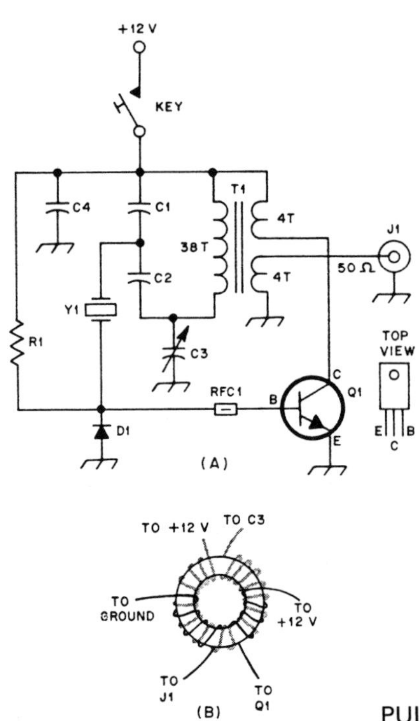

PULSE WIDTH MODULATOR—Gives variable output pulses depending upon input signal. $f_c < f_n < f$ unity gain, where $f_c = 1/2\pi R_2 C_1$ and $f_n = 1/2\pi R_1 C_1$.—"Intersil Data Book," Intersil, Cupertino, CA, 1981, p. 5-84.

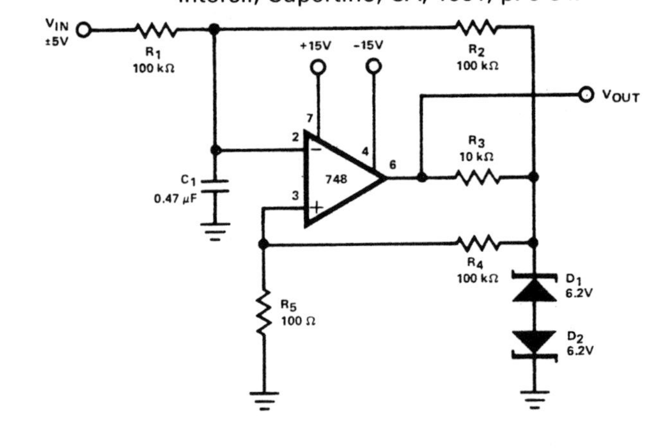

MICROPHONE EQUALIZER—Operates on 200- to 4000-Hz range of audio frequencies by equalizing microphone response to all frequencies in that range and rejecting frequencies outside it. It may be used with all types of radiotelephone transmitting equipment as well as recording and public address systems if proper shielding is used. RFC1 is a miniature ferrite core choke between 300 and 500 μH.—C. Nouel, Build Your Own Microphone Equalizer, *CQ,* July 1984, pp. 24–26.

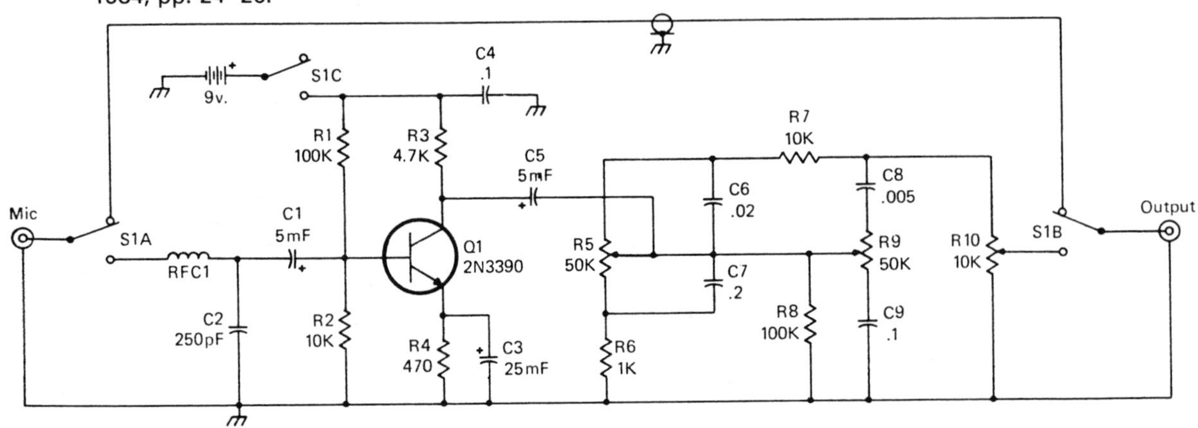

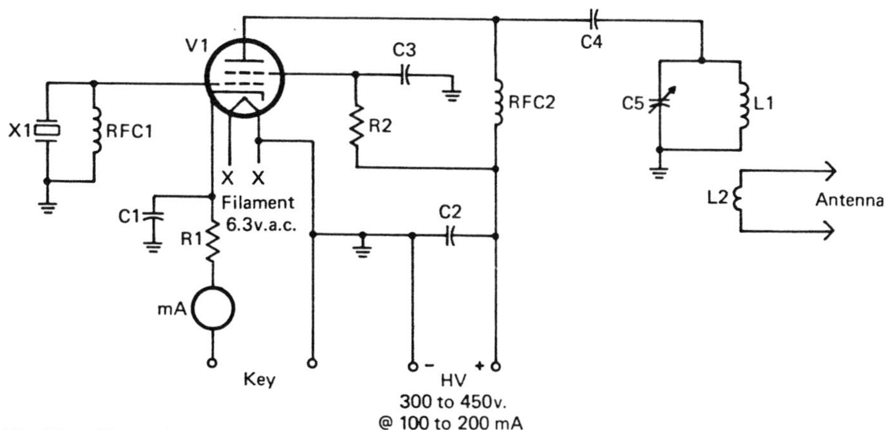

X1: 40 or 30 meter crystal
C1, C2, C3: .001 mFd paper
R1: 150 ohm, 2 watt
R2: 15K ohm, 2 watt
C4: .01 mFd mica
C5: 250 pFd variable
L1: 40 meters—17 turns No. 18 enamel
30 meters—12 turns No. 18 enamel
L2: 6 turns No. 18 enamel on cold end of L1
All coils wound on 2 inch form.
V1: 6L6G or 6V6G tube
RFC1, RFC2: 2.5 MHz 250 ma choke
Remember: use authentic parts and build on oak board!
ma: 0 to 200 d.c. ma meter

ONE-TUBE CW TRANSMITTER—Classic one-tube CW transmitter popular with amateur radio operators in past. The choice of the crystal and L1 permits operation in the 7000- to 7300- and 10,100- to 10,150-kHz amateur bands. Careful tune-up is important to avoid TVI.—D. Ingram, Collecting, Restoring, and Using Vintage Gear, *CQ,* September 1983, pp. 70–73.

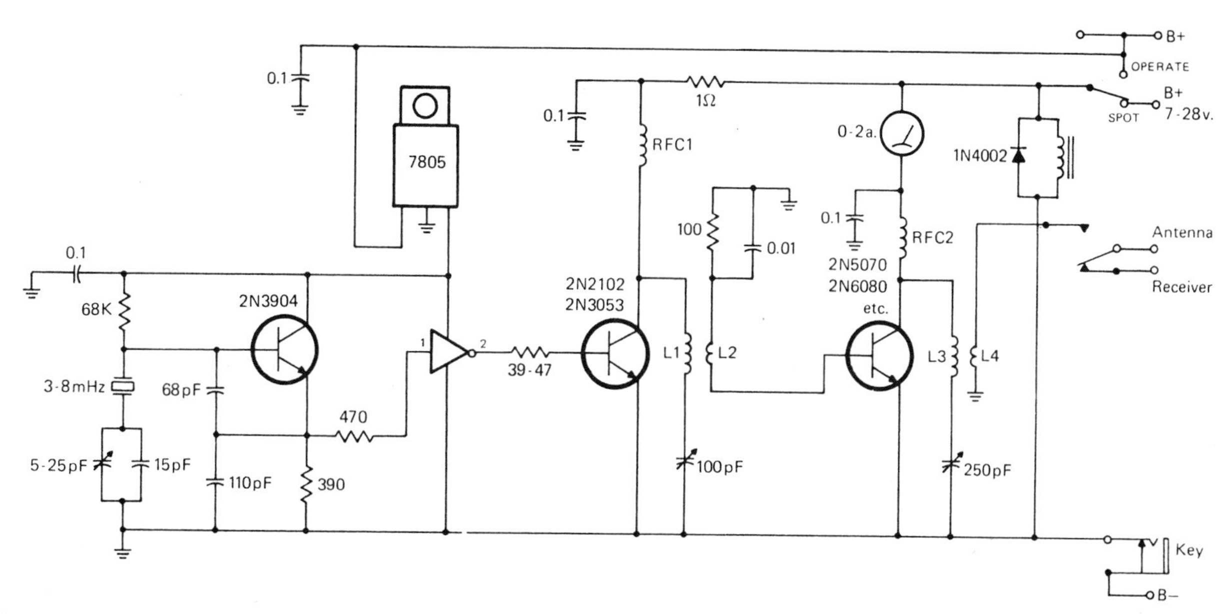

MINIATURE CW QRP TRANSMITTER—Output of TTL clock oscillator is amplified to approximately 5 W output for CW work in 80-meter (3500- to 4000-kHz) amateur radio band. The unit can be constructed inside a recipe file box. RFC is 20 turns of No. 22 or 26 wire on a small powdered iron core or 1-W resistor. L1 is 20 turns of No. 26 or 28 wire on a 1.5-in-diameter cardboard tube. L2 is one or two turns of No. 26-28 wire on the collector end of L1. L3 is 20 turns of No. 22 wire on a 1.5-in-diameter cardboard tube. L4 is three or four turns of No. 22 wire on the collector end of L3.—E. Fruitman, Roll Your Own TTL QRPp Transmitter in a Card File Box, *CQ,* June 1983, pp. 26–30.

SUBAUDIBLE TONE GENERATOR—Generates subaudible tone in range 52.6 to 152.2 Hz whenever microphone button is pressed to unlock tone-encoded squelch. The IC used is the American Microsystems S2559B/D digital tone generator. C1 is a 0.01-μF 50-V disc, C2 is a 10-F 16-V tantalum, C3 is a 120-pF 1000-V DX disc, C4 is a 330-pF 1000-V disc, C5 is a 1-μF 25-V tantalum, and C6 is a 4.7-F 16-V tantalum. D1 is a 1N4001, D2 is a 1N4739, and D3 and D4 are each 1N914. FB is a ferrite bead. R1 is 220 Ω, R2 is 10 M, R3 is 8.2 K, R4 is 470 Ω, R5 is a 10-K potentiometer, R6 is 47 K, and R7 is 10 K. X1 is a ceramic resonator; how to determine the exact frequency is explained in the original article.—R. Harold, Two Dollars a Tone, *73 Magazine,* August 1980, pp. 74–78.

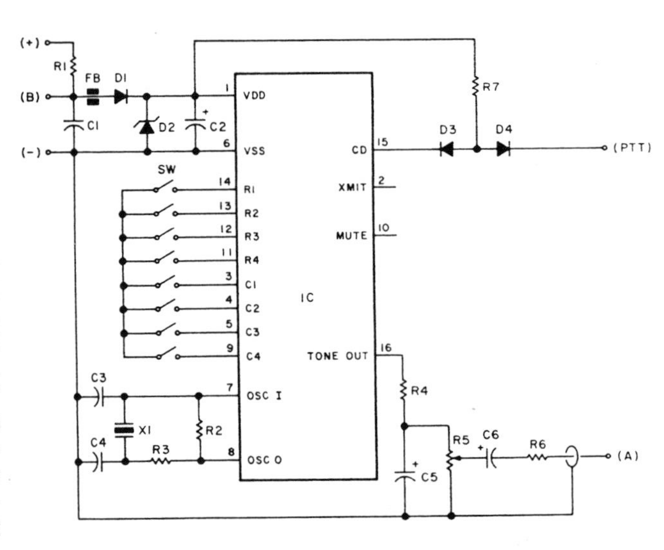

TRANSMIT-RECEIVE SWITCH—Useful from HF through VHF and can switch powers up to 400 W. D1 and D2 are both UM-9401 PIN diodes. RFC1 is five turns of No. 22 wire wound 0.25 in in diameter and to a length of 1 in. The transmitter/receiver isolation is typically 30–40 dB.—I. Ridpath, T-R Switching with PIN Diodes, *QST,* March 1981, pp. 19–21.

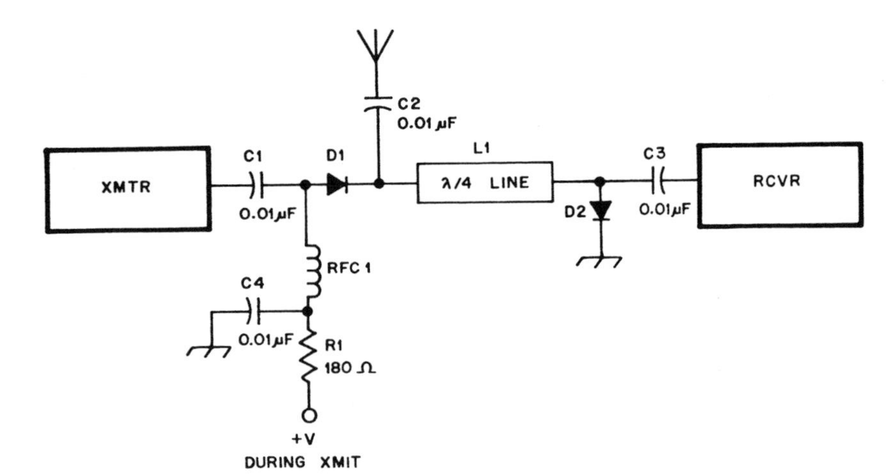

30-METER CW TRANSMITTER—Operates from 10,100 to 10,150 kHz with CW output of 1.5 W. L1 is 13 turns of No. 22 wire on a T68-6 toroid form and L2 is 30 turns of No. 28 wire on an FT37-63 toroid form.—P. Hoffman, A Two Transistor Transmitter for 30 Meters, *QST,* February 1984, pp. 46–47.

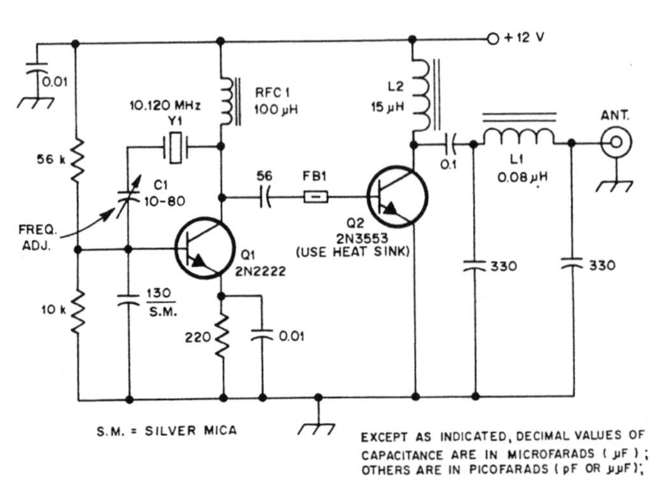

14- TO 28-MHz TRANSMITTING CONVERTER—Designed for use with low-power transmitters supplying 3 W or less of drive signal; output is approximately 1 W. R1 is used to null the 14-MHz signal feedthrough. C9, C10, C11, L2, and L3 form a low-pass filter to reduce harmonic content in the output signal. Q1 is a 2N3866. L1 is nine turns of No. 23 on a T-50-6 core. L2 and L3 is formed from five turns of No. 24 on a T-50-6 core. T1 is formed from eight twisted trifilar turns of No. 28 wire on an FT-37-61 form. T2 and T3 form a 16 : 1 broadband impedance matching network.—J. Pitts, A QRP Transmitting Converter, *QST,* April 1981, pp. 35–37.

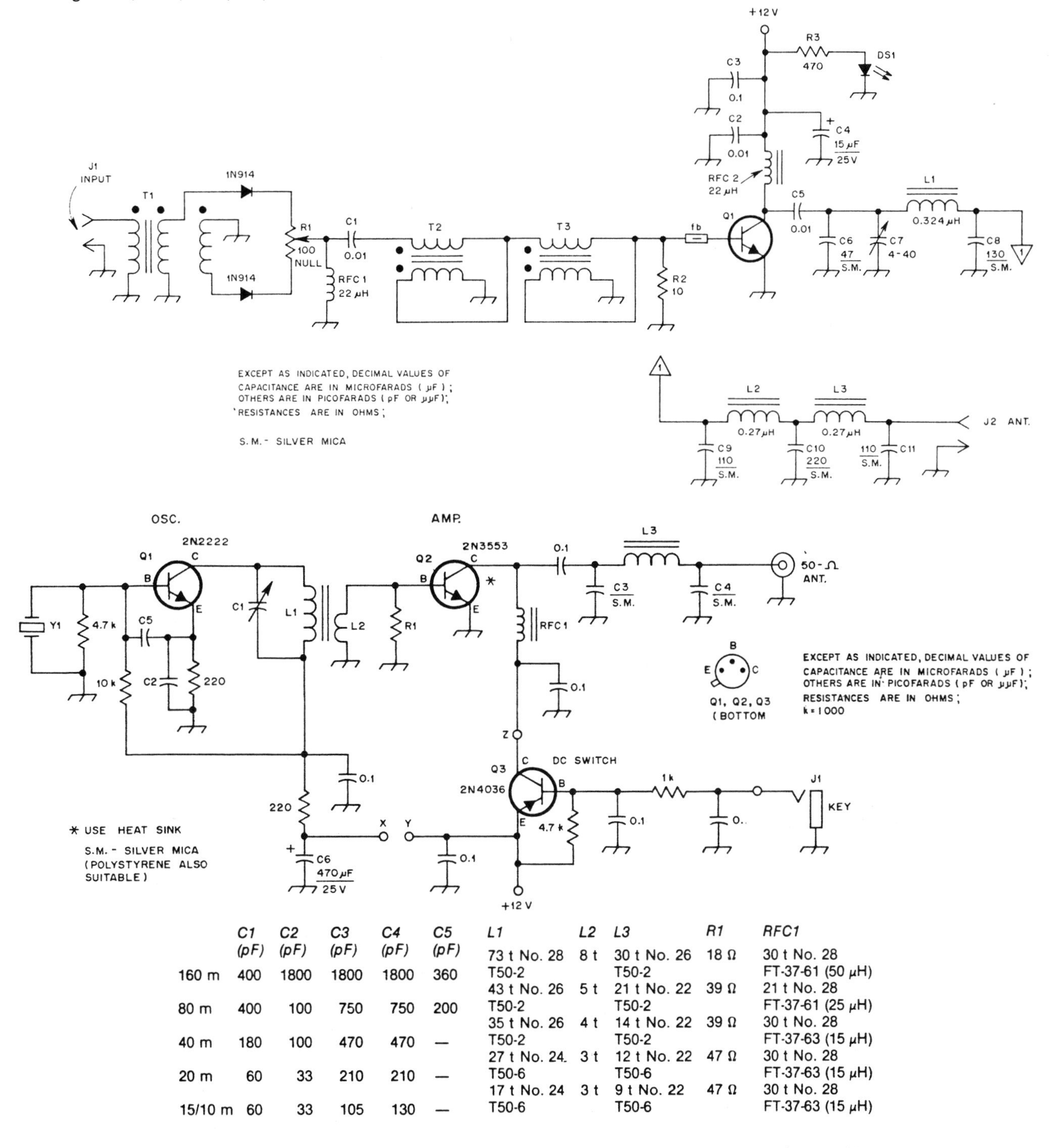

	C1 (pF)	C2 (pF)	C3 (pF)	C4 (pF)	C5 (pF)	L1	L2	L3	R1	RFC1
160 m	400	1800	1800	1800	360	73 t No. 28 T50-2	8 t	30 t No. 26 T50-2	18 Ω	30 t No. 28 FT-37-61 (50 μH)
80 m	400	100	750	750	200	43 t No. 26 T50-2	5 t	21 t No. 22 T50-2	39 Ω	21 t No. 28 FT-37-61 (25 μH)
40 m	180	100	470	470	—	35 t No. 26 T50-2	4 t	14 t No. 22 T50-2	39 Ω	30 t No. 28 FT-37-63 (15 μH)
20 m	60	33	210	210	—	27 t No. 24. T50-6	3 t	12 t No. 22 T50-6	47 Ω	30 t No. 28 FT-37-63 (15 μH)
15/10 m	60	33	105	130	—	17 t No. 24 T50-6	3 t	9 t No. 22 T50-6	47 Ω	30 t No. 28 FT-37-63 (15 μH)

Toroid cores are used in L1, L2 and L3. These are powdered-iron cores available from Amidon Associates and Palomar Engineers (T50-2, etc.). RFC1 is wound on a small ferrite core (FT-37-67), and so on), available from same suppliers. The letter "t" signifies the number of wire turns in the winding.

"UNIVERSAL" CW QRP TRANSMITTER—Delivers up to 1.5 W of output when operated with 12- to 14-V supply. The circuit can operate on all amateur radio bands from 160 meters (1800–2000 kHz) to 10 meters (28000–29700 kHz) through the proper selection of components as detailed in the table that accompanies the drawing.—D. DeMaw, Experimenting for the Beginner, *QST,* September 1981, pp. 11–15.

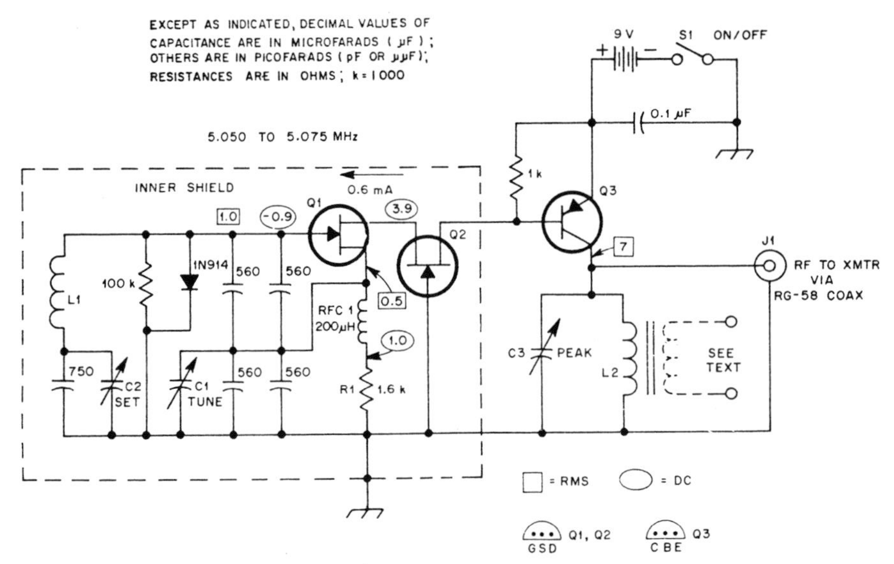

30-METER VFO—Series-tuned Clapp oscillator provides coverage of 10,100- to 10,150-kHz amateur radio band. C1 is a 4- to 50-pF air variable capacitor, C2 is a 2- to 17-pF air trimmer capacitor, C3 is a 180-pF mica compression capacitor, L1 is a 2.9-μH air-wound inductor, L2 is 32 turns of fine gauge wire on a T50-2 core, Q1 is a 2N5484 JFET, Q2 is a 2N5486 JFET, Q3 is an MPS3640, and RFC1 is a 200-μH molded choke.—D. Monticelli, A Battery-Powered 30-Meter VFO, *QST,* May 1984, pp. 35–37.

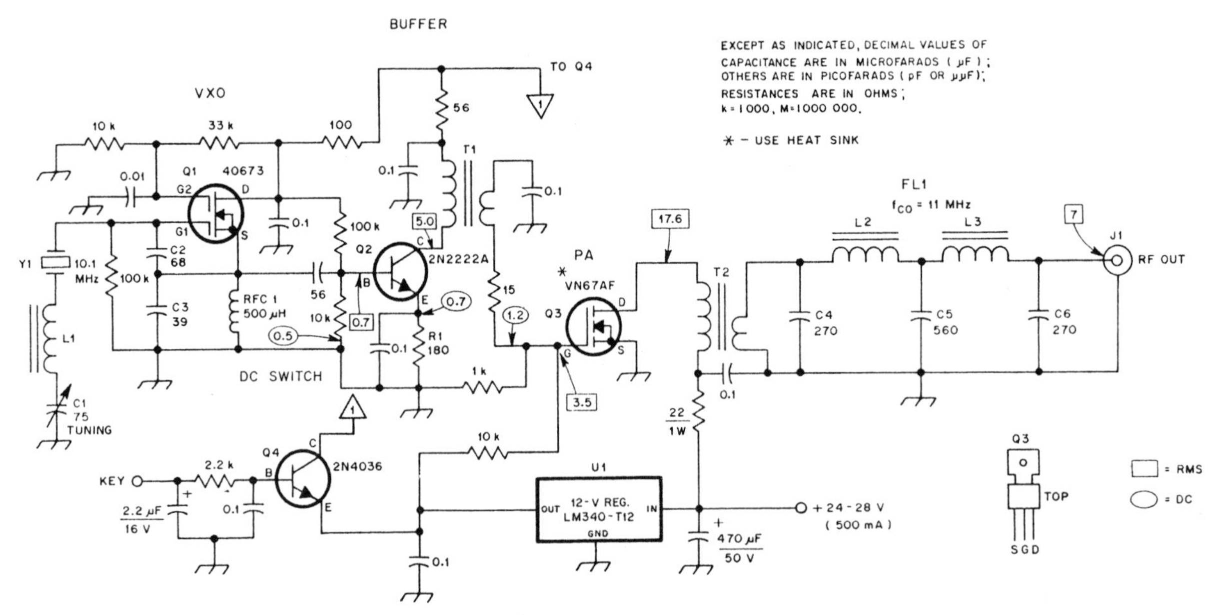

30-METER LOW-POWER CW TRANSMITTER—Delivers 1 to 2 W in the 10,100- to 10,150-kHz amateur radio band. L1 is a 12-μH inductor made of 50 turns of No. 28 wire on a T50-2 toroid form. Both L2 and L3 are 0.72-μH inductors made of 13 turns of No. 24 wire on a T50-6 toroid form. T1 is a broadband transformer made of No. 26 wire wound on an FT50-43 ferrite toroid form, 12 turns for the primary and six turns for the secondary. T2 is a broadband transformer made of No. 24 wire wound on an FT50-43 ferrite toroid form, 15 turns for the primary and six turns for the secondary. The value of R1 varies with the power output desired.—D. DeMaw, A VXO CW Rig for 30 Meters, *QST,* November 1983, pp. 31–34.

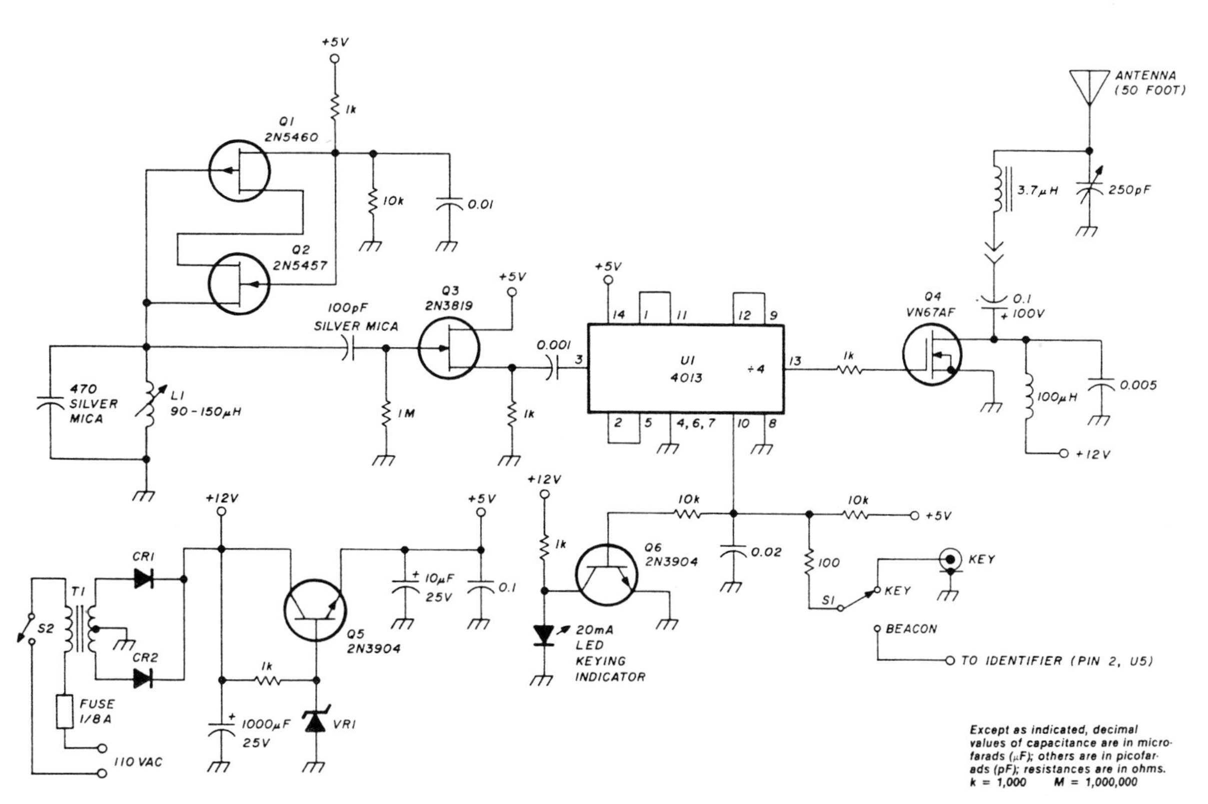

160- TO 190-kHz CW TRANSMITTER—Produces 1-W input in CW mode in 160- to 190-kHz range. Current Federal Communications Commission regulations permit unlicensed operation in the 160- to 190-kHz range so long as the input power is 1 W or less and the antenna used is 50 feet or less in length. Communications have taken place at distances of up to 700 miles with such transmitters. The circuit may be keyed manually or with a programmable keyer for beacon operation. CR1 and CR2 are each 1N4002. VR1 is a 1N52 zener rated at 5.2 V. T1 has a 110-V AC primary, 24-V center-tapped secondary, and a current rating of 300 mA.—S. DeFrancesco, VMOS on 1750 Meters, *Ham Radio,* October 1983, pp. 71–73.

VLF TRANSMITTER—Gives 1 W for CW in 160- to 190-kHz range. It may be used without a license in the United States if the antenna used (including its feedline) is less than 50 feet long. L2 is 440 turns of No. 26 wire on a 0.75-in-diameter PVC pipe about 10 in long; L1 is 30 turns of insulated hookup wire wound over L2.—L. Jack, Become a Low-Band Pioneer, *73 Magazine,* October 1983, pp. 34–36.

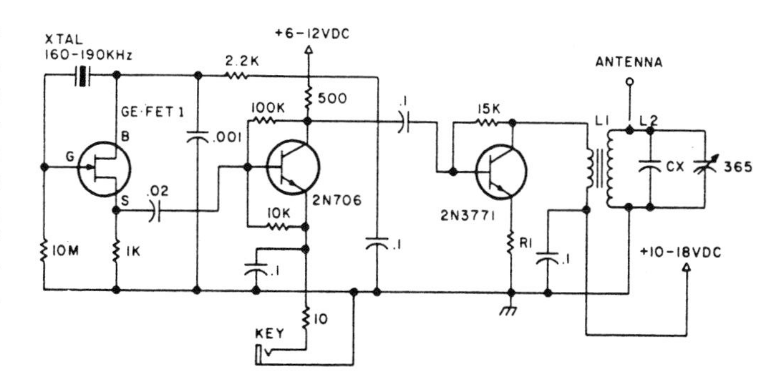

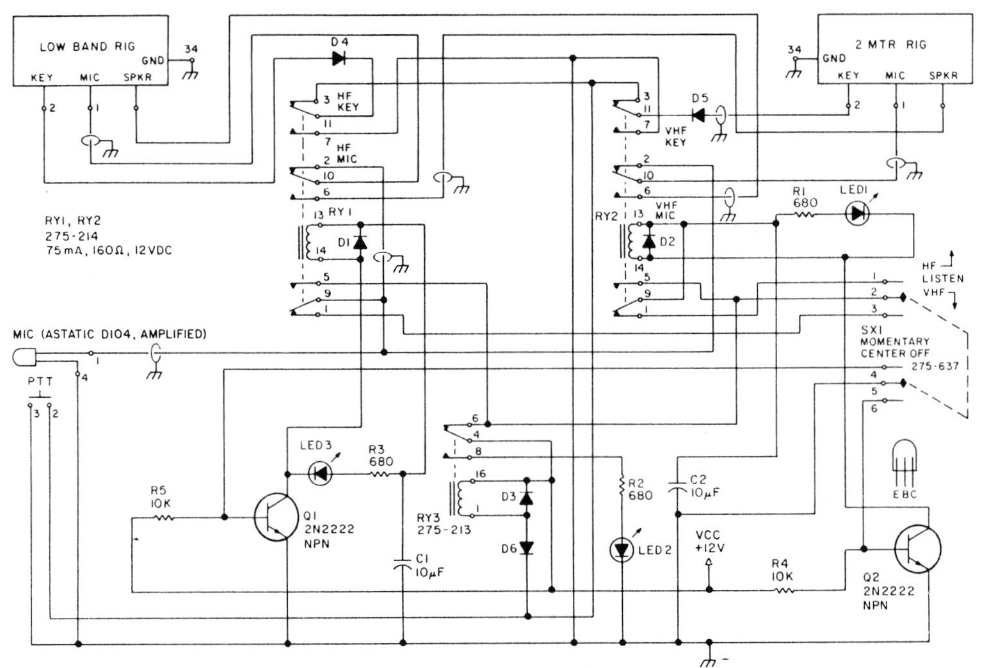

CROSSLINKING LOGIC BOX—Allows linking of VHF and HF transceivers together so that transmissions on VHF may be relayed on HF and vice versa. The operation is entirely manual; a control operator decides which transmissions are to be retransmitted on another band. RY1 and RY2 are 4PDT 12-V DC relays. RY3 is an SPDT 12-V DC relay.—D. Allen, Breakthrough in Boston: The Birth of Crosslinking, *73 Magazine,* January 1984, pp. 10–14.

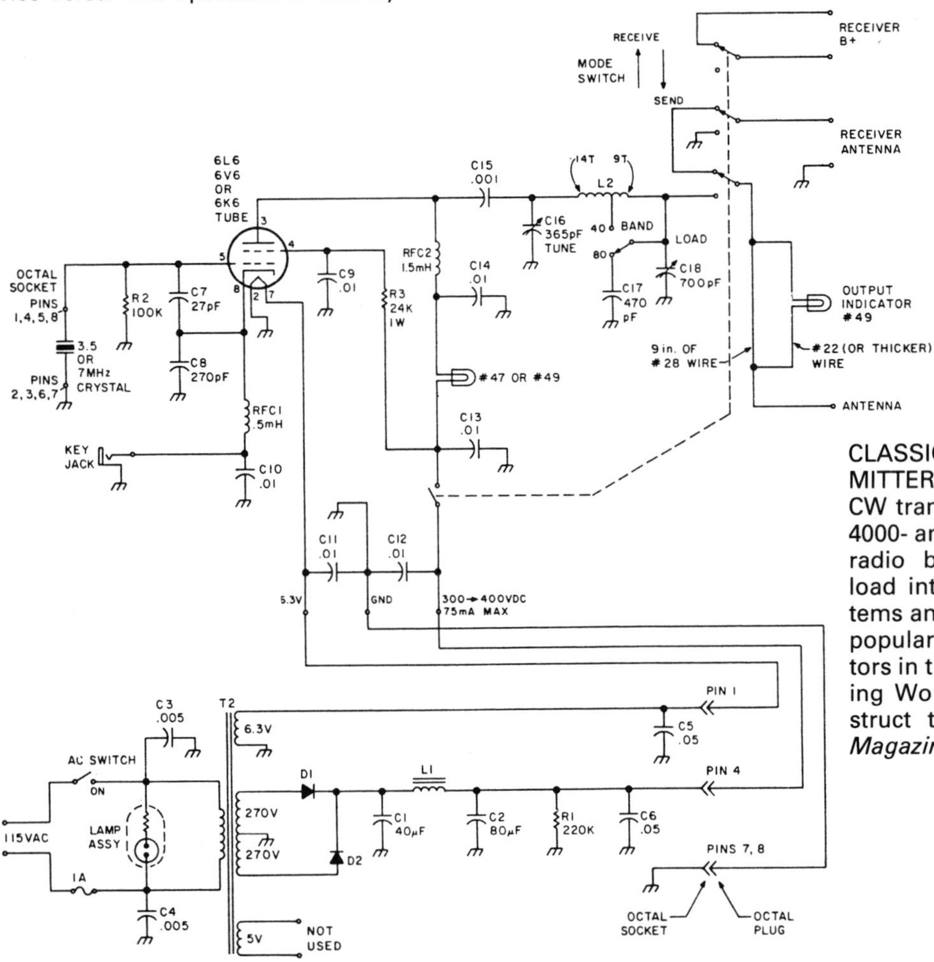

CLASSIC AMATEUR CW TRANSMITTER—Crystal controlled 15-W CW transmitter operates in 3500- to 4000- and 7000- to 7300-kHz amateur radio bands. The transmitter can load into a variety of antenna systems and is a replica of a transmitter popular with amateur radio operators in the years immediately following World War II.—P. Clower, Construct this Classic Transmitter, *73 Magazine,* May 1983, pp. 14–21.

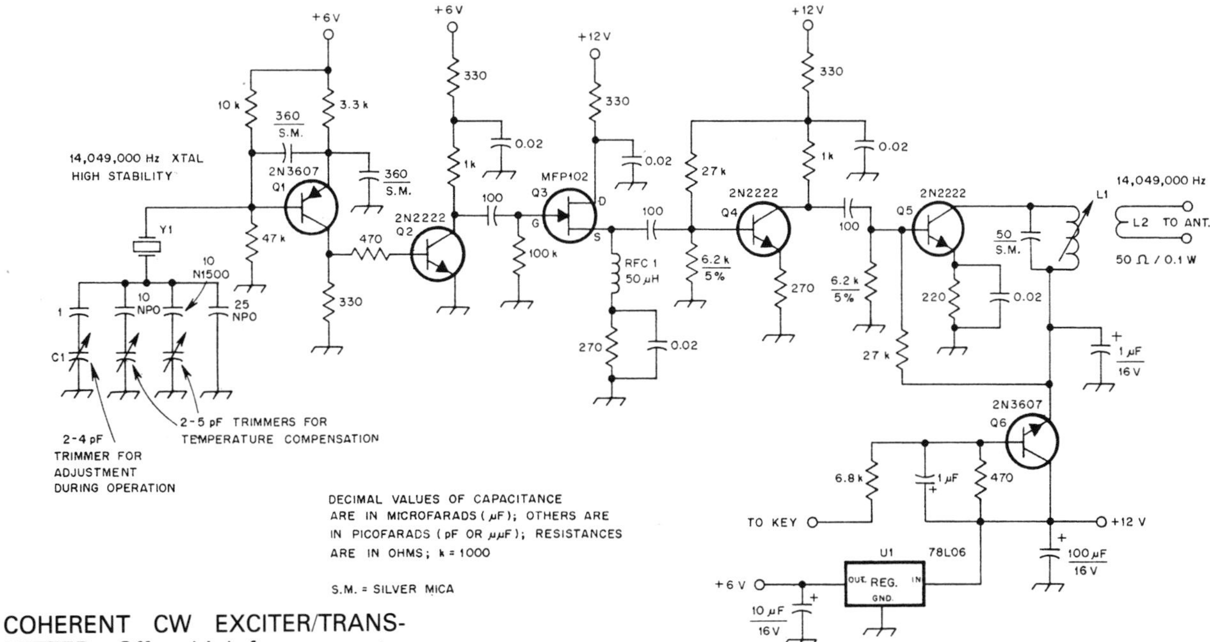

COHERENT CW EXCITER/TRANSMITTER—Offers high-frequency stability (within 1–2 Hz) when keyed for coherent CW operation. The stability is achieved by using high-quality crystal oscillators which are not keyed and which are followed by several stages of amplifiers and buffers to minimize the loading effects of the keying. The output of the circuit is approximately 100 mW. L1 is a Miller 4404 inductor having a value of approximately 2.5 μH. L2 is two turns of No. 18 wire over the "cold" end of L1.—C. Woodson, Coherent CW—The Practical Aspects, *QST,* June 1981, pp. 18–23.

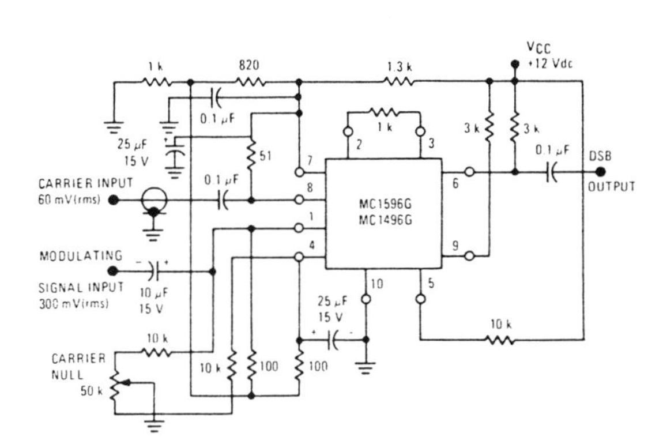

DSB BALANCED MODULATOR—Produces DSB output with suppressed carrier. The 50-K potentiometer is used to eliminate the carrier. The basic function of the MC1496 is DSB-suppressed carrier modulation. Carrier suppression is rated at −50 dB at 10 MHz.—"Motorola Linear Integrated Circuits Databook," Motorola, Phoenix, AZ, 1979, p. 6-98.

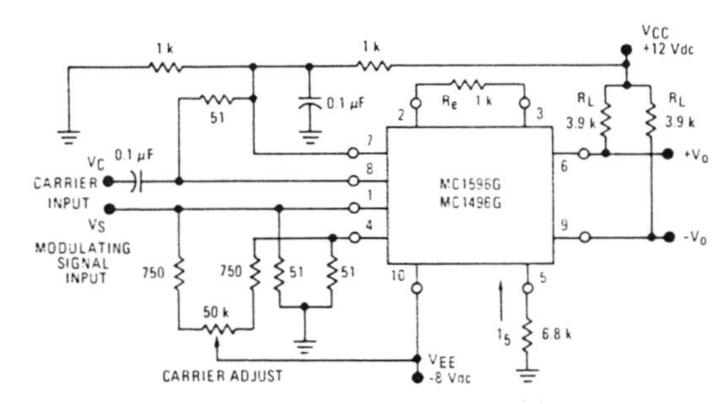

AM MODULATOR—Basically DSB suppressed carrier modulator but carrier null adjustment circuitry has sufficient range to insert full carrier for normal AM. The carrier level may be reduced but still left at sufficient level to allow the detection by AM receivers without the BFO.—"Motorola Linear Integrated Circuits Databook," Motorola, Phoenix, AZ, 1979, p. 6-98.

27

VHF/UHF Circuits

UHF OSCILLATOR—Designed to operate in range from 1100 to 1200 MHz. The frequency range may be adjusted by altering the position of the tuning diode on the line. The oscillator output may be used to drive the frequency multipliers as well. Part *b* gives additional details for the oscillator assembly. The RF coupling loop is formed from 0.1875-in shim stock.—N. Foot, Simplifying the Multipurpose UHF Oscillator, *Ham Radio,* September 1981, pp. 26–31.

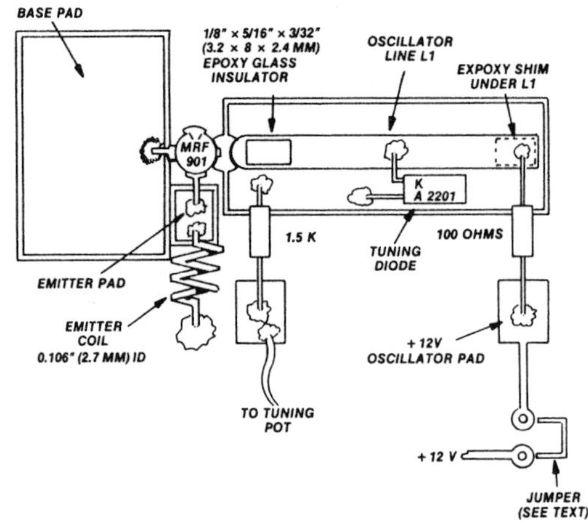

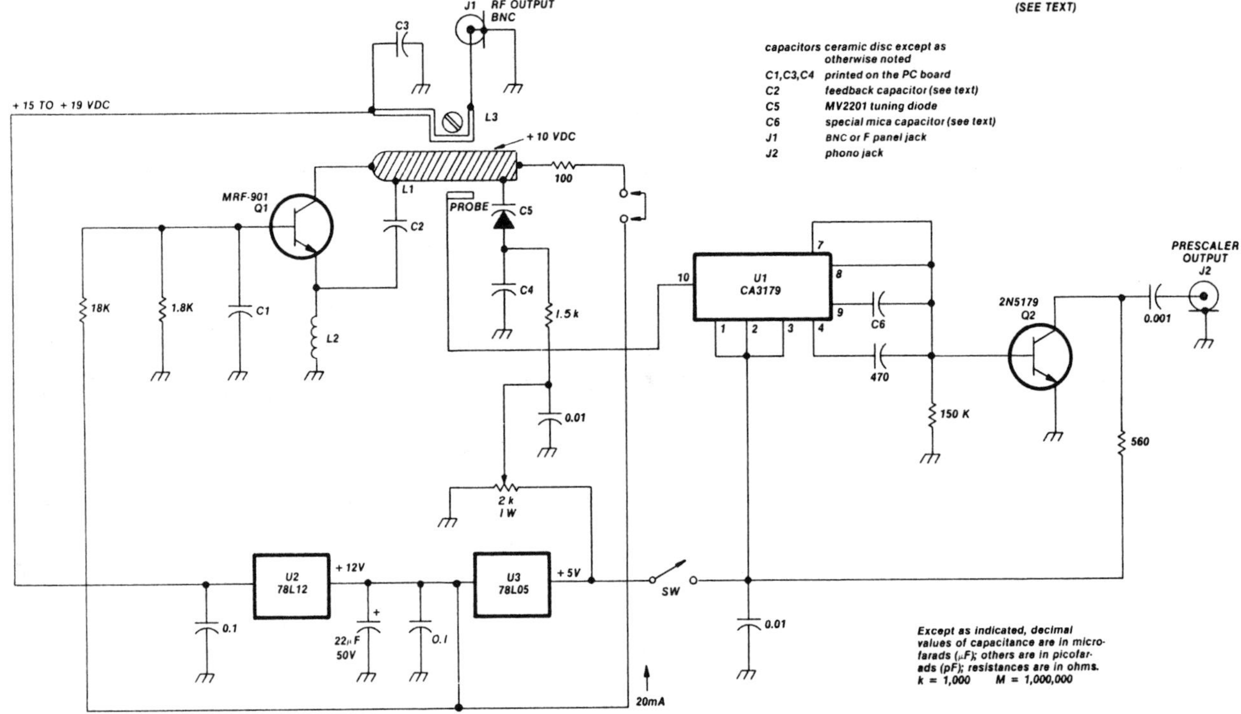

UHF TEST OSCILLATOR—Tunes 340–510 MHz with no major output level variations. With C12 set to its maximum capacity, C11 tunes the oscillator from 340 to 420 MHz. With C12 at its minimum, C11 tunes 410–510 MHz. C11 and C12 are 9-pF subminiature variable types (Johnson 9M11 or equivalent). L15 and L16 are formed from No. 14 copper wire 5.75 in long tapped 2 in from the hot (C) end. RFC2 is 15 turns of No. 28 wire wound 0.75 in long on a 0.25-in-diameter plastic form.—J. Reed, Make Mine Modular: Easy-To-Build Receiving Converter and Test Equipment for 435 MHz, *QST,* March 1983, pp. 11–15.

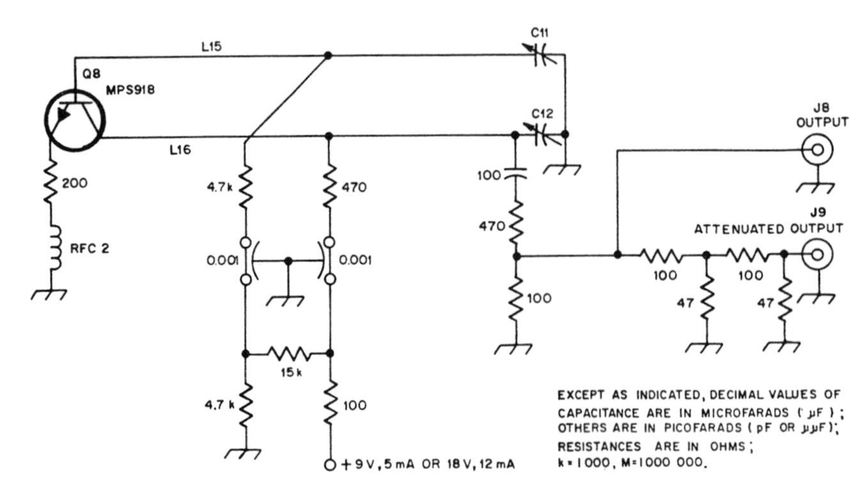

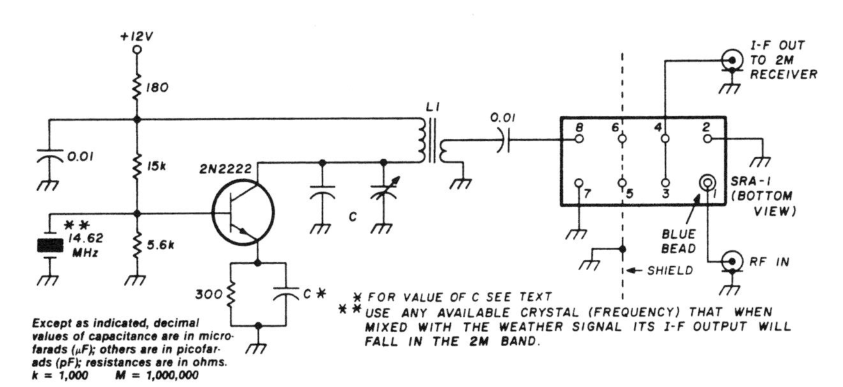

WEATHER FREQUENCY TO 2-METER CONVERTER—Converts 162-MHz weather broadcast frequency to frequency in 144- to 148-MHz amateur radio band. In the circuit shown, the 162.40-MHz signals mix with a 16-MHz local oscillator to produce a signal at 146.40 MHz. The primary of L1 should be selected so that its reactance at the crystal frequency is about 200 Ω. In the circuit shown, L1 is 20 turns of wire on a T50-2 toroid form for the primary, while the secondary is four turns. The fixed and variable capacitors at the collector must resonate with L1; in the circuit shown, a 20-pF fixed and 7- to 45-pF variable were used in the collector and a 39-pF fixed in the emitter circuit.—L. Nagurney, Two-Meter Weather Converter, *Ham Radio,* December 1983, p. 87–88.

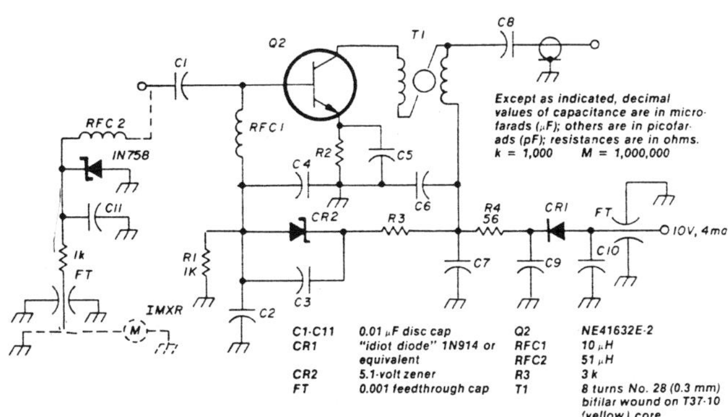

10- TO 60-MHz PREAMP—Designed for use in VHF and microwave receiving systems by providing gain at IF frequencies in 10- to 60-MHz range commonly used by such systems. The source impedance is approximately 200 Ω. R1 can be adjusted for either maximum gain or minimum noise.—G. Krauss, Low-Noise, Low-Cost 10-60 MHz Preamp, *Ham Radio,* May 1981, pp. 65–68.

144- TO 148-MHz RECEIVING PREAMPLIFIER—Designed to be mounted at antenna. The solenoid (Ledex No. 12180133-REV A or equivalent) switches the preamp out of the antenna circuit during transmit. Careful construction and tuneup is necessary for stability.—R. Brossman, Wheeling and Dealing with Preamps, *73 Magazine,* April 1984, pp. 84–88.

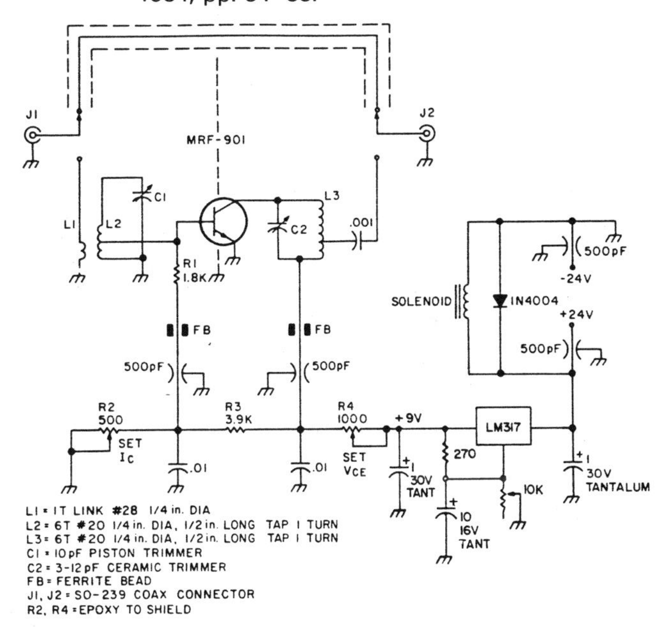

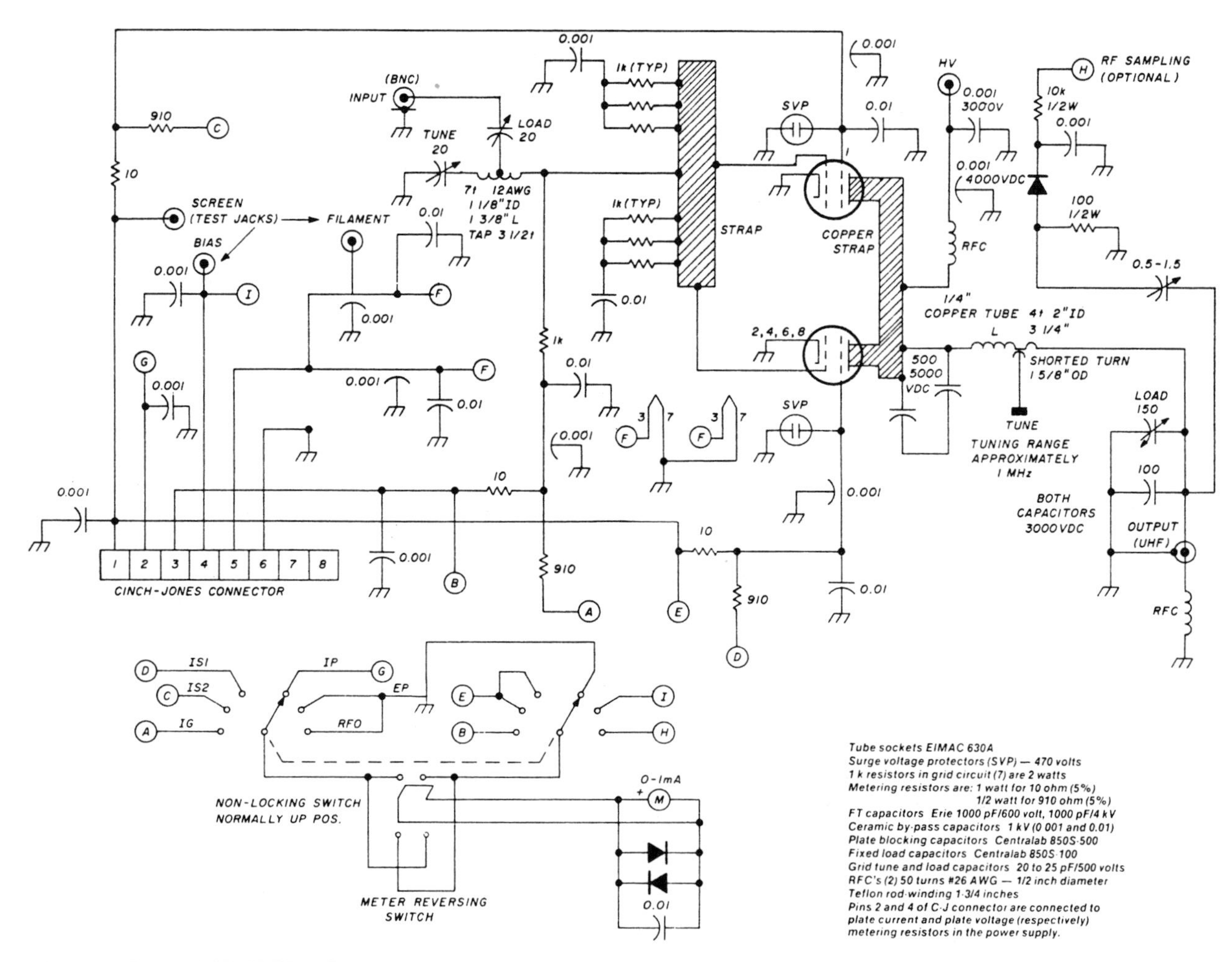

50-MHz LINEAR AMPLIFIER—Capable of delivering approximately 800-W output from 10-W drive. Parallel tetrodes (4CX250, 8930, etc.) have their grids connected by a copper strap between the sockets. The two anodes are paralleled by a brass or copper plate assembly. The plate coil is wound from 0.25-in copper tubing in four turns, 2 in long. The original article includes extensive details on the construction and proper placement of parts.—F. Merry, 6-Meter Amplifier, *Ham Radio,* April 1983, pp. 72–82.

1296-MHz GaAs FET PREAMPLIFIER—Q1 may be NE24483, MGF1400, or Dexel 3501. The circuit produces 17 dB of gain at 1296 MHz with a noise figure of 0.7 dB. The input and output capacitors should be high-quality chip capacitors. L1 and L2 are both Microstrip lines 0.25 in wide, 0.8 in long, and 0.2 in above ground. C1 through C4 are 0.8- to 10-pF air dielectric piston trimmers. C5 through C7 are 470- to 1000-pF feedthrough capacitors. RFC1 and RFC2 are both six turns of No. 20 wire, 0.125 in in diameter.—B. Atkins, The New Frontier, *QST,* January 1981, p. 68.

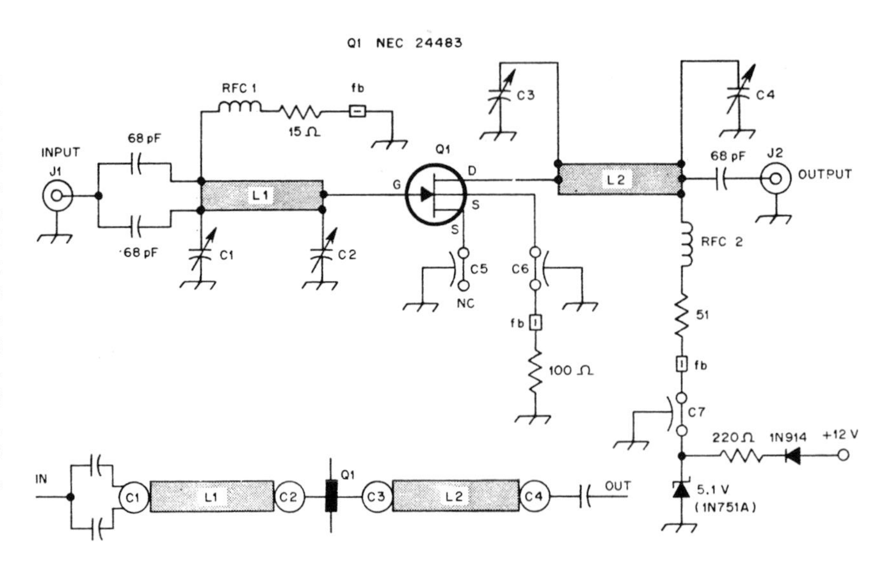

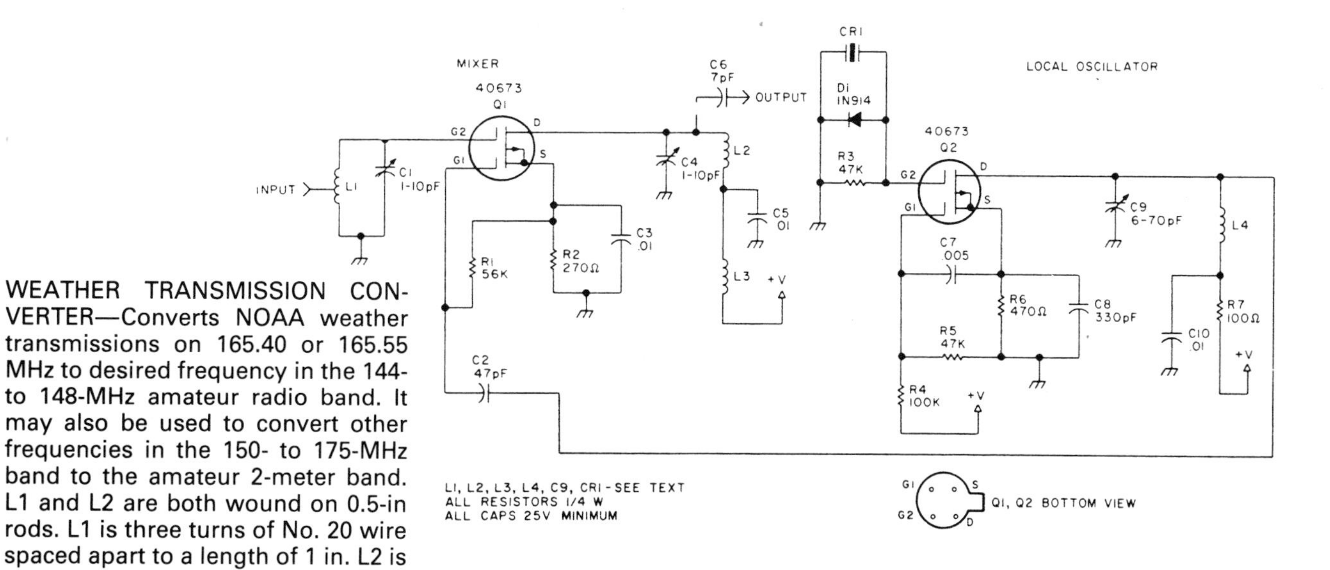

WEATHER TRANSMISSION CONVERTER—Converts NOAA weather transmissions on 165.40 or 165.55 MHz to desired frequency in the 144- to 148-MHz amateur radio band. It may also be used to convert other frequencies in the 150- to 175-MHz band to the amateur 2-meter band. L1 and L2 are both wound on 0.5-in rods. L1 is three turns of No. 20 wire spaced apart to a length of 1 in. L2 is three turns of No. 20 wire spaced to 1.25 in long. L3 has a value of 10 μH. L4 is 12 turns of No. 24 wire on a T50-2 form and resonates with C9, a 6- to 70-pF variable. CR1 depends upon the operating frequencies desired.—P. Danzer, Build the Weather-Grabber, *73 Magazine,* June 1983, pp. 84–88.

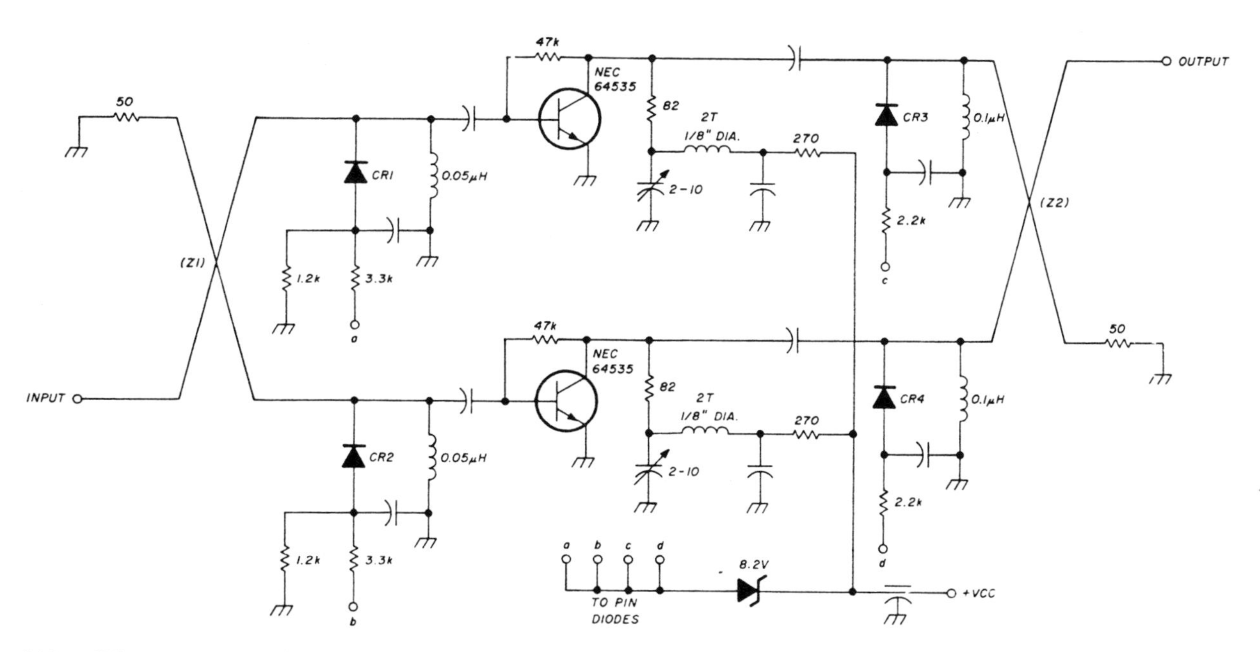

200- TO 400-MHz GAIN CONTROLLED AMPLIFIER—Useful in receivers or preamplifiers operating at 200–400 MHz. The gain with VCC = 7 V is 20 dB. All unmarked bypass capacitors are 0.01 μF, and CR1 through CR4 are PIN diodes. Z1 and Z2 are both constructed from Wireline® brand coaxial cable using the formula of one-quarter wavelength multiplied by 0.65. The original article has mounting details for Z1 and Z2.—H. Cross, Low-Noise Preamplifiers with Good Impedance Match, *Ham Radio,* November 1982, pp. 36–40.

50- TO 54-MHz LINEAR AMPLIFIER—Produces 100-W PEP into 50-Ω load from 8- to 10-W drive in SSB/CW service. Harmonic suppression is good but additional filtering is required to use the unit on the air. C1 is a 9- to 180-pF mica compression trimmer, while C2, C4, and C5 are all 50- to 380-pF mica compression trimmers. C3 is a 80- to 480-pF mica compression trimmer. D1 is a 15-A, 40-V rectifier diode. RFC1 is a 6.8-μH inductor (Nytronics SWD 6.8 or equivalent). RFC2 is made from No. 16 wire wound full length over a 330-Ω 2-W resistor. L1 is two turns of No. 14 bare wire, 13/32 in in diameter and 3/16 in long. L2 is a half-loop of No. 14 bare wire, 19/32 in high and 13/32 in long. L3 is two turns of No. 14 bare wire 13/32 in in diameter and 0.25 in long.—T. Tammaru, A Solid-State 6-Meter Linear Amplifier You Can Build, *QST,* May 1982, pp. 11–14.

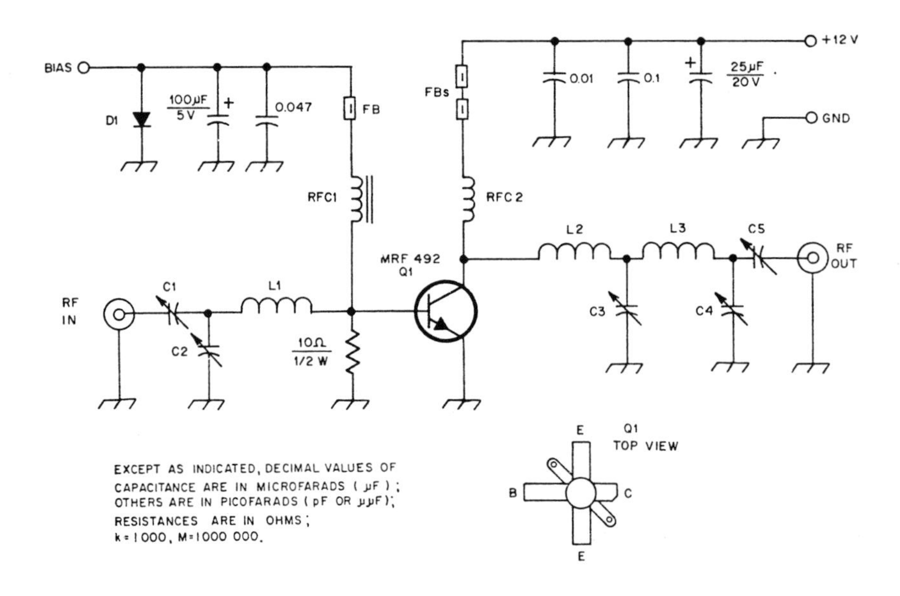

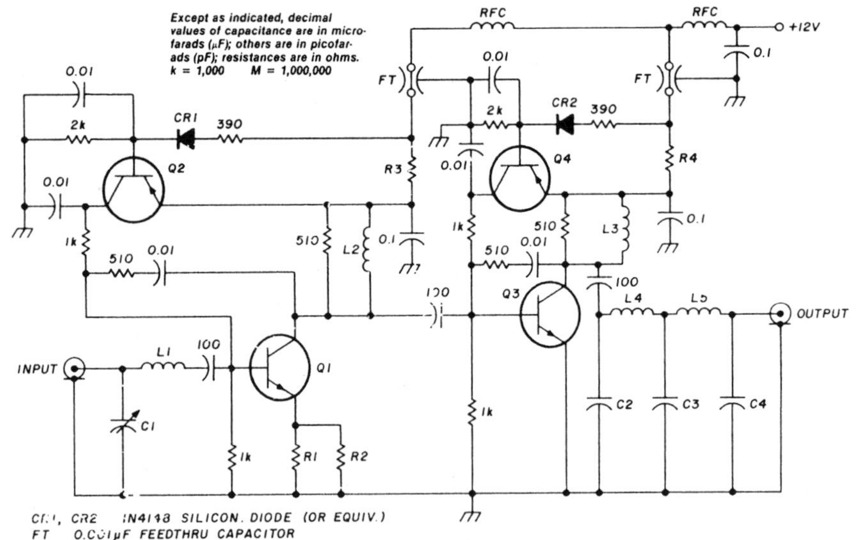

CR1, CR2 1N4148 SILICON DIODE (OR EQUIV.)
FT 0.001μF FEEDTHRU CAPACITOR
L2, L3 5T NO. 20 AWG ON 1/4" (6.4mm) DIAMETER x 0.5" (12.7mm) LONG
Q1, Q3 SEE TEXT
Q2, Q4 2N2905 OR EQUIV. HIGH GAIN PNP TRANSISTOR
RFC 1.0 MICROHENRY RF CHOKE. VALUE NOT CRITICAL
R1, R2 10 OHMS, 1/4 WATT
R3 39 OHM, 1/4 WATT
R4 13 OHM (THREE 39 OHM, 1/4 WATT, RESISTORS IN PARALLEL)

	144 MHz	220 MHz	432 MHz
C1	NOTE 1	NOTE 1	1-10pF
C2, C4	24pF	18pF	NOTE 1
C3	47pF	36pF	NOTE 1
L1	NOTE 1	NOTE 1	1/2" (12.7mm) LEAD ON THE 100pF CAPACITOR
L4, L5	6T, NO. 24 AWG ON 0.1" (2.5mm) DIAMETER 0.25" (6mm) LONG	3T, NO. 24 AWG ON 0.1" (2.5mm) DIAMETER 0.125" (3mm) LONG	NOTE 1

VHF/UHF LINEAR AMPLIFIER STAGE—Offers 10- to 15-dB gain per stage at frequencies from 144 to above 432 MHz. The circuit has good linearity with typical outputs of 100–250 mW on SSB and 0.5 W on CW. Q1 and Q3 can be Microwave Semiconductors 82091, Acrian CD1899, NEC NE74020, or their equivalents. The table accompanying the drawing gives values for various components to allow operation on the 144-, 220-, or 432-MHz amateur radio bands.—J. Reisert, VHF/UHF World, *Ham Radio,* April 1984, pp. 84–88.

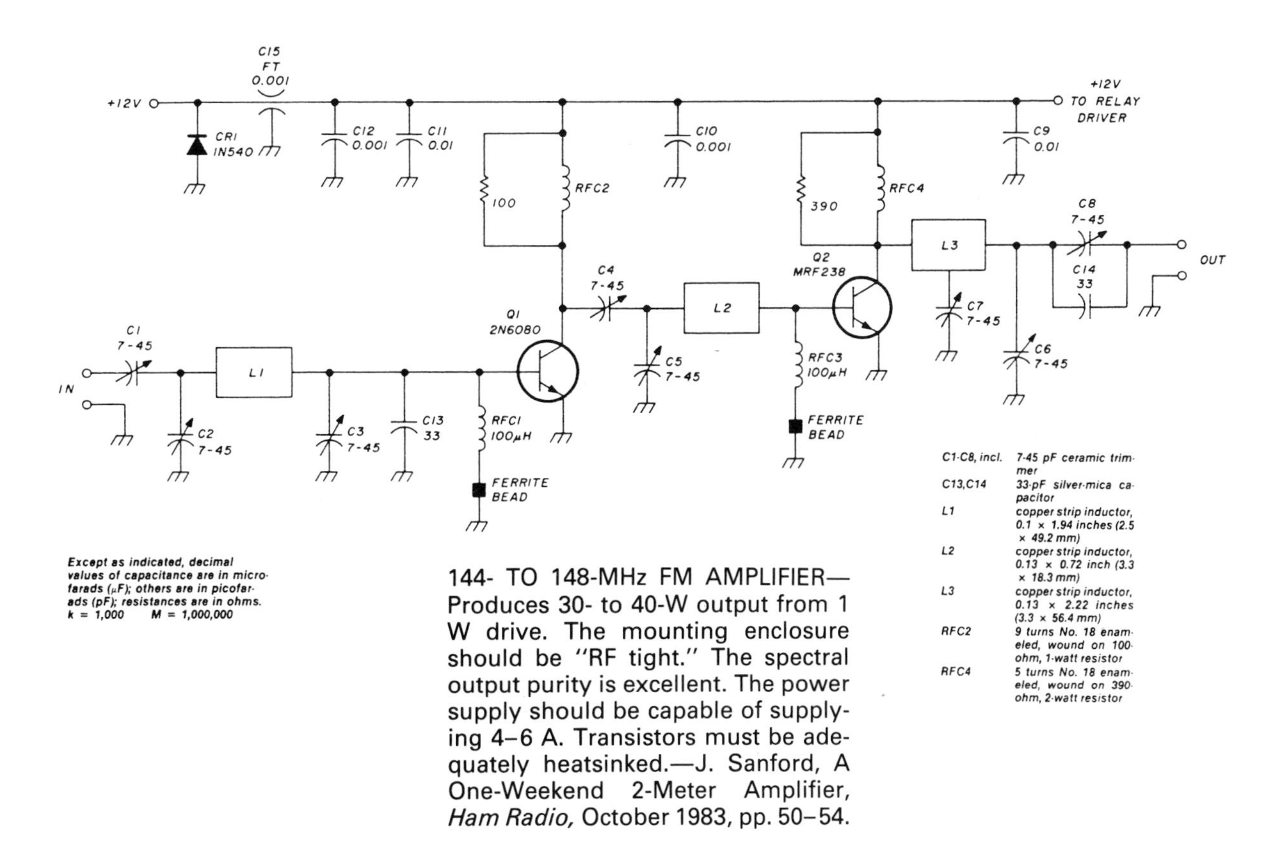

144- TO 148-MHz FM AMPLIFIER—Produces 30- to 40-W output from 1 W drive. The mounting enclosure should be "RF tight." The spectral output purity is excellent. The power supply should be capable of supplying 4–6 A. Transistors must be adequately heatsinked.—J. Sanford, A One-Weekend 2-Meter Amplifier, *Ham Radio,* October 1983, pp. 50–54.

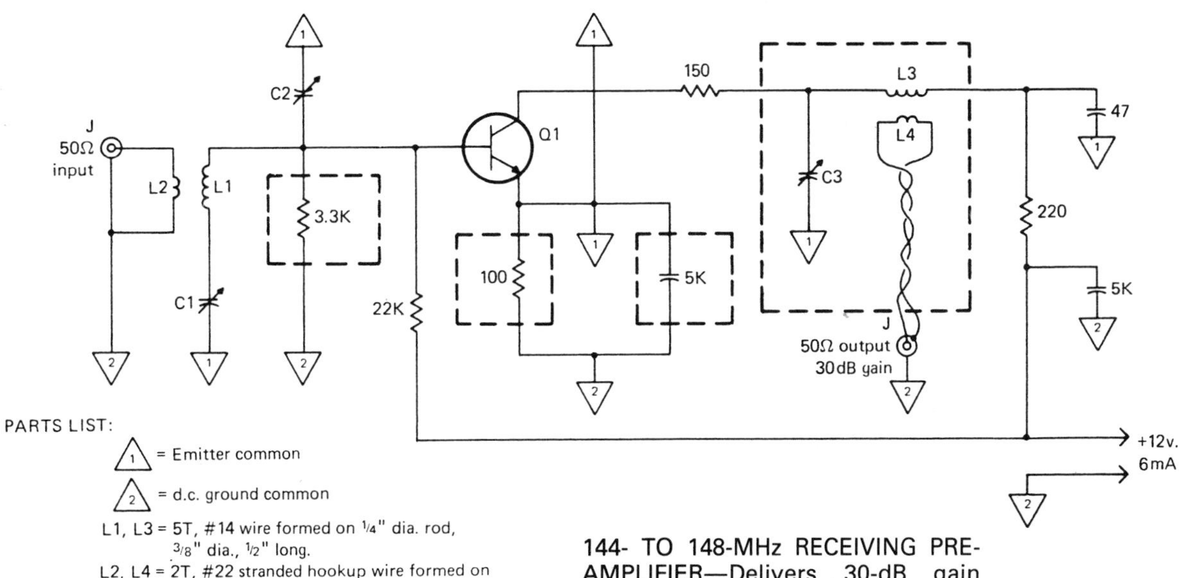

PARTS LIST:

1 = Emitter common

2 = d.c. ground common

L1, L3 = 5T, #14 wire formed on 1/4" dia. rod, 3/8" dia., 1/2" long.
L2, L4 = 2T, #22 stranded hookup wire formed on L1/L3 end.
Q1 = MRF901 (276-2044)
C1, C2, C3 = 5-60pF (272-1340)
J = BNC chassis connectors (278-105)
All fixed capacitors = pF/50v. disc ceramic.
All resistors = 1/4w., 5% (Radio Shack catalog numbers)

144- TO 148-MHz RECEIVING PRE-AMPLIFIER—Delivers 30-dB gain with noise figure of 1.5 dB. The device is built on a double-sided PC board (common ground and common emitter sides). All components enclosed in dashed lines are mounted on the common ground side.—J. Reed, An Easy to Build Two Meter Preamp and Gated Noise Source, *CQ,* July 1984, pp. 52–54.

50- TO 54-MHz VFO—Based upon MC145109 PLL IC, circuit is variable crystal oscillator covering any 500-kHz segment between 50 and 54 MHz. The project was built on separate circuit boards for the PLL, oscillator, display, and power supply.—E. Miller, The Six-Meter VFO that Won't Quit, *73 Magazine,* November 1983, pp. 48–56.

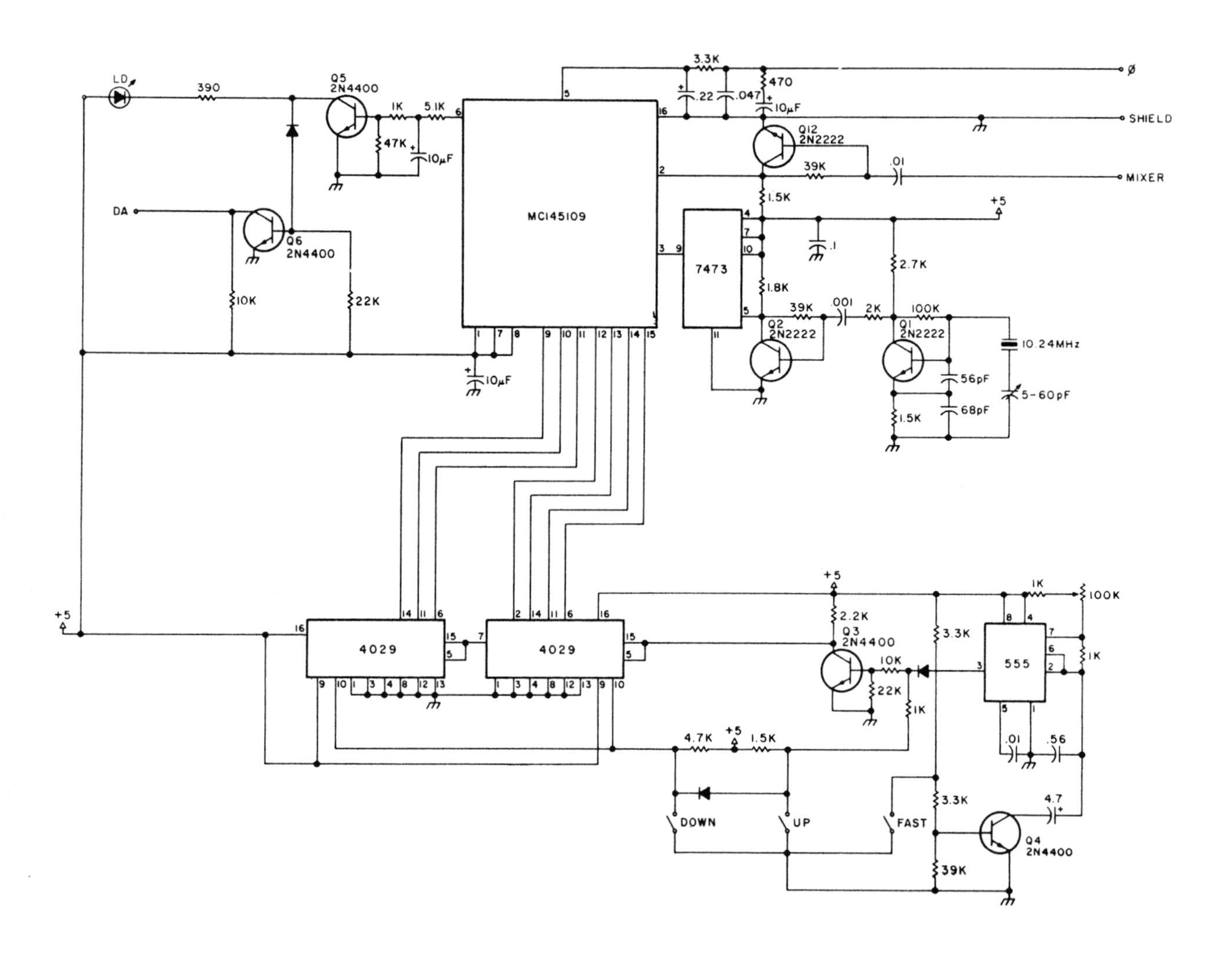

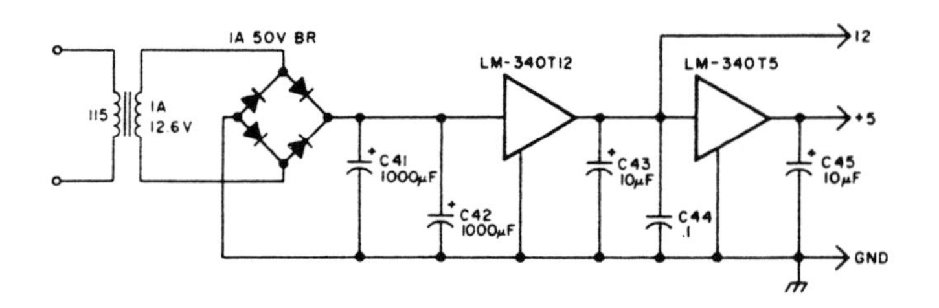

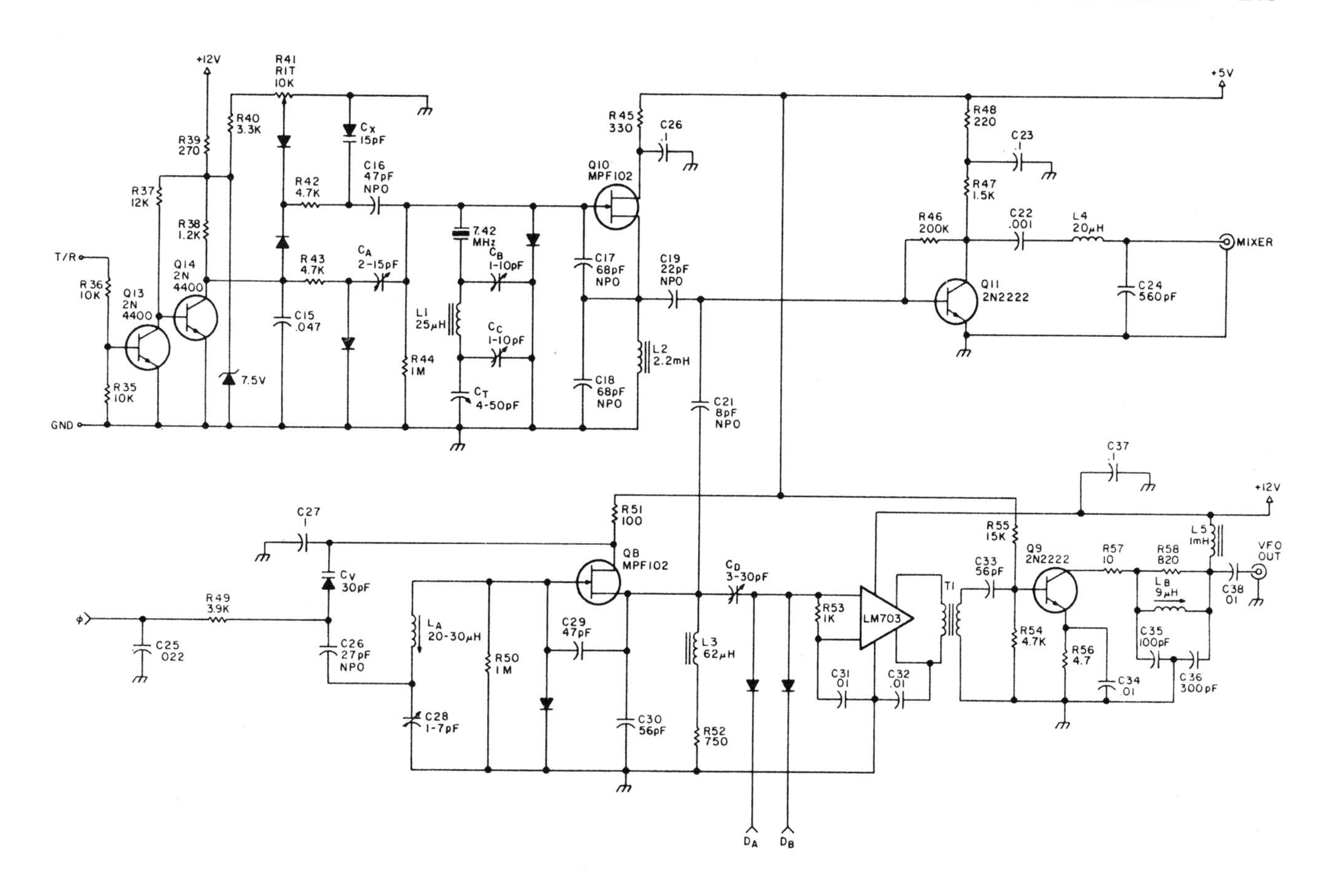

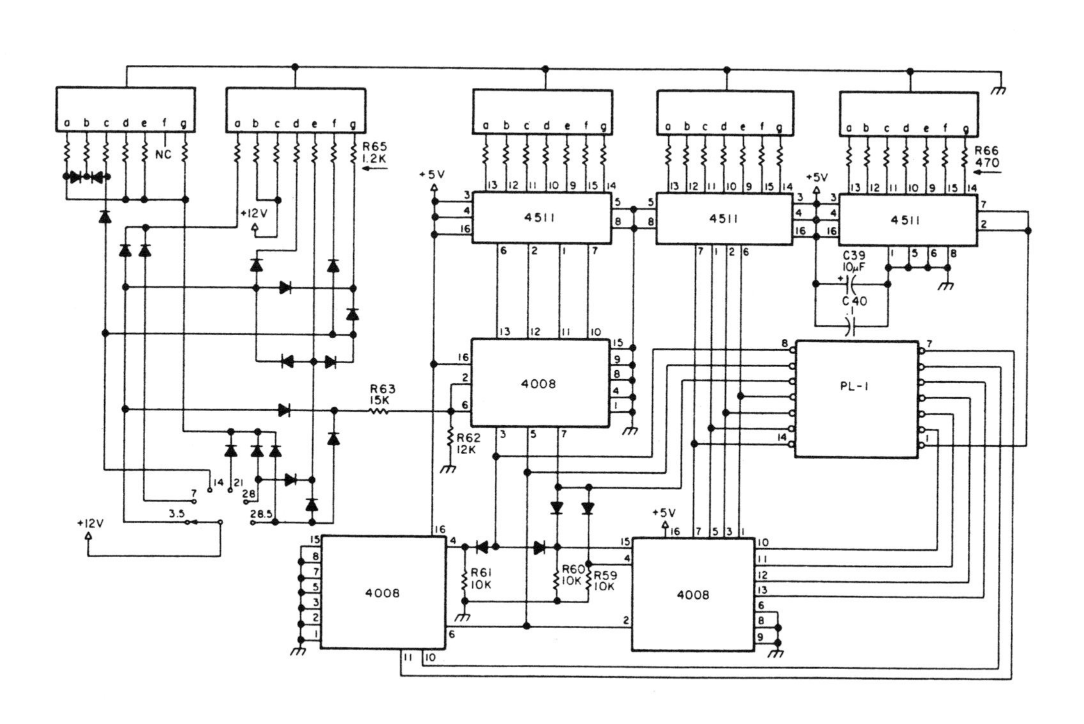

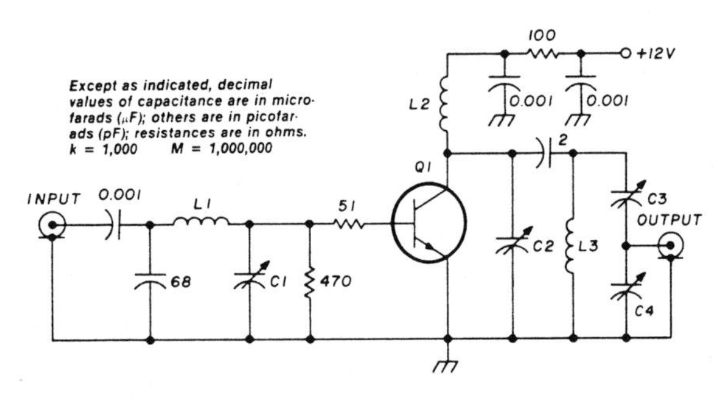

FREQUENCY DOUBLER—Produces a 190- to 220-MHz output at 10–20 mW from a 95- to 110-MHz input at 5–10 mW. High quality tuning capacitors with short leads should be used. Properly constructed, the circuit will not require a spectrum analyzer for alignment; the circuit is merely peaked for maximum output power.—J. Reisert, VHF/UHF World, *Ham Radio,* March 1984, pp. 45–46.

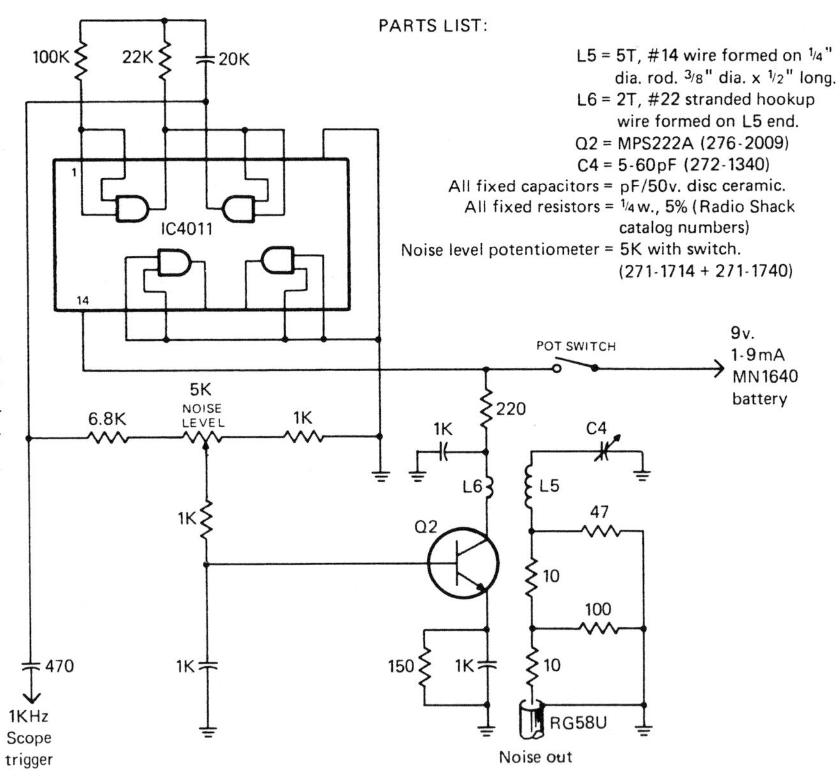

GATED NOISE SOURCE—Generates 1-kHz square wave output audible as random noise on AM receiver tuned to frequency in range of 140–175 MHz. It is useful for adjusting receivers and preamps operating at VHF.—J. Reed, An Easy to Build Two Meter Preamp and Gated Noise Source, *CQ,* July 1984, pp. 52–54.

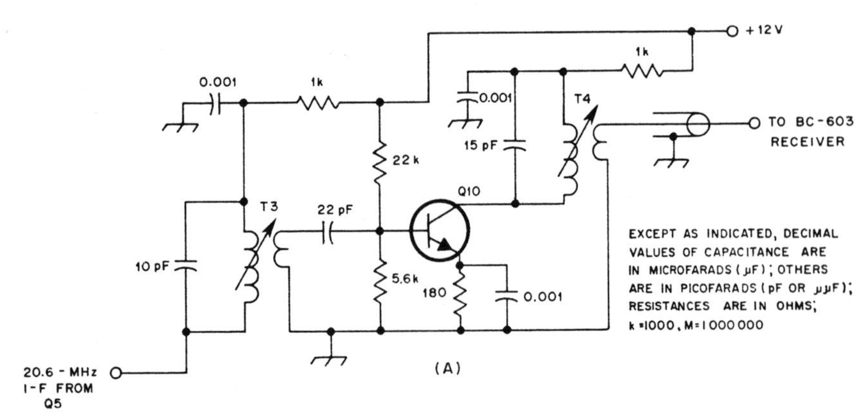

SATELLITE RECEIVER IF AMPLIFIER—Designed for use in receiver to receive signals from weather satellites operating at 1691 MHz. The operating frequency is 20.6 MHz. T3 and T4 consist of a primary made from 16 turns of No. 24 wound on a slug-tuned form 6 mm in diameter and a secondary of a four-turn link of No. 18 wire wound over the cold end of the primary winding. Q10 is a BF 155.—G. Emilani and M. Righini, An S-Band Receiving System for Weather Satellites, *QST,* August 1980, pp. 28–33.

OVERTONE CRYSTAL OSCILLATOR—Capable of covering 90–125 MHz with approximately 10-mW output. Harmonic suppression is in excess of 25 dB, and spurious responses are minimal. The crystal used should be a high-quality fifth or seventh overtone series-resonant type.—J. Reisert, VHF/UHF World, *Ham Radio,* March 1984, p. 45.

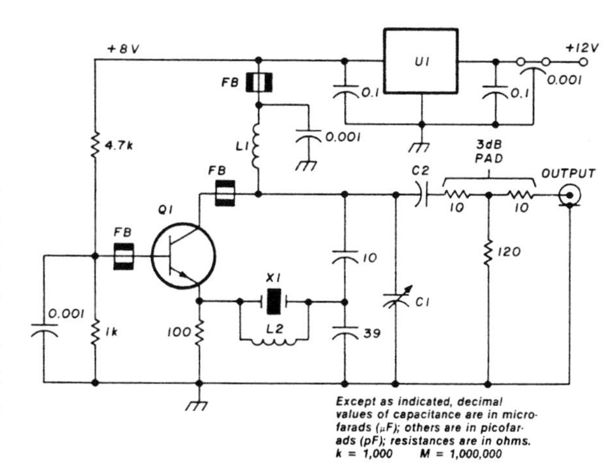

C1 2-20 pF small trimmer capacitor
C2 5 pF capacitor
FB ferrite bead - Ferroxcube No. 56 590 65/3B or equivalent
L1 8 turns No. 24 AWG enamelled wire close wound on 0.1 inch (2.5 mm) diameter
L2 0.39 μH RF choke or 10 turns No. 28 AWG on T-25-6 toroid core
Q1 2N5179, NE73432 or equivalent
U1 78L08 3-terminal 8-volt regulator
X1 5th or 7th overtone series resonant 90-125 MHz crystal

IF PREAMPLIFIER FOR MICROWAVE RECEIVERS—Offers 3.5-dB noise figure and constant gain from below 10 MHz to over 90 MHz. The input and output impedances are approximately 50 Ω. The circuit stability is excellent. The circuit was designed for use with receivers designed for the 10-GHz amateur radio band, where 30 MHz has become the de facto standard receiver IF frequency. Q1 should be a low-noise NEC transistor. T2 is seven bifilar turns on a FT23-75 core.—D. Petersen, The Care and Feeding of Gunnplexers, *QST,* April 1983, pp. 14–18.

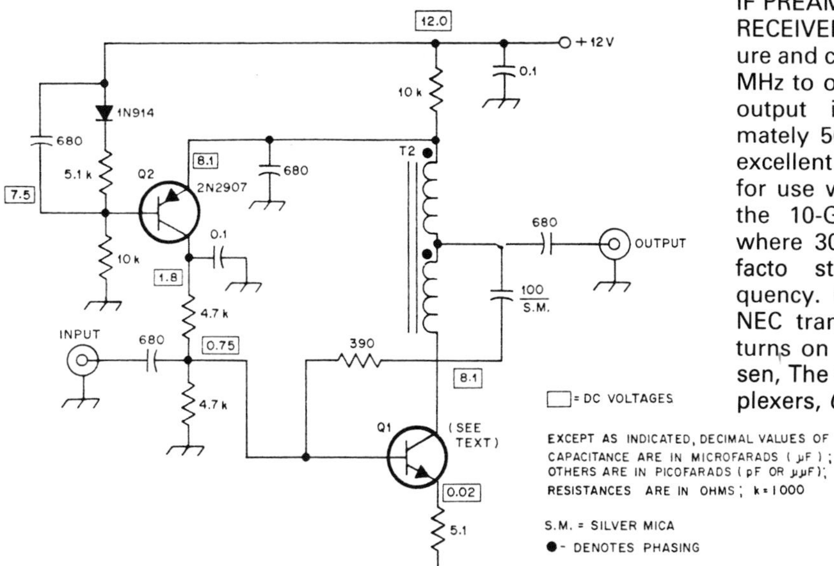

UHF INSTRUMENTATION PREAMPLIFIER—Accepts input signals from 100 MHz to 1 GHz. The gain ranges from 15 dB at the lower end of the frequency range to 5 dB at the upper end. D1 through D4 are HP5082-2811 Schottky barrier diodes. A1 is a hybrid amplifier (Amperex ATF417 or Philips OM185). D5 and D6 are 1-W, 5.1-V zener diodes. R6 and R10 should be rated at 5 W. C1 through C12 are 1-nF ceramic types. C13 is 0.33 μF.—"Plessey Applications IC Handbook," Plessey Semiconductor, Irvine, CA, 1982, p. 123.

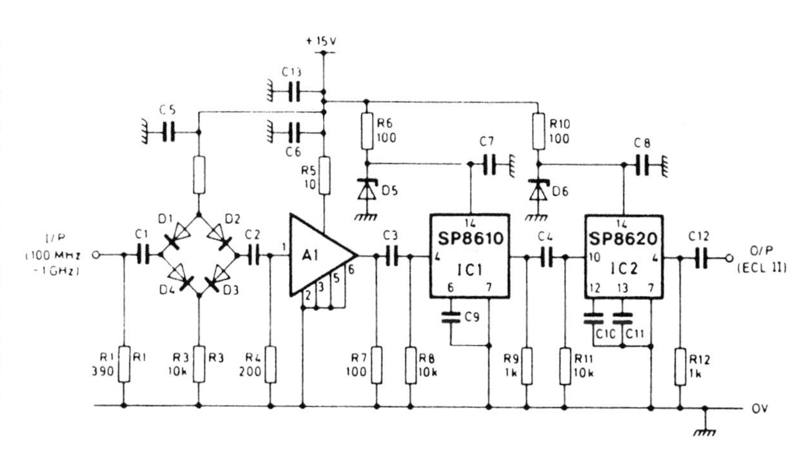

28

Voltage Regulation Circuits

5-V, 4-A REGULATOR—Uses one amplifier of the MC3401 quad op amp to deliver 5 V at 4 A. R (10 Ω) is used to bias the 1N3824. V_O is equal to V_Z + 0.6 V.—"Motorola Linear Integrated Circuits Databook," Motorola, Phoenix, AZ, 1979, p. 3-147.

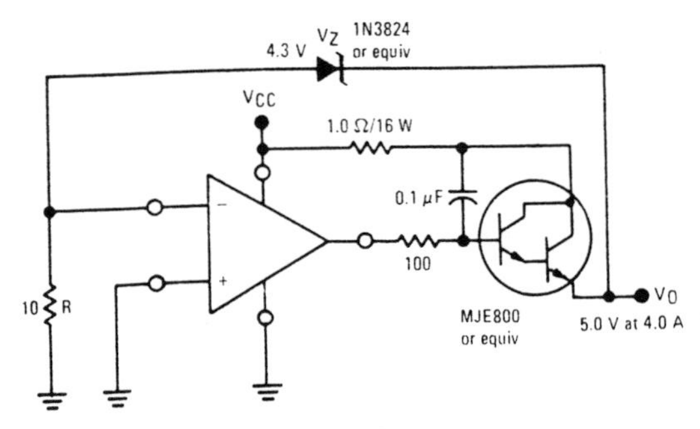

5-V, 10-A REGULATOR—Delivers 5-V output at up to 10 A from 10- to 30-V input. It includes short-circuit current limiting for safe-area protection of pass transistors.—"Motorola Linear Integrated Circuits Databook," Motorola, Phoenix, AZ, 1979, p. 4-15.

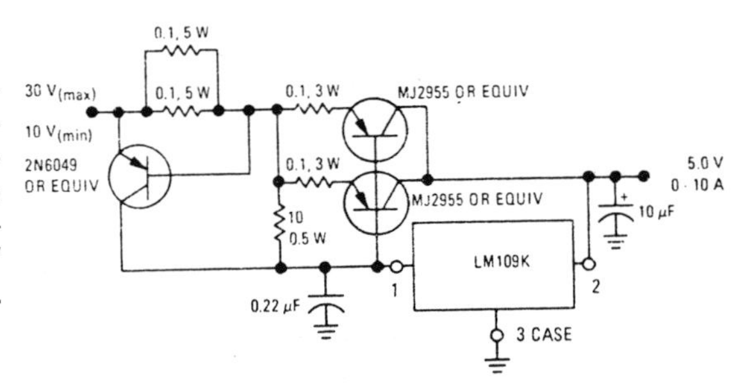

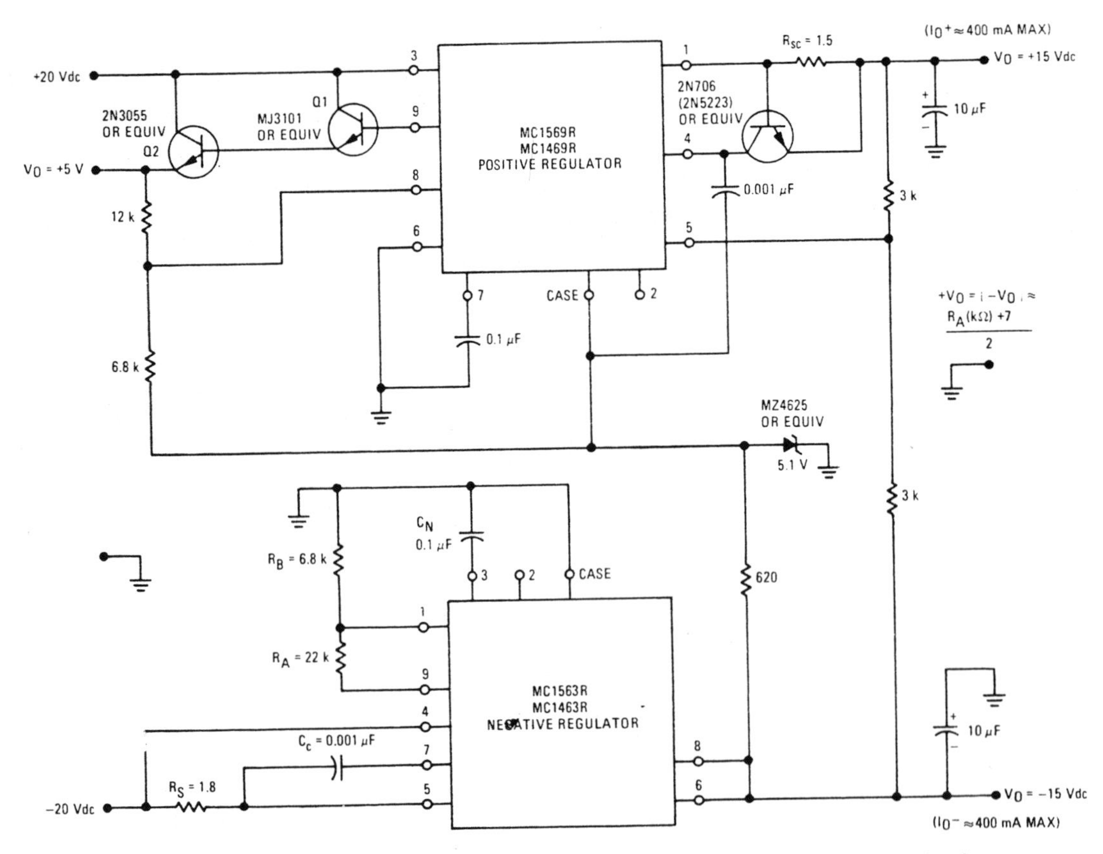

COMPLEMENTARY TRACKING REGULATOR WITH AUXILIARY 5-V SUPPLY—Provides positive and negative 15-V supplies at 400 mA maximum. Both supplies arrive at the voltage level simultaneously, and variations in the two voltages also track. An auxiliary 5-V supply is boosted up to 2 A by Q1 and Q2. It varies less than 5 mV.—"Motorola Linear Integrated Circuits Databook," Motorola, Phoenix, AZ, 1979, p. 4-94.

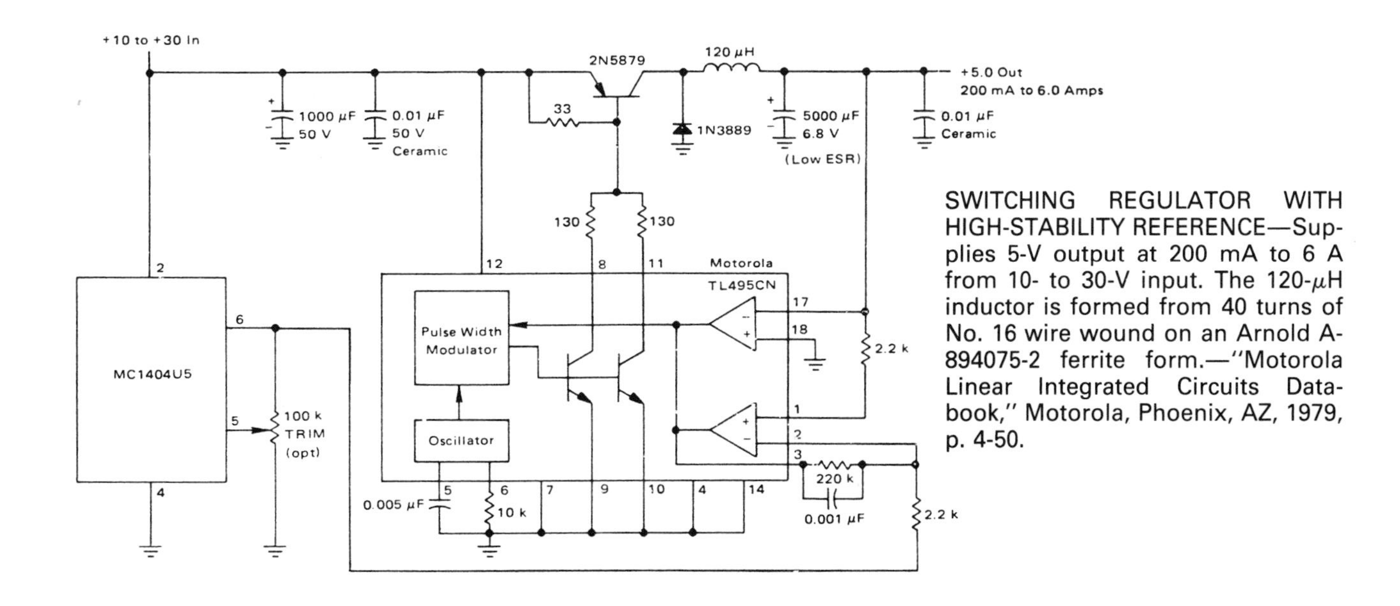

SWITCHING REGULATOR WITH HIGH-STABILITY REFERENCE—Supplies 5-V output at 200 mA to 6 A from 10- to 30-V input. The 120-μH inductor is formed from 40 turns of No. 16 wire wound on an Arnold A-894075-2 ferrite form.—"Motorola Linear Integrated Circuits Databook," Motorola, Phoenix, AZ, 1979, p. 4-50.

4-A SWITCHING REGULATOR—Based on LM100 voltage regulator device. L1 is 60 turns of No. 20 wire on an Arnold Engineering A930157.2 molybdenum permalloy core.—"Intersil Data Book," Intersil, Cupertino, CA, 1981, p. 5-14.

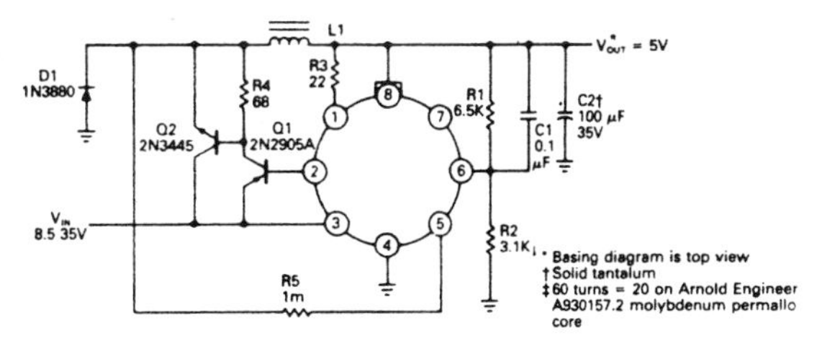

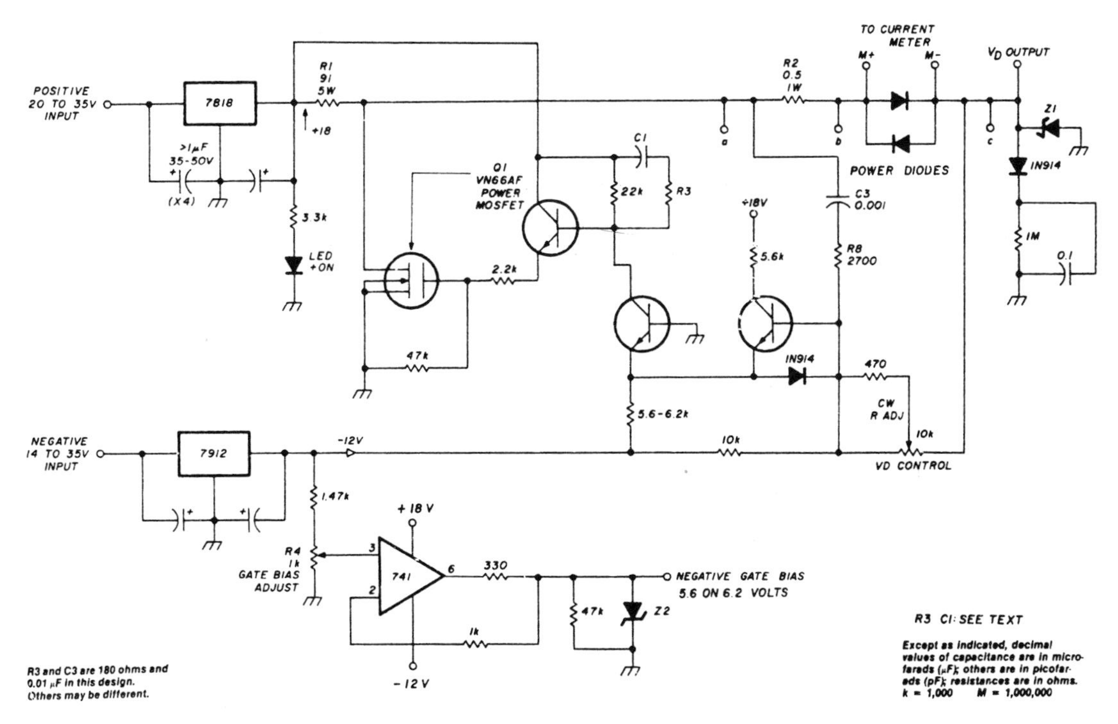

GENERAL-PURPOSE REGULATOR—Can provide up to 198 mA at 12 V and also provides an adjustable −5-V supply. R1 sets the maximum available load power, which in this case is 1.011 W. C1 and R3 control the transient response of the shunt regulator; the values used in the prototype were 0.01 μF and 180 Ω.—H. Cross, Protection for Your Solid-State Devices, *Ham Radio,* March 1981, pp. 52–56.

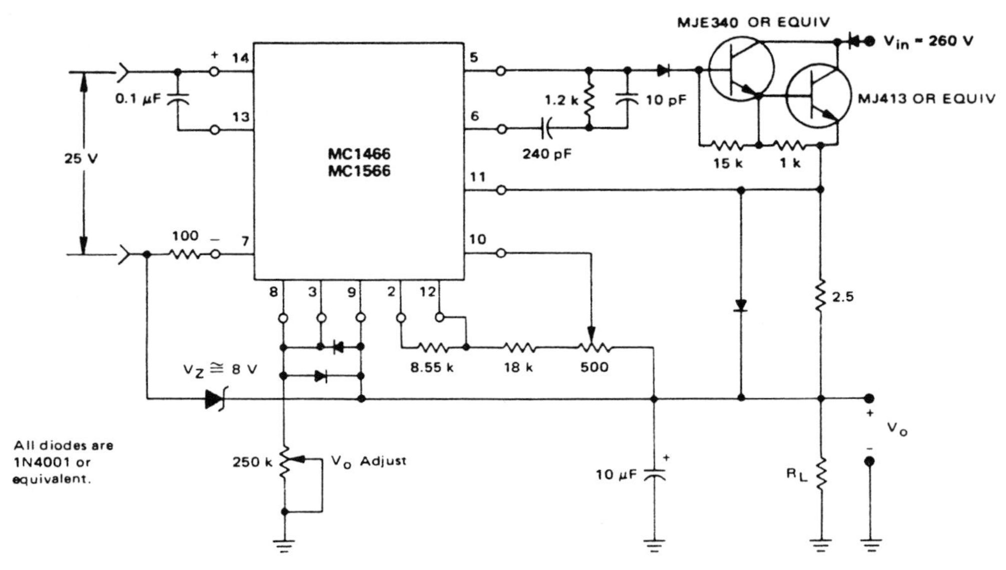

0- TO 250-V, 0.1-A REGULATOR—Based on MC1466 monolithic voltage and current regulator. All diodes are 1N4001 or equivalent. The resistor and the zener diode between pins 9 and 7 minimize the possibility of transients on the output.—"Motorola Linear Integrated Circuits Databook," Motorola, Phoenix, AZ, 1979, p. 4-76.

LATCHING REMOTE SHUTDOWN REGULATOR—Uses latch capability of NE550 regulator. The circuit is operated by TTL gates with separate inputs for shutdown and unlatch (or reset). In normal operation, the shutdown line is high, so the output of the shutdown gate is low, and regardless of the state of the unlatch gate, V_{OUT} is set at the normal level.—"Signetics Analog Applications Manual," Signetics, Sunnyvale, CA, 1979, pp. 63–64.

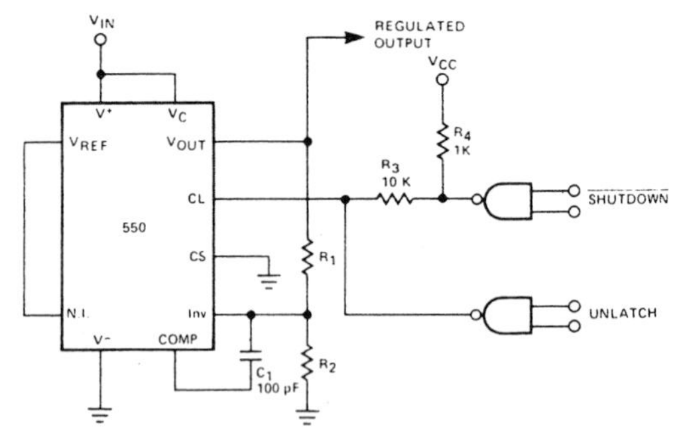

Index

ABOUT THE AUTHOR

HARRY HELMS is a Consulting Editor at McGraw-Hill. Formerly an editor at Prentice-Hall, he was also a technical writer for Radio Shack and Texas Instruments. Mr. Helms has written numerous magazine articles, and is the author of *Computer Language Reference Guide, The BASIC Book, The McGraw-Hill Computer Handbook,* and *Electronics Applications Sourcebook.*